T0319338

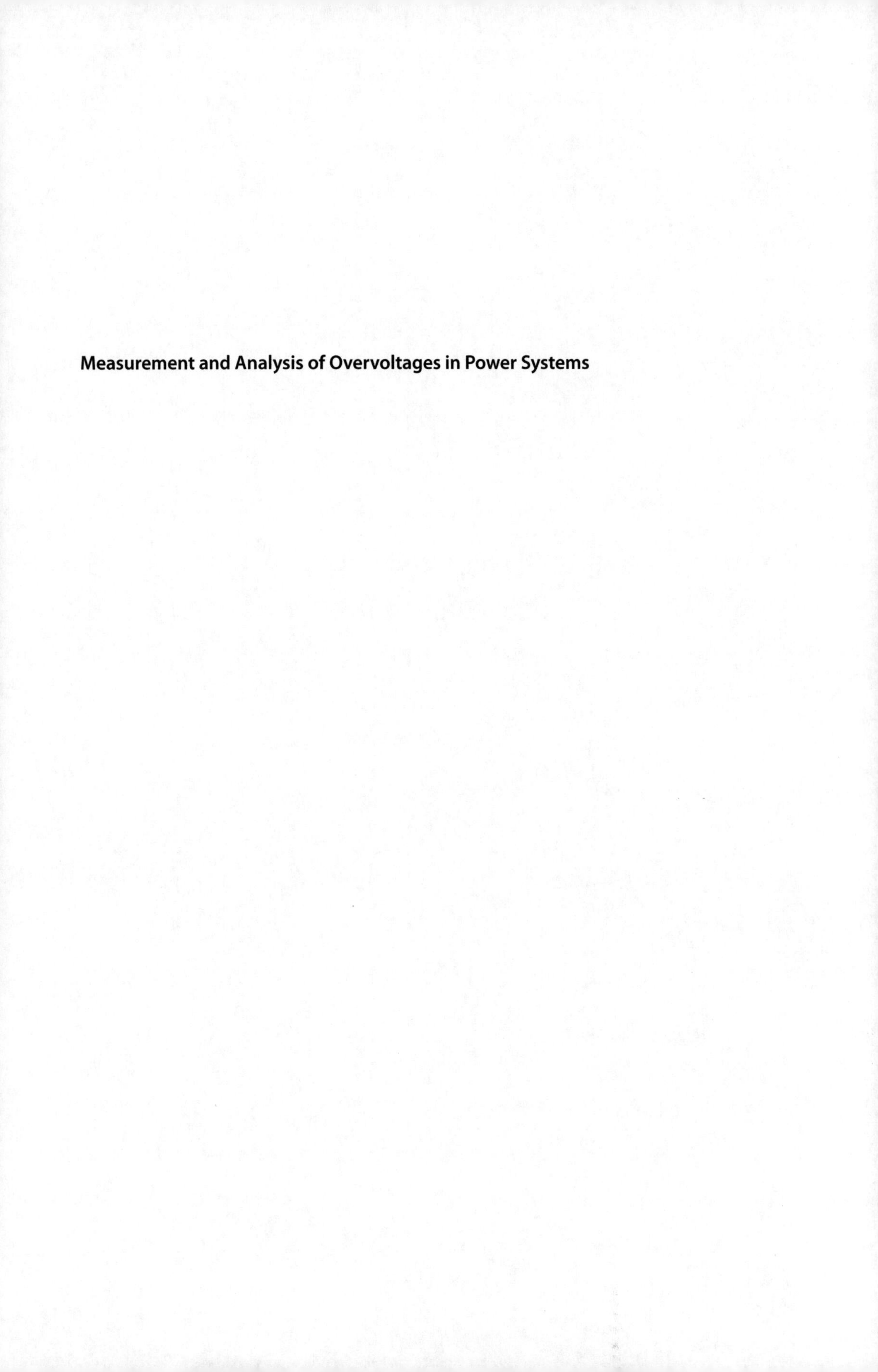

Measurement and Analysis of Overvoltages in Power Systems

Measurement and Analysis of Overvoltages in Power Systems

Measurement and Analysis of Overvoltages in Power Systems

Jianming Li
Sichuan Electric Power Research Institute (SEPRI)
State Grid Corporation of China (SGCC)
Sichuan, China

This edition first published 2018

Registered Offices
John Wiley & Sons, Inc., 111 River Street, Hoboken, NJ 07030, USA
John Wiley & Sons Singapore Pte. Ltd, 1 Fusionopolis Walk, #07-01 Solaris South Tower, Singapore 138628

Editorial Office
The Atrium, Southern Gate, Chichester, West Sussex, PO19 8SQ, UK

For details of our global editorial offices, customer services, and more information about Wiley products visit us at www.wiley.com.

Wiley also publishes its books in a variety of electronic formats and by print-on-demand. Some content that appears in standard print versions of this book may not be available in other formats.

Library of Congress Cataloging-in-Publication Data

Names: Li, Jianming, 1952- author.
Title: Measurement and analysis of overvoltages in power systems / by Jianming Li.
Description: Singapore ; Hoboken, NJ : John Wiley & Sons, 2018. | Includes bibliographical references and index. |
Identifiers: LCCN 2017041391 (print) | LCCN 2017055744 (ebook) | ISBN 9781119129059 (pdf) | ISBN 9781119129042 (epub) | ISBN 9781119128991 (cloth)
Subjects: LCSH: Overvoltage.
Classification: LCC TK7870 (ebook) | LCC TK7870 .L4825 2018 (print) | DDC 621.31/7–dc23
LC record available at https://lccn.loc.gov/2017041391

Cover design by Wiley
Cover image: © Pobytov/Gettyimages

Set in 10/12pt Warnock by SPi Global, Chennai, India

Printed in Singapore by C.O.S. Printers Pte Ltd

10 9 8 7 6 5 4 3 2 1

Contents

Preface

With the rapid development of computer technology and power electronic technology, the methods of measuring and analyzing overvoltages in power systems have changed significantly. In this book, by adopting digital simulation and entity dynamic simulation and by using a substantial amount of field-measured data, overvoltage parameters are extracted and overvoltage patterns recognized. The new concepts and methods of overvoltage measurement in this book can be used to instruct companies and organizations about monitoring, recording and analyzing overvoltages so as to ensure secured operation of the power system.

The book consists of eight chapters: Chapter 1 briefly introduces the overvoltage mechanism; Chapter 2 relates to the acquisition system of overvoltage monitoring devices; Chapter 3 deals with the transmission and recording system of overvoltage monitoring devices; Chapter 4 studies the transient response of lightning waves; Chapter 5 is concerned with typical field tests and waveform analysis for the UHVDC transmission system; Chapter 6 is mainly about digital simulation; Chapter 7 relates to entity dynamic simulation for overvoltages on transmission lines; and Chapter 8 provides details on overvoltage pattern recognition. I am the chief editor of the book, but experiments involved in this book, together with compilation and proofreading were conducted cooperatively by Chen Shaoqing, Zhang Yu, He Xiangyu, Xu Wen, Qin Dahai, Ren Xiaohua, Bi Yanqiu, Jiang Yuhan, Xie Shijun, Li Shuqi, Ouyang Renle, Li Guoyi, Huang Yidan, Zhang Luo, Liu Yuqing, Chen Xin, Li Hewei, Zhou Yue, Yang Hailong and Luo Yiqiao.

My heartfelt appreciation also goes to Professor Zeng Rong from Tsinghua University, Professor Yang Qing from Chongqing University and Senior Engineer Cao Yongxing from Sichuan Electric Power Research Institute, State Grid Corporation of China. All of them have provided a considerable amount of valuable advice related to the manuscript.

Errors and deficiencies may be found in the book due to my limited ability. Advice and criticism are welcome!

April 2017 *Jianming Li*

1

Overvoltage Mechanisms in Power Systems

1.1 Electromagnetic Transients and Overvoltage Classification

1.1.1 Electromagnetic Transients in Power System

In the case of system faults or switching operations, the operating parameters of the system will change sharply. The system may transit from one operation state to another, or it may be damaged partially or entirely with operating parameters considerably deviated from the normal values. If countermeasures are not taken, the system is hard to restore to normal operation, which may have dramatic consequences on the national economy and on people's livelihoods.

Changes in operation states cannot be completed instantaneously, so there will be a transition, known as the transient process. Transient processes in power systems are generally divided into the electromagnetic transient process and the electromechanical transient process. The electromagnetic transient process refers to the changing dynamics of the electric field and the magnetic field as well as the corresponding voltage and current of each component in the power system, while the electromechanical transient process refers to the dynamics of mechanical movement of electric machine rotors caused by electromagnetic torque changes in generators and motors.

Although the electromagnetic and electromechanical transient processes take place simultaneously and are interrelated, it is difficult and complicated to perform a unified analysis because a lot of influencing factors have to be taken into account due to the constantly expanding scale and increasingly complex structure of modern power systems. Also, the changing rates of the two processes vary widely. For example, the speed of the rotating machinery such as generators and motors will not change immediately because of the inertia in electromagnetic transient analysis. In this case, the transient process mainly depends on the electromagnetic parameters of each system component; the speed changes of the generator and the motor are therefore generally not considered, so electromechanical transients can be ignored. In electromechanical transient analysis such as the analysis of static stability and transient stability, the rotating machinery speed has already changed, thus the transient process depends on both the electromagnetic and electromechanical parameters (speed and angular displacement). In this case, the electromagnetic transient process is often ignored. Comprehensive consideration of both electromagnetic and electromechanical transients is only necessary when analyzing problems such as sub-synchronous resonance phenomena caused by

Measurement and Analysis of Overvoltages in Power Systems, First Edition. Jianming Li.

generator shaftings or calculating transient torque of the generator shafting after major disturbances.

The main purpose of electromagnetic transient analysis is to analyze and calculate transient overvoltages and overcurrents which might occur in the case of system faults or switching operations for reasonable and feasible design of electrical equipment. Generally, the equipment insulation levels are decided by overvoltages arising from power system electromagnetic transients, and standards for high voltage tests are made accordingly, to determine the possibility of the safe operation of existing equipment while restrictive and protective measures are studied at the same time. Electromagnetic transient analysis is also needed for a number of issues including operating principles and operation conditions regarding novel high-speed protection devices, fault location and positioning methods and electromagnetic interferences. In addition, electromagnetic transient calculation and simulation is indispensable for investigating accident causes and finding countermeasures, calculating overvoltage occurrence probability and predicting accident rates, checking equipment actuation performances (such as transient recovery voltage and zero offset of circuit breakers) and examining responses concerning protective relays and automatic safety devices.

1.1.2 Characteristics and Research Methods of Electromagnetic Transients

Electromagnetic transients in power systems are characterized by wide frequency range, including travelling wave process and distributed parameters. To reveal the physical mechanism of the wave process on single-conductor lines, the equivalent circuit of a differential-length line segment is used, as shown in Figure 1.1, where R_0, L_0, C_0 and G_0 refers to the resistance, inductance, capacitance and conductivity per unit length, respectively.

In Figure 1.1, the equation for the voltage drop between Nodes ① and ② is

$$u - \left(u + \frac{\partial u}{\partial x} \cdot dx\right) = R_0 \cdot dx \cdot i + L_0 \cdot dx \cdot \frac{\partial i}{\partial t} \tag{1.1.1}$$

Applying Kirchhoff's current law, the equation for Node ② becomes

$$i - \left(i + \frac{\partial i}{\partial x} \cdot dx\right) = G_0 \cdot dx \cdot \left(u + \frac{\partial u}{\partial x} \cdot dx\right) + C_0 \cdot dx \cdot \frac{\partial \left(u + \frac{\partial u}{\partial x} \cdot dx\right)}{\partial t} \tag{1.1.2}$$

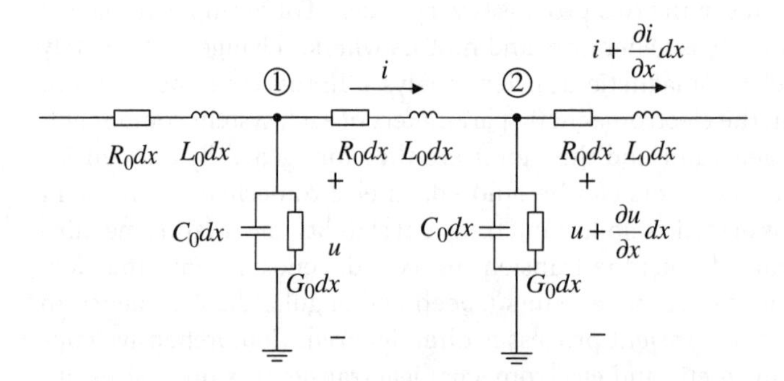

Figure 1.1 Equivalent circuit of a differential-length line segment.

Omitting the second-order infinitely small item $(dx)^2$ gives

$$-\frac{\partial u(x,t)}{\partial x} = R_0 \cdot i(x,t) + L_0 \cdot \frac{\partial i(x,t)}{\partial t} \tag{1.1.3}$$

$$-\frac{\partial i(x,t)}{\partial x} = G_0 \cdot u(x,t) + C_0 \cdot \frac{\partial u(x,t)}{\partial t} \tag{1.1.4}$$

where $u(x,t)$ and $i(x,t)$ are the binary functions for the circuit with regard to time and space. The sending end of the line can be deemed as the origin of the x-axis, and the positive direction is, by definition, towards the receiving end.

In transient calculation for high-voltage or ultra-high-voltage lines, G_0 can be neglected, and R_0 is also generally neglected. In this case, lines can be regarded as lossless with distributed parameters. The equivalent circuit of a simplified lossless single-conductor line of a differential length is as shown in Figure 1.2.

The equation for the lossless single-conductor line is given by

$$-\frac{\partial u(x,t)}{\partial x} = L_0 \cdot \frac{\partial i(x,t)}{\partial t} \tag{1.1.5}$$

$$-\frac{\partial i(x,t)}{\partial x} = C_0 \cdot \frac{\partial u(x,t)}{\partial t} \tag{1.1.6}$$

where signs are based on the precondition that the sending end of the line is the origin of the x-axis, and the positive direction is towards the receiving end.

Solving the above set of partial differential equations using Laplace transformation or by separating variables, we obtain

$$u = u_q\left(t - \frac{x}{v}\right) + u_f\left(t + \frac{x}{v}\right) \tag{1.1.7}$$

$$i = \frac{1}{z}\left[u_q\left(t - \frac{x}{v}\right) - u_f\left(t + \frac{x}{v}\right)\right] \tag{1.1.8}$$

with

$$v = \frac{1}{\sqrt{L_0 C_0}} \tag{1.1.9}$$

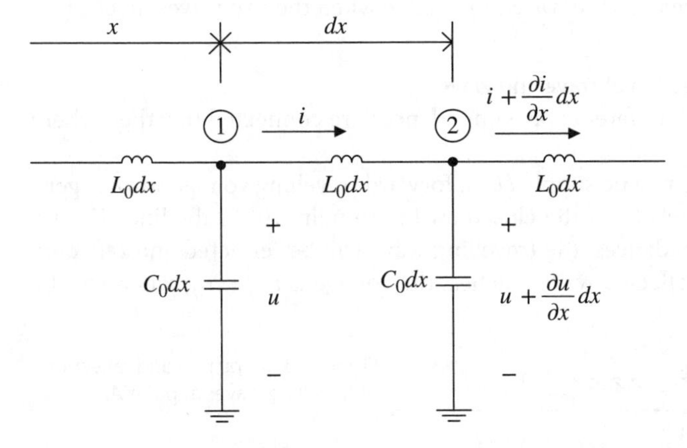

Figure 1.2 Equivalent circuit of a simplified lossless single conductor line of a differential length.

which refers to the electromagnetic wave velocity, and

$$Z = \sqrt{\frac{L_0}{C_0}} \tag{1.1.10}$$

which refers to the wave (characteristic) impedance of the line.

For overhead transmission lines, the wave velocity v is close to the speed of light, i.e. $v = 3 \times 10^8\ \text{m/s}$, while for cables, the wave velocity v is ½ – ⅓ of the speed of light, as the capacitance C_0 is larger and the dielectric constant of the medium is larger than that of air. For overhead single-conductor lines, $Z \approx 500\ \Omega$; for bundle conductor lines, $Z \approx 300\ \Omega$, and for cables, Z ranges from several to dozens of ohms.

In the above two equations, $u_q(t - \frac{x}{v})$ is the wave travelling along the line in the positive x-direction, usually known as the forward travelling wave voltage, while $u_f(t + \frac{x}{v})$ is the wave travelling along the line in the negative x-direction, usually known as the backward travelling wave voltage. By definition, $i_q = \frac{1}{z} \cdot u_q(t - \frac{x}{v})$ is the forward travelling wave current and $i_f = \frac{1}{z} \cdot u_f(t + \frac{x}{v})$ is the backward travelling wave current.

For forward and backward travelling waves, the relations between the voltages and the currents are given by

$$u_q(x, t) = Zi_q(x, t) \tag{1.1.11}$$

$$u_f(x, t) = -Zi_f(x, t) \tag{1.1.12}$$

According to Equation (1.1.11) and Equation (1.1.12), the voltages and the currents of travelling waves are related based on the wave impedance $Z = \sqrt{L_0/C_0}$.

Three differences between the wave impedance and the resistance are: (1) the wave impedance refers to the ratio between the voltage and current waves propagating in the same direction, which is independent of the line length, while the resistance increases with the line length, (2) from the perspective of power, the wave impedance does not consume energy and only determines how much energy the conductor absorbs or releases, while the resistance converts electric energy to thermal energy, (3) if both the forward and backward travelling waves coexist on lines, the ratio of the total voltage to the total current no longer equals the wave impedance when the two waves meet.

1.1.2.1 Refraction and Reflection of Travelling Waves

In Figure 1.3, two lines with different wave impedances are connected together, where A is the connecting point.

When connecting line Z_1 to a dc supply U_0, a forward travelling voltage wave is generated $u_{1q} = U_0$, which travels from the closed switch to point A. As the lines Z_1 and Z_2 have different wave impedances, the travelling wave will be refracted and reflected at point A. Wherever the reflected wave reaches, the voltage is $u_{1q} + u_{1f}$; wherever the

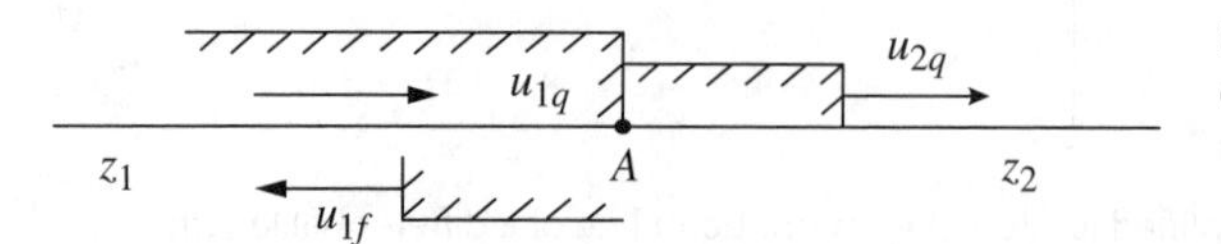

Figure 1.3 Refraction and reflection of travelling waves at point A.

refracted wave reaches, the voltage is u_{2q}. In addition,

$$u_{2q} = \frac{2Z_2}{Z_1 + Z_2} U_0 = \alpha U_0 = \alpha u_{1q} \tag{1.1.13}$$

$$u_{1f} = \frac{Z_2 - Z_1}{Z_1 + Z_2} U_0 = \beta U_0 = \beta u_{1q} \tag{1.1.14}$$

with

$$\alpha = \frac{2Z_2}{Z_1 + Z_2} \tag{1.1.15}$$

$$\beta = \frac{Z_2 - Z_1}{Z_1 + Z_2} \tag{1.1.16}$$

In this expression, α and β are the refraction and reflection coefficients of point A respectively, and $\alpha = \beta + 1$. Besides, $0 \leq \alpha \leq 2$, which means that α is always positive, hence the refraction and incoming voltage waves will always be of the same polarity. The reflection coefficient β may either be a positive or negative value, $-1 \leq \beta \leq 1$.

1.1.2.2 Peterson Principle

When the wave impedances of semi-infinitely long lines 1 and 2 are different, calculate the current and the voltage of point A where the forward wave $u_{1q}(t)$ with arbitrary waveforms arrives.

Line 1 is equivalent to a voltage source, whose emf is double the amplitude of the incoming forward wave voltage, i.e. $2u_{1q}(t)$, and the internal resistance is Z_1, the wave impedance of line 1; line 2 is equivalent to the resistance Z_2, with its value being Z_2, the wave impedance of line 2. The calculation based on the equivalent circuit is one using the Peterson principle, and the original and equivalent circuits are shown in Figure 1.4.

The Peterson principle can also be expressed in the form of current sources, as depicted in Figure 1.5.

By converting the distributed parameter circuit to an equivalent lumped parameter circuit, i.e. converting differential equations to algebraic equations, the Peterson principle simplifies the calculation process. Note that the Peterson principle can only be applied under the following conditions: (1) the incoming wave travels to point A along a uniform lossless line, (2) no backward wave exists on line 2 or the backward wave has not yet reached point A. In the equivalent calculation, Z_2 may either be a line or any network composed of resistors, reactors and capacitors.

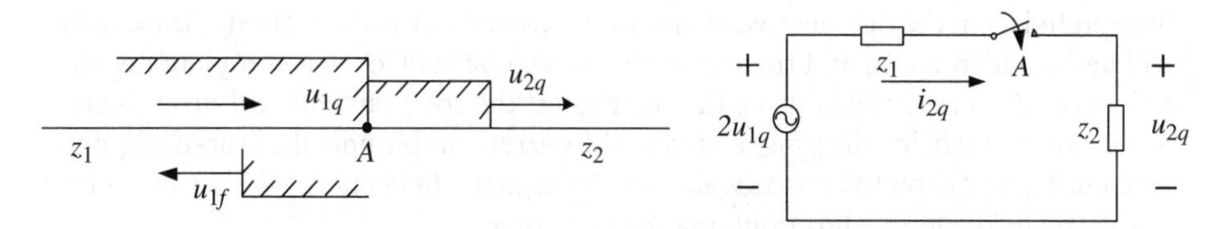

(a) Refraction and reflection of the incident wave voltage u_{1q} at point A;
(b) Peterson equivalent circuit of the refracted voltage u_{2q} on Z_2.

Figure 1.4 Peterson principle.

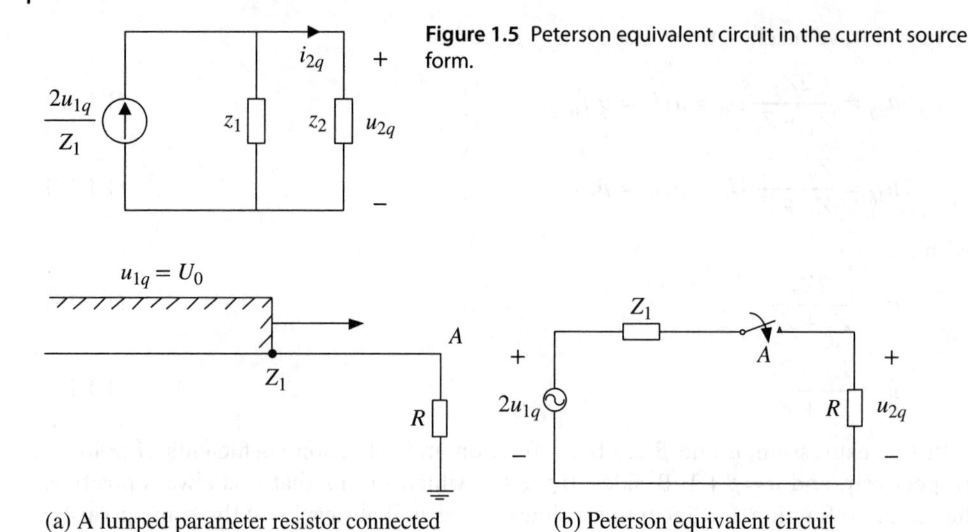

Figure 1.5 Peterson equivalent circuit in the current source form.

(a) A lumped parameter resistor connected at the receiving end

(b) Peterson equivalent circuit

Figure 1.6 Peterson equivalent circuit for calculating refraction voltage.

The following are conclusions made by adopting the Peterson principle in applications where a resistor is connected to the receiving end of the line or where a series inductor or a parallel capacitor connects two lines with different wave impedances.

Case A: When the receiving end of the line is connected to a lumped parameter resistor Point A is connected to a resistor Z_2, as shown in Figure 1.6.
When the incoming wave is a rectangular wave,

1. If the resistance R equals the wave impedance Z_1, the reflection coefficient of the voltage wave $\beta_u = 0$, which means that neither refraction nor reflection will occur when u_{1q} reaches R.
2. If the resistance $R \to \infty$, i.e. the receiving end of line Z_1 becomes an open circuit, $\beta_u = 1$, $u_{1f} = u_{1q}$, $u_2 = 2u_{2q}$ (for the voltage waves) and $i_{1f} = -u_{1f} / Z_1 = -u_{1q} / Z_1 = -i_{1q}$, $i_2 = i_{2q} = i_{1q} + i_{1f} = 0$ (for the current waves), which means that u_{1q} is reflected at the terminal, and the reflected voltage wave equals the incoming voltage wave. The terminal voltage is doubled while the terminal current drops to zero. When the reflected voltage wave travels backwards, the voltage will be doubled and the current will drop to zero wherever it reaches. This phenomenon can be interpreted from the perspective of energy: at the end of the open circuit, the current will be zero identically, and the current wave will be subject to a complete negative reflection. As a result, wherever the wave goes, the total current will drop to zero. As the magnetic field energy is related to the current amplitude, the stored magnetic field energy will drop to zero as well, and all magnetic field energy will be converted to electric field energy, thus doubling the line voltage.
3. If the resistance $R = 0$, i.e. the receiving end of the line is short-circuit grounded, then the voltage $\alpha_u = 0$, $\beta_u = -1$, $u_{1f} = -u_{1q}$, $u_2 = u_{2q} = 0$, which means that u_{1q} is reflected when reaching the receiving end. The reflected voltage wave equals the negative incoming voltage wave, and there is no refraction wave, since alternative

lines do not exist. Where the reflected voltage wave reaches, the voltage drops to 0, and $i_{1f} = -u_{1f} / Z_1 = u_{1q} / Z_1 = i_{1q}$, thus, $i_2 = i_{1q} + i_{1f} = 2i_{1q}$, meaning that the terminal current is doubled. Where the reflection voltage wave arrives, all electric field energy is converted to magnetic field energy, thus doubling the current.

4. If the resistance $R < Z_1$, then $\beta_u < 0$, the reflected voltage $u_{1f} < 0$, and the reflected wave will result in a line voltage drop.
5. If the resistance $R > Z_1$, then $\beta_u > 0$, the reflected voltage $u_{1f} > 0$, and the reflected wave will result in a line voltage rise.

Case B: Series inductors connected between the lines

In practical projects, series inductors are often used to connect lines, as depicted in Figure 1.7, where an inductor is connected in series with Z_1 and Z_2.

It can be shown that the refracted voltage wave u_{2q} travelling along line Z_2 is expressed by

$$u_{2q} = i_{2q} \cdot Z_2 = \frac{2Z_2}{Z_1 + Z_2} u_{1q}(1 - e^{-\frac{t}{T}}) = \alpha u_{1q}(1 - e^{-\frac{t}{T}}) \tag{1.1.17}$$

where $T = L / (Z_1 + Z_2)$, the time constant of the circuit; $\alpha = 2Z_2 / (Z_1 + Z_2)$, the refraction coefficient of the voltage wave.

$$u_{1f} = \frac{Z_2 - Z_1}{Z_1 + Z_2} u_{1q} + \frac{2Z_1}{Z_1 + Z_2} u_{1q} e^{-\frac{t}{T}} \tag{1.1.18}$$

For line Z_1, as the current flowing across the inductor cannot change instantly, the inductor is equivalent to open circuit at instant $t = 0$, and $u_{1f} = u_{1q}$. Since all the magnetic field energy is converted to electric field energy, the voltage is doubled, following an exponential change; when $t \to \infty$, $u_{1f} \to \beta u_{1q}$ ($\beta = (Z_2 - Z_1)/(Z_1 + Z_2)$).

For line Z_2, the refracted voltage u_{2q} increases exponentially with time. When $t = 0$, $u_{2q} = 0$; when $t \to \infty$, $u_{2q} \to \alpha u_{1q}$ ($\alpha = 2Z_2 / (Z_1 + Z_2)$). It means that after passing through the inductor, the infinitely long rectangular wave becomes a travelling wave with an exponential front, and the series inductor plays a role in reducing the front steepness of the incoming wave.

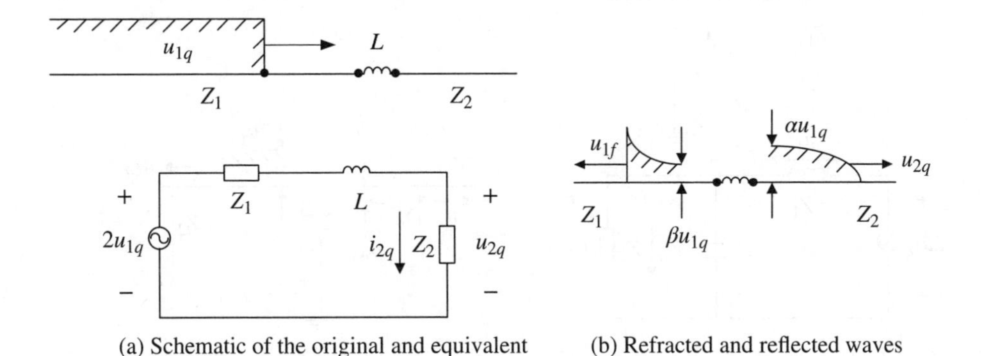

Figure 1.7 Travelling wave passing through an series inductor.

Differentiating Equation (1.1.17) with respect to t, we obtain the maximum steepness at instant $t = 0$:

$$\left(\frac{du_{2q}}{dt}\right)_{max} = \left.\frac{du_{2q}}{dt}\right|_{t=0} = \frac{2u_{1q}Z_2}{L} \tag{1.1.19}$$

According to Equation (1.1.19), the maximum steepness is independent of Z_1, and is determined solely by Z_2 and L. The larger the L value is, the more the steepness reduces.

Case C: Parallel capacitors connected between the lines

In practical projects, parallel capacitors are often used in line applications, as depicted in Figure 1.8, where a capacitor is connected in parallel with Z_1 and Z_2.

It can be shown that the refracted voltage wave u_{2q} propagating along line Z_2 is given by

$$u_{2q} = i_{2q} \cdot Z_2 = \frac{2Z_2}{Z_1 + Z_2} u_{1q}\left(1 - e^{-\frac{t}{T}}\right) = \alpha u_{1q}\left(1 - e^{-\frac{t}{T}}\right) \tag{1.1.20}$$

where $T = Z_1Z_2 / (Z_1 + Z_2)\, C$, the time constant of the circuit; $\alpha = 2Z_2/(Z_1 + Z_2)$, the refraction coefficient of the voltage wave.

$$u_{1f} = \frac{Z_2 - Z_1}{Z_1 + Z_2} u_{1q} - \frac{2Z_2}{Z_1 + Z_2} u_{1q} e^{-\frac{t}{T}} \tag{1.1.21}$$

For line Z_1, when $t = 0$, $u_{1f} = -u_{1q}$. This is because the voltage across the capacitor cannot change instantaneously, the capacitor appears as if it were short-circuited to ground. All the electric field energy is converted to magnetic field energy, and the voltage changes exponentially according to the time constant; when $t \to \infty$, $u_{1f} \to \beta u_{1q}$ ($\beta = (Z_2 - Z_1) / (Z_1 + Z_2)$).

For line Z_2, the refracted voltage u_{2q} increases exponentially with time. When $t = 0$, $u_{2q} = 0$; when $t \to \infty$, $u_{2q} \to \alpha u_{1q}$ ($\alpha = 2Z_2 / (Z_1 + Z_2)$). The infinitely long rectangular wave becomes a travelling wave with an exponential front after passing through the capacitor. The parallel capacitor plays the same role as the series inductor, reducing the incoming wavefront steepness or smoothing the wavefront.

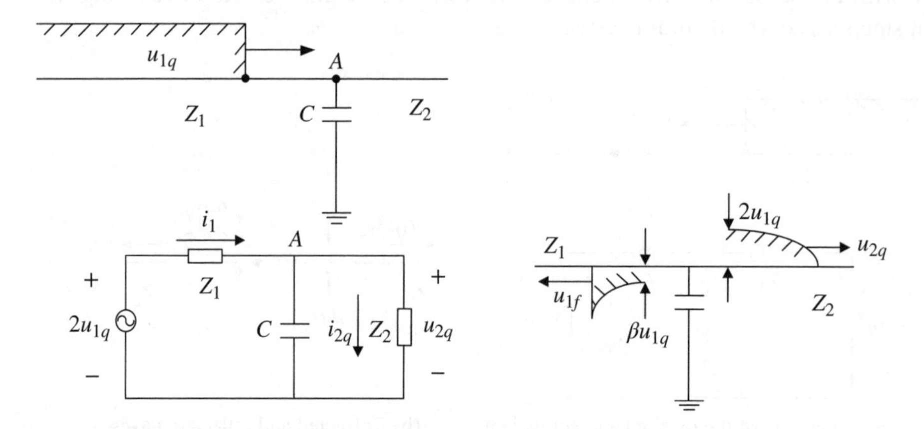

Figure 1.8 Travelling wave passing through a parallel capacitor.

Differentiating Equation (1.1.20) with respect to t, we obtain the maximum steepness at instant $t = 0$ given by

$$\left(\frac{du_{2q}}{dt}\right)_{\max} = \left.\frac{du_{2q}}{dt}\right|_{t=0} = \frac{2u_{1q}}{Z_1 C} \tag{1.1.22}$$

According to Equation (1.1.22), the maximum steepness is independent of Z_2, and is determined solely by Z_1 and C. The larger the C value is, the more the steepness reduces.

According to the above analysis, both parallel capacitors and series inductors are capable of reducing incoming wavefront steepness. Impulse corona, which may also have the same effect, will be introduced next.

Impulse corona is generated on the lines by lightning and by switching impulse waves. The presence of corona sleeves is equivalent to increasing conductor radial dimensions, which may change the capacitance between conductors and increase coupling coefficients. Corona may result in sparks and heating of conductors, consuming the impulse wave energy. Consequently, the occurrence of impulse corona will attenuate and distort travelling waves. Typical attenuation and distortion of travelling waves resulting from impulse corona are illustrated in Figure 1.9.

In the figure, curve 1 is the original waveform of the incoming wave, and curve 2 shows the wave attenuated and distorted by impulse corona after propagating some distance. Impulse corona is beneficial for equipment safety as it can reduce the amplitude and the steepness of the incoming surge.

1.1.2.3 Multiple Refraction and Reflection of Travelling Waves

In a practical power system, a transmission line often consists of many kinds of conductors. For example, the steel reinforced aluminum conductor may be used at the sending and receiving ends, while cables may be used for the transmission part, as shown in Figure 1.10.

In the figure, the first voltage wave u_{0f} reflected from point B not only turns the voltage of line 2 into $u_{0q} + u_{0f}$ but will also give rise to a new round of refractions and reflections when it reaches point A. Similarly, more refractions and reflections will continue to occur. The discussions concerning this case are presented next.

As shown in Figure 1.10, it is assumed that a short conductor with length of l_0 and wave impedance of Z_0 is connected between infinitely long lines Z_1 and Z_2 with different wave impedances, and the two nodes are labelled points A and B. An infinitely long rectangular wave U_0 travels from line Z_1 to point A and is refracted and reflected there. The refracted wave $\alpha_1 U_0$ continues to propagate along the line and reaches point B, while the first refracted wave $\alpha_1\alpha_2 U_0$ generated at point B continues to travel along

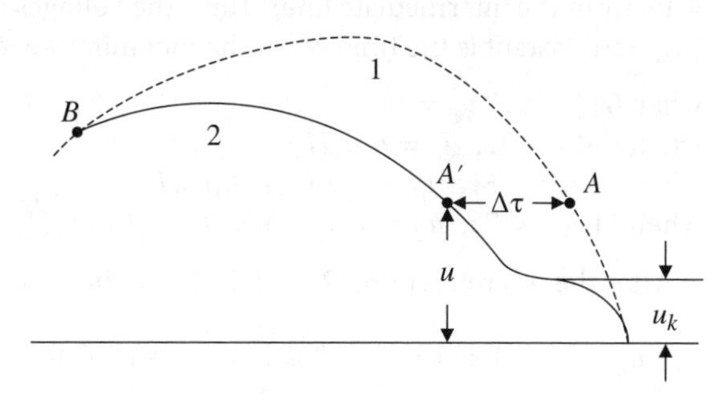

Figure 1.9 Travelling wave attenuated and distorted by corona.

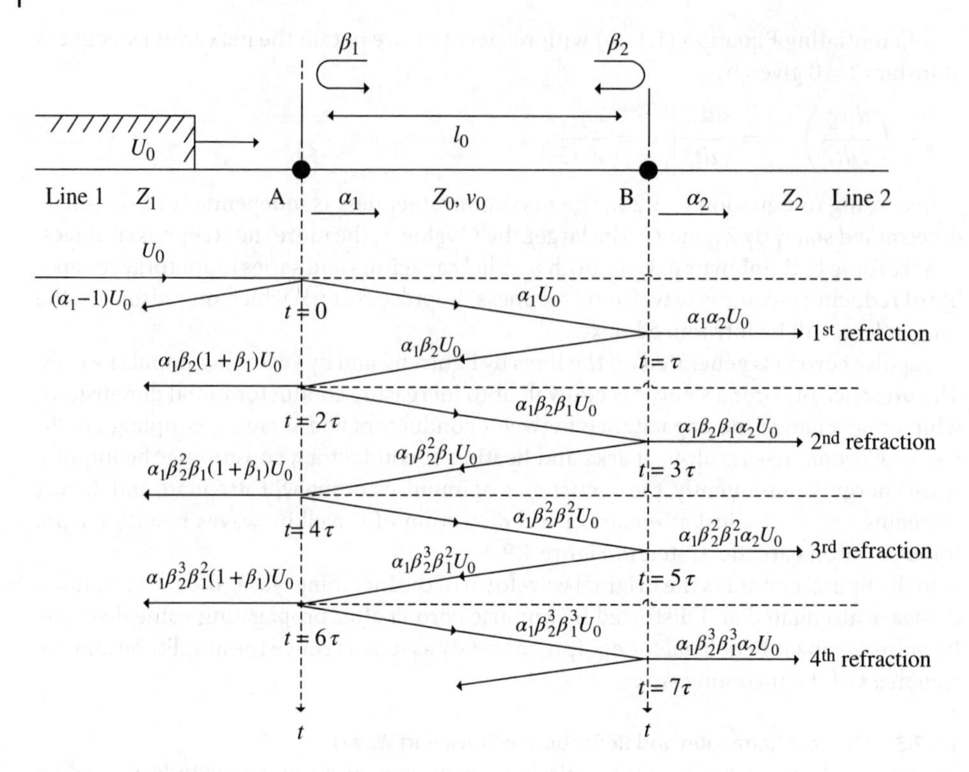

Figure 1.10 Multiple refraction and reflection of travelling waves.

line Z_2, and the first reflected wave $\alpha_1\beta_2 U_0$ travels back to point A, where the reflected wave $\alpha_1\beta_2\beta_1 U_0$ travels along line Z_0 until it reaches point B. The second refracted wave $\alpha_1\beta_2\beta_1\alpha_2 U_0$ at point B travels along line Z_2, while the second reflected wave $\alpha_1{\beta_2}^2\beta_1 U_0$ travels towards point A, and so forth. The expressions for calculating the refraction coefficients α_1, α_2 and the reflection coefficients β_1, β_2 are given by

$$\alpha_1 = \frac{2Z_0}{Z_1 + Z_0}, \quad \alpha_2 = \frac{2Z_2}{Z_0 + Z_2}, \quad \beta_1 = \frac{Z_1 - Z_0}{Z_0 + Z_1}, \quad \beta_2 = \frac{Z_2 - Z_0}{Z_0 + Z_2} \tag{1.1.23}$$

Each point on the line may be calculated via superposition of all the refracted and reflected waves, but attention should be paid to the arrival time sequence of each refracted and reflected wave. And note that the time required for the wave to travel through the intermediate line is calculated according to $\tau = l_0 / v_0$ (where v_0 is the wave velocity of the intermediate line). Thus, the voltages of point B at different time instants (the zero instant is the time when the incoming wave first arrives at point A) are:

when $0 \le t < \tau,\ u_B = 0$;
when $\tau \le t < 3\tau,\ u_B = \alpha_1\alpha_2 U_0$;
when $3\tau \le t < 5\tau,\ u_B = \alpha_1\alpha_2(1 + \beta_1\beta_2)U_0$;
when $51 \le t < 7\tau,\ u_B = \alpha_1\alpha_2(1 + \beta_1\beta_2 + (\beta_1\beta_2)^2)U_0$.

After the nth refraction, $(2n - 1)\tau \le t < (2n + 1)\tau$, the voltage of point B is given by

$$u_B = \alpha_1\alpha_2(1 + \beta_1\beta_2 + (\beta_1\beta_2)^2 + \cdots + (\beta_1\beta_2)^{n-1})U_0 = U_0\alpha_1\alpha_2\frac{1 - (\beta_1\beta_2)^n}{1 - \beta_1\beta_2} \tag{1.1.24}$$

When $t \to \infty$, $n \to \infty$, $(\beta_1\beta_2)^n \to 0$, the above equation is simplified to

$$u_B = U_0\alpha_1\alpha_2\frac{1}{1-\beta_1\beta_2} \tag{1.1.25}$$

Substituting Equation (1.1.23) into the above equation, the amplitude of the ultimate voltage of point B is

$$U_B = \frac{2Z_2}{Z_1+Z_2}U_0 = \alpha U_0 \tag{1.1.26}$$

where α is the voltage refraction coefficient when the wave travels directly from line Z_1 to line Z_2. It is indicated by the equation that the ultimate voltage entering line Z_2 is only related to Z_1 and Z_2, and is independent of the intermediate line Z_0, which, however, determines the waveform and wavefront properties of u_B. A detailed discussion will be stated as follows.

1. When $Z_0<Z_1$ and $Z_0<Z_2$ (for instance, connecting a cable between different overhead lines), both β_1 and β_2 are positive, all the refracted waves are thus positive, and u_B increases gradually until reaching the ultimate voltage U_B, as shown in Figure 1.11(a). In particular, when $Z_0 \ll Z_1$ and $Z_0 \ll Z_2$, the intermediate line exhibits small inductance and large capacitance to ground. The inductance may thus be ignored and the intermediate line may be substituted by a parallel capacitor that can reduce the incoming wave steepness.
2. When $Z_0>Z_1$ and $Z_0>Z_2$ (for instance, connecting an overhead line between different cables), both β_1 and β_2 are negative, all the refracted waves are thus negative, but the product of β_1 and β_2 is positive. u_B increases gradually until reaching the ultimate voltage U_B with the same waveform as shown in Figure 1.11(a). In particular, when $Z_0 \gg Z_1$ and $Z_0 \gg Z_2$, the intermediate line exhibits large inductance and small capacitance to ground. The capacitance may thus be ignored and the intermediate line may be substituted by a series inductor that can reduce the incoming wave steepness.

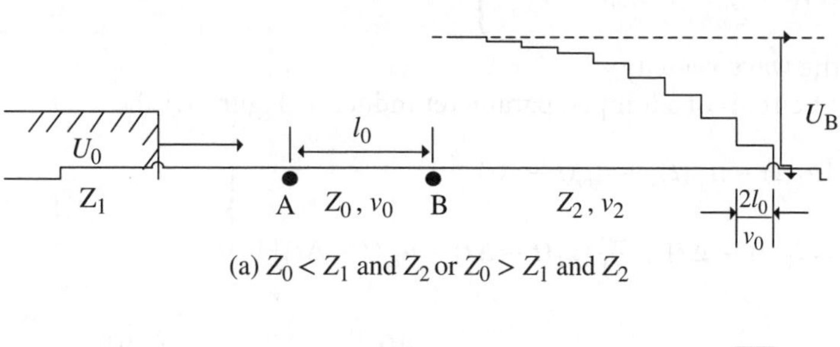

(a) $Z_0 < Z_1$ and Z_2 or $Z_0 > Z_1$ and Z_2

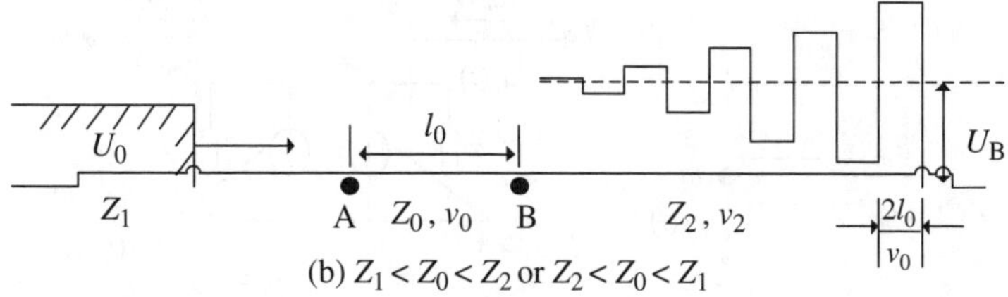

(b) $Z_1 < Z_0 < Z_2$ or $Z_2 < Z_0 < Z_1$

Figure 1.11 Diagram of u_B waveforms under different wave impedance combinations.

3. When $Z_1 < Z_0 < Z_2$, β_1 is negative, β_2 is positive, so the product $(\beta_1\beta_2)$ is negative, and u_B will increase gradually until reaching the ultimate voltage U_B during the oscillation, as shown in Figure 1.11(b). The final voltage $U_B > U_0$.
4. When $Z_2 < Z_0 < Z_1$, β_1 is positive, β_2 is negative, so the product $(\beta_1\beta_2)$ is negative, and u_B will increase gradually until reaching the ultimate voltage U_B during the oscillation with the same waveform shown in Figure 1.11(b). The final voltage $U_B < U_0$.

1.1.2.4 Evaluation of Overvoltages Using the Bergeron Method

As there are thousands of components and nodes in a practical power system, manual analysis is very complicated and may be impossible. In general, such complicated evaluations are completed by computers. Many methods are used for overvoltage evaluation, but the Bergeron method is the only widely used one that can realize EMTP transient simulation. Using computers for calculating power system overvoltages, the Bergeron method has the advantages of high calculation speed, convenient parameter modification and high accuracy. A brief introduction to the Bergeron method will be given here.

The core concept of the Bergeron method is to first use the equivalent impedance circuit with a current source to replace components through which travelling waves flow and then solve the equations based on circuit and matrix theories. The Bergeron equivalent circuits of various power system components are presented next.

1. Bergeron equivalent circuit of a lossless single-conductor line (Figure 1.12)

$$\left.\begin{aligned} i_{km}(t) &= \frac{u_k(t)}{Z} + I_{km}(t-\tau) \\ i_{mk}(t) &= \frac{u_m(t)}{Z} + I_{mk}(t-\tau) \\ I_{km}(t-\tau) &= -\frac{u_m(t-\tau)}{Z} - i_{mk}(t-\tau) \\ I_{mk}(t-\tau) &= -\frac{u_k(t-\tau)}{Z} - i_{km}(t-\tau) \end{aligned}\right\} \tag{1.1.27}$$

with $\tau = l / v$ (v is the wave velocity).

2. Bergeron equivalent circuit of a lumped parameter inductor (Figure 1.13)

$$\left.\begin{aligned} i_{km}(t) &= \frac{\Delta t}{2L}[u_k(t) - u_m(t)] + I_{km}(t-\Delta t) \\ I_{km}(t-\Delta t) &= i_{km}(t-\Delta t) + \frac{\Delta t}{2L}[u_k(t-\Delta t) - u_m(t-\Delta t)] \end{aligned}\right\} \tag{1.1.28}$$

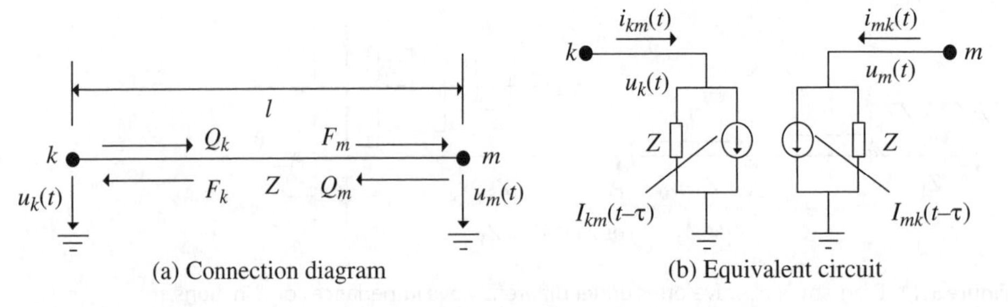

Figure 1.12 Bergeron equivalent circuit of a lossless single-conductor line.

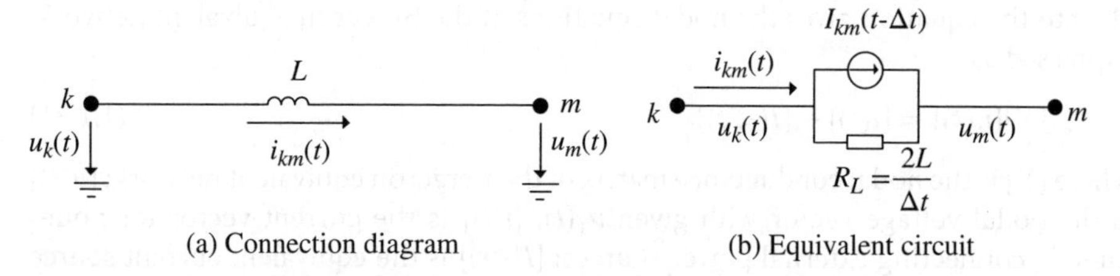

Figure 1.13 Bergeron equivalent circuit of a lumped parameter inductor.

where Δt is the time increment.

3. Bergeron equivalent circuit of a lumped parameter capacitor (Figure 1.14)

$$\left.\begin{aligned} i_{km}(t) &= \frac{2C}{\Delta t}[u_k(t) - u_m(t)] + I_{km}(t - \Delta t) \\ I_{km}(t - \Delta t) &= -i_{km}(t - \Delta t) - \frac{2C}{\Delta t}[u_k(t - \Delta t) - u_m(t - \Delta t)] \end{aligned}\right\} \tag{1.1.29}$$

4. Bergeron equivalent circuit of a lumped parameter resistor (Figure 1.15)
The resistance of the equivalent circuit remains the same as that of the original circuit, where

$$i_{km}(t) = \frac{1}{R}[u_k(t) - u_m(t)] \tag{1.1.30}$$

When applying the Bergeron method, the variables of each node need to be calculated first at time instant $t = 0$, and then at time constants $t = \Delta t$, $2\Delta t$..., until the required time constant. The calculation sequence at a particular time constant t is as follows:

1. Based on the previous calculation before the time constant t, and combined with Equations (1.1.27–29), equivalent current sources $I(t - \tau)$ and $I(t - \Delta t)$ have to be calculated.
2. According to Kirchhoff's current law, the equation is written based on the condition that the current sum of the node is zero. Substituting the current source obtained in

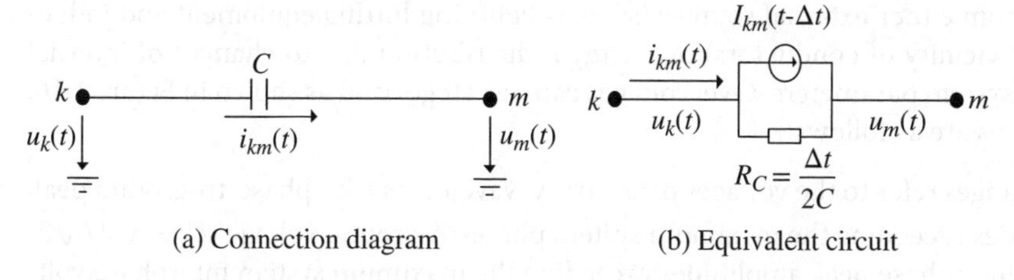

Figure 1.14 Bergeron equivalent circuit of a lumped parameter capacitor.

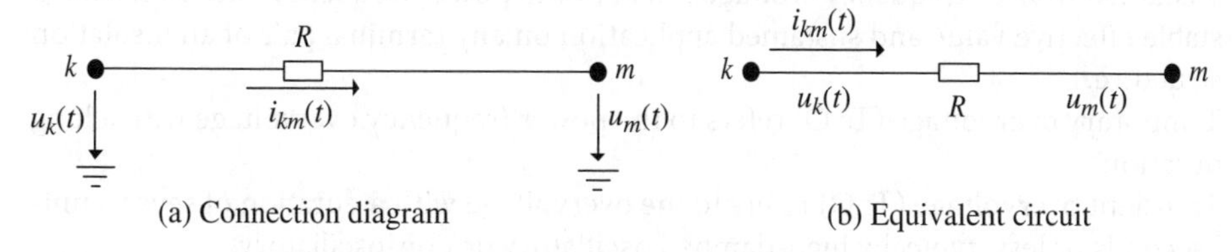

Figure 1.15 Bergeron equivalent circuit of a lumped parameter resistor.

(1) into the equation gives the nodal equations of the Bergeron equivalent network, expressed as

$$[Y][u(t)] = [i(t)] - [I(< t)] \tag{1.1.31}$$

where $[Y]$ is the nodal conductance matrix of the Bergeron equivalent network; $[u(t)]$ is the nodal voltage vector with given $u_1(t)$; $[i(t)]$ is the current vector for nodes directly connecting external power sources; $[I(<t)]$ is the equivalent current source vector determined by the voltage and current values before time constant t. The value of $[u(t)]$ can be obtained from the above equation.

3. By combining Equations (1.1.27–30), all the current values $i(t)$ at time constant t can be calculated.
4. By once again using Equations (1.1.27–30), the equivalent current source $I(t)$ at time constant t can be calculated as the preparation for the next step.
5. Suppose $t = t + \Delta t$, return to step (1).

1.2 Overvoltage Classification in Power Systems

Under normal conditions, damage or short-time aging will not occur in electrical equipment under long-term operating voltages. Nevertheless, equipment may experience much higher voltages for various reasons, resulting in insulation damage, aging or even breakdown. The voltage level that endangers the insulation is referred to, in this book, as the power system overvoltage.

There are different categories of overvoltages in a power system. It is essential to distinguish between overvoltages as they are different in amplitude, frequency and duration. Basically, the present standards mainly classify overvoltages based on their sources and causes. The specific classification of overvoltages will be detailed next.

1.2.1 Overvoltage Classification

Overvoltages are generally caused by sudden changes of the electromagnetic energy resulting from either external sources (such as lightning hitting equipment and induction in the vicinity of conductors) or energy redistribution due to changes of internal states and system parameters. Overvoltages can be categorized as shown in Figure 1.16.

Definitions are as follows:

- Overvoltages refer to the voltages of arbitrary waveforms with phase-to-ground peak amplitudes exceeding the maximum system phase-to-ground voltage ($U_s \times \sqrt{2}/\sqrt{3}$) or with interphase peak amplitudes exceeding the maximum system interphase voltage ($\sqrt{2}\,U_s$).
- Sustained (power frequency) voltage refers to the power frequency voltage having a stable effective value and sustained application on any terminal pair of an insulation structure.
- Temporary overvoltage (Te.O) refers to the power frequency overvoltage with a long duration.
- Transient overvoltage (Tr.O) refers to the overvoltage with a duration of several milliseconds or less, typically high-damped oscillatory or non-oscillatory.

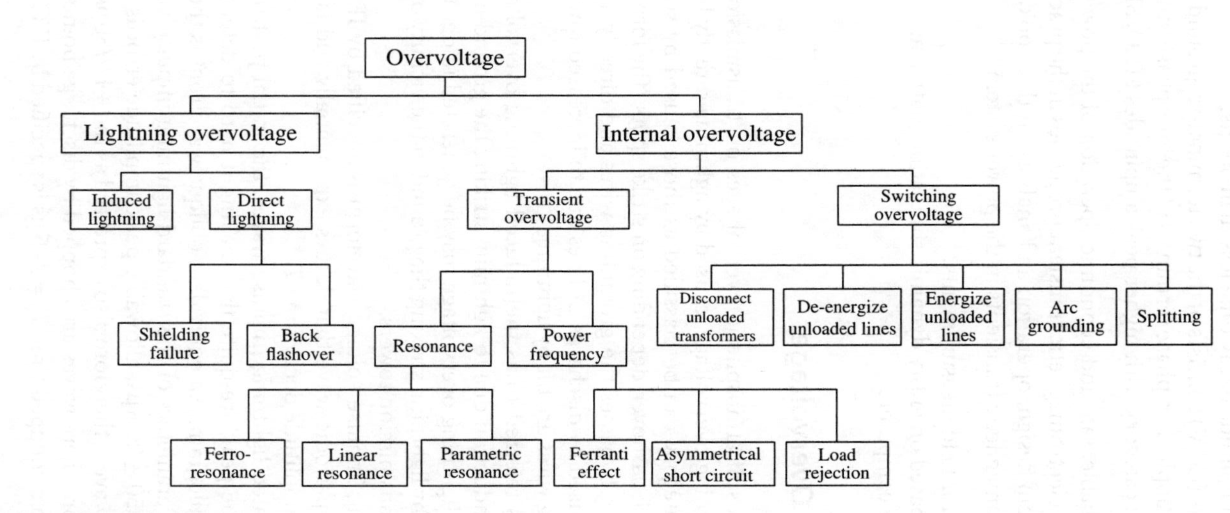

Figure 1.16 Overvoltage classification.

Transient overvoltages are subdivided into the following categories:

- Slow-front overvoltage (SFO) refers to the transient overvoltage with a zero-to-peak time of 20–5000 μs and a half-amplitude decay time less than 20 ms.
- Fast-front overvoltage (FFO) refers to the transient overvoltage with a front time of 0.1–20 μs and a half-amplitude decay time less than 300 μs.
- Combined overvoltage (COV) consists of the two terminal-to-ground voltage components simultaneously applied on phase terminals of the interphase (minor) insulation. It is classified into the category with higher peak amplitudes (Te.O, SFO, FFO).

In order to ensure reliable, safe and economic operation of the power system, comprehensive insights into electromagnetic transient properties in the practical system are essential regarding system design, operation and analysis. In this book, three methods are used to study electromagnetic transients in the power system.

1. Systematic methods for field measurements;
2. Physical simulation based on entity dynamic simulation platform;
3. Digital simulation by computers.

1.3 Atmospheric Overvoltages

There are two categories of atmospheric overvoltages on transmission lines, one of which is the direct lightning overvoltage, caused by lightning directly hitting the lines. Overvoltages of this category can be classified as those caused by shielding failures and those caused by back flashover, depending on strike spots. The former subcategory is caused when lightning bypasses the grounding wires and directly strikes the lines, whereas the latter is caused by discharge of lines due to increased tower potentials as tower tops or grounding wires are hit by lightning.

The other category is referred to as the induced lightning overvoltage, originating from electromagnetic induction due to lightning striking the ground in the vicinity of the lines. This induced lightning overvoltage may be generated under two conditions, induced on lines either by lightning striking the ground in the vicinity of the lines or by lightning striking the transmission towers.

The standard lightning impulse voltage waveform is specified by IEC, as shown in Figure 1.17. It is a non-periodic wave that decays exponentially and is determined by front time T_1 and half-amplitude decay time T_2.

For laboratory-generated lightning impulse waves, the initial part of the front and the area surrounding the peak are quite flat, making it hard to determine the origin and the peak in the oscillogram. An equivalent oblique wavefront is therefore used, as shown in Figure 1.17. Parameters of the standard lightning impulse voltage wave are $T_1 = 1.2 \pm 30\%$ μs; $T_2 = 50 \pm 20\%$ μs; allowable peak amplitude error is ±3%. The standard lightning impulse wave is therefore also expressed as the ±1.2/50 μs wave. In order to simulate the condition where waves are chopped by discharge points on the lines, a chopped waveform with a chopped time $T_c = 2–5$ μs is specified by IEC as well.

1.3.1 Lightning Discharge

Lightning discharge, by its physical nature, is a kind of spark discharge with ultra-long air gaps. It is similar to the process of long air gap breakdown, while the difference

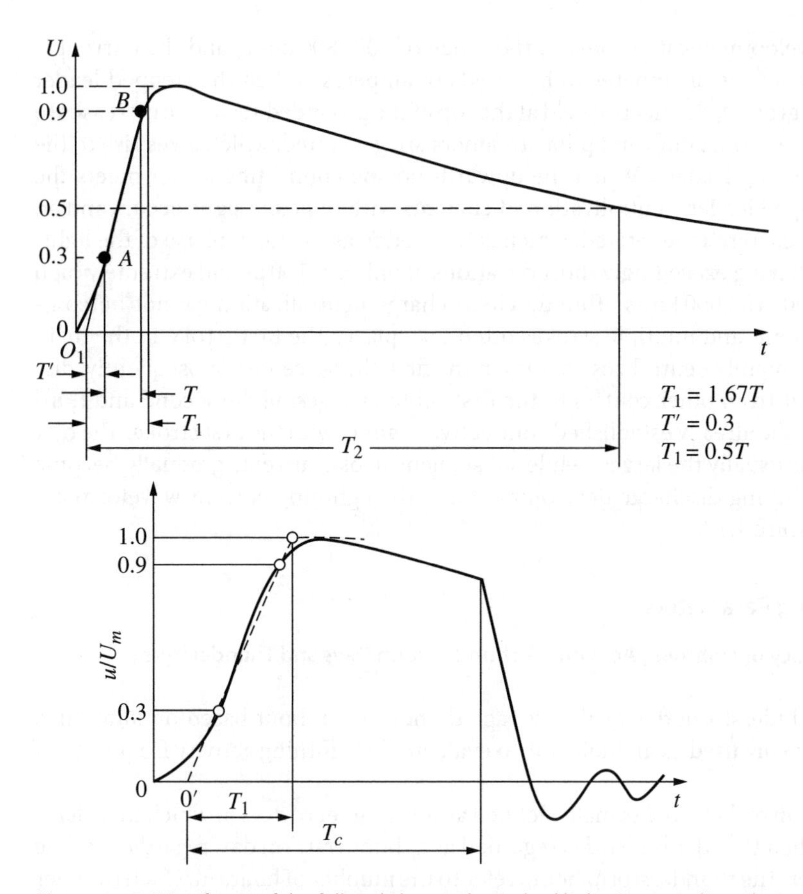

Figure 1.17 Waveforms of the full and chopped standard lightning impulse voltage.

is that it is caused by lightning discharging electrodes rather than metal electrodes. The process and characteristics of lightning discharge will be introduced next.

When thunderclouds appear in the sky, they ascend or descend with the airflow. Generally, the thundercloud bottom carries negative charges, hence most lightning strikes are of negative polarity, and positive charges will be induced on the ground. In this case, high electric field will be generated between the thundercloud and the ground or between two clouds carrying charges of different polarities, giving rise to larger potential differences of several megavolts. The average field intensity, however, is not large owing to the long distances. Lightning discharge may be initiated once the electric field is sufficiently high to set off electron avalanches and corona. And lightning strokes are usually found in linear, flaky and spherical shapes. Although there are others, the linear lightning stroke is the cause of most lightning accidents in power systems, so this book will primarily focus on linear lightning strokes.

The field intensity initiating the discharge usually appears at the bottom of the thundercloud, giving rise to pilot streamers that are followed by downward moving stepped leaders. The leader progresses downwards with a step length of about 25–50 m and a propagation rate of about 10^4 km/s; as the time interval between steps is 30–90 μs, the average development rate is only in the range of 100–800 km/s, and the current is not large, from dozens of amperes to hundreds of amperes. When the stepped leader approaches the ground, the electric field at the top of the grounded structure increases to such a level that air ionization and pilot streamers are generated, which gives rise to the so-called connecting leaders. When the upward-moving connecting leader meets the downward stepped leader, neutralization of currents with opposite signs occurs and an extremely large current is generated, which is referred to as the main phase of the lightning discharge having exceedingly short durations of only 50–100 μs and extremely high speeds of 20,000–150,000 km/s. Thundercloud charge neutralization cannot be completed in one event, and multiple strokes often take place. The first stroke in the main discharge phase mainly neutralizes charges in the first charge center. Subsequently, discharge can occur from other centers to the first center, triggering the second and third strokes through the already established conductive channel with the first stroke. The first stroke current is usually the largest while subsequent stroke currents gradually become smaller. The lightning discharge development and the lightning current waveform are illustrated in Figure 1.18.

1.3.2 Lightning Parameters

1.3.2.1 Frequency of Lightning Activities – Thunderstorm Days and Thunderstorm Hours

The average thunderstorm day or the average thunderstorm hour based on long-term statistics is normally used as an indicator to evaluate the lightning activity frequency of a region.

The thunderstorm days are normally defined as the number of days in which thunder is heard. A day when thunder is heard is regarded as a thunderstorm day, regardless of the times of thunder. The thunderstorm hours refer to the number of hours in which thunder is heard. If thunder is heard within an hour, that hour is regarded as one thunderstorm

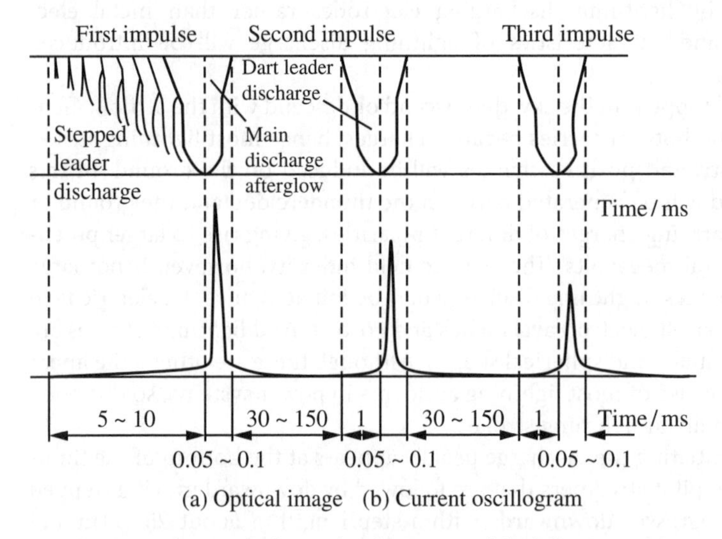

Figure 1.18 Lightning discharge development and the lightning current waveform.

hour. The thunderstorm days and the thunderstorm hours vary greatly between regions around the world as they are related to factors such as climate and terrain.

1.3.2.2 Ground Flash Density

Unable to distinguish between cloud-to-cloud discharge and cloud-to-ground discharge, the thunderstorm days and thunderstorm hours only indicate the frequency of lightning activities. However, from the prospective of lightning protection, the times of cloud-to-ground discharge is of the greatest concern, so ground flash density is introduced to indicate the average number of lightning flashes that strike the ground per square kilometer in a thunderstorm day.

1.3.2.3 Lightning Current Amplitude

The lightning current amplitude is an index indicating lightning intensity, and also the cause for lightning overvoltages. It is the most important lightning parameter, which also receives the most attention. According to long-term measurements, the probability that the lightning current amplitude exceeds I in a normal region is calculated by

$$\lg P = -\frac{I}{88} \tag{1.3.1}$$

where I is the lightning current amplitude, kA; P is the probability that the lightning current amplitude is larger than I.

1.3.2.4 Front Time, Front Steepness and Wavelength of the Lightning Current

According to actual measurements, the front time of the lightning current T_1 is in the range of 1–4 μs, and the average front time is about 2.6 μs; the wavelength of the lightning current T_2 is in the range of 20–100 μs, with the majority around 40 μs. In general, the 2.6/40 μs waveform is used in lightning protection design. The front steepness, an important parameter for lightning evaluation and protection, is determined by the lightning current amplitude and front time. The relation between the front steepness and the amplitude I is expressed by

$$\mathrm{A} = \frac{I}{2.6}(\mathrm{kA}/\mu\mathrm{s}) \tag{1.3.2}$$

1.3.2.5 Lightning Current Waveforms for Calculation

Parameters of the lightning current such as amplitude, front time, steepness and wavelength vary over a wide range, and the lightning current wave is a non-periodic impulse wave. Different waveforms have to be adopted to satisfy various demands in calculations for lightning protection. Several common waveforms for calculation are given in Figure 1.19.

1. Double exponential wave

$$i = I_0(e^{-\alpha t} - e^{-\beta t}) \tag{1.3.3}$$

where I_0 is a particular current value larger than lightning current amplitude I.

2. Oblique wave

$$i = at \tag{1.3.4}$$

where a is the front steepness, kA/ μs.

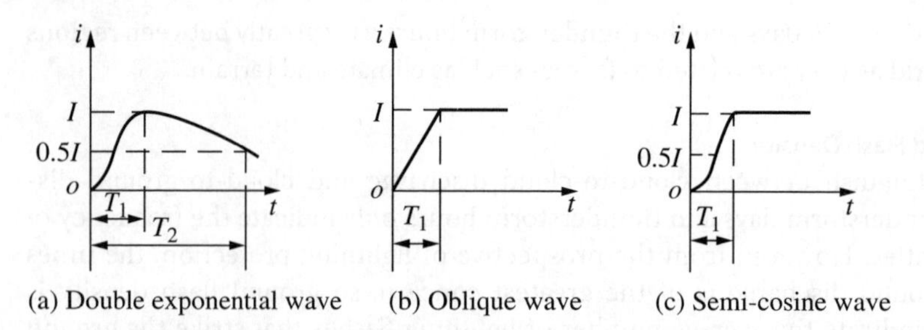

Figure 1.19 Equivalent lightning current waveforms for calculation purposes.

3. Semi-cosine wave

$$i = \frac{I}{2}(1 - \cos \omega t) \tag{1.3.5}$$

1.3.3 Induced Lightning Overvoltages

1.3.3.1 Induced Lightning Overvoltages on the Line When Lightning Strikes the Ground Near the Line

When lightning strikes the ground in vicinity of the line, induced overvoltages will be generated on the conductor because of electromagnetic induction. The generation of induced overvoltages is as shown in Figure 1.20.

In the initial phase of thundercloud discharge, there exists a discharge process in which stepped leaders move towards the ground. As the line is in the electric field generated by the thundercloud and the leader channel, the electric field component E_X along the conductor will attract positive charges at the terminals to the segment close to the

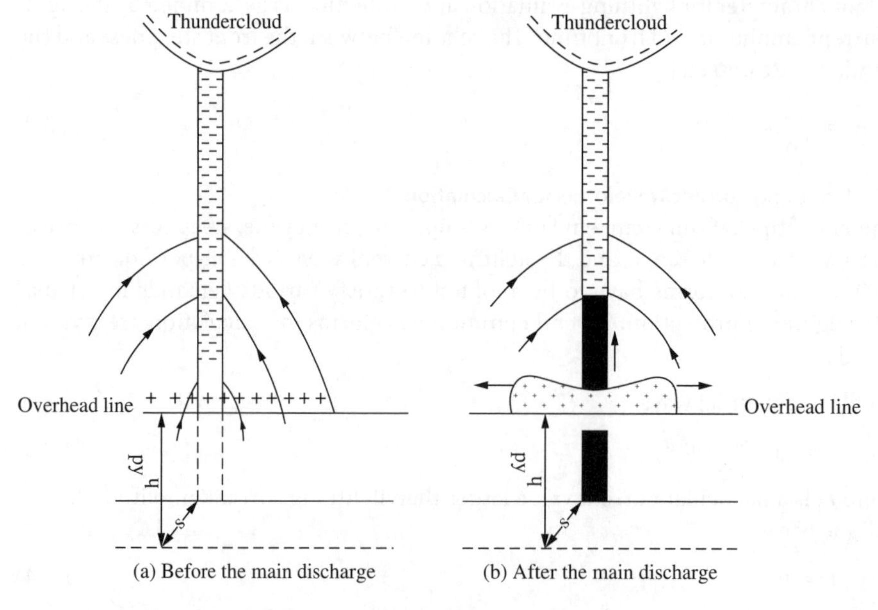

Figure 1.20 Schematic of the generation of induced overvoltages.

leader channel due to electrostatic induction and this will result in bound charges, and the negative charges on the conductor will flow into the ground. When discharge occurs between the cloud and the ground near the line, negative charges in the leader channel are neutralized quickly, and so the resulting electric field also disappears quickly; the released bound charges travel towards the conductor terminals, giving rise to induced overvoltages. The induced overvoltage amplitude occasionally reaches 500–600 kV, high enough for discharge to take place in a 60–80 cm air gap. Therefore, induced overvoltages can endanger transmission lines rated 35 kV and below, where special care has to be taken.

According to theoretical analysis and actual measurements, it is recommended by relevant specifications that when the distance between the lightning strike spot and the line S > 65 m, the maximum induced lightning overvoltage U_g on the conductor is given by

$$U_g = 25\frac{I_L \times h_d}{S}\text{kV} \tag{1.3.6}$$

where I_L is the lightning current amplitude (kA); P_y is the average height of the conductor (m); S is the distance between the lightning strike spot and the line.

As can be concluded from this, the induced lightning overvoltage and the lightning current are of opposite polarities.

As seen from Equation (1.3.6), the induced lightning overvoltage is proportional to the lightning current amplitude I_L and the average line height h_d; a higher h_d will result in a smaller conductor-to-ground capacitance and a higher induced voltage. The induced lightning overvoltage is inversely proportional to the distance S between the lightning strike spot and the line; a larger S will result in a smaller induced overvoltage.

Induced lightning overvoltages are present on three-phase conductors simultaneously without interphase potential differences, hence only flashover to ground can happen. If ground wires hang above the conductor, induced charges and the induced overvoltage on the conductor will be reduced due to the shielding effect. Assuming that the average heights to ground of the conductor and the ground wire are h_d and h_b respectively, if the ground wire is not grounded, then the induced overvoltages on the ground wire and on the conductor $U_{g,b}$ and $U_{g,d}$ can be solved based on Equation (1.3.6), expressed as

$$U_{g.b} = 25\frac{I_L \times h_b}{S} \tag{1.3.7}$$

$$U_{g.d} = 25\frac{I_L \times h_d}{S} \tag{1.3.8}$$

Therefore,

$$U_{g.b} = U_{g.d}\frac{h_b}{h_d} \tag{1.3.9}$$

As the ground wire is actually grounded through each line tower, a potential $U_{g,b}$ is assumed on the ground wire to keep the ground wire at zero potential. Because of the coupling effect between the ground wire and the conductor, $U_{g,b}$ will generate a coupling voltage $kU_{g,b}$ on the conductor, where k is the coupling coefficient between the ground wire and the conductor.

In this way, the potential on the conductor $U'_{g.d}$ can be expressed as

$$U'_{g.d} = U_{g.d} - KU_{g.b} \approx U_{g.d}(1 - K) \tag{1.3.10}$$

According to the above equation, the induced overvoltage on the conductor is reduced from $U_{g,d}$ to $U'_{g,d}$ with the help of ground wires. The larger the coupling coefficient K, the lower the induced overvoltage on the conductor.

1.3.3.2 Induced Overvoltages on the Line When Lightning Strikes the Line Tower

Equation (1.2.6) is valid only when $S > 65$ m, and lightning with shorter distances from the line will directly strike the line due to the lightning attraction effect. When lightning strikes the line tower, voltage of the polarity opposite to the lightning current will be induced on the conductor due to rapid changes in electromagnetic field generated by the lightning channel. Presently, in the case of lines of a general height (approximately below 40 m) without equipping ground wires, the maximum induced overvoltage is given by

$$U_{g.d} = ah_d \tag{1.3.11}$$

where a is the overvoltage coefficient (with unit of kV/m), and its value is equal to the average steepness of the lightning current $\alpha = I_L/2.6$ (kA/μs). When the ground wire is installed, considering the shielding effect, Equation (1.2.11) becomes

$$U'_{g.d} = ah_d(1 - K) \tag{1.3.12}$$

where K is the coupling coefficient.

When induced lightning occurs on transmission lines, the travelling current waves are induced current, and the three-phase wave shapes resemble each other. In the case of back flashover, electromagnetic coupling current appears before insulator breakdown and instantly transforms into direct stroke current after the breakdown. In the case of shielding failures, the currents flowing through the lines are direct stroke current components. Figures 1.21–1.23 demonstrate the travelling current waves at the strike spot on the 220 kV transmission line.

For induced lightning overvoltages, the wave shapes of its three-phase induced current are still similar after transmission and attenuation on the lines. In the case of overvoltages caused by back flashover, the presence of electromagnetic coupling components having small gradient and long rise time can be found. After the insulator

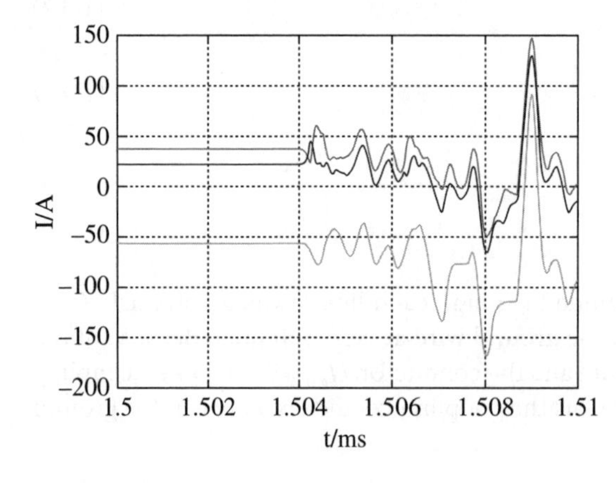

Figure 1.21 Induced lightning overvoltages.

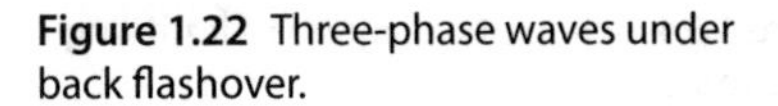

Figure 1.22 Three-phase waves under back flashover.

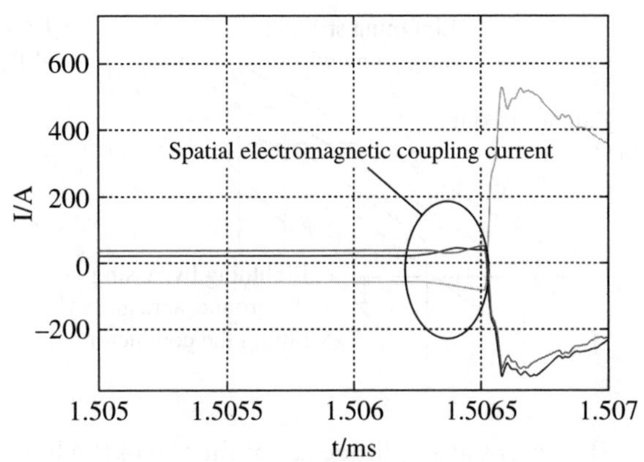

Figure 1.23 Three-phase current waves under shielding failures.

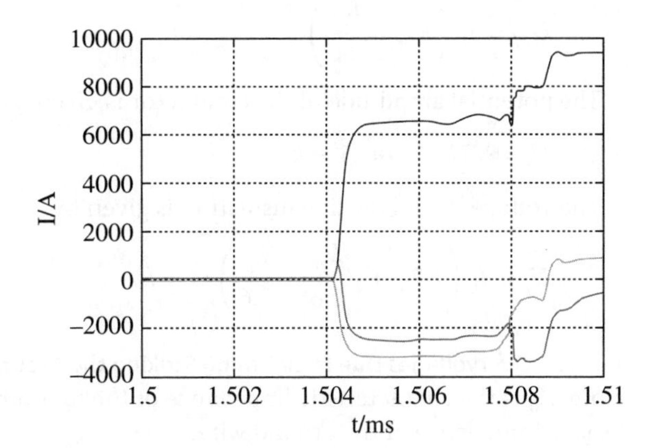

string breakdown, there comes a surge in current amplitude and an increase in current gradient due to the large stroke current injected into the conductor. In the case of overvoltages caused by shielding failures, the current surges, shortly after the occurrence of lightning, as there is no electromagnetic coupling current. Therefore, the similarity between three-phase current waves and the existence or non-existence of the electromagnetic coupling current are adopted as the parameters to identify overvoltage categories.

1.3.4 Direct Lightning Overvoltages

There are three situations (Figure 1.24) for lightning directly striking the line: lightning striking the tower top, lightning striking the ground wire at midspan and lightning bypassing the ground wire and striking the conductor (also referred to as the shielding failure).

1.3.4.1 Overvoltages Due to Lightning Striking the Tower Top

In approximate calculation in engineering, line towers and ground wires are often replaced by lumped parameter inductors L_{gt} and L_b, respectively, as illustrated in Figure 1.25, which shows the equivalent circuit of lightning striking the line tower.

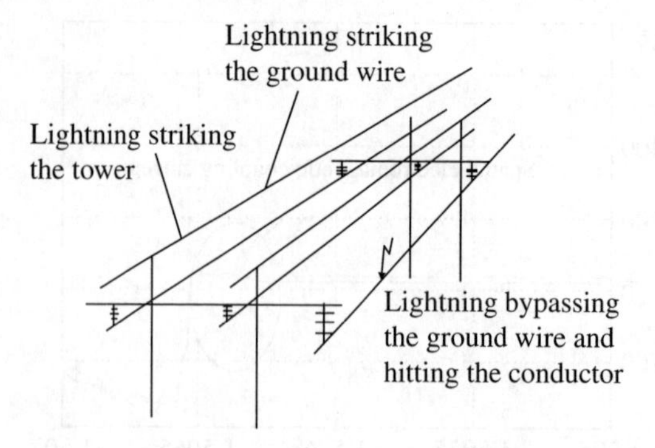

Figure 1.24 Schematic diagram of lightning strike spots.

The potential amplitude U_{td} at the top of the line tower is given by

$$U_{td} = I_L \left(R_{ch} + \frac{L_{gt}}{2.6} \right) \quad (1.3.13)$$

The potential amplitude of the conductor is given by

$$U_d = kU_{td} - ah_d(1 - k) \quad (1.3.14)$$

The voltage across the line insulation is given by

$$U_j = I_L \left(\beta R_{ch} + \beta \frac{L_{gt}}{2.6} + \frac{h_d}{2.6} \right) (1 - k) \quad (1.3.15)$$

1.3.4.2 Overvoltages Due to Lightning Striking the Ground Wire at Midspan

Note: Z_0 is the characteristic impedance of the lightning channel, and Z_b is the characteristic impedance of the ground wire.

Using the parameters in Figure 1.26, the maximum voltage at the strike spot is

$$U_A = a \times \frac{l}{v_b} \times \frac{Z_0 Z_b}{2Z_0 + Z_b}(1 - k) \quad (1.3.16)$$

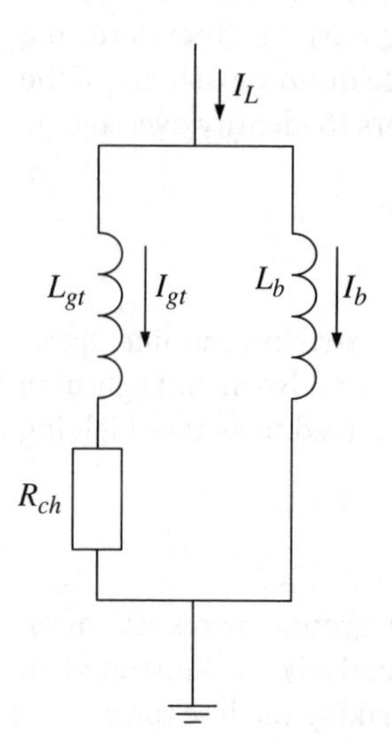

Figure 1.25 Equivalent circuit of lightning striking the line tower.

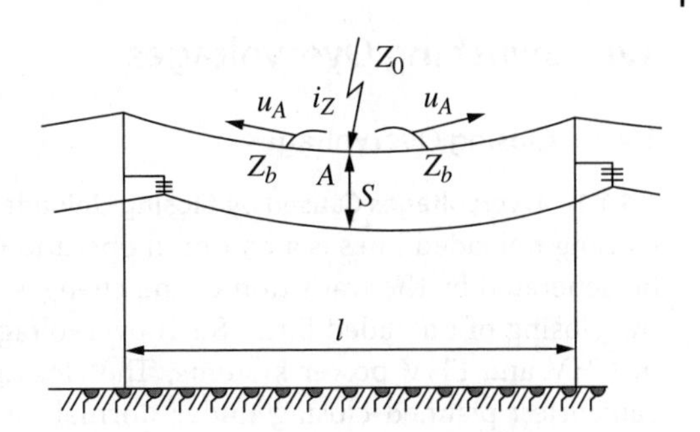

Figure 1.26 Lightning striking the ground wire at midspan.

The coupling voltage kU_A will be generated on the conductor due to coupling between the ground wire and the conductor, and hence the maximum voltage U_s across the air gap S between the ground wire and the conductor at the strike spot is

$$U_s = U_A(1-k) = a \times \frac{l}{v_b} \times \frac{Z_0 Z_b}{2Z_0 + Z_b}(1-K) \tag{1.3.17}$$

1.3.4.3 Overvoltages Due to Shielding Failures

The installation of ground wires is able to protect three-phase conductors. However, there is still the possibility that lightning bypasses the ground wire and directly hits the conductor. The probability of such a shielding failure is referred to as the shielding failure probability p_α.

In plain areas,

$$l_g p_\alpha = \frac{\alpha\sqrt{h}}{86} - 3.9 \tag{1.3.18}$$

and in mountainous regions,

$$l_g p_\alpha = \frac{\alpha\sqrt{h}}{86} - 3.35 \tag{1.3.19}$$

where a is the shielding angle and h is the tower height (m).

As seen in the above equation, the shielding failure probability in the mountainous region is twice to three times as high as that of the plain area, equivalent to an 8° increase in the shielding angle.

Overvoltages caused by shielding failures are calculated as follows. In the case of shielding failures, the impedance of the strike spot is $Z_d/2$ (Z_d is the wave impedance of the conductor), and the lightning current wave i_z flowing through the strike spot is

$$i_z = \frac{i_L}{1 + \frac{Zd/2}{Z_0}} \tag{1.3.20}$$

The amplitude of the voltage across the conductor is

$$U_d = I_L \frac{Z_0 Zd}{2Z_0 + Zd} \tag{1.3.21}$$

1.4 Switching Overvoltages

1.4.1 Closing Overvoltages

1.4.1.1 Overvoltages Caused by Closing Unloaded Lines

Closing unloaded lines is a common operation in the power system. Overvoltages will be generated by the transition of the changes, as the line voltage changes suddenly at the closing of unloaded lines. Such overvoltages are the main switching overvoltages in UHV and EHV power systems. The closing of unloaded lines is divided into two categories: planned closing under normal operation (such as putting overhauled or newly constructed lines into operation as per the plan) and automatic reclosing after clearing line faults that may cause higher overvoltages due to the non-zero initial conditions.

1.4.1.2 Overvoltages Caused by Planned Closing

Before closing the unloaded line, no residual voltage and grounding fault are present on the line; the three phases are symmetric; the line is under a zero initial state. The single-phase line is studied without considering the electromagnetic coupling between the conductors. A T-type equivalent circuit for the line is adopted, which is a LC oscillating circuit, as illustrated in Figure 1.27. Due to high oscillating frequency ($f_0 = 1/2\pi\sqrt{L_s C_T}$), the supply amplitude can be regarded as a constant value E in a short time. E is related to the closing voltage phase angle. In the worst case, the supply voltage amplitude E_m is given by the differential equation

$$E_m = L_s \frac{di}{dt} + u_c \tag{1.4.1}$$

with $i = C_T \frac{du_c}{dt}$, which gives

$$L_s C_T \frac{d^2 u_c}{dt^2} + u_c = E_m \tag{1.4.2}$$

This is a second-order differential equation with its solution given by

$$u_c = A \sin \omega_0 t + B \cos \omega_0 t + E_m \tag{1.4.3}$$

where $\omega_0 = 1/\sqrt{L_s C_T}$; A and B are integration constants. Under the initial condition that $t = 0$, $u_c = 0$, $i = C_T \frac{du_c}{dt} = 0$, it can be derived that $A = 0$, $B = -E_m$, which gives

$$u_c = E_m(1 - \cos \omega_0 t) \tag{1.4.4}$$

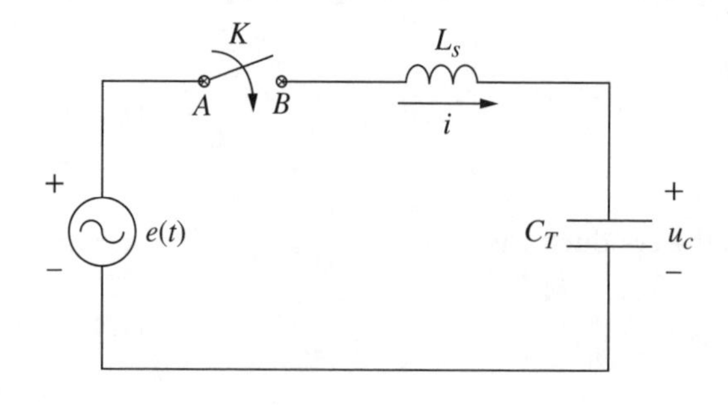

Figure 1.27 Equivalent circuit of closing unloaded lines.

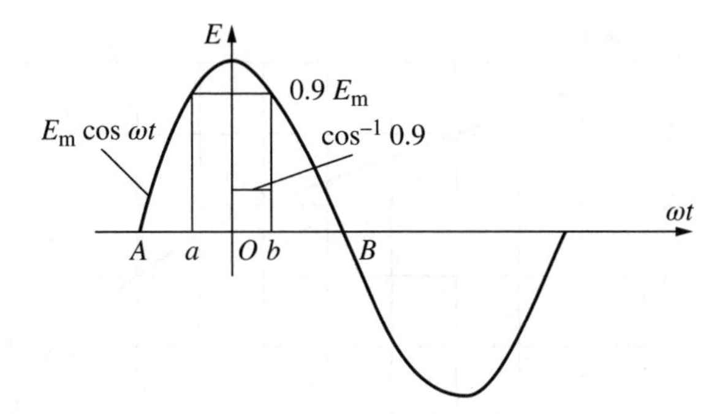

Figure 1.28 Schematic of closing probability calculation.

When $\omega_0 t = \pi$, u_c has the maximum value of

$$u_{cm} = 2E_m \tag{1.4.5}$$

When closing the unloaded line, the phase angle φ of the supply voltage is a random variable over the range of a cycle (see Figure 1.28), and uniform distribution is generally adopted. At closing the line, the supply voltage $|E|$ is usually lower than E_m, and the closing overvoltage is usually lower than $2E_m$ as well.

When closing unloaded lines, resistance loss and corona loss will reduce the overvoltage amplitude. Considering the voltage rise caused by the capacitive effect of transmission lines, the measured overvoltage is 1.9–1.96 E_m.

The above analysis is based on the worst condition: the supply voltage amplitude equals E_m at the closing of the line. However, according to Figure 1.25, when calculating the distribution of closing overvoltage probability, the closing phase angle value has to be distributed uniformly over a range of the length 2π.

Assuming that the per unit amplitude of the supply voltage $|E|/E_m = 0.9$, the probability that the supply voltage $|E|/E_m \geq 0.9$ when closing the line is

$$P(|E|/E_m \geq 0.9) = \frac{ab}{AB} = \frac{2\arccos 0.9}{\pi} = 28.7\%$$

which means that the probability that the supply voltage is equal to or larger than $0.9E_m$ when closing the line is 28.7%.

Generally speaking, the probability that $|E|/E_m \geq k$ (k ranges between 0 and 1) is given by

$$\varphi = \frac{2\arccos k}{\pi} \tag{1.4.6}$$

According to Equation (1.4.4), when closing unloaded lines, the maximum overvoltage $2E = 2kE_m$. The line overvoltage multiple $K_0 = 2\beta K_c K$, if considering the attenuation coefficient β and the power frequency voltage rise coefficient k_c (which is in the range 1.1–1.15 for receiving ends of 220 kV lines, and 1.15–1.30 for receiving ends of 330 kV lines).

Taking all the mentioned factors into account, the probability for the overvoltage to exceed K_0 is given by

$$\varphi = \frac{2}{\pi}\arccos\frac{K_0}{2\beta K_c} \tag{1.4.7}$$

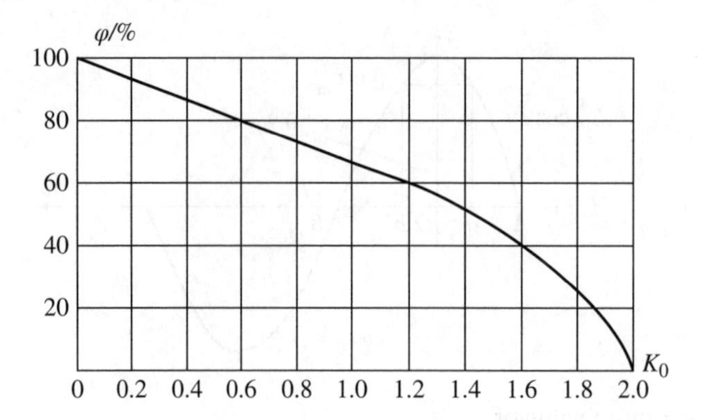

Figure 1.29 The probability curve of the closing overvoltage.

Figure 1.29 displays the cumulative probability distribution of overvoltages caused by closing unloaded lines. In the calculation, $\beta = 0.9$, $K_c = 1.1$ and the extremum of this overvoltage is $2\beta K_c E_m$.

The above analysis utilizes the LC model of the lumped parameter circuit. If the distributed parameter model of multi-phase transmission lines is used, the overvoltage multiple may be larger than 2 under the condition that hundreds of times of closing operations are performed on long transmission lines and electromagnetic coupling between the conductors are taken into account. Such a result is impossible when adopting a lumped parameter circuit model.

1.4.1.3 Overvoltages Caused by Automatic Reclosing

Overvoltages caused by automatic line reclosing are relatively higher. When phase C of the transmission line is grounded, K_2 trips first, then K_1 is actuated and connected to the power supply, as shown in Figure 1.30.

Currents flowing through non-fault phases A and B are capacitive, and the tripping of K_1 is equivalent to switching off unloaded lines. According to the analysis, the arc at the non-fault phase contact of K_1 is extinguished at the zero-crossing point of the capacitive current. At this particular moment, the supply voltage reaches its maximum value. Considering factors such as single-phase grounding and the Ferranti effect, the average line residual voltage $u_r = 1.3E_m$. Before the automatic reclosing of K_1, residual charges of the line flow into the ground via conductor leakage resistance, which includes leakage of the insulator string conductance and air conductance (see Figure 1.31), which is

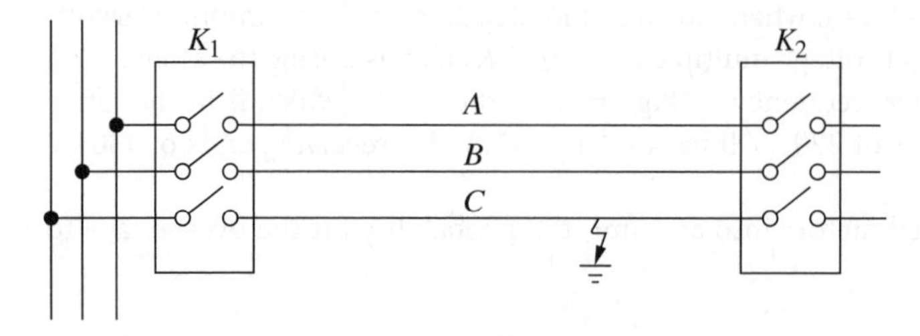

Figure 1.30 Schematic diagram of automatic reclosing.

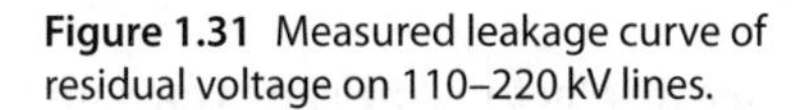
Figure 1.31 Measured leakage curve of residual voltage on 110–220 kV lines.

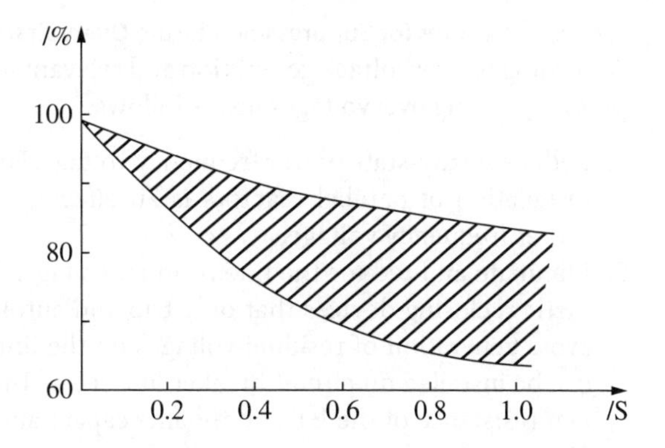

related to surface pollution of insulator strings and climate conditions such as humidity, rain and snow. The power supply is attenuated over a wide range. As seen in the figure, the residual voltage is reduced to 10–30% of the initial value within 0.5 s. Generally, the time constant of line discharge is larger than 0.1 s.

Assuming that before the automatic reclosing of K_1, the residual line voltage has dropped by 30%, then $u_r = (1 - 0.3) \times 1.3E_m = 0.91E_m$.

The worst closing time is the instant when the polarity of the power supply is opposite to that of the line residual voltage, and the power supply amplitude attains the maximum value $-E_m$. Such a situation is similar to switching off unloaded lines. At this moment, high frequency oscillation will occur on lines with a stable value of $-E_m$, an initial value of $0.91E_m$, and the maximum value $u_{0v} = -E_m + (-E_m - 0.91E_m) = -2.91E_m$. If the charge leakage of the line is ignored, the overvoltage may be even higher. However, the supply voltage may not be $-E_m$ at the reclosing instant. In this case, the overvoltage will be relatively lower.

1.4.1.4 Factors Influencing Closing Overvoltages

According to the above analysis, overvoltages caused by closing unloaded lines depend on supply voltage phase angles at the closing moment, and the phase angle is a random value following statistical regularity. Pre-breakdown phenomena will generally occur at the closing of circuit breakers: as the contacts are getting closer during the closing process, the potential difference between the contacts has already resulted in dielectric breakdown, enabling electrical connection prior to physical touch of the mechanical contacts. According to statistics of oil circuit breakers, the closing phase angle is mostly located $\pm 30°$ surrounding the maximum value. However, in the case of high-speed air circuit breakers, the closing phase angle is distributed fairly uniformly.

The overvoltage amplitude is also significantly affected by the polarity and amplitude of the residual voltage, which is a key characteristic of three-phase reclosing overvoltages. The insulator surface leakage may affect the residual voltage as well. When the reclosing duration is 0.3–0.5s, the residual voltage usually drops by 10–30%. When single phase reclosing operation is utilized in the system, there is no residual voltage on the lines.

Besides, overvoltages caused by closing unloaded lines are related to such factors as power network structures, line parameters, three-phase synchronization at the closing of circuit breakers, the number of busbar outgoing lines and conductor corona.

1.4.1.5 Measures for Suppressing Closing Overvoltages

According to overvoltage generation and relevant influencing factors, measures for suppressing closing overvoltages are as follows:

1. Reduce steady-state power frequency voltage. For long power frequency lines, proper installation of parallel reactors is an effective measure to reduce the steady-state power frequency voltage.
2. Eliminate and reduce the line residual voltage. The utilization of single-phase automatic reclosing devices that only trip and automatically reclose the fault phase can avoid generation of residual voltages on the lines. In addition, electromagnetic PTs can be installed on circuit breaker line sides. The equivalent inductance and equivalent resistance of the PT and the line capacitance form a damped oscillating circuit. Since the series resistor consumes energy, it can be seen as a damped factor of oscillation, releasing line residual charges within several power frequency cycles before automatic reclosing so that reclosing overvoltages will not double the original voltage.
3. Use circuit breakers with parallel resistors. When closing auxiliary contacts, the higher the parallel resistance, the lower the overvoltage; when closing main contacts, the lower the parallel resistance, the lower the overvoltage. To sum up, the reclosing overvoltage multiple is in a V-shape relation with the parallel resistance. An appropriate resistor has to be selected according to line parameters as well as the manufacture difficulty.
4. Adopt synchronous closing. Closing operations carried out by special devices to enable the same polarity and even the same potential for breaker contact terminals are able to considerably reduce or even eliminate closing and reclosing overvoltages.
5. Use high performance arresters. Installing ZnO or magnetic blowout arresters at the starting and receiving ends of the lines (at breaker line sides) can restrict overvoltages generated when closing unloaded lines. Arresters have to be able to operate reliably under high-amplitude overvoltages resulting from parallel resistor faults or other accidents, so as to control the overvoltages within the allowable range, so arresters are deployed as backup protection.

1.4.2 Opening Overvoltages

1.4.2.1 Overvoltages Caused by De-Energizing Unloaded Lines

Switching off unloaded lines is a common operation in the power system under normal operation. In China, a high insulation level is provided for power systems rated 35–220 kV, but accidents such as flashover and breakdown still take place sometimes, due to de-energizing unloaded lines. Arc reignition is the root cause for such an overvoltage having a high amplitude and a duration as long as 0.5 to 1.0 power frequency cycles. Besides, this overvoltage is used as the design basis for determining the switching impulse level of power systems rated 220 kV or below.

For unloaded lines, the current flowing through the breaker is virtually capacitive and is related to such factors as voltage grade, line length and line structure. Generally, the current is only several to hundreds of amperes, much lower than the short-circuit current in the high-voltage system. A circuit breaker capable of interrupting large short-circuit currents may not be able to interrupt capacitive current arising from de-energizing unloaded lines. This is because the high recovery voltage between the breaker contacts can cause gap breakdown and arc reignition. If arc reignition occurs in the gap, the capacitive circuit connected to the de-energized line where electromagnetic oscillation is generated will continue to absorb energy from the power supply and produce overvoltages. For this reason, a large capacity to interrupt short-circuit current and an unloaded line de-energization experiment are required for high-voltage circuit breakers. Switching off capacitor banks will give rise to overvoltages of similar kinds.

1.4.2.2 Physical Process

An unloaded line can be represented by a T-type equivalent circuit, where L_T is the line inductance, C_T is the line-to-ground capacitance, L_e is the sum of the leakage inductance of the generator and the transformer and the bus capacitance is ignored (Figure 1.32).

In the common case of switching off unloaded lines, the amplitude of the maximum operational phase voltage of the power supply $e(t)$ is U_{xg}. When faults are cleared, the transient potential of the generator or the system has to be used for calculation. In Figure 1.32(b), the schematic of simplified equivalent circuit, it is assumed that the supply potential is

$$e(t) = E_m \cos \omega t \tag{1.4.8}$$

Before switching off the line, the inductance is small, and the current is given by

$$i(t) \approx \frac{E_m}{X_c - X_s} \cos(\omega t + 90^\circ) \tag{1.4.9}$$

where X_c and X_s are the capacitive reactance of C_T and the inductive reactance of L_s, respectively.

The overvoltage caused by switching off unloaded lines develops as illustrated in Figure 1.33.

Given that inductance L_s may be negligible, the line voltage is $u(t)$, that is, capacitor voltage $u_c(t)$ equals the supply voltage $e(t)$ before disconnecting the circuit breaker, i.e. prior to time instant t_1 (Figure 1.33). In the worst case in which the circuit breaker is disconnected at time instant t_1 when the line voltage is $-E_m$, the power frequency current of the line would just pass the zero-crossing point, and the arc of the circuit breaker would extinguish. In practice, if the circuit breaker is disconnected in the first half cycle before time instant t_1 with the presence of the power supply, the arc between breaker

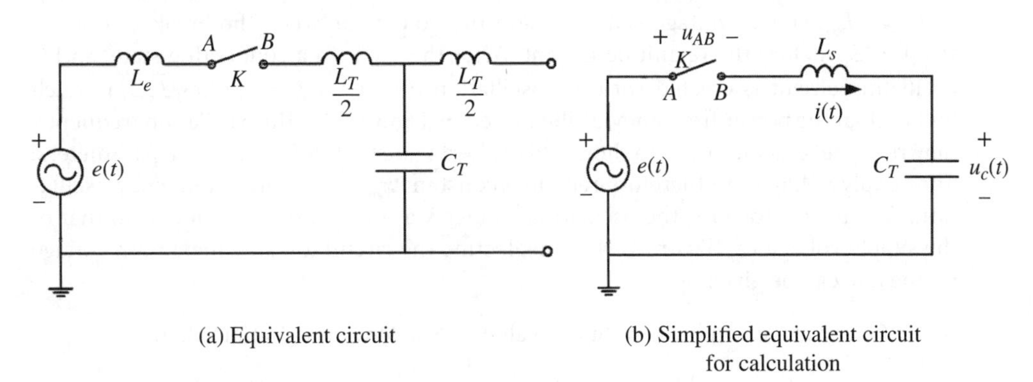

Figure 1.32 Equivalent circuit of switching off unloaded lines.

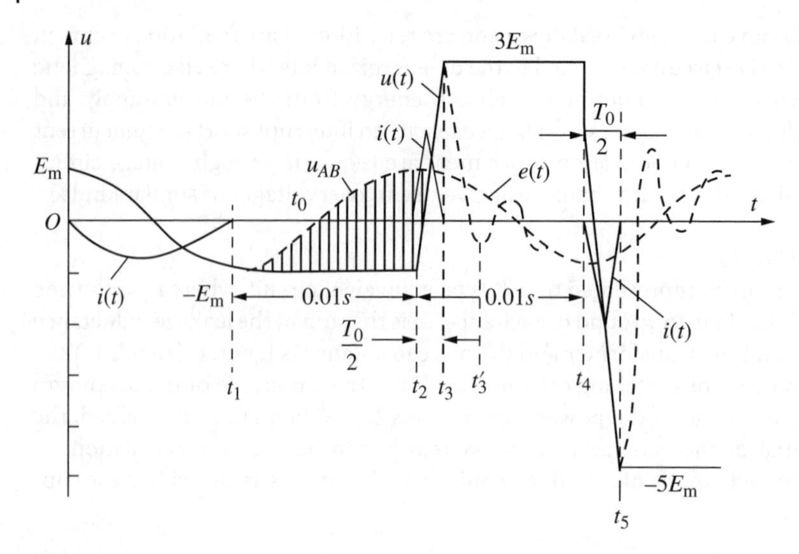

Figure 1.33 Development process of the overvoltage caused by de-energizing unloaded lines.

contacts would be extinguished at time instant t_1. The line voltage remains $-E_m$ in view of line-to-ground insulation, but the potential at breaker contact A i.e. the dotted line after t_1, changes according to the cosine curve. The recovery voltage increasing gradually between the breaker contacts is given by

$$u_{AB} = u_A - u_B = e(t) - (-E_m) = E_m(1 + \cos \omega t) \tag{1.4.10}$$

If the insulation strength between the breaker contacts recovers faster than the rising rate of recovery voltage u_{AB}, the arc between the contacts will not be reignited, and line de-energization will not result in overvoltage. However, if the circuit breaker performs poorly, the insulation strength will recover slowly; $|u_{AB}|$ may reach $2E_m$ at t_1, and arc reignition is likely to occur within the time interval t_1–t_2. Depending on the randomness of the occurrence and the duration of arc reignition, the overvoltage magnitude is thus random as well.

The most severe arc reignition occurs at time instant t_2 when capacitor voltage $u_c(t_2) = -E_m$, supply voltage $e(t) = E_m$, and the voltage between the breaker contacts $|u_{AB}| = 2E_m$ before the reigniting instant. After the arc is in a conducting state, an LC oscillating circuit is created with the oscillation frequency $f_0 = 1/(2\pi\sqrt{L_s C_T})$, much higher than the power frequency, as illustrated by Figure 1.33. The oscillation frequency may reach several hundreds to thousands of hertz due to different network parameters. The supply voltage can therefore remain a constant E_m during high-frequency oscillation. Due to resistor loss, the oscillating voltage waveform would be similar to that of the supply voltage E_m (Figure 1.34). If neglecting the circuit loss, the maximum voltage on the line can be given by

$$\text{Overvoltage amplitude} = \text{Stable value} + (\text{Stable value} - \text{Initial value})$$

In Figure 1.34, the stable value $= E_m$, the initial value $= -E_m$, and the overvoltage amplitude $= E_m + [E_m - (-E_m)] = 3E_m$.

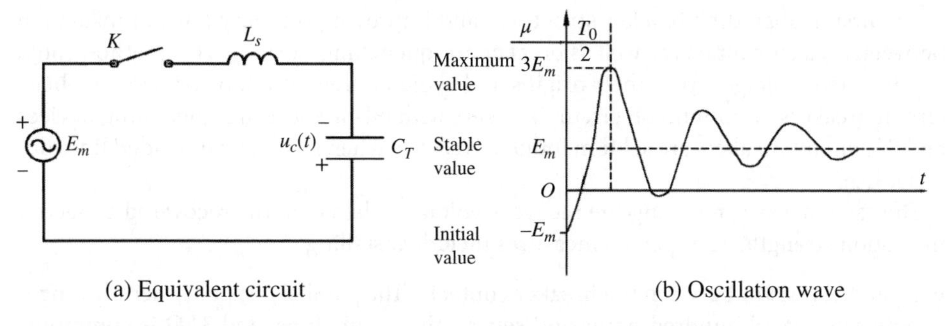

Figure 1.34 Equivalent circuit and the oscillation wave during arc reignition.

According to the oscillation wave, electromagnetic oscillation in the LC circuit is similar to the mechanical oscillation of a swing, moving from the initial position to the static position, and then to the other extreme position with the same distance.

Under the most dangerous condition, the overvoltage is developed as $3E_m$, $-5E_m$, $7E_m$, … In practical cases, arc reignition cannot increase infinitely due to a number of complicated factors.

1.4.2.3 Influencing Factors

The overvoltage caused by de-energizing unloaded lines is generated by arc reignition between the contacts when disconnecting circuit breakers. To reduce the occurrence probability of such overvoltages, the arc-quenching performances of circuit breakers need to be improved. If breaker restrike does not occur when the supply voltage reaches its maximum value, and arc-quenching does not occur when the high frequency current crosses the zero position, the generated overvoltage magnitude will be lower. Besides, the recovery capacity of breaker insulation will be considerably improved with the increase in opening time and gap distance, and hence probability of a second restrike will be reduced as well.

When a system bus connects alternative outgoing lines, a large capacitor regarded as connected in parallel with the supply side of the equivalent circuit would give rise to charge redistribution along with the equivalent line capacitance C_T at the breaker restriking instant. As a result, the initial line voltage would approach the supply voltage value at that instant, reducing oscillation magnitudes and overvoltages. In addition, active loads of the outgoing lines play the role of damped oscillation, which can also reduce overvoltages.

The neutral grounding method would also significantly impact overvoltages caused by unloaded line de-energization. If the neutral point is ungrounded, asynchronous operation of three-phase circuit breakers will introduce instantaneous asymmetric components and generate neutral displacement voltage, thus increasing the overvoltage.

In addition, high-amplitude overvoltages caused by de-energizing unloaded lines can be reduced due to corona-caused losses.

1.4.2.4 Measures

The most fundamental measure to limit overvoltages caused by switching off unloaded lines is to eliminate circuit breaker restrike, which could be achieved in two ways.

The first is upgrading breaker structures and improving recovery rates of insulation between breaker contacts as well as breaker arc-quenching capacity. This method could suppress overvoltages from their origins and avoid arc reignition. At present, air-blast circuit breakers, minimum oil circuit breakers with oil-pressure arc-quenching devices and SF_6 circuit breakers are seldom found to restrike when switching unloaded lines out of service.

The second way is reducing the recovery voltage to less than the recovered dielectric insulation strength. The specific measures include installing:

- a parallel resistor between the breaker contacts. The parallel resistor generally ranges between several hundred ohms and several thousand ohms, and 3 kΩ is commonly selected to suppress overvoltages caused by de-energizing unloaded lines.
- electromagnetic PTs at the breaker line sides. If the transmission line is equipped with an electromagnetic PT, residual charges of the line will be discharged via the PT when the circuit breaker is disconnected. The leakage will accelerate attenuation in the transition period; residual charges trapped in the line will be completely released within several power frequency periods, enabling a fast decline in recovery voltage between breaker contacts and thus avoiding or reducing overvoltages caused by breaker restrike.
- parallel reactors for EHV transmission lines. When disconnecting the breaker, the parallel reactor and the line capacitance constitute an oscillation circuit with its natural frequency approaching the power frequency; the line voltage then becomes the oscillating power frequency voltage, which considerably reduces the rising rate of the recovery voltage across the breaker contacts. In this way, arc reignition will be avoided and the probability of a high-amplitude overvoltage is minimized.
- zinc oxide arresters with high performance. These are intended for backup protection for overvoltages due to disconnecting unloaded lines.

1.4.2.5 Switching Off Unloaded Transformers

Switching off unloaded transformers is a common operation in the power system. During normal operation, an unloaded transformer appears as an excitation inductor, and disconnecting an unloaded transformer is like interrupting a small inductive load. The original electromagnetic energy stored in the interrupted inductive component will be transformed and released in the circuit constituted by the inductor and the capacitance to ground, generating oscillation and high switching overvoltages on the transformer and the breaker. Similarly, the presence of such overvoltages can also be found when interrupting such inductive components as parallel reactors, arc-quenching coils and motors. The cause, influencing factors and restrictive measures concerning such overvoltages are illustrated in the following sections, by analyzing the disconnection of unloaded transformers.

1.4.2.6 Cause and Physical Process

Such overvoltages are generated due to energy stored in the inductor as the inductor current is forced to be interrupted by the circuit breaker prior to the natural zero-crossing. When interrupting alternating current of over 100 A, the arc between the breaker contacts could generally be extinguished at the natural zero-crossing point; at this particular moment, the equivalent inductance stores no magnetic energy and overvoltages would not be generated. However, when switching off unloaded transformers, the excitation current is small, generally 0.5–4% of the rated current with its effective value ranging from several amps to dozens of amps. Under the condition of intense breaker deionization, current chopping will occur due to forced arc-quenching before the current crosses the zero position, and electromagnetic energy will be stored in the inductive component. When the magnetic energy of the inductor is transformed into electric energy of the capacitor, overvoltages will appear because the small capacitance leads to a high voltage.

The simplified equivalent circuit of disconnecting unloaded transformers is used to analyze the physical process, as shown in Figure 1.35, where L_T is the excitation inductance of the unloaded transformer; C_T is the parallel capacitance of the stray capacitance to ground of the transformer and the capacitance to ground of the transformer-side lead; L_S is the equivalent inductance of the bus-side supply; QF is the circuit breaker. Given that $i_c \ll i_{L_T}$ under the application of power frequency voltage, the approximation $i = i_{L_T} + i_c \approx i_{L_T}$ can be adopted. Thus, the unloaded current lags the supply voltage.

Assuming that the unloaded current is chopped when $i = I_0 = I_m \sin\alpha$ (α is the phase angle at the current-chopping instant), the supply voltage $U_0 = U_m \sin(\alpha+90°)\cos\alpha$. When the transformer excitation current is suddenly cut off by the breaker, the current of the circuit has a large changing rate, i.e. $\frac{di}{dt}$ is large, the voltage across the transformer winding inductance $L\frac{di}{dt}$ is large, thus resulting in overvoltages. In terms of energy conversion, the original magnetic energy stored in the excitation inductance and the original electric energy stored in the excitation capacitance to ground at the instant of current chopping are $W_L = \frac{1}{2}L_T I_0^2$ and $W_C = \frac{1}{2}C_T U_0^2$, respectively. This is followed by electromagnetic oscillation in the circuit constituted by L_T and C_T. The maximum voltage $U_{\max}$ will appear across the capacitor when all the energy is stored in the capacitor, as given by the following equations based on the energy conservation law.

$$\frac{1}{2}C_T U_{max}^2 = \frac{1}{2}L_T I_0^2 + \frac{1}{2}C_T U_0^2 \tag{1.4.11}$$

$$U_{\max} = \sqrt{U_{\mathrm{m}}^2 \cos^2\alpha + \frac{L_{\mathrm{T}}}{C_{\mathrm{T}}} I_{\mathrm{m}}^2 \sin^2\alpha} \tag{1.4.12}$$

With $I_m \approx \frac{E_m}{2\pi f L}$ and the natural frequency $f_0 = \frac{1}{2\pi\sqrt{L_T C_T}}$,

$$U_{max} = U_m \sqrt{\cos^2\alpha + \left(\frac{f_0}{f}\right)^2 \sin^2\alpha} \tag{1.4.13}$$

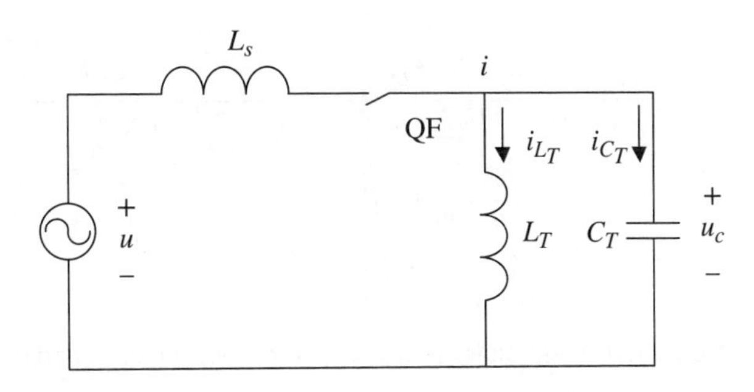

Figure 1.35 Equivalent circuit of disconnecting unloaded transformers.

Therefore, the overvoltage multiple is represented by

$$K = \frac{U_{max}}{U_m} = \sqrt{\cos^2\alpha + \left(\frac{f_0}{f}\right)^2 \sin^2\alpha} \tag{1.4.14}$$

Given that energy losses like hysteresis and eddy current losses of the iron core and conductor copper loss exist in the practical circuit during energy conversion, the loss coefficient η_m ($\eta_m < 1$) is introduced with its value related to the material of the iron core of the transformer winding and the circuit oscillating frequency. A smaller loss coefficient corresponds to a higher frequency, and the typical range is 0.3–0.5. By introducing such a revision, the overvoltage multiple K is expressed as

$$K = \sqrt{\cos^2\alpha + \eta_m \left(\frac{f_0}{f}\right)^2 \sin^2\alpha} \tag{1.4.15}$$

As can be seen from this equation, the overvoltage multiple attains its maximum when $\alpha = 90°$, i.e. the unloaded excitation current is cut off at the peak, and the maximum overvoltage multiple is given by

$$K = \sqrt{\eta_m}\frac{f_0}{f} \tag{1.4.16}$$

1.4.2.7 Waveform Characteristics

Current chopping usually takes place in the descending portion of the curve. As an example, if $K = \sqrt{\eta_m}\frac{f_0}{f}$ is positive, the voltage is certainly negative, based on the voltage-current phase relation. When the arc is suddenly quenched, the excitation inductor current is unable to change instantly, and would continue to charge the capacitance to ground, increasing the voltage across the capacitor continuously in the negative position. Subsequently, attenuated oscillation would happen due to energy losses. The oscillation frequency is shown in Figure 1.36.

The circuit oscillation frequency f_0 is related to such factors as voltage rating, capacity, structure, external wiring and stray capacitance of the transformer. Statistics have shown that the maximum oscillation frequency of HV transformers is generally around 10 times the power frequency, while large-capacity UHV transformer has a relatively lower

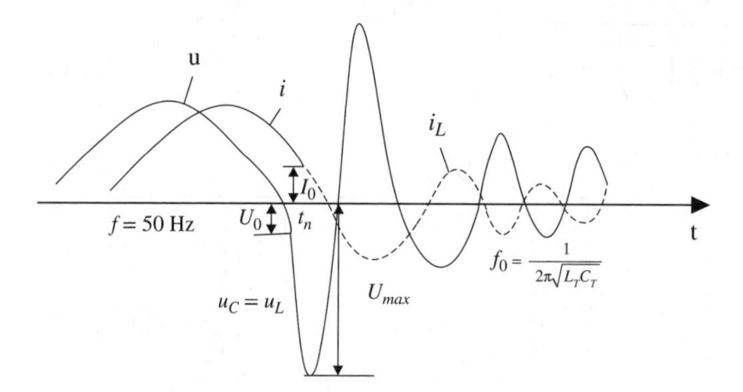

Figure 1.36 Overvoltages caused by disconnecting unloaded transformers.

oscillation frequency, only several times the power frequency, and the corresponding overvoltage amplitude is lower as well.

Field data has indicated that the overvoltage multiple in the case of disconnecting unloaded transformers is typically 2–3 times; the probability of exceeding 3.5 times is only 10%, and on very rare occasions can it reach or exceed 4.5–5.0 times.

Electromagnetic connections are present between transformer windings. Hence, switching operations at the transformer MV or LV side could generate overvoltages of the same multiple at the high-voltage side, posing a threat to the insulation at the HV side.

1.4.2.8 Influencing Factors

Cut-off value or breaker arc-quenching performances – The above analysis has shown that the amplitude of such an overvoltage is approximately proportional to the current cutoff value I_0, which is associated with the breaker arc-quenching performances. The current cutoff value of each circuit breaker is largely scattered, but a maximum potential value $I_{0(max)}$ that is virtually stable could be utilized. In general, bulk-oil circuit breakers are relatively weak in terms of arc-quenching capability; the arc would not be quenched before the zero-crossing point. In the case of compressed-air circuit breakers, oil-pressed minimum oil circuit breakers and vacuum circuit breakers, the excitation current could be interrupted before its natural zero position or even near its peak. Hence, circuit breakers with good arc-quenching performances could interrupt no-load currents at higher amplitudes, meaning opening overvoltages of higher amplitudes as well.

Excitation current or excitation inductance – The excitation current I_{L_T} or the inductance L_T of the unloaded transformer would also exert some influence on overvoltage amplitudes. If the excitation current amplitude $I_{L_T} \leq I_{0(max)}$, the overvoltage amplitude would increase with I_{L_T}, and the maximum overvoltage amplitude would appear when $i_{L_T} = I_{L_T}$, while the maximum overvoltage amplitude would appear when $i_{L_T} = I_{0(max)}$ if $I_{L_T} > I_{0(max)}$.

Circuit oscillation frequency – The magnitudes of overvoltages caused by switching off unloaded transformers are related to oscillation frequency of the circuit: the higher the oscillation frequency, the higher the corresponding overvoltage amplitude. The oscillation frequency of the circuit is related to the rated voltage, capacity, structural type, stray capacitance and external wiring of the transformer. For example, if a transformer is equipped with a cable with a large lead capacitance, the equivalent capacitance of the transformer side are accordingly large, thus reducing the overvoltage amplitude. Note also that the magnetic linkage generated by the oscillating current in the winding passes through the entire iron core when unloaded transformers are switched off; hence, capacitance to ground of the winding on the other side is also involved in the oscillation, and computation has to be performed on the basis of the transformer ratio. If this particular side is connected to a long lead, the capacitance to ground will increase accordingly, which is beneficial to overvoltage suppression. Moreover, such overvoltages can also be limited via approaches designed to increase the capacitance to ground C_T, such as the tangled winding method and electrostatic shielding enhancement.

Grounding methods of transformer neutrals – The amplitude of overvoltages due to disconnecting unloaded transformers is also affected by the grounding method of transformer neutrals. For three-phase transformers with ungrounded neutrals, displacement of transformer neutrals occurs due to interphase electromagnetic connections when the three phases of the breaker act asynchronously. In this way, the maximum amplitude of overvoltages caused by disconnecting three-phase unloaded transformers is 50% higher than that of disconnecting single-phase unloaded transformers.

Arc reignition between breaker contacts – The previous analysis is based on the assumption that arc reignition does not occur between breaker contacts after the current is cutoff. In practice, arc reignition is prone to happen as high-frequency oscillation in the transformer circuit enables a fast rise in the recovery voltage across the breaker gap after current interruption. Unlike general conditions where arc reignition result in overvoltage amplitude increase, reignition in this case can suppress overvoltages. The reason is that charges in the capacitance to ground of the transformer produce high frequency discharge via the power supply circuit; as a result, stored energy in the electric field is released rapidly; the capacitance voltage value drops to the supply voltage value; the arc across the breaker gap is extinguished due to declined recovery voltage. Since the inductor current cannot change instantly during reignition, the inductor will charge the capacitor after arc-quenching, the recovery voltage across the gap will rise, which may again lead to reignition and release of the stored energy. As described, the stored energy will gradually decrease and arc reignition will no longer exist when recovery strength of the breaker gap is greater than the maximum recovery voltage. In this regard, overvoltages are suppressed.

1.4.2.9 Restrictive Measures

Install arresters – Characterized by short duration and low energy, such overvoltages can be limited without much difficulty. For instance, installing arresters is an option for protection. Note that arresters used to limit overvoltages caused by switching off unloaded transformers has to be connected in parallel with the transformer sides of breakers so as to ensure that the transformer is still under arrester protection after the breaker is switched off. Unlike the guide against lightning overvoltages, such arresters are not allowed to quit operation even on occasions other than the thunderstorm season. If the transformer high-voltage side has the same neutral grounding method with the low-voltage side, arresters can be installed in the low-voltage side to limit high-voltage side overvoltages caused by switching unloaded transformers in consideration of economy.

Install resistors in parallel with circuit breakers – Such an overvoltage can also be suppressed by installing a resistor in parallel with the breaker main contact; this plays a role in achieving sufficient damping effects and limiting excitation current. The resistance, approximating to the excitation impedance of the inductor, is tens of thousands of ohms. The resistor can thus be regarded as having high resistance compared with that used for limiting overvoltages due to switching unloaded lines.

1.4.3 Arc Grounding Overvoltages

More than 60% of the faults in the power system can be ascribed to single-phase grounding, based on practical operation data. When a single-phase metallic grounding fault occurs in the grid with ungrounded neutral points, non-fault phase voltage will increase to the line voltage while the size and symmetry of the three-phase line voltages remain unchanged. In order to ensure power supply reliability, lines need not be disconnected immediately; they can continue to operate with faults for another two hours. Capacitive current appears at the grounding point when a single-phase grounding fault occurs because of line-to-ground capacitance. If large enough, the capacitive current will cause an arc that quenches and reignites alternately at the zero-crossing position of power frequency current. Such intermittent arc will lead to electromagnetic energy oscillation between inductive and capacitive elements, thus causing severe overvoltages on both fault and non-fault phases, referred to as arc grounding overvoltages.

In most cases, insulation of fine-quality electrical equipment that conforms to the standards will not be damaged by arc grounding overvoltages. However, arc grounding overvoltages may greatly endanger: (a) existing electrical equipment with poor insulation properties in the system, (b) equipment that has degraded insulation strength, (c) equipment whose potential insulation faults have not been found in the preventive test. In particular, distribution networks have higher probability of single-phase grounding occurrence, and are prone to unstable grounding arc that may result in arc grounding overvoltages of long duration and extensive impact range. Therefore, corresponding measures need to be taken to limit arc grounding overvoltages in systems with ungrounded neutral points.

1.4.3.1 Cause and Formation

Such overvoltages are caused by electromagnetic energy oscillation between inductive and capacitive elements, due to changes in system structure and parameters at the instants of arc striking and reignition.

The development and amplitude of the arc grounding overvoltage are influenced by arc-quenching instants. The arc grounding current includes two current components: the high-frequency component and the power (fundamental) frequency component. And there exist two possible arc-quenching instants: the instant when high-frequency oscillating current crosses the zero-crossing position, and the instant when power frequency current crosses the zero-crossing position. In fact, arc-quenching is determined by recovery strength and recovery voltage of the gap at the zero-crossing of the current. The mechanism of overvoltage development will be illustrated by the case where the arc is quenched at the zero-crossing of the power frequency current.

Simplifications here are made to facilitate analysis: (a) the influence of interconductor capacitance is ignored, (b) conductors of different phases possess equal capacitance to ground. The resulting equivalent circuit is illustrated in Figure 1.37.

Assuming that a grounding fault occurs at phase A when the phase voltage just becomes positive, the potential of phase A conductor will instantly drop to zero; the potential of the neutral point will become the phase voltage value, and phase-to-ground voltages of phase B and phase C will rise to the line voltage. u_A, u_B, u_C are used to represent the three-phase supply voltages, while u_1, u_2, u_3 are used to represent the conductor-to-ground voltages, i.e. voltages across the capacitance to ground. The overvoltage development is presented graphically in Figure 1.37.

Provided that the grounding arc occurs at phase A when $t = t_1$ ($u_A = +U_m$, where U_m is the amplitude of the phase voltage of the supply), the voltages across the three-phase capacitances at the time prior to the arcing instant are the three-phase supply voltages,

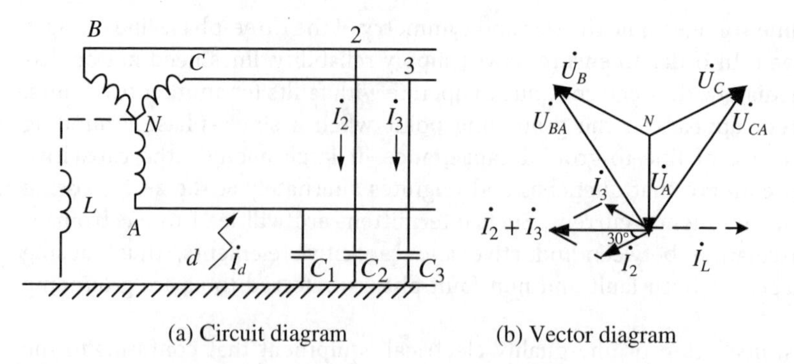

Figure 1.37 Circuit and vector diagrams for single-phase grounding.

respectively given by the following set of equations.

$$\left.\begin{aligned} u_1(t_1^-) &= +U_\varphi \\ u_2(t_1^-) &= u_3(t_1^-) = -1.5U_\varphi \end{aligned}\right\} \tag{1.4.17}$$

Right after the arcing, charges stored in the capacitance to ground of phase A, C_1, flow into the ground through the arc, and its voltage is reduced to zero; the capacitances to ground of the two non-fault phases C_2 and C_3 are charged by line voltage u_{BA} and u_{CA} via supply inductance with their voltages transiting from the initial value $-0.5U_\varphi$ to the instantaneous value of u_{BA} and u_{CA} $-1.5U_\varphi$. The frequency of the high-frequency oscillation is determined by supply inductance and conductor-to-ground capacitance. The stable values of the voltages across the three-phase conductors are given by

$$\left.\begin{aligned} u_1(t_1^+) &= 0 \\ u_2(t_1^+) &= u_3(t_1^+) = -1.5U_\varphi \end{aligned}\right\} \tag{1.4.18}$$

Therefore, the maximum voltages across C_2 and C_3 during the oscillation are expressed as

$$u_{2m}(t_1) = u_{3m}(t_1) = 2(-1.5U_\varphi) - (-0.5U_\varphi) = -2.5U_\varphi \tag{1.4.19}$$

At the end of the transition, u_2 and u_3 will be equal to u_{BA} and u_{CA}, respectively. The arc at the faulty site includes a fundamental frequency component and a high-frequency oscillating component that decays rapidly. Provided that the arc is not quenched at the zero-crossing of the high-frequency component, then the arc will be quenched at the zero-crossing of the power current component (marked as t_2) after burning for another half power frequency cycle; at this particular moment, $u_A = -U_\varphi$. A new transition will take place again after arc-quenching, and the initial voltage values of the three-phase conductors are respectively given by

$$u_1(t_2^-) = 0,\ u_2(t_2^-) = u_3(t_2^-) = +1.5U_\varphi \tag{1.4.20}$$

Now we will move on to calculate the stable values after arc-quenching. As the neutral point is ungrounded, the initial charges stored in each conductor capacitance are retained in the system, and equally redistributed over the three-phase capacitances. Each conductor thus has the same conductor-to-ground voltage. Consequently, a dc displacement voltage to ground $U_N(t_2)$ will be generated across the neutral point, as represented by

$$U_N(t_2) = \frac{0 \times C_1 + 1.5U_\varphi C_2 + 1.5U_\varphi C_3}{C_1 + C_2 + C_3} = U_\varphi \tag{1.4.21}$$

Therefore, after arc-quenching, the voltage across capacitance for each phase is overlaid by the corresponding supply voltage component and a dc component. The stable values are thus respectively given by

$$\left.\begin{aligned} u_1(t_2^+) &= u_A(t_2) + U_N(t_2) = -U_\varphi + U_\varphi = 0 \\ u_2(t_2^+) &= u_B(t_2) + U_N(t_2) = 0.5U_\varphi + U_\varphi = 1.5U_\varphi \\ u_3(t_2^+) &= u_C(t_2) + U_N(t_2) = 0.5U_\varphi + U_\varphi = 1.5U_\varphi \end{aligned}\right\} \tag{1.4.22}$$

As seen from the above equations, the new stable values of the three-phase capacitance voltages are equal to the initial values. It can therefore be concluded that no oscillation occurs during the arc-quenching transition.

Half a power frequency cycle later, i.e. when $t_3 = t_2 + T/2$, the voltage of the fault phase conductor reaches its maximum value $2U_\varphi$. If arc reignition of the fault-phase occurs at this particular moment, u_1 will drop to zero, and the transition process will take place again. For voltages across the three-phase capacitances, the initial values prior to reignition and the stable values after reignition are respectively given by the equation set below.

$$\left.\begin{aligned} u_1(t_3^-) &= 2U_\varphi \\ u_2(t_3^-) &= u_3(t_3^-) = -0.5U_\varphi + U_\varphi = 0.5U_\varphi \end{aligned}\right\} \tag{1.4.23}$$

$$\left.\begin{aligned} u_1(t_3^+) &= 0 \\ u_2(t_3^+) &= u_{BA}(t_3) = -1.5U_\varphi \\ u_3(t_3^+) &= u_{CA}(t_3) = -1.5U_\varphi \end{aligned}\right\} \tag{1.4.24}$$

The highest possible voltages across C_2 and C_3 during the oscillation are given by

$$u_{2m}(t_3) = u_{3m}(t_3) = 2(-1.5U_\varphi) - 0.5U_\varphi = -3.5U_\varphi \tag{1.4.25}$$

The subsequent "quenching-reignition" process follows the same pattern.

1.4.3.2 Overvoltage Characteristics and Waveforms

Theoretically, the maximum overvoltage multiples for fault and non-fault phases are 2.0 and 3.5, respectively.

The maximum overvoltage occurs at the instant of arc reignition of the fault phase. No oscillation is ever noted in the fault-phase voltage, while high-frequency oscillation occurs to non-fault phases at the instants of arcing and reignition of the fault phase, and no difference is noted in waveform polarity of the overvoltages.

The oscillation frequency of such overvoltages is determined by the power supply inductance and the capacitance to ground of conductors, ranging from hundreds to thousands of hertz. Generated at arc reignition, such overvoltages will last for a long time if not suppressed. Arc grounding overvoltages caused by arc-quenching at the instant of the zero-crossing of the power frequency current are shown in Figure 1.38.

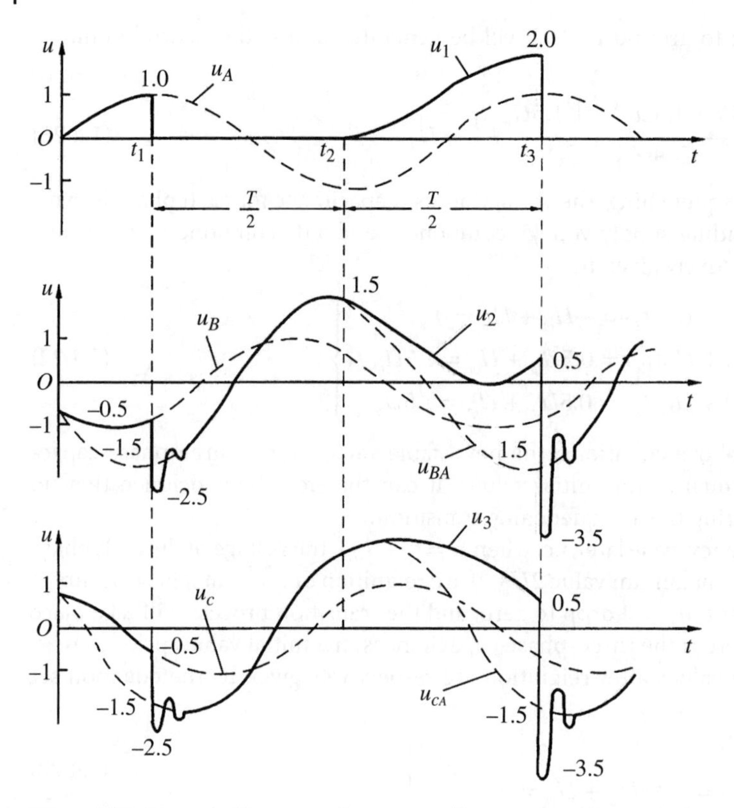

Figure 1.38 Arc grounding overvoltage caused by arc-quenching at the instant of zero-crossing of the power frequency current.

1.4.3.3 Influencing Factors

1. Phase position at arc-quenching and reignition – Analysis of the worst case for overvoltage multiples is based on the assumption that both arcing and arc reignition take place when the fault-phase voltage attains its peak and when arc-quenching occurs at the zero-crossing instant of the power frequency current. In practical cases, arc-burning and arc-quenching, affected by the surrounding medium and atmospheric conditions of the arcing position, are random.
2. Relevant parameters of the system – Such overvoltages appear lower under the same conditions if taking interphase capacitance into account. The reason is that the phase-to-ground capacitance of the non-fault phase connects in parallel with the interphase capacitance between the fault and non-fault phase after arcing, and the voltages of the two phases are different, charges are thus redistributed before the oscillation, narrowing the gap between the initial value and stable value of the non-fault phase voltage, reducing oscillation amplitude and thus lowering the overvoltage.
   The overvoltage amplitude during the oscillation is also reduced due to losses caused by resistance of the power supply, the line and the arc.
3. Grounding methods of neutrals – If the neutral point is directly grounded, once the single-phase grounding fault occurs, a large short-circuit current will flow through

the grounding point; the circuit breaker will trip, taking the power supply out of service and thus thoroughly eliminating the arc grounding overvoltage.

1.4.3.4 Restrictive Measures

As arc grounding overvoltages are caused by an intermittent arc, the fundamental method of limiting such overvoltages is to limit the intermittent arc, for instance, by altering the grounding method of neutrals.

1. Adopting effective neutral grounding methods – When adopting effective neutral grounding methods, the single-phase ground fault generates a large single-phase short-circuit current, and the circuit breaker trips at once, to clear the fault. In a short time interval, the arc at the fault point is quenched, and automatic reclosing is performed. Power supply will be regained if the operation succeeds; otherwise, the circuit breaker will trip again. In this way, no intermittent arc will appear, thus limiting the arc grounding overvoltage. This method is applicable to power grids rated 110 kV or higher.
2. Neutral grounding via arc suppression coils – Adopting effective neutral grounding may cause breaker tripping in the case of single-phase grounding faults, thus significantly reducing power supply reliability. For lines rated 66 kV and below, ineffective neutral grounding methods are commonly used to increase distribution reliability as it will not bring apparent economic benefits regarding lowering insulation levels. When a single-phase grounding fault occurs and the current of capacitance to ground of the line is too large to self-extinguish, an inductance coil can be installed at the neutral point. Reactive current is generated in the inductance coil by the neutral-to-ground potential, thus compensating for the capacitive current at the grounding point and making the overall current flowing though the fault point small. Therefore, intermittent arc generation is prevented and such overvoltages are suppressed.

Functions of arc suppression coils – The arc suppression coil is an adjustable inductance coil with air gaps on the iron core. Its volt–ampere characteristics are not prone to saturation, and it has to be connected between the neutral point and the ground. Figure 1.34(a) can be used to analyze how overvoltages are suppressed by arc suppression coils:

Assume that a single-phase grounding fault occurs to phase A, the current of the grounding point includes the vector sum of the currents flowing through the capacitances to ground of non-fault phases $(\dot{I}_2 + \dot{I}_3)$ and the inductive current flowing through the arc suppression coil $\dot{I}_L$; according to the vector diagram, the grounding point current is $\dot{I}_d = \dot{I}_2 + \dot{I}_3 + \dot{I}_L$. Appropriate selection of the inductance L for arc suppression coils (i.e. appropriate selection to compensate the inductive current $\dot{I}_L$) can ensure sufficiently small grounding current so that the grounding arc quenches rapidly and is unlikely to reignite, thus suppressing arc grounding overvoltages.

Compensation degree of arc suppression coils – The compensation degree of arc suppression coils, represented by K, is defined as the ratio in percentage between the inductive current of the arc suppression coil and the current of capacitance to ground. And $1 - K$, represented by v, is referred to as the out-of-resonance degree. The compensation degree of the arc suppression coil K is expressed as

$$K = \frac{I_L}{I_C} = \frac{U_\varphi/\omega L}{\omega(C_1 + C_2 + C_3)} = 1/\omega^2 L(C_1 + C_2 + C_3)$$
$$= \frac{(1/\sqrt{L(C_1 + C_2 + C_3)})^2}{\omega^2} = \frac{\omega_0{}^2}{\omega^2} \tag{1.4.26}$$

with $\omega_0 = \frac{1}{\sqrt{L(C_1+C_2+C_3)}}$, the angular frequency of the circuit natural oscillation.

Arc suppression coils can operate under the following three states based on variation in compensation degree.

1. Under-compensation – $I_L < I_C$, $\omega L > 1/[\omega(C_1+C_2+C_3)]$, i.e. the inductive current of the arc suppression coil is not sufficient to completely compensate for the capacitive current. Grounding point current is capacitive under this circumstance, which corresponds to $K < 1$, $v > 0$.
2. Full compensation – $I_L = I_C$, $\omega L = 1/[\omega(C_1+C_2+C_3)]$, i.e. the inductive current of the arc suppression coil can exactly compensate for the capacitive current. At this particular moment, the arc suppression coil and the capacitance to ground are in a state of parallel resonance, and the current flowing through the fault point is a small resistive current. Full compensation corresponds to $K = 1$, $v = 0$.
3. Over-compensation – $I_L > I_C$, $\omega L < 1/[\omega(C_1+C_2+C_3)]$. In this case, the fault point current is inductive, and over-compensation corresponds to $K > 1$, $v < 0$.

If the compensation degree of arc suppression coils is too small, large residual current flowing though the fault point and rapid growth rate of recovery voltage at the fault point will block arc-quenching. With the compensation degree increasing, the growth rate of the recovery voltage at the fault point declines, and arc-quenching is easier. Considerable displacement voltage will appear at neutral points during normal operation in the state of full compensation. Lines have to be transposed to make the three-phase capacitances to ground equal so as to avoid danger imposed by potential rise of neutral points. In fact, symmetry of capacitance to ground can never be fully realized; so arc suppression coils cannot be completely operated in the state of full compensation.

In general, arc suppression coils adopt an over-compensation degree of 5–10% due to capacitive current increase in power grid development. The compensation degree will be decreased and cannot play the role in arc-quenching if under-compensation is adopted, while arc suppression coils presently operated in the over-compensation state can still continue to play a role when the grid further develops. Additionally, if under-compensation is adopted, full compensation may occur when lines are taken out of service, generating large potential displacement in the neutral and causing severe ferro-resonance overvoltages in zero-sequence networks.

1.4.4 Power System Splitting Overvoltages

When the power grid is unstable for some reason (e.g. grounding faults on lines), the emfs of the power supplies at the line terminals will generate relative swing (out of step) of generally low frequency. System splitting must be performed at a high speed to prevent the accident from expanding. In theory, such splitting could happen at any swing angle, i.e. the relative power angles δ of the power supplies at the line terminals could

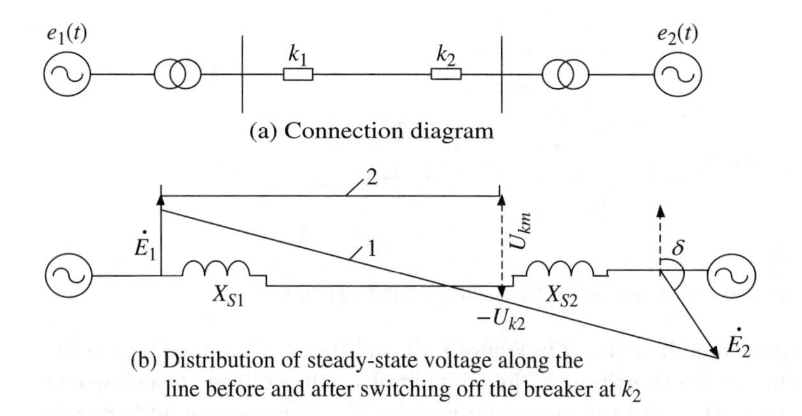

Figure 1.39 Schematic of splitting overvoltages due to supply out of step.

be any value. In the worst case where trip occurs at or near phase reversing ($\delta = 180°$), high-amplitude overvoltages will be produced. Besides, if single-phase grounding faults occur, the faults will be cleared by circuit breakers, and overvoltages will also be generated due to oscillation.

Figure 1.39 shows the connection diagram when the emfs of the supplies at the transmission line ends are out of step. As the frequency of the out-of-step swing is low, the system is actually in a stable power frequency state when the circuit breaker is switched off. Assuming that the power angle difference δ between the potential E_1 at the sending end and the potential E_2 at the receiving end is greater than 90° at the opening of the circuit breaker (at site k_2), the distribution of the steady-state power frequency voltage along the line before switching off the breaker is shown by curve 1 in Figure 1.39(b). At this particular moment, the power supplies at the ends are approximately reversed in polarity, the voltage along the line is linearly distributed, and the voltage at terminal k_2 is $-U_{k2}$. When the breaker at k_2 trips and system splitting occurs, the steady-state power frequency voltage along the line is distributed according to the cosine law, as illustrated by curve 2 in Figure 1.39(b), and the steady-state voltage at the line end is U_{km}. After the breaker at k_2 is switched off, high-frequency oscillation occurs along the line, and the terminal voltage oscillates from $-U_{k2}$ to U_{km}. The maximum overvoltage at the line end during the transition is given by

$$U_{2m} = 2U_{km} - (-U_{k2}) = 2U_{km} + U_{k2} \tag{1.4.27}$$

It could be concluded from the above analysis that the splitting overvoltage amplitude depends on the power angle difference δ between the potentials of the power sources at the line ends at the splitting instant. When the power angle difference δ is larger than 90°, the maximum splitting overvoltage may exceed 2 p.u., and when the power angle difference δ is smaller than 90°, the splitting overvoltage is smaller than 2 p.u.

When single-phase grounding occurs in the system and the breaker trips to clear the fault, a transition will occur that causes transient overvoltages. Figure 1.39(a) shows the connection diagram. Assume that a grounding fault occurs at the output end of k_2. The voltage is then basically linearly distributed over the line, as shown in Figure 1.40, and the voltage at the terminal k_2 (the grounding fault point) is zero. When the breaker trips to clear the fault, the steady-state voltage is distributed over the line, based on

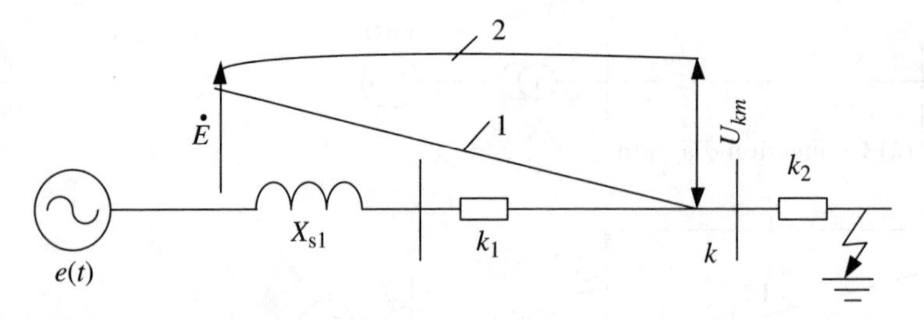

Figure 1.40 Schematic of overvoltages caused by clearing grounding faults.

the cosine law, represented by curve 2 in Figure 1.40, and the maximum voltage at the line end is U_{km}. During the transition of the tripping of k_2, the voltage at the line end oscillates from zero to U_{km}, and the maximum possible overvoltage amplitude during this process is given by

$$U_{2m} = 2U_{km} \tag{1.4.28}$$

It is further demonstrated that the oscillation waveform is constituted by the superposition of the steady-state power frequency voltage and the harmonics of each order. The maximum overvoltage multiple can reach 2.0 when all the harmonics are of the same sign. In practice, the overvoltage multiple is generally 1.5–1.7 due to the attenuation of oscillation and the differences between the time when each harmonic attains the peak.

If grounding faults are cleared in a series compensating system, higher transient overvoltages will be generated. The connection diagram is shown in Figure 1.41, where C is the series compensating capacitance, generally in the state of under-compensation; therefore the voltage drop U_c is in reversal with the supply voltage when the grounding fault occurs. After k_2 trips to clear the fault, the voltage at point k transits from $-U_c$ to U_{km}, and the maximum possible overvoltage in this process is expressed as

$$U_{2m} = 2U_{km} + U_C \tag{1.4.29}$$

In Figure 1.41, the oscillating overvoltage at point n is even higher, given by

$$U_{nm} = U_{2m} + U_C = 2(U_{km} + U_C) \tag{1.4.30}$$

It could be concluded from the above analysis that, in this case, the maximum overvoltage depends on the value of U_c, and the specific value of U_c is determined by the compensation degree and the position of the fault point.

To conclude, the power angle difference δ between the emfs at the two ends is the primary factor influencing splitting overvoltages. Other influencing factors include the

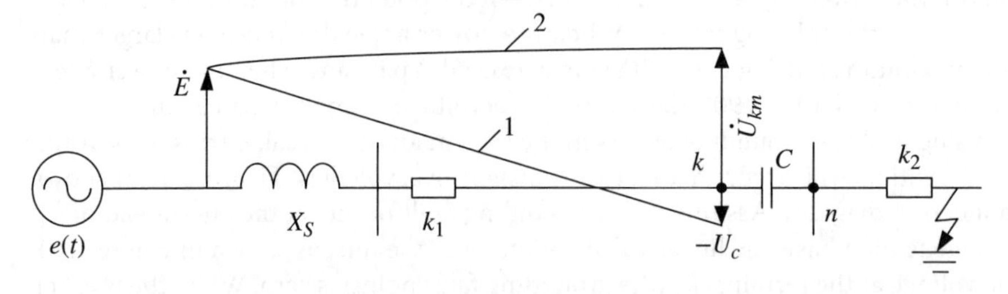

Figure 1.41 Schematic of overvoltages due to fault clearing in series compensating systems.

line length, the capacity of the supply that still serves the lines after splitting and the system splitting site. Splitting overvoltages will be very high under adverse conditions, such as small power capacity, long lines and splitting time at $\delta = 180°$. In practical situations, such possibilities are, however, very small, and the above adverse conditions are even less likely to occur simultaneously.

Using metal oxide arresters is a practical way to limit splitting overvoltages. And a better option is to adopt automatic devices that can control the opening of the breaker within a certain scope that the swing of emfs at the two ends will not exceed. In this manner, overvoltages will be limited at the origin.

1.5 Power Frequency Overvoltages

According to the statistics, faults in power systems can be found in a variety of forms, of which asymmetrical short-circuit faults are the most common. In a system where the neutral point is directly grounded, the potential of the neutral point has been made equal to the ground potential. Therefore, the voltages of non-fault phases will not rise when a single-phase grounding fault occurs. However, for systems with ungrounded neutral points, the voltages of non-fault phases will rise if a single-phase grounding fault occurs. In the case of two-phase grounding faults, the voltage of the non-fault phase rises as well. Comparatively speaking, single-phase grounding contributes to greater voltage rises. The power frequency overvoltage is defined as the overvoltage with an amplitude exceeding the maximum operating phase voltage and a frequency equal to or close to the power frequency (50 Hz) in the case of normal operation or grounding faults. The following causes mainly contribute to the formation of power frequency overvoltages: Ferranti effect of unloaded long lines, voltage rise of non-fault phases due to asymmetrical short-circuit faults and rise of the power frequency voltage due to load rejection.

Power frequency overvoltages are characterized by the following properties.

1. The extent of rise of power frequency voltage has a direct influence on the actual amplitude of the switching overvoltage.
2. The extent of rise of the power frequency voltage can influence effects and operating conditions of protection devices.
3. Long duration of power frequency voltage rise has a significant influence on insulation and operational performances of the equipment.

In terms of the overvoltage multiple, power frequency overvoltages pose no threats to electrical equipment with normal insulation. In ultra-high voltage systems, the determination of insulation levels of electrical equipment will play an increasingly major role.

Power frequency voltage rise has gained much attention in UHV transmission systems with small insulation margins for the following reasons.

1. Power frequency voltage rise mostly occurs in unloaded or lightly loaded conditions, under which switching overvoltages of various types are likely to happen as well. Therefore, power frequency overvoltages and switching overvoltages may simultaneously occur and overlay. Combined effects of these overvoltages must be taken into account in high voltage insulation design.

2. Power frequency voltage rise is the major basis for the determination of operating conditions of overvoltage protection devices, meaning that it can directly influence protection features of arresters and insulation levels of electrical equipment.
3. As power frequency voltage rise is a non-attenuated or weakly attenuated phenomenon with a long duration, insulation and operating conditions of the equipment could also be greatly influenced.

Experience shows that power frequency voltage rise poses no threats to equipment with normal insulation because of its small amplitude in most cases. However, attention still needs to be paid under the following conditions:

1. Long duration of (or long-term) power frequency voltage rise could significantly impact insulation and operational performances of the equipment.
2. The magnitude of power frequency voltage rise can influence protective effects and operating conditions of protection devices.
3. The magnitude of power frequency voltage rise can influence the amplitude of switching overvoltages.

1.5.1 Power Frequency Overvoltages due to the Ferranti Effect

It is known that transmission lines can be represented by distributed elements. A π-type equivalent circuit of lumped elements can be used to replace transmission lines whose distance is not too long, as shown in Figure 1.42.

According to operational experience, as the inductive reactance is much smaller than the capacitive reactance for most lines, the following equation can be given based on Kirchhoff's voltage law (KVL) under the applied voltage at the sending end in the case of unloaded lines ($\dot{I}_2 = 0$):

$$\dot{U}_1 = \dot{U}_2 + \dot{U}_R + \dot{U}_L = \dot{U}_2 + R\dot{I}_{C2} + jX_L\dot{I}_{C2} \tag{1.5.1}$$

If $\dot{U}_2$ is taken as the reference vector, the vector diagram depicted in Figure 1.43 can be obtained.

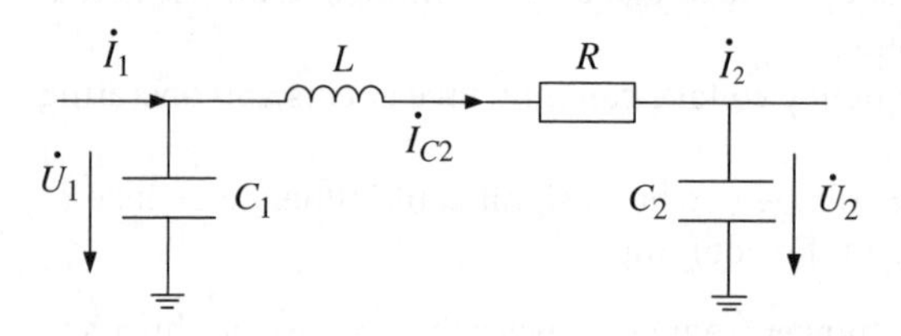

Figure 1.42 Equivalent circuit.

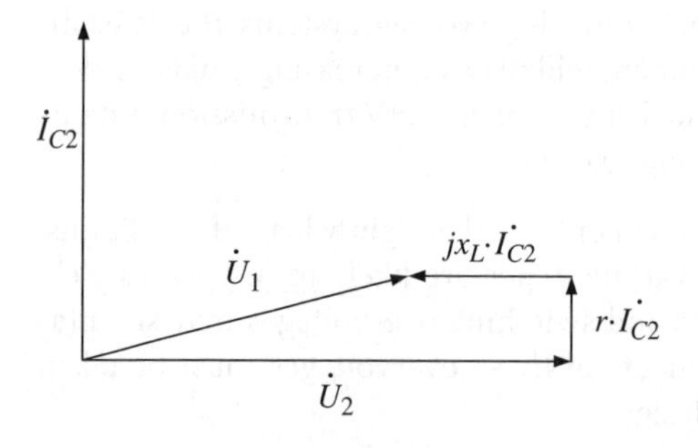

Figure 1.43 Vector diagram.

It is known that, for unloaded lines, the power frequency inductive reactance X_L is smaller than the power frequency capacitive reactance X_C, and from operational experience R is far less than X_L and X_C in most cases. Therefore, when applying the emf of the power source E, the voltage drop $\dot{U}_L$ across X_L caused by the capacitive current through the line makes voltage $\dot{U}_C$ across X_C higher than E. It could be thus concluded that the voltage across the capacitance is higher than the emf of the power source, i.e. the power frequency voltage rise is caused by the capacitance effect of unloaded lines (also referred to as the Ferranti effect). Due to the presence of line-to-ground capacitance, the line voltage is the vector sum given by $\dot{U} = \dot{U}_R + \dot{U}_L + \dot{U}_C$, where a phase difference of 90° exists between the neighboring vectors of $\dot{U}_R$, $\dot{U}_L$ and $\dot{U}_C$. The phase difference between $\dot{U}_L$ and $\dot{U}_C$ is 180°, meaning that they are opposite in phase. Consequently, the larger $\dot{U}_C$ is, the smaller the original $\dot{U}_L + \dot{U}_C$ will be, and the smaller voltage division thus leads to voltage rise at the receiving end.

Distributed parameter circuits have to be adopted when analyzing power frequency overvoltages caused by the capacitance effect of unloaded long lines due to the increase in transmission distances and transmission voltage amplitudes. The π-type equivalent circuit depicted in Figure 1.44 is adopted in this book.

As can be learned from Figure 1.44, the voltage at distance X from the open end of the line is given by

$$\dot{U}_X = \frac{\dot{E}\cos\theta}{\cos(\alpha l + \theta)}\cos\alpha x \tag{1.5.2}$$

$$\theta = \tan^{-1}\frac{X_S}{Z} \tag{1.5.3}$$

$$Z = \sqrt{\frac{L_0}{C_0}} \tag{1.5.4}$$

$$\alpha = \frac{\omega}{v} \tag{1.5.5}$$

where: $\dot{E}$ is the supply voltage; X_s is the equivalent reactance of the supply; Z is the wave impedance of the conductor; ω is the angular frequency of the supply; v is the velocity of light.

The following conclusions can be drawn:

1. Power frequency voltage is distributed in accordance with the cosine law from the receiving end to the sending end along the line, given by

$$\dot{U}_x = \dot{U}_2 \cos\alpha x \tag{1.5.6}$$

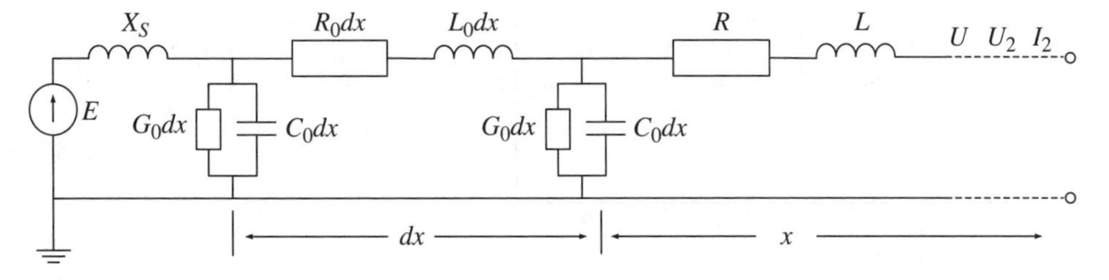

Figure 1.44 π-type equivalent circuit.

It can thus be seen that the voltage at the receiving end of the line is the highest.

2. The length of the line could influence the extent of the voltage rise at the receiving end of the line.

$$\dot{U}_1 = \frac{\dot{E}\cos\theta}{\cos(\alpha l + \theta)}\cos\alpha x\bigg|_{x=l} = \dot{U}_2 \cos\alpha l \tag{1.5.7}$$

$$U_B = U_C = aU_{xg}, \; \frac{\dot{U}_2}{\dot{U}_1} = \frac{1}{\cos\alpha l} \tag{1.5.8}$$

It can be seen that longer lines result in more severe power frequency voltage rise at the receiving end. According to operational experience, power frequency voltage rise will not exceed 2.9 times in practical conditions due to the limitation from losses of line resistance and corona.

3. Power frequency voltage rise is influenced by the supply capacity.

$$\dot{U}_x = \frac{\dot{U}_1}{\cos\alpha l}\cos\alpha x \tag{1.5.9}$$

The presence of the power supply leakage reactance can be regarded as increasing the line length, thus improving the voltage rise at the receiving end of the unloaded long line. It is recommended to estimate the most severe power frequency voltage rise on the basis of X_s under the minimum operation mode in single-supply systems. In the case of long lines powered at the ends, the following steps for breaker operation should be followed: when closing the line, the side with the larger supply capacity must be closed before that with the smaller power capacity; when disconnecting the line, the side with the smaller power capacity must be taken out of service before the side with the larger power capacity.

4. The distribution of voltage at each point on the unloaded line is shown in Figure 1.45.

$$\dot{U}_x = \frac{\dot{U}_1}{\cos\alpha l}\cos\alpha x \tag{1.5.10}$$

The parallel capacitor is commonly used to solve the problem of power frequency voltage rise based on the root cause. The capacitive current is reduced via compensation, thus limiting the power frequency voltage rise due to the Ferranti effect of unloaded long lines.

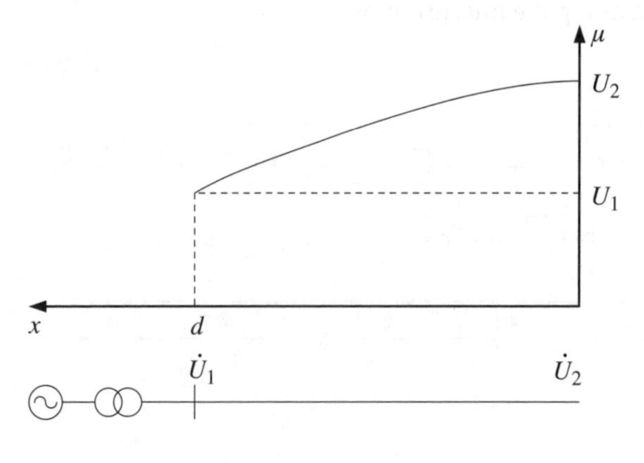

Figure 1.45 Voltage distribution over the unloaded line.

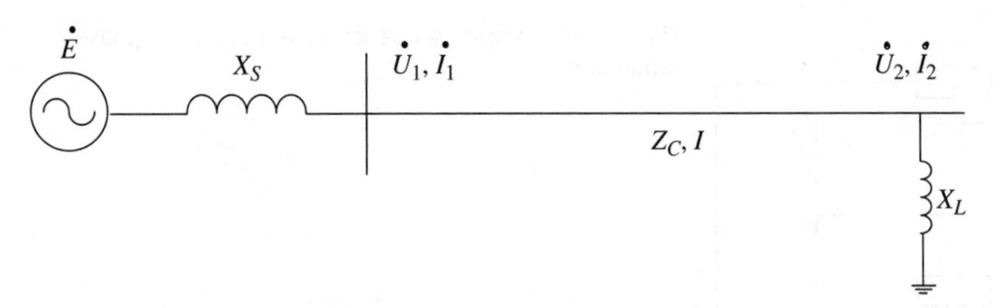

Figure 1.46 Parallel reactor connected at the receiving end of the line.

Assuming that a reactor is connected in parallel with the receiving end of the line, the connection diagram is shown in Figure 1.46.

Based on the circuit fundamentals, the following equation is given

$$\dot{U}_2 = \frac{\dot{E}}{\left(1+\frac{X_S}{X_L}\right)\cos\alpha l + \left(\frac{Z_C}{X_L} - \frac{X_S}{Z_C}\right)\sin\alpha l} \tag{1.5.11}$$

As seen from this equation, the terminal voltage decreases with the increase in reactor capacity. The reason is that the inductance of the parallel reactor can compensate the capacitance to ground of the line and reduce the Ferranti effect. Parallel reactors not only play a role in controlling the power frequency voltage rise but also in such areas as system stability, self-excitation and reactive power balance.

1.5.2 Power Frequency Voltage Rise Due to Asymmetrical Short-Circuit Faults

Three-phase short-circuit is symmetrical, while single-phase grounding, two-phase short-circuit and three-phase short-circuit with one phase grounded via impedance are all asymmetrical. The asymmetrical short-circuit fault is the most common fault for transmission lines. The zero-sequence component of the short-circuit current can cause power frequency voltage rise on non-fault phases (often called the asymmetrical effect). The single-phase grounding fault is the most common of the asymmetrical short-circuit faults, and can result in the most severe power frequency voltage rise.

When an asymmetrical fault occurs in the power system, current and voltage at the fault point of each phase are asymmetrical. Symmetrical components and sequence networks are generally adopted as the analytical means to facilitate calculation. Besides, arc-quenching voltage of valve arresters generally depends on the power frequency voltage rise in the case of single-phase grounding. Therefore, only single-phase grounding are discussed in this chapter.

In Figures 1.47 and 1.48, assume that phase A is grounded,

$$\dot{U}_A = 0, \dot{I}_B = \dot{I}_C = 0$$

hence,

$$\dot{U}_B = \frac{(a^2-1)Z_0 + (a^2-a)Z_2}{Z_1+Z_2+Z_0}\dot{E}_A \tag{1.5.12}$$

$$\dot{U}_C = \frac{(a-1)Z_0 + (a^2-a)Z_2}{Z_1+Z_2+Z_0}\dot{E}_A \tag{1.5.13}$$

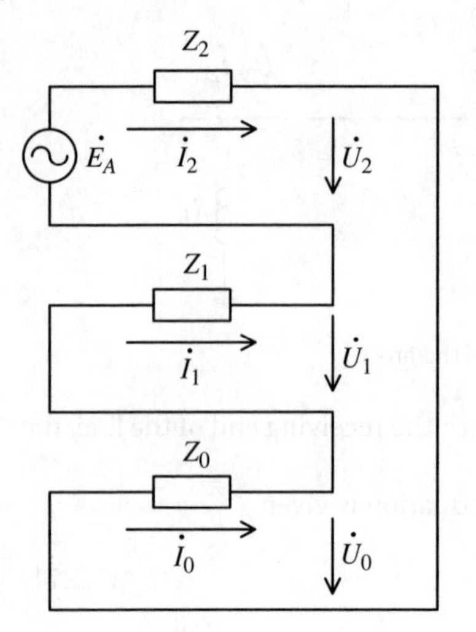

Figure 1.47 Complex sequence network of single-phase grounding.

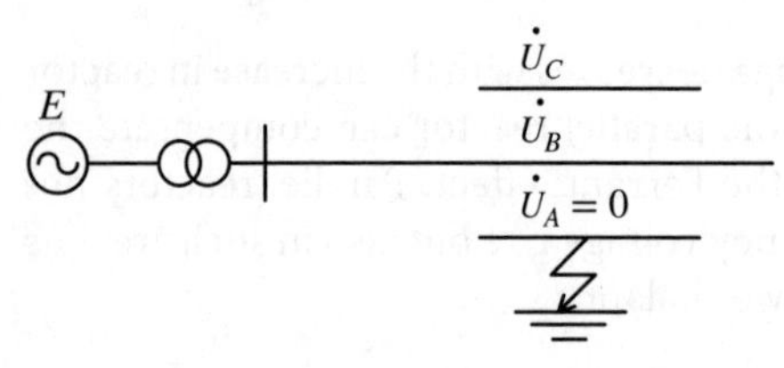

Figure 1.48 Wiring of single-phase grounding.

where $\dot{E}_A$ is the voltage at the fault point during normal operation; Z_0, Z_1, Z_2 are positive sequence, negative sequence and zero-sequence impedance of the grid seen from the fault point, respectively.

$$a = e^{j\frac{2\pi}{3}} \tag{1.5.14}$$

For a power supply of large capacity, the percentage of the generator reactance compared to the entrance impedance is small, and hence $Z_1 \approx Z_2$. Ignoring the resistive component of the impedance gives

$$\dot{U}_B = \left[-\frac{1.5\frac{X_0}{X_1}}{2+\frac{X_0}{X_1}} - j\frac{\sqrt{3}}{2} \right] \dot{E}_A \tag{1.5.15}$$

$$\dot{U}_C = \left[-\frac{1.5\frac{X_0}{X_1}}{2+\frac{X_0}{X_1}} - j\frac{\sqrt{3}}{2} \right] \dot{E}_A \tag{1.5.16}$$

The module values of $\dot{U}_B$ and $\dot{U}_C$ are expressed as

$$U_B = U_C = aU_{xg} \tag{1.5.17}$$

with

$$a = \sqrt{3}\frac{\sqrt{1+k+k^2}}{k+2} \tag{1.5.18}$$

Here, α, the grounding coefficient, is defined as the ratio between the effective value of the highest phase-to-ground power frequency voltage of the non-fault phases and the effective value of the phase-to-ground voltage of the non-fault phase when single-phase grounding fault occurs. The grounding coefficient is closely related to the zero-sequence impedance, particularly in the simplified form where its magnitude depends on the ratio $k = X_0/X_1$.

The relationships between X_0/X_1 and the power frequency voltage rise of the non-fault phases are shown in Figure 1.49 and Figure 1.50 when a grounding fault occurs to phase A.

The extent of voltage rise of non-fault phases under different grounding methods of neutrals will be discussed in detail. In the case of grids with ungrounded neutral points, X_0 depends upon the capacitive reactance of the line and is thus negative. In the case of single-phase grounding, the power frequency voltage rise of the non-fault phases is about 1.1 times the rated (line) voltage U_n. And the arc-quenching voltage of arresters is selected as 110% U_n, and therefore referred to as the "110% arrester".

In the case of 35–60 kV grids with neutral points grounded via arc suppression coils, X_0 takes a considerable positive value when the system operates in the

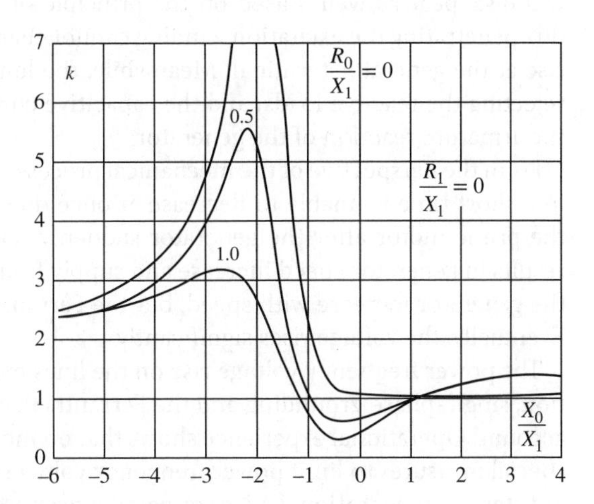

Figure 1.49 Power frequency voltage rise of phase B when $R_1/X_1 = 0$

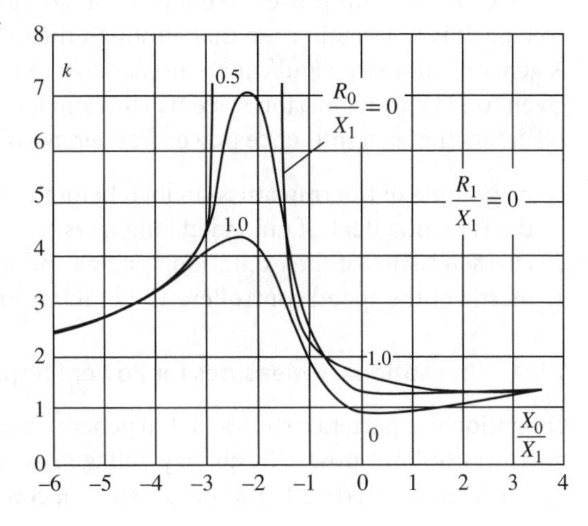

Figure 1.50 Power frequency voltage rise of phase C when $R_1/X_1 = 0$.

over-compensation state, and the non-fault phase voltage approaches the rated voltage U_n. Thus "100% arresters" are used.

In the case of 110–220 kV grids with neutrals effectively grounded, X_0 is a small positive value, and $X_0/X_1 \leq 3$. The voltage rise of the non-fault phases when single-phase grounding occurs will not exceed 0.8 times the maximum system voltage, thus, "80% arresters" are adopted.

1.5.3 Power Frequency Voltage Rise Due to Load Rejection

For large capacity transmission lines, a series of electromechanical transient processes will be generated in prime motors and generators when load rejection is caused by sudden breaker trip for some reason. Load rejection is one of the reasons behind power frequency voltage rise.

When the generator loses part of or even all the loads instantaneously, the current through the inductive grid load will demagnetize the armature reaction of the generator. And when the load is suddenly rejected, the demagnetization for the armature reaction will disappear as well. Based on the principle of flux linkage conservation, magnetic flux penetrating the excitation winding cannot change suddenly, thus generating voltage rise at the generator terminal. Meanwhile, the long line appears to be capacitive after rejecting the reactive loads, and the capacitive current would assist the magnetism for the armature reaction of the generator.

From the perspective of the mechanical process, power inputted into the prime motor in a short time is unable to decrease at once due to inertia of the speed controller of the prime motor after the generator suddenly rejects part of the active loads, which results in generator speed increase and supply frequency rise. Not only does the emf of the generator increase with speed, but the Ferranti effect of the line is also aggravated. Eventually, the voltage rises significantly.

The power frequency voltage rise on the lines may be very large if sudden load rejection, single-phase grounding and the Ferranti effect of unloaded lines are all taken into account. Operational experience shows that on most occasions, there is no need to take special measures to limit power frequency voltage rise in grids rated 220 kV and below, but static compensation devices or parallel reactors have to be used in UHV grids rated 330–500 kV to limit power frequency voltage rise to less than 1.3–1.4 times the phase voltage. It is thus again seen that although the multiple of power frequency voltage rise is generally not large and causes no damage to power system insulation, care has to be given to EHV transmission systems with small margins.

Factors that can influence power frequency voltage rise include:

1. magnitude of the transmission load before the trip of circuit breakers;
2. the Ferranti effect of unloaded long lines;
3. characteristics of the excitation systems and the voltage regulators of generators;
4. inertia of the speed controllers and braking equipment of prime motors.

1.5.4 Precautionary Measures for Power Frequency Overvoltages

Operational experience shows that in general conditions there is no need to take special measures to limit power frequency voltage rise in grids rated 220 kV and below, while for UHV grids rated 330–500 kV, parallel reactors or static compensation devices need to be used to limit power frequency voltage rise to lower than 1.3–1.4 times the phase voltage, for example:

1. use high-voltage parallel reactors to compensate for the Ferranti effect of unloaded lines;
2. use static var compensators (SVC) to compensate for the Ferranti effect of unloaded lines;
3. use transformers with neutrals directly grounded to lower power frequency voltage rise caused by asymmetrical grounding faults;
4. equip generators with excitation or voltage regulators of high performance to suppress the effect of the capacitive current on the optimized magnetism for the armature reaction in the case of sudden load rejection, thereby preventing the generation and development of overvoltages.

1.6 Resonance Overvoltages

Electrical equipment includes components of an inductive or capacitive nature. Inductances of power transformers, instrument transformers, generators, arc suppression coils, reactors, conductors, etc. can be regarded as inductive components, while conductor-to-ground capacitance and interphase capacitance of lines, compensating capacitors, stray capacitance of high-voltage equipment, etc. can be regarded as capacitive components. During system operation or when fault occurs, oscillation circuits constituted by these inductive and capacitive components may generate resonance phenomena under certain conditions, thus resulting in overvoltages on certain parts (or equipment) of the system; these are also known as resonance overvoltages.

Resonance is a periodic or quasi-periodic operating state, which appears in circuits whose natural frequency is close to the harmonic frequency of the power supply due to harmonics generated by nonlinear characteristics of some equipment and loads in power grids or circuits. The duration of resonance overvoltages is much longer than that of switching overvoltages, and resonance overvoltages may exist stably until the resonance is suppressed. It is the long duration of resonance overvoltages that leads to significant damage. Resonance overvoltages pose threats to insulation of electrical equipment, and lasting overcurrent due to resonance may burn inductive components with small capacity; this may influence the selection of protective measures.

The resonance circuit consists of inductance, capacitance and resistance. The capacitance C and resistance R in the system are generally deemed as linear components while the inductive components are divided into three kinds based on different properties: linear inductance, nonlinear inductance and periodically varying inductance. In power systems with different voltage ratings and structures, resonance is classified into the following categories based on the inductance in the circuit: linear resonance, nonlinear resonance and parametric resonance.

In UHV and EHV systems, as transformer neutrals are directly grounded or grounded via a low impedance, the potential of neutrals is basically fixed; therefore in low-voltage systems, resonance overvoltages due to saturation of electromagnetic PTs will never occur. However, UHV and EHV systems are generally characterized by parallel reactors connected to the terminal due to long distance, which may increase occurrence possibility of resonance. For example, power frequency resonance could take place at the switching of non-full-phase parallel reactors; frequency-dividing resonance may occur in series and parallel compensation networks; high-frequency resonance is likely to happen at the closing of unloaded capacitive lines connecting transformers-generator units.

1. Linear resonance – Linear resonance is the simplest mode of resonance in power systems. Parameters of the linear resonance circuit are all constant, and do not vary with the voltage or current. These circuits mainly consist of inductive components without iron cores (line inductance and transformer leakage inductance) or with iron cores whose excitation characteristics are close to linear characteristics (arc suppression coils) and capacitive components in the system. Under the effect of the ac power supply, linear resonance is likely to happen when the natural frequency of the system equals or approaches the power supply frequency.
   Linear resonance that can occur during operation of the power system includes: resonance due to the Ferranti effect of unloaded long lines, resonance caused by asymmetrical faults in the system with non-effective grounded neutrals, resonance attributed to full compensation of arc suppression coils and resonance of transferring overvoltages.
   Apart from changing the parameters of the circuit, measures for limiting overvoltages mainly include increasing resistance R of the circuit, suppressing current, increasing circuit losses and reducing overvoltages. In power system design and operation, it is advisable to avoid resonance conditions to eliminate such linear resonance overvoltages.
2. Ferro-resonance (nonlinear resonance) – Nonlinear resonance will occur in LCR series circuits containing inductance with iron cores, if certain conditions are satisfied. Saturation will be noted in inductive components with iron cores in the circuit; at this particular moment, the inductance is no longer a constant, and varies with the current or magnetic flux. Ferro-resonance will be generated when certain resonance conditions are satisfied; this is very different from linear resonance in terms of characteristics and properties. Ferro-resonance in power systems generally occurs in the state of non-full phase operation, such as line breakage, unloaded transformers, etc.
3. Parametric resonance – Parametric resonance is attributed to periodic changes of parameters of inductive components in the system under the effect of external forces. Generally, for circuits consisting of inductive components and capacitive components with periodical changes, the changing inductance periodically inputs system energy into the resonance circuit under improper cooperation of system parameters, thus forming resonance overvoltages.

As actual power systems are complicated in structure, and faults as well as operation are wide-ranging in generation mode, there are many connection ways that can lead to resonance. Priority will be given to qualitative analysis of the physical process of resonance overvoltages and comparison between characteristics of various resonance overvoltages. Measures for preventing and limiting such overvoltages are presented in the following section as well.

1.6.1 Linear Resonance Overvoltages

In LC series linear circuits, under the effect of ac power supply, series resonance is likely to happen when the natural frequency of the system is equal or near to the power supply

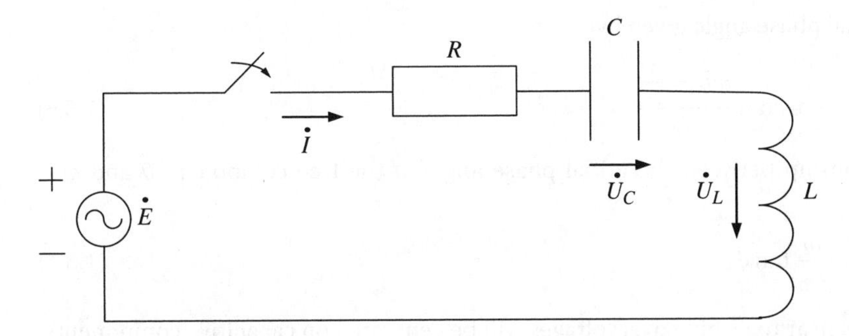

Figure 1.51 Series linear resonance circuit.

frequency, which will generate very high overvoltages on the inductance or capacitance. Therefore, series resonance is also called voltage resonance. Figure 1.51 shows the series resonance circuit, which is typical in the transition due to switching or faults.

Assume that the power supply voltage is $\sqrt{2}E\sin(\omega t+\varphi)$, and R is the damping resistance of the circuit, and $\xi = R/2L$ is the circuit resistivity. As R is small, $\xi/\omega_0 \ll 1$, influence of the resistance on the angular frequency of natural oscillation can be neglected, then $\omega_0 = 1/\sqrt{LC}$. When the initial value of the current of the inductor and the voltage on the capacitor in the circuit are zero, the voltage across the capacitor C during the transition is given by

$$u_c(t) = \sqrt{2}U_c\left[-\cos(\omega t+\varphi) + \sqrt{\left(\frac{\omega}{\omega_0}\right)^2 \sin^2\varphi + \cos^2\varphi}\, e^{-\xi t}\cos(\omega_0 t+\theta)\right] \quad (1.6.1)$$

where U_c and φ are the effective value and the initial phase angle of the steady-state component of the voltage across the capacitor, respectively, which can be obtained through calculation of the steady-state circuit. In the steady state, the impedance angle φ of the circuit is given by

$$\varphi = \arctan\frac{\omega L - \frac{1}{\omega c}}{R} \quad (1.6.2)$$

In the circuit, the effective values of the current, voltage across the capacitor and voltage across the inductor are respectively given by

$$I = \frac{E}{\sqrt{R^2 + \left(\omega L - \frac{1}{\omega C}\right)^2}} \quad (1.6.3)$$

$$U_c = \frac{I}{\omega C} = \frac{E}{\sqrt{\left[1-\left(\frac{\omega}{\omega_0}\right)^2\right]^2 + (R\omega C)^2}} = \frac{E}{\sqrt{\left[1-\left(\frac{\omega}{\omega_0}\right)^2\right]^2 + \left(\frac{2\xi}{\omega_0}\frac{\omega}{\omega_0}\right)^2}} \quad (1.6.4)$$

$$U_L = \omega L I = \frac{E}{\sqrt{\left[1-\left(\frac{\omega_0}{\omega}\right)^2\right]^2 + \left(\frac{R}{\omega L}\right)^2}} = \frac{E}{\sqrt{\left[1-\left(\frac{\omega_0}{\omega}\right)^2\right]^2 + \left(\frac{2\xi}{\omega_0}\frac{\omega_0}{\omega}\right)^2}} \quad (1.6.5)$$

with the initial phase angle given by

$$\phi = \phi_0 - \arctan\frac{\omega L - \frac{1}{\omega C}}{R} \tag{1.6.6}$$

The relationship between the initial phase angle of the free component θ and φ is expressed as

$$\tan\theta = \frac{\omega}{\omega_0}\tan\phi \tag{1.6.7}$$

When ω_0 is near to ω, high overvoltages will be generated on capacitive components. In the following, overvoltage amplitude of two different states will be discussed: when the circuit is in resonance and when circuit is near to resonance.

1. Circuit parameters satisfy $\omega L = \frac{1}{\omega C}$, namely $\omega = \omega_0$. At this moment, the current in the circuit is limited only by the resistance R; the current in the circuit is $I = E/R$; the voltage across the inductor is equal to that across the capacitor, represented by

$$U_L = U_c = I \cdot \frac{1}{\omega C} = \frac{E}{R}\sqrt{\frac{L}{C}} \tag{1.6.8}$$

   When the circuit resistance is relatively small, extremely high resonance overvoltages will be generated.
2. When $\omega < \omega_0$, that is $\frac{1}{\omega C} > \omega L$, the circuit is in the capacitive operating state. When the circuit resistance R is very small that it can be ignored, $U_L = U_C - E$. The voltage on the capacitor is given by

$$U_c = \frac{E}{1 - \left(\frac{\omega}{\omega_0}\right)^2} \tag{1.6.9}$$

   The voltage U_c across the capacitor is always larger than the supply voltage E. Such a power frequency voltage rise in the state of non-resonance is referred to as the capacitance effect, or the Ferranti effect.
3. When $\omega > \omega_0$, that is $\frac{1}{\omega C} < \omega L$, the circuit is in the reactive operating state. If the circuit resistance is ignored, $U_C = U_L - E$. According to the formula, the voltage on the capacitance is given by

$$U_c = \frac{E}{1 - \left(\frac{\omega}{\omega_0}\right)^2} \tag{1.6.10}$$

   When $\frac{\omega}{\omega_0} \leq \sqrt{2}$, voltage on the capacitor is equal to or larger than the supply voltage E, and the overvoltage declines rapidly with the increase of $\frac{\omega}{\omega_0}$.

Strict parameter coordination is required for linear resonance in the circuit. Resonance overvoltages could be avoided by avoiding the resonance range in the actual power system design or operation.

Meanwhile, linear resonance may occur in compensation networks of arc suppression coils when asymmetrical grounding faults occur or during non-full phase operation.

1.6.2 Ferro-Resonance Overvoltages

Ferro-resonance occurs in circuits containing components with iron cores. The inductance of the inductive component with iron core varies with the voltage and current, therefore nonlinear characteristics are noted in its inductance, thus resulting in a series ferro-resonance properties. That is why ferro-resonance is also called nonlinear resonance.

Figure 1.52 illustrates the simplest series circuits of R, C and the inductor with an iron core L. Under normal operating conditions, the inductive reactance is larger than the capacitive reactance (i.e. $\omega L > \frac{1}{\omega C}$); at this moment, resonance conditions are not satisfied. When the iron core reaches saturation and its inductive reactance declines $\omega L > 1/\omega C$ (i.e. $\omega = \omega_0 = \frac{1}{\sqrt{LC}}$), the conditions of series resonance are satisfied, and resonance will take place; the overvoltage is generated on the inductor and the capacitor. This phenomenon is called ferro-resonance.

As the inductance in the resonance circuit is not constant, the resonance frequency in the circuit is not fixed. When the resonance frequency f_0 equals the power frequency, fundamental resonance occurs in the circuit; when the resonance frequency is an integral multiple of power frequency (such as 2 times, 3 times, 4 times…), high-order harmonic resonance occurs in the circuit; when the resonance frequency is the fractional multiple of the power frequency (such as 1/2 time, 1/3 time, 1/4 time……), fractional harmonic resonance occurs in the circuit. Even in the case of fundamental resonance, high-order harmonics are likely to be generated as well as the fundamental component. Ferro-resonance therefore features the possibility of containing harmonic resonance of different orders.

Figure 1.53 provides the volt–ampere curves for the capacitor and the inductor with an iron core. The voltages and currents are both expressed in effective values. Obviously, U_C has to be a straight line ($U_C = I/\omega C$). As for the inductor with an iron core, U_L is basically a straight line before the iron core is saturable; the inductance is reduced after the iron core is saturable, and U_L is no longer a straight line. Therefore, the two volt–ampere curves will intersect. It is seen that $\omega L > 1/\omega C$ is a prerequisite for the occurrence of ferro-resonance.

When the circuit resistance is ignored, the sum of the voltage drops on L and C will be balanced with the supply potential: $\dot{E} = \dot{U}_L + \dot{U}_C$. As $\dot{U}_L$ and $\dot{U}_C$ are reversed in phase, the equilibrium equation becomes $E = \Delta U$, while $\Delta U = |U_L - U_C|$, as illustrated by the curve of ΔU. In the figure, the curve of ΔU and the dotted line intersect at α_1,α_2 and α_3, all of which can meet the condition of potential balance $E = \Delta U$, and are thus called the

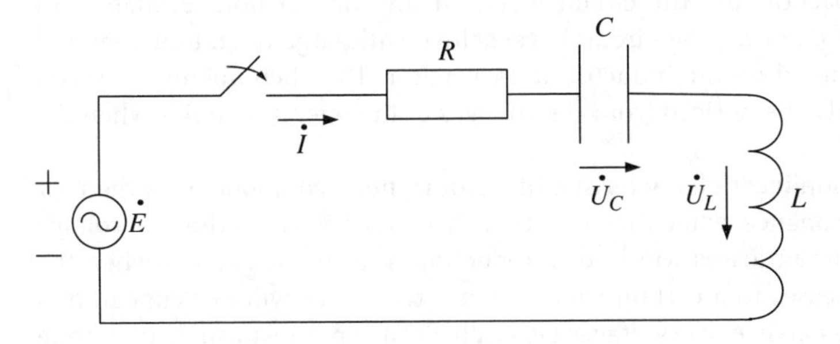

Figure 1.52 Series ferro-resonance circuit.

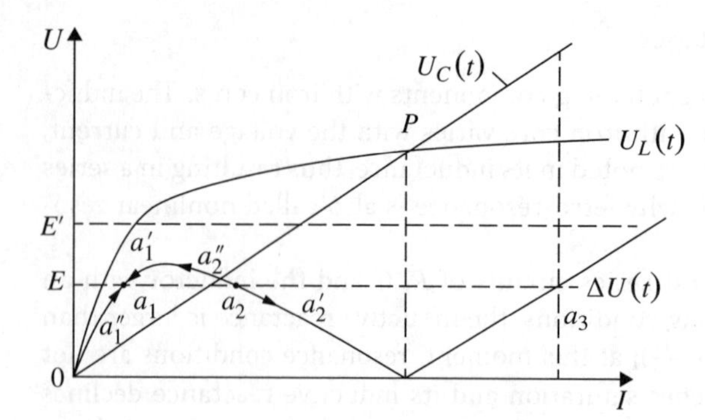

Figure 1.53 Volt-ampere characteristic of the series ferro-resonance circuit.

equilibrium point. Based on its physical conception, an equilibrium point must satisfy conditions for potential balance, but may not necessarily satisfy conditions for stability, which hinders the point from becoming an actual operating point. "Slight disturbance" is generally used to estimate the stability of an equilibrium point. That is, the equilibrium point will be deemed as stable and can become an actual operating point if the circuit state is able to automatically return to the original balance after deviating from the equilibrium point due to the applied "slight disturbance". Otherwise, the balance point will be deemed as unstable and cannot become an actual operating point if the circuit state gradually deviates from the equilibrium point after applying the slight disturbance.

On the basis of this principle, the following judgment can be made, that point α_1 is a non-resonance operating point of the circuit; the circuit is inductive with small circuit current and small voltage across the inductor and capacitor. However, as for point α_2, if a slight increase in the current of the circuit is caused by a certain new disturbance and the current deviates from α_2 to α_2' when $E > \Delta U$, then the current will continue to increase until reaching the new equilibrium point α_3. Otherwise, if the disturbance leads to a slight decrease in the circuit current, and ΔU deviates from point α_2 to point α_2' when $E < \Delta U$, then the circuit current will continue to decrease until reaching the equilibrium point α_1. It is thus clear that the equilibrium point α_2 is unstable and cannot withstand any slight disturbance. Point α_3 can be determined as stable by adopting the same method.

Thus two potential stable operating states exist – point α_1 and point α_3 – for the ferro-resonance circuit under the effect of applied E. In operating state of point α_1, $U_L > U_C$; the circuit is inductive with small circuit current and small voltage across the inductor and capacitor, and the circuit works in the state of non-resonance. In the operating state of point α_3, the circuit is capacitive with large circuit current and overvoltages will be noted on the inductor and capacitor. The phenomenon of series ferro-resonance can also be noticed from the changes of the operating point when the potential E increases.

In conclusion, the nonlinear characteristic of ferromagnetic components is the root cause behind ferro-resonance, while the saturation characteristic limits the overvoltage amplitude. Besides, circuit losses will lead to reduction in overvoltage, and when the circuit resistance increases to a certain value, intense resonance will not appear. It is indicated that ferro-resonance overvoltages generally occur on transformers operating in unloaded or lightly loaded conditions.

When the iron cores of such inductive components as transformers, reactors and instrument transformers are saturated, inductive parameters will alter with the changes of current or flux. If the supply frequency and circuit parameters meet the requirement $\omega = \frac{1}{\sqrt{L_0 \times C_0}}$, ferro-resonance will be generated, often accompanied by phase reversal. For distribution systems with ungrounded or non-effectively grounded neutrals, the ferro-resonance caused by core saturation of electromagnetic PTs is one of the most frequent overvoltages causing the greatest harm. Depending on the main frequency component, ferro-resonance is divided into fractional-frequency resonance, fundamental resonance and high-frequency resonance. Figure 1.54 presents the wave shapes of ferro-resonance of the three categories.

The above discussion concerns the basic properties of fundamental ferro-resonance overvoltages. It is demonstrated by practical operation and experimental analysis that for resonance circuits containing inductive components with iron cores, continuous resonance of other frequencies may appear if certain conditions are satisfied. The resonance frequency may be an integral or fractional multiple of the power frequency, respectively referred to as high-order harmonic resonance and frequency-dividing resonance. Resonance of two or more frequencies may simultaneously appear in some particular conditions.

Ferro-resonance overvoltages in power systems generally occur in the state of non-full phase operation where the inductance may be the excitation inductance of unloaded or lightly loaded transformers, of arc suppression coils and of electromagnetic PTs, and the capacitance may be conductor-to-ground capacitance, interphase capacitance and the stray capacitance between the inductance coil and the ground.

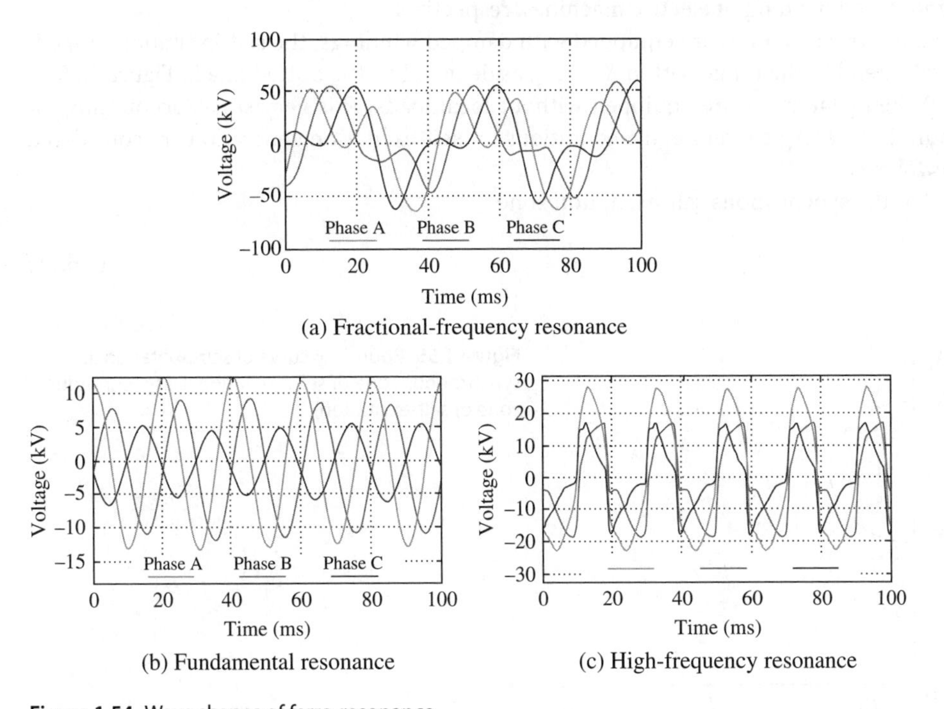

Figure 1.54 Wave shapes of ferro-resonance.

1.6.3 Parametric Resonance Overvoltages

The self-excitation overvoltage due to periodic changes of parameters of inductive components and capacitive components is referred to as the parametric resonance overvoltage. When synchronous generators are connected to capacitive loads (such as a segment of the unloaded line), the reactance periodically changes within the range of x'_d to x_q. Owing to the effect of optimized magnetism of capacitive current, parametric resonance may be triggered if parameter cooperation is improper. Small or even zero excitation current can lead to rises in the voltage and current of the generator terminals, making it impossible to operate in parallel with other electric machines. Such a phenomenon is called the self-excitation of electric machines, and the resulting self-excitation overvoltage is called the self-excitation overvoltage. In terms of physical nature, self-excitation of electric machines is attributed to parametric resonance due to the capacitance and the inductance with periodic changes during rotation of the electric machines. Prior to official operation, the design department will check the self-excitation overvoltage parameters to avoid the resonance point; therefore, parametric resonance is not likely to take place in normal conditions.

When synchronous generators are equipped with unloaded long lines, synchronous and asynchronous self-excitation, which are different by nature, may appear, as shown in Figure 1.55.

Self-excitation will be generated with parameter cooperation of X_c and R within the boundary of the curve. Of the two semicircular curves, range I is the synchronous self-excitation zone, while range II is the asynchronous self-excitation zone, where the dotted and solid lines represent the self-excitation zone of the damped and the undamped winding of electric machines, respectively.

If electric machines are equipped with damped windings, the self-excitation range II will extend to the range within $X_c < X'_d$, as depicted by the dotted line in Figure 1.55.

When generators are equipped without local loads, R is very small, accounting for only 2–3% of X_c; therefore, the conditions for self-excitation generation are considered as follows.

For the synchronous self-excitation zone

$$X_q < X_c < X_d \tag{1.6.11}$$

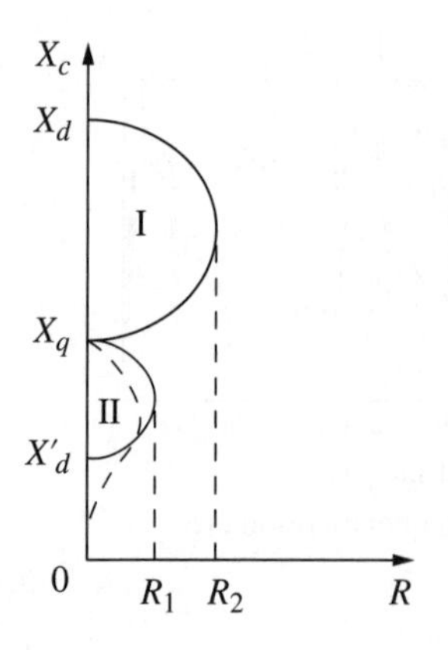

Figure 1.55 Boundary curve of self-excitation (I: synchronous zone of self-excitation II: asynchronous zone of self-excitation).

For the asynchronous self-excitation zone

$$X_c < X_d' < X_q \quad (1.6.12)$$

As can be seen from Equation (1.6.11), the synchronous self-excitation zone is produced by the prerequisite $X_d \neq X_q$, which can only be met by water turbine generator sets. Overvoltages produced by synchronous self-excitation are generally slow in speed, which are easy to eliminate by automatic excitation regulators. For asynchronous self-excitation with rapid speed of rise, even very fast excitation regulators are unable to control it in time; therefore, the latter voltage is more dangerous. In an actual power system, if the generator is equipped with a long unloaded line, then X_c will be relatively small, and the circuit may be in the scope of self-excitation as the loss of system is generally also small. If parallel resistors or reactors are used for compensation, X_c can be regarded as being increased, thus avoiding self-excitation overvoltages.

In order to eliminate the phenomenon of self-excitation, the following measures may be taken: using fast automatic excitation regulators to eliminate synchronous self-excitation; installing parallel reactors in UHV grids to compensate line capacitance to allow the equivalent capacitive reactance to be greater than X_d' and X_q, thus eliminating resonance; temporarily connecting large resistors in series to generator stator windings to increase the damped resistance of the circuit to the extent greater than R_1 and R_2 illustrated in Figure 1.55.

To sum up, parametric resonance overvoltages possess the following characteristics.

1. The energy required for parametric resonance is provided by the prime motor, which changes the parameters, and individual power supply is not needed. The presence of a certain amount of remanence or a small quantity of residual charge on the capacitor will lead to development of parametric resonance if parameters are mismatched.
2. Since there are losses in the circuit, only when the energy produced by parametric variation is enough to compensate for circuit losses, can the development of resonance be ensured. Therefore, for a certain loop resistance, there is a certain corresponding self-excitation scope. Upon the occurrence of resonance, saturation of inductance would make the circuit automatically deviate from the resonance condition, preventing self-excitation overvoltages from growing.

2

Transducers for Online Overvoltage Monitoring

Overvoltage signal acquisition, generally achieved by voltage transducers, is the key feature of online lightning overvoltage monitoring. The design of voltage transducers is thus a technical challenge in online overvoltage monitoring. Overvoltages on transmission lines are characterized by high magnitude and high frequency, which are very demanding for transducers. Outstanding performance in stability, linearity and transient response characteristics is needed for transducers to acquire accurate lightning overvoltage signals. Transducers used for online overvoltage monitoring to date are categorized into three kinds depending on overvoltage acquisition methods: the contact overvoltage transducer, the non-contact overvoltage transducer and the fiber-optic overvoltage transducer. Specific transducers of the three categories will be introduced next, including voltage transducers at transformer bushing taps, gapless metal oxide arrester (MOA) voltage transducers, transducers for transmission lines coupled by capacitance and all-optic voltage transducers.

2.1 Overvoltage Transducers at Transformer Bushing Taps

Presently, HV voltage dividers, characterized by simple structure, high measuring accuracy and good transient response, are usually used to acquire voltage signals in the field of online overvoltage monitoring. We will consider a number of issues such as long-term operation reliability, heating, impedance matching and ac impulse, especially those concerning the safety of personnel and measuring devices when voltage dividers are connected in parallel to systems of higher rating. To ensure safety while using minimum primary equipment, this chapter will present a method to acquire real-time overvoltage signals from the transformer capacitive bushing tap by a voltage-dividing system consisting of a specially made voltage transducer and a transformer capacitive bushing.

2.1.1 Design

2.1.1.1 Structural Design of the Main Body

The voltage transducer is installed at the transformer capacitive bushing tap, as shown in Figure 2.1.

The schematic of the voltage transducer installed at the bushing tap is shown in Figure 2.2, where C_S is the voltage-dividing capacitance, R_S is the voltage-dividing resistance, R_P is the matched resistance and P_S is the protection unit.

Measurement and Analysis of Overvoltages in Power Systems, First Edition. Jianming Li.

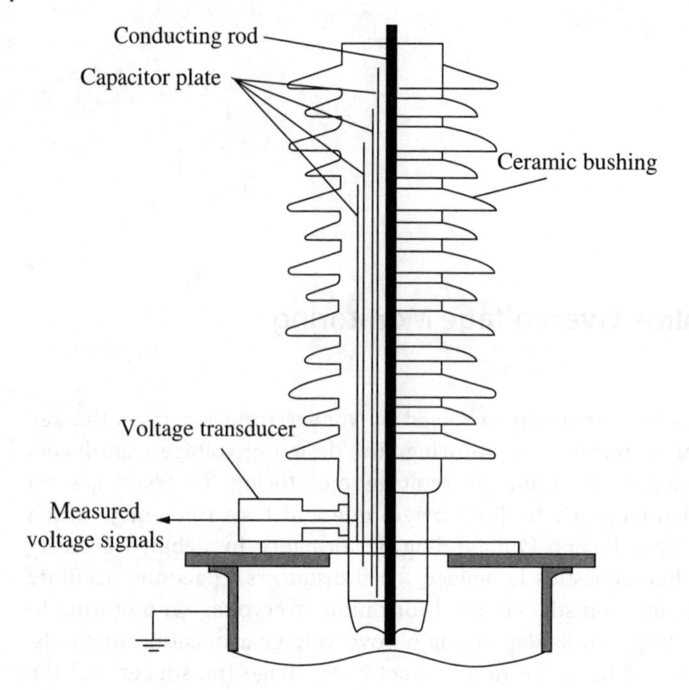

Figure 2.1 Installation scheme of transducers.

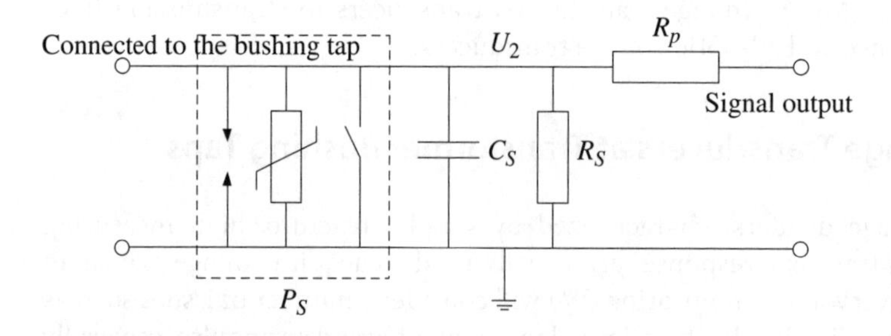

Figure 2.2 Schematic of a voltage transducer at the bushing tap.

Figure 2.3 shows the structure of the transducer, consisting of a voltage-dividing unit, a protection unit and a signal output cable. The voltage-dividing unit is composed of a voltage-dividing capacitance and a voltage-dividing resistance, which constitute high-frequency and low-frequency response circuits, respectively. The coaxial cylinder structure is adopted to reduce magnetic coupling between the elements of the voltage-dividing unit and the output circuit, and to minimize the effect of residual inductance on the response characteristics. The matched resistance is connected to the signal cable interface through the transducer center. Resistive and capacitive voltage-dividing elements are connected in parallel, and distributed uniformly over the circumference and symmetrically to the center. The grounded metal shell, as the transducer outer layer, achieves electromagnetic shielding for the core components.

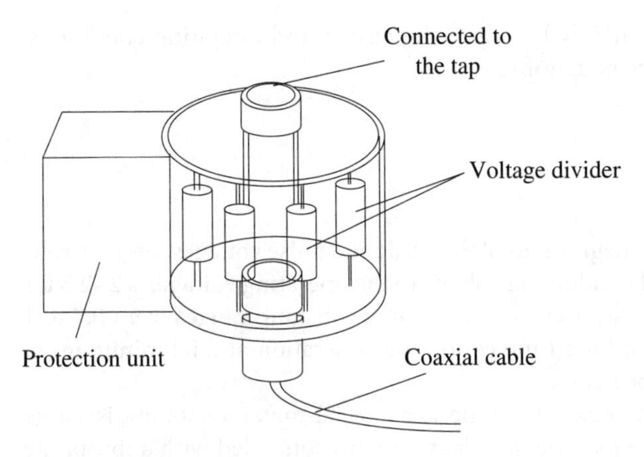

Figure 2.3 Transducer structure.

2.1.1.2 Protection Unit Design

A protection unit circuit is designed to prevent disconnection of the bushing tap circuit and to suppress detrimental overvoltages to the secondary side. In the case of transducer failures, this circuit can ensure a secured variation range of the voltage on the clamp measuring terminal, reliable operation of the protection unit and reliable grounding of the bushing tap. The protection unit is a hybrid protection circuit consisting of varistors, discharge tubes and relays, and will be detailed in this section.

Varistor – The ZnO varistor, a sensitive element with nonlinear volt–ampere characteristics, is often used. It acts as a small capacitor under normal voltage; when the applied voltage reaches beyond its critical value, its internal resistance declines dramatically, leading to current conduction in a very short time (with a response time of nanoseconds). The order of the operating current increases to several magnitudes higher, and the overvoltage is absorbed by the varistor in the form of discharge current, equivalent to a partial short circuit. Therefore, the other elements in the circuit are effectively protected from overvoltages. Figure 2.4 shows the volt–ampere characteristic of the varistor that can ensure that the measuring terminal voltage is within the safety range.

Figure 2.4 Typical varistor volt–ampere characteristic.

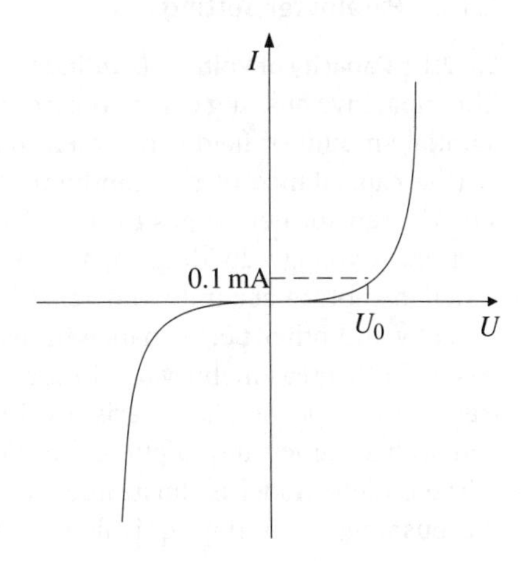

The selection of varistors is mainly based on the circuit and operating conditions, usually given by one of the three equations.

$$
\begin{aligned}
V &= V_{AC} \times K_P \times T_S \\
V &= V_{DC} \times K_P \times T_S \\
V &= V_{CP} \times K_P \times T_S
\end{aligned}
\tag{2.1.1}
$$

where V_{AC}, V_{DC} and V_{CP} are the magnitudes of the ac, dc and pulse voltages, respectively; K_P, the multiple of the practical conducting voltage to the operating voltage, is 2–2.5 for ac or dc circuits and 2 for impulse circuits; T_S, the time constant, ranges from 0.7 to 1 and takes its maximum value in long-time continuous operation and its minimum in short-time or discontinuous operation.

Gas discharge tube – The gas discharge tube is a special metal–ceramic discharge element made of highly reactive material for electron emission, filled with appropriate gas medium in the discharge gap and equipped with discharge and ignition structures; it is mainly used for transient overvoltage protection.

With a high impedance ($\geq 1000\,\mathrm{M\Omega}$) and a low capacitance ($\leq 10\,\mathrm{pF}$), the discharge tube would virtually cause no damage to the circuit in normal operation when connected as a protection unit. When the tube encounters harmful transient overvoltages, breakdown and discharge of the tube take place first, and this is followed by a fast impedance drop (almost appearing as a short-circuit). By transferring harmful current to the ground via the grounding wire and by restricting the voltage within its arc voltage (approximately 20 V), the tube eliminates the detrimental transient overvoltages and overcurrents, and thus protects the circuit elements. It will return to the high impedance state rapidly when the overvoltage extinguishes, and the circuit will continue to operate under normal conditions.

Protective relay unit – The protective relay unit is the executive body of the control circuit. Controlled by the control circuit, it can realize rapid and reliable short circuit of the bushing tap grounding wire under open circuit of the bushing tap. The relay will be disconnected when the transducer is in operation, and will be closed to prevent bushing tap open circuit and meanwhile send out fault signals to inform operation personnel in the case of transducer failures or loss of the control supply.

2.1.2 Parameter Setting

2.1.2.1 Capacity of Voltage-Dividing Capacitance

The capacitive bushing can be regarded as a pure capacitance as a simplification. A substantial amount of field experiment data has demonstrated that the main capacitance C (the capacitance of the conducting rod of the HV bushing tap to ground) of the 110 kV transformer ranges from 200 pF to 300 pF, and takes the value of 280 pF in this section. A voltage-dividing capacitance is connected in series with the bushing tap in order to capture accurate and reliable overvoltage signals. Adequate selection of the capacity and other performance indicators of the voltage-dividing capacitance C_2 can ensure both a reasonable variation range of the transducer output voltage and excellent frequency response characteristics. In this way, the signal-to-noise ratio is improved, and both accuracy and safety of measurement are ensured.

The circuit model is illustrated in Figure 2.5, where C_1 is the main capacitance of the bushing; C_2 is the equivalent voltage-dividing capacitance; R_1 is the equivalent

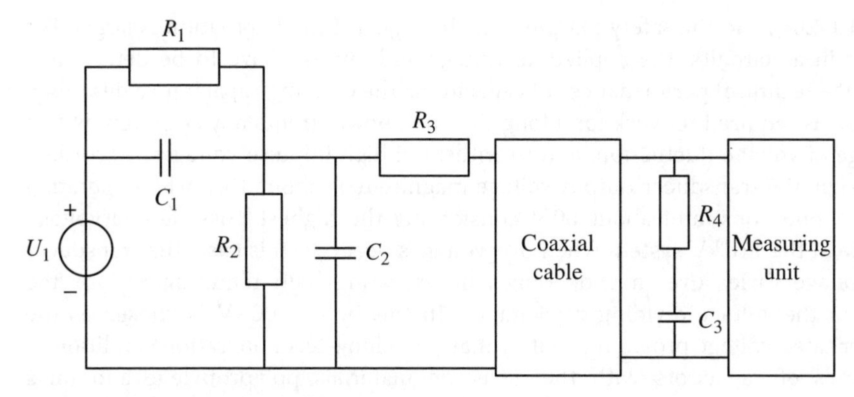

Figure 2.5 Circuit model of the entire system.

insulation resistance of the bushing rod to the tap; R_2 is the equivalent voltage-dividing resistance. R_1 and R_2 are usually high, and the input resistance of the measuring circuit is in the order of a megohm.

The coaxial cable is matched at both ends, and the values of the matched resistances R_3 and R_4 are equal to the cable wave impedance of the cable Z. C_C, the distributed capacitance of the cable, is about several thousand pF and can thus be considered negligible. The wave entering the cable is given by

$$U_1 \frac{C_1}{C_1 + C_2} \frac{Z}{Z + R_3} = U_1 \frac{C_1}{C_1 + C_2} \frac{Z}{Z + Z} = \frac{1}{2} \frac{C_1 U_1}{C_1 + C_2} \tag{2.1.2}$$

The receiving end of the coaxial cable connects to R_3 and C_3; R_4 equals the wave impedance of the cable Z, $C_3 \approx C_2$. As there is no reflection when the wave reaches the end, the initial voltage-dividing ratio is thus given by

$$\left. \frac{U_1}{U_2} \right|_{t=0} = \frac{2(C_1 + C_2)}{C_1} \tag{2.1.3}$$

The final steady voltage-dividing ratio is given by

$$\left. \frac{U_1}{U_2} \right|_{t=\infty} = \frac{(C_1 + C_2 + C_3 + C_C)}{C_1} \tag{2.1.4}$$

The initial voltage-dividing ratio equals the final voltage-dividing ratio if $C_1 + C_2 = C_3 + C_C$ is satisfied. When primary bus voltage equals the rated voltage, and the designed transducer output voltage to the signal conditioning circuit is 13 V, the voltage-dividing ratio can be calculated as

$$k = \frac{U_1}{U_2} = \frac{115 \times 10^3}{13\sqrt{3}} = 5107 \tag{2.1.5}$$

where C_2 is 0.71 μF according to Equation (2.1.3).

2.1.2.2 Voltage Rating

Capacitor breakdown or failure could occur when the circuit voltage is higher than the capacitor rated voltage, hence a safety margin is needed for the selection of capacitor voltage ratings. The operating voltage of the circuit is usually lower than the rated value by 10–20%, and the safety margin must be higher if the fluctuation is larger. For capacitors in ac circuits, the applied ac voltage and current have to be determined based on the technical performances of capacitors. The dividing capacitance described in this book is required to work for a long time in a power frequency ac system with a wide range of voltage fluctuation, and to withstand high impulse voltages. According to the design, the transducer output voltage magnitude is about 13 V when operating in normal conditions, and about 60 V considering the highest possible overvoltage magnitude in the 110 kV system when overvoltages occur. In this case, the transducer output voltage varies over a wide range, necessitating high requirements for the selection of the voltage-dividing capacitance. In this book, 500 kV is chosen as the transducer rated voltage providing a safety margin in long-term operation conditions.

Two kinds of capacitors with the dielectric materials polypropylene and mica are compared through tests. Both of them have good high-frequency performance. However, the volume of the polypropylene capacitor is much smaller than that of mica for the same capacitance. Therefore the polypropylene capacitor is adopted as the voltage-dividing capacitor in the design to decrease the transducer size.

2.1.3 Feasibility Analysis

2.1.3.1 Error Analysis and Dynamic Error Correction

According to Equations (2.1.2) and (2.1.3), when the two ends are completely matched, the voltage-dividing ratio k of the capacitive dividing system is given by

$$k = \frac{U_1}{U_2} = \frac{2(C_1 + C_2)}{C_1} \tag{2.1.6}$$

If the variation of the main capacitance during system operation ΔC is small, the two ends can still be deemed as meeting the requirement $C_1 + \Delta C + C_2 = C_3 + C_C$, and the new voltage-dividing ratio k' can be represented by

$$k' = \frac{2(C_1 + \Delta C + C_2)}{C_1 + \Delta C} \tag{2.1.7}$$

The voltage-dividing ratio error Δk then is given by

$$\begin{aligned}\Delta k = k - k' &= \frac{2(C_1 + C_2)}{C_1} - \frac{2(C_1 + \Delta C + C_2)}{C_1 + \Delta C}\\ &= \frac{2C_1C_1 + 2C_1C_2 + 2C_1\Delta C + 2C_2\Delta C - 2C_1C_1 - 2C_1\Delta C - 2C_1C_2}{C_1(C_1 + \Delta C)}\\ &= \frac{2C_2\Delta C}{C_1(C_1 + \Delta C)}\end{aligned} \tag{2.1.8}$$

and the relative voltage-dividing ratio $\Delta k/k$ is expressed as

$$\frac{\Delta k}{k} = \frac{2C_2\Delta C}{C_1(C_1 + \Delta C)} \cdot \frac{C_1}{2(C_1 + C_2)} = \frac{C_2\Delta C}{(C_1 + \Delta C)(C_1 + C_2)} \tag{2.1.9}$$

Considering variations of the transducer dividing ratio with the bushing capacitance which contribute to larger measurement errors, this system designs the function of dynamic error correction with the help of software. The measuring system takes the secondary voltage from PT as the correction voltage at a regular time, calculates the average voltage $U_{B\text{-}A}$, $U_{B\text{-}B}$, $U_{B\text{-}C}$ as the references, and simultaneously collects the three-phase bus voltages U_A, U_B, U_C to identify voltage saltation. Provided no voltage saltation occurs,

$$\begin{aligned}\Delta U_A &= U_A - U_{b-A}\\ \Delta U_B &= U_B - U_{b-B}\\ \Delta U_C &= U_C - U_{c-C}\end{aligned} \tag{2.1.10}$$

are taken as the dynamic error correction values of the measuring system.

2.1.3.2 Impulse Response Characteristics Tests

The adopted supply in the impulse response test is a 2400 kV/210 kJ impulse voltage generator with the standard rated voltage of ±2400 kV, the rated energy of 210 kJ, 8-class charging capacitors with the main capacitance of each reaching 0.655 μF/150 kV, and with the highest tolerant charging voltage of ±300 kV. The generated standard output lightning wave has a front time of 1.2 ± 0.36 μs, a tail time of 50 ± 10 μs, and the output efficiency no less than 90%, whereas the standard switching voltage wave has a front time of 250 μs, a tail time of 2500 ± 60 μs and the output efficiency no less than 75%. The impulse voltage divider is the capacitive voltage divider TCF2000/0.0012 with the rated voltage of 2000 kV and rated voltage-dividing ratio of 3000:1.

The bushing in the test is a capacitive oil-filled paper bushing with the rated voltage of 363 kV and the main capacitance of 473 pF. The equivalent capacitance of the voltage transducer is 0.64 μF. Considering that the signal cable is not long, the test circuit adopts head matching and the calculated voltage-dividing ratio is 1353:1. The wire connection for the test is shown in Figure 2.6.

In the test, the wave front/tail time of the lightning impulse voltage is 1.56/51.2 μs, and that of the switching impulse voltage is 219/2650 μs, which are both within the standard error range specified by GB/T16927.1-1997 and IEC60060.1. The test curve of the lightning impulse test is depicted in Figure 2.7(a), and the waveform of the lightning impulse voltage measured by the transducer is shown in Figure 2.7(b).

The test curve of the switching impulse voltage is shown in Figure 2.8(a), and the waveform of the switching impulse voltage measured by the transducer is illustrated in Figure 2.8(b). It can be seen that the output curve of the transducer is virtually the same as that of the standard voltage divider.

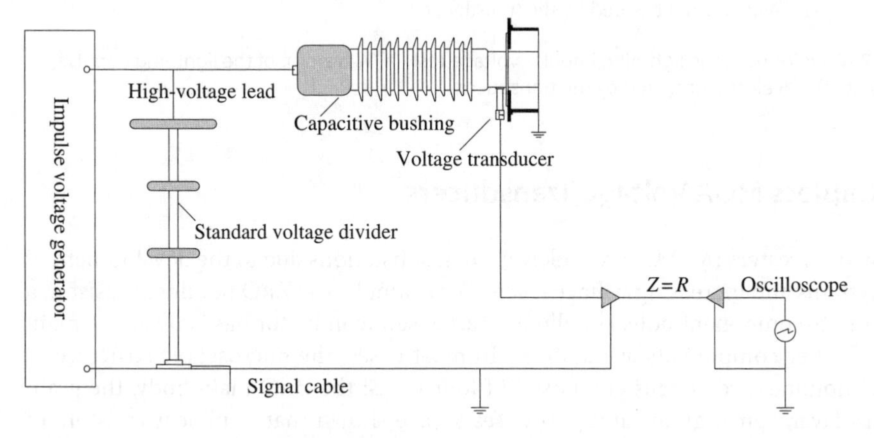

Figure 2.6 Wire connection for the impulse response test.

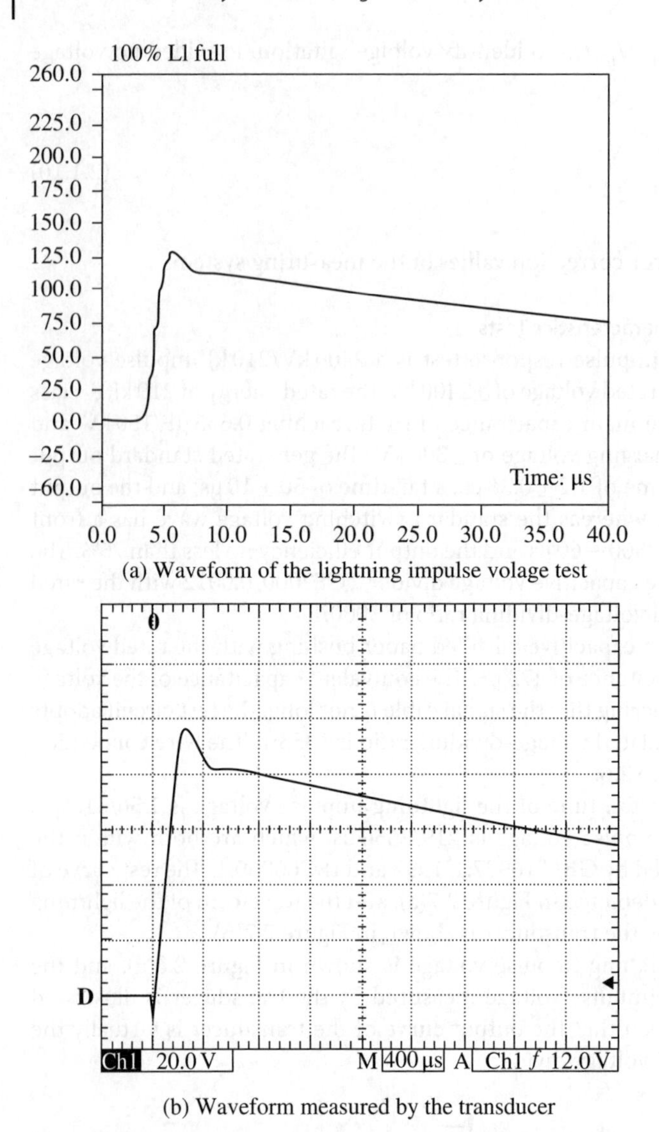

Figure 2.7 Waveforms of the lightning impulse voltage test. (a) Waveform of the lightning impulse voltage test (b) Waveform measured by the transducer.

2.2 Gapless MOA Voltage Transducers

Metal oxide arresters (MOAs) are widely used in substations due to the development of power systems and increasing voltage levels. For example, the ZnO nonlinear resistor, a kind of multi-component polycrystalline ceramic semiconductor has ZnO as the main body and other components as additives. In most cases, the microscopic structure of the ZnO nonlinear resistor is composed of four parts: the ZnO main body, the grain boundary layer, spinel grains and pores. Tests have shown that nonlinear resistors of

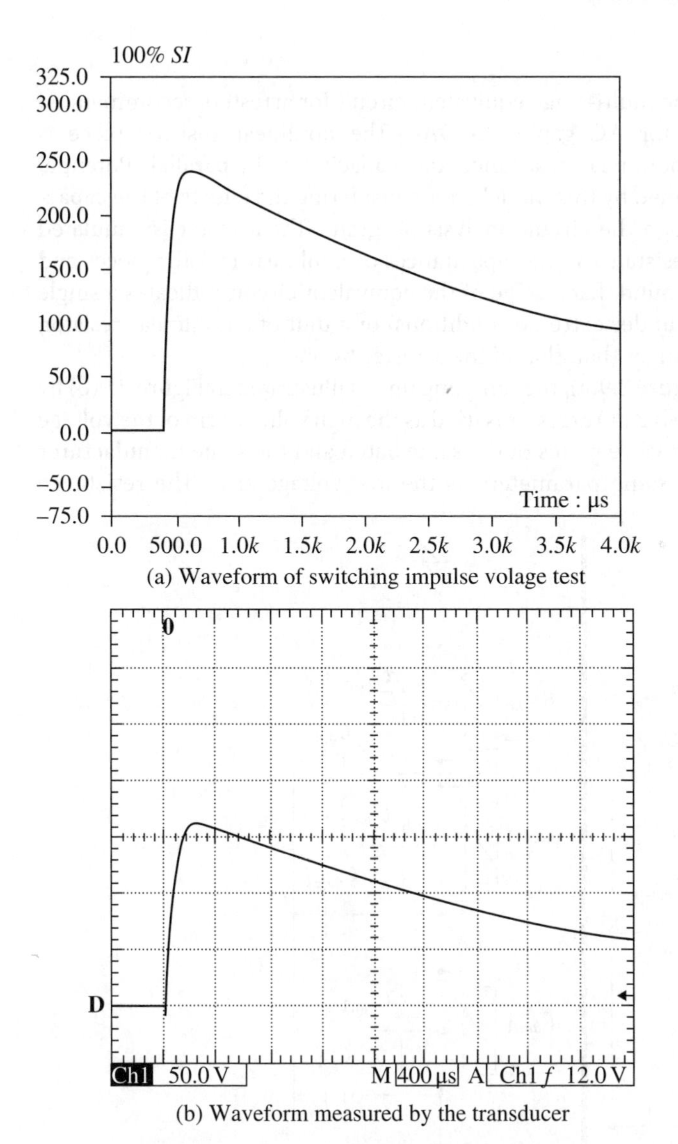

Figure 2.8 Waveforms of the switching impulse voltage test. (a) Waveform of switching impulse voltage test (b) Waveform measured by the transducer.

the same manufacturer and the same batch have only tiny differences in performance due to the same manufacturing technique and additive compositions. Moreover, the improved manufacturing techniques of ZnO nonlinear resistors enable high thermal and electric stability of the grain boundary layers of valve plates, thereby ensuring their reliable operating properties and stable volt–ampere characteristics.

The above characteristics of the ZnO valve plates makes it feasible to manufacture of the voltage transducers using valve plates with MOAs as voltage dividers. More details about the design, feasibility analysis and practical applications of this kind of transducer will be presented next.

2.2.1 Design

Shown in Figure 2.9(a) is the multi-stage equivalent circuit for arresters recommended by the national standard for AC gapless MOAs. The nonlinear resistor piece is represented by a pair of nonlinear resistance and capacitance in parallel. Potential distribution can be determined by this model after considering the effects of the capacitance and resistance through the circuit analysis program. The arrester is simulated using the voltage-related resistances, the capacitances of nonlinear resistor pieces and the stray capacitances to ground. Each stage of the equivalent circuit indicates a single metal oxide resistor piece (under extreme conditions) or a unit of a nonlinear resistor, whose length must be no longer than that of the arrester by 3%.

Based on the model in Figure 2.9(a), the sampling unit is illustrated in Figure 2.9(b) by the dashed frame. The gapless ZnO arrester is used as the high voltage arm of the voltage transducer, and the arrester valve plates of the same batch and the same manufacturer or with approximately the same parameters as the low voltage arm. The resistance

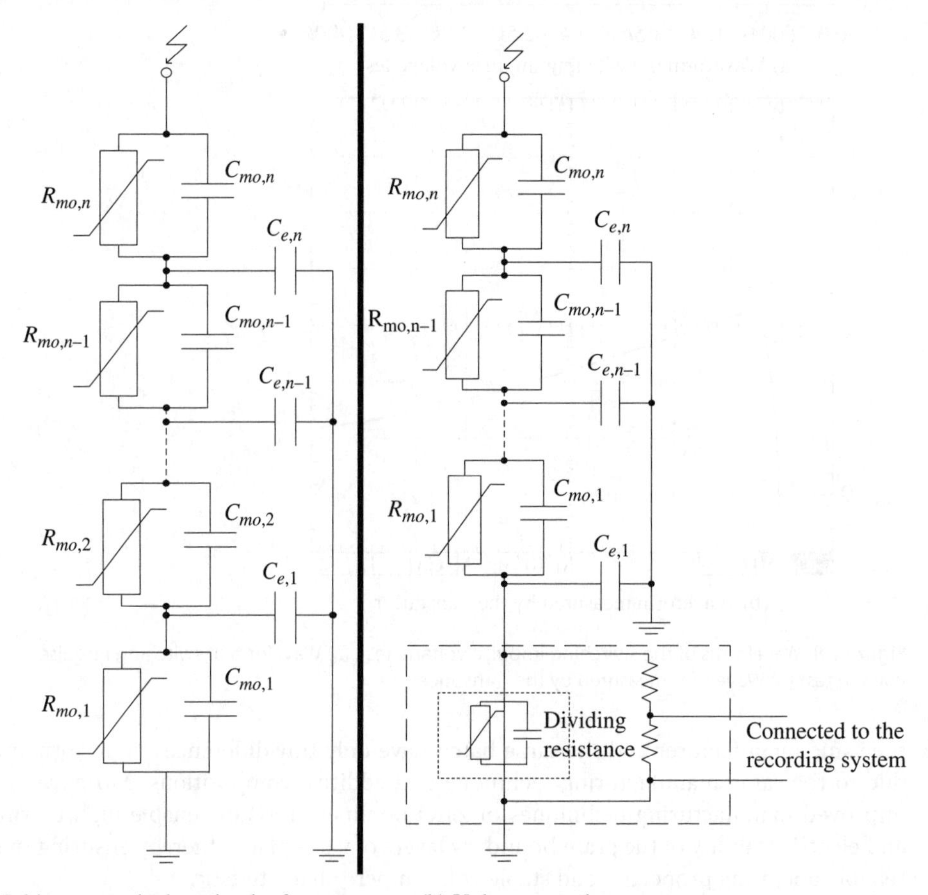

Figure 2.9 Sampling structure of arrester valve plates. ($R_{mo,x}$ – resistance related to the voltage in the unit x; $C_{mo,x}$ – capacitance of the unit x; $C_{e,x}$ – stray capacitance to ground of the node x; n – number of units) (a) Multi-stage equivalent circuit of arrestors (b) Voltage transducer structure.

in parallel to the low-voltage arm valve plate have to be sufficiently high so that the current could still flow into the ground through the valve plate, thus not affecting normal operation of the arrester. Moreover, parallel resistors have to be divided into multiple groups for the purpose of voltage division so as to avoid the impact of high overvoltages on the back-end sampling and recording system. In order to ensure acquisition data accuracy, a digital signal conversion module is also needed prior to the sampling and recording system, which, by converting acquisition signals into digital ones in the front end, can minimize the impact of complex field electromagnetic interferences.

Generally, in order to avoid the interference of arrester stray capacitance on online monitoring, the stray capacitance to ground has to be determined first by calculating capacitance and electric field, and potential distribution can then be calculated via circuit analysis that introduces the resistance characteristics. Due to the effect of temperature on the resistance, iterative calculation programs need to be run. However, for the ZnO arrester that fails to meet the requirement of balanced voltage distribution, a grading ring mounted at the high-voltage terminal is an option to laterally compensate for the effect of the stray capacitances to ground of each ZnO element. For this reason, the stray capacitance effect is ignored in this model. In practical applications, errors caused by stray capacitances can be calibrated and compensated through the sampling capacitance of the low-voltage arm. In this way, the ideal voltage-dividing precision can be achieved.

2.2.2 Operating Properties and Feasibility Analysis

The microstructure and properties of ZnO varistors mainly depend on the fabrication process and additive compositions. The conducting properties of the micro-unit of ZnO particles can be represented by the series-parallel circuit of resistors and capacitors, as shown in Figure 2.10.

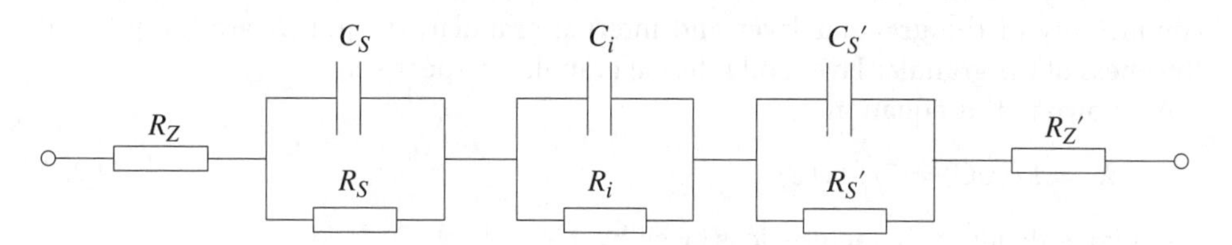

Figure 2.10 Equivalent circuit of the conducting properties of ZnO micro-unit.

where R_z and R_z' refer to the volume resistance of ZnO; R_i and C_i refer to the resistance and capacitance of the granular layer, respectively; C_s, R_s, C_s' and R_s' refer to the capacitance and resistance between the crystals and the dielectric.

2.2.2.1 Working in the Small Current Section

For the valve plate working in the small current section, breakdown or conduction would not occur, the voltage divider is equivalently capacitive, as shown in Figure 2.11.

$$C_1 = e\frac{s}{d_1}; C_2 = e\frac{s}{d_2} \tag{2.2.1}$$

$$\varepsilon = \varepsilon_{obs}\varepsilon_0 \tag{2.2.2}$$

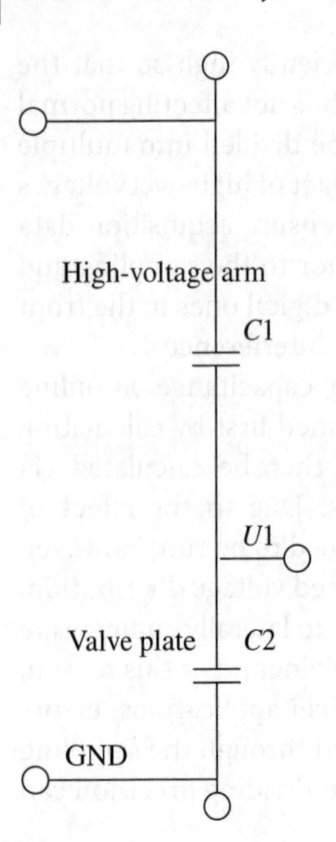

Figure 2.11 Equivalent circuit for small current range.

where ε_{obs} and ε_0 are the relative dielectric constant and the vacuum dielectric constant respectively; C is the inter-electrode capacitance of the valve plate; S is the surface electrode area of the valve plate; d is the valve plate thickness.

Among others, the relative dielectric constant of the valve plate is expressed as

$$\varepsilon_{obs} = [d_1 d_2(\sigma_1 - \sigma_2)^2 / (d_1\sigma_2 - d_2\sigma_1)^2]\varepsilon_b \tag{2.2.3}$$

where ε_b is the dielectric constant of the valve plate material; σ_1 and σ_2 are the conductivity of the granular layer and internal granular; d_1 and d_2 are the internal thickness of the granular layer and internal granular, respectively.

As shown in this equation:

$$X_c = 1/(\omega C) = 1/(2\pi f C) \tag{2.2.4}$$

the voltage-dividing characteristic is given by

$$U_1 = U_0 \frac{X_{C2}}{X_{C1} + X_{C2}} = U_0 \frac{d_2}{d_1 + d_2} \tag{2.2.5}$$

2.2.2.2 Working in the Large Current Range

The nonlinear volt–ampere characteristics of ZnO valve plates can be described by

$$U = 10^{A_0} \cdot I^{(1+A_0+A_2 \lg I)} \tag{2.2.6}$$

where A_0, A_1 and A_2 are constants calculated according to the current–voltage test of the sample valve plate.

The nonlinear resistivity of the ZnO element is given by

$$\alpha_v = \frac{1}{K} \cdot 10^{[1+A_1+A_2(\lg I)^2]} \tag{2.2.7}$$

where K is the ratio between the polarity area and the polarity distance of the ZnO element.

The nonlinear resistance of the ZnO valve plate is therefore

$$R = \alpha_v \cdot d/s \tag{2.2.8}$$

When the valve plate breaks down and becomes conducting in the large current section, the equivalent voltage-dividing circuit is as shown in Figure 2.12.

$$R_1 = R_{v1} = \alpha_v \frac{d_1}{S};\ R_2 = R_{v2} = \alpha_v \frac{d_2}{S} \tag{2.2.9}$$

where α_v is the nonlinear resistivity of the ZnO elements in high- and low-voltage arms.

The voltage-dividing ratio of the voltage transducer in the large current range is

$$U_1 = U_0 \frac{R_2}{R_1 + R_2} = U_0 \frac{d_2}{d_1 + d_2} \tag{2.2.10}$$

It can be seen from Equations (2.2.5) and (2.2.10) that the ratio between the input and output voltage is equal to the ratio between the total thickness of the valve plate and the thickness of the sampling plate whatever the range is. That is to say, the voltage transducer amplitude ratio has a good linearity over a wide range. The simulated current using the impulse supply is shown in Figure 2.13, and the simulated voltage is shown in Figure 2.14.

Figure 2.13 shows that when the impulse current is applied and the conduction is not initiated, the current flowing through the arrester is mainly the current of the ZnO volume capacitance (curve a), and the voltage-dividing ratio is determined by the inter-electrode capacitive reactance between the ZnO elements of the high-voltage and low-voltage arms. After the conduction, the current flowing through the nonlinear resistor increases rapidly (curve b), and the voltage-dividing ratio is determined by

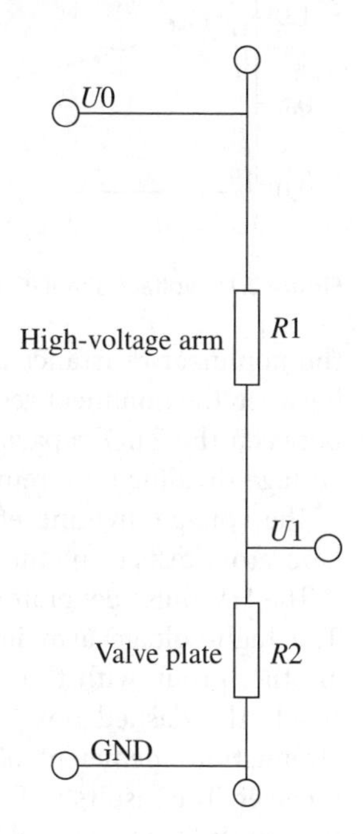

Figure 2.12 Equivalent circuit for large current range.

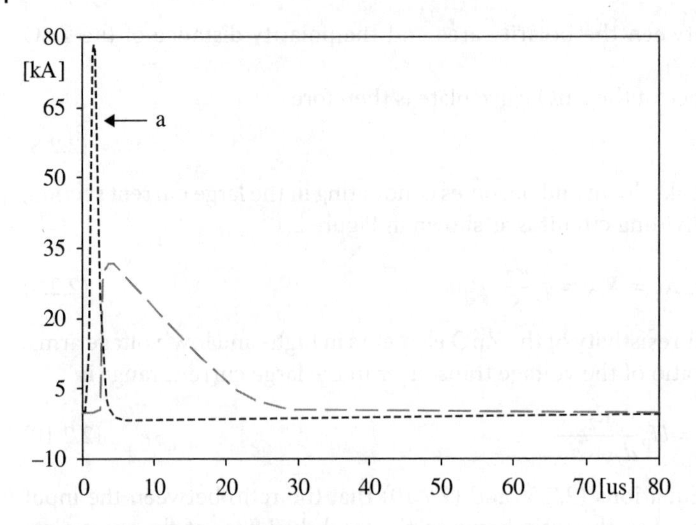

Figure 2.13 Current simulation result.

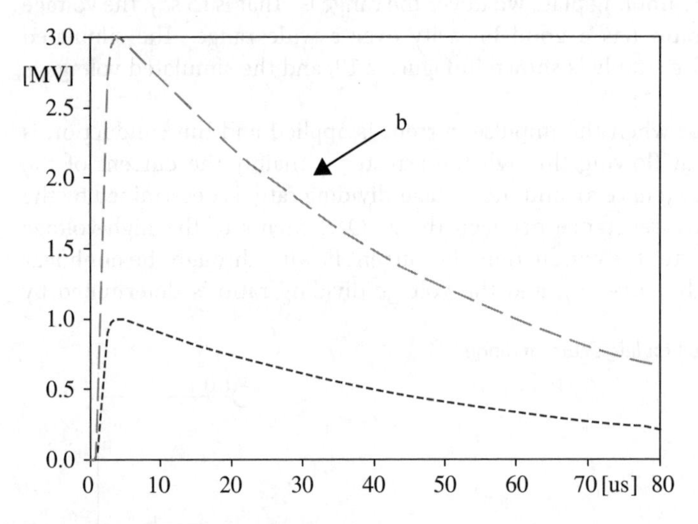

Figure 2.14 Voltage simulation result.

the nonlinear resistance between the high-voltage and low-voltage arms. As the ratio between the nonlinear resistances of the high-voltage and low-voltage arms equals that between the ZnO capacitive reactance of the high-voltage and low-voltage arms, the voltage-dividing ratio remains unchanged irrespective of the conducting state.

The voltage-dividing effects of four groups of valve plates have been analyzed in laboratory tests using the principles shown in Figure 2.15.

The HV pulse generator or power frequency voltage is adopted as the signal source. The high-voltage arm in the test is simulated using three groups of valve plates; another group with the same kind of valve plate is regarded as the low-voltage arm. Inside the dashed box is the attenuator, utilizing resistive voltage-dividing with an attenuation coefficient of 10. An oscilloscope is used at the terminal as the signal receiver. The results in Figure 2.16 illustrate the voltage-dividing effect when applying the power frequency and impulse voltage sources.

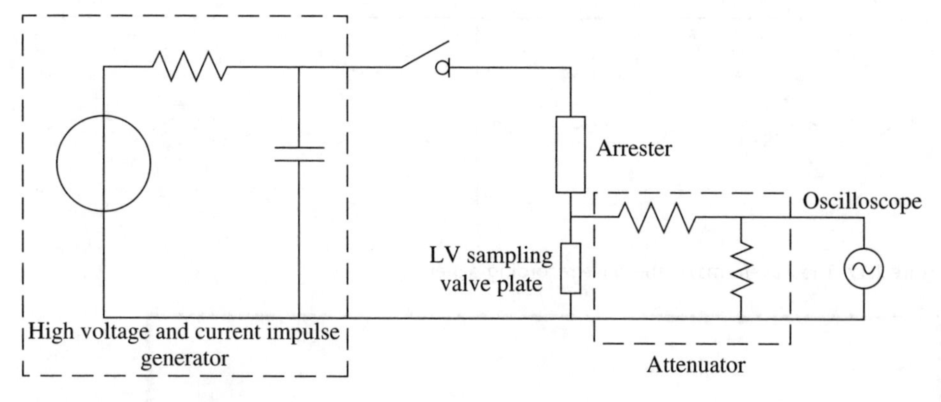

Figure 2.15 Test principles.

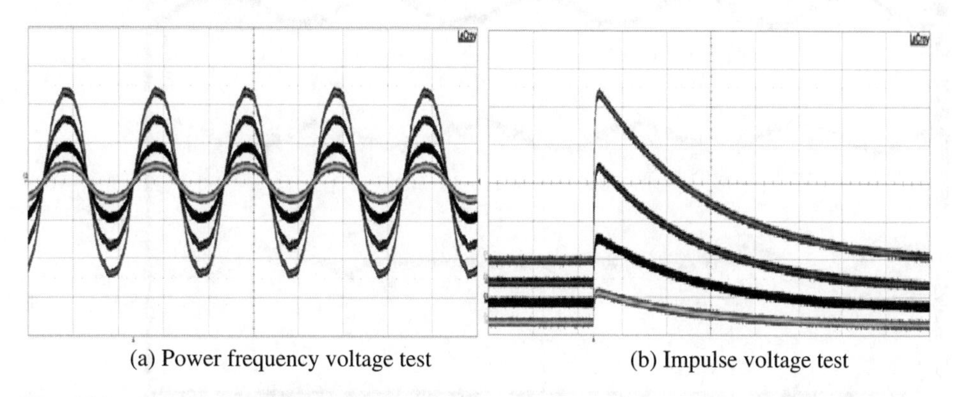

(a) Power frequency voltage test
(b) Impulse voltage test

Figure 2.16 Waveforms in laboratory tests.

In the power frequency voltage test, the supply rms value is 220 V, and the peak value is 311 V. The acquired wave has a magnitude of 7.17 V with the calculation accuracy of 99.14% and a wave frequency of 50 Hz, and the waveform appears smooth. These all prove that under power frequency voltage the measuring accuracy and frequency response are highly applicable. In addition, when an impulse voltage of 1500 V is applied, data errors are controlled within ±1% by comparing the acquisition data and input data.

2.2.3 Analysis of Field Applications

With online overvoltage monitoring transducers it is hard to achieve the same accuracy as that of laboratory simulation and tests owing to complex onsite conditions. Therefore, the practicability of transducers in the field needs to be verified. Figure 2.17 and Figure 2.18 are waveforms acquired from the in-station wave recording panel and the ZnO arrester voltage transducer in the case of a single-phase grounding fault simulated at a substation incoming line.

It can be observed from the above plots that during power frequency pre-triggering, the sampling waveforms of the voltage transducer strongly agree with that on the recording panel, and the errors are controlled within ±2% by data comparison. When the single-phase grounding fault is simulated, the recording panel data, sampled by the CVT, has shortcomings in measuring accuracy and frequency response; for example, the waveform breakpoint and high-frequency data loss are clearly noticeable.

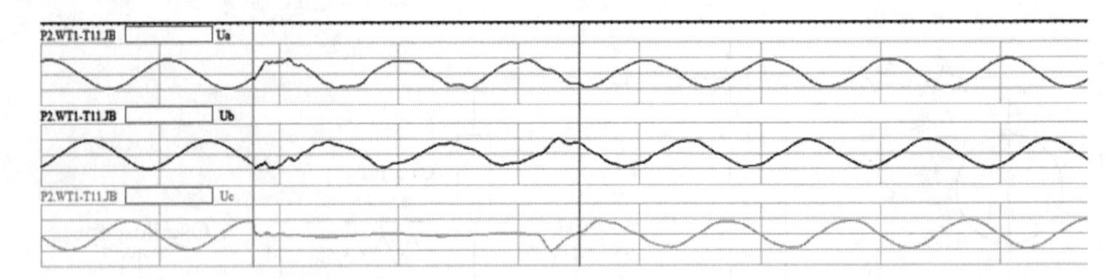

Figure 2.17 The waveforms on the wave recording panel.

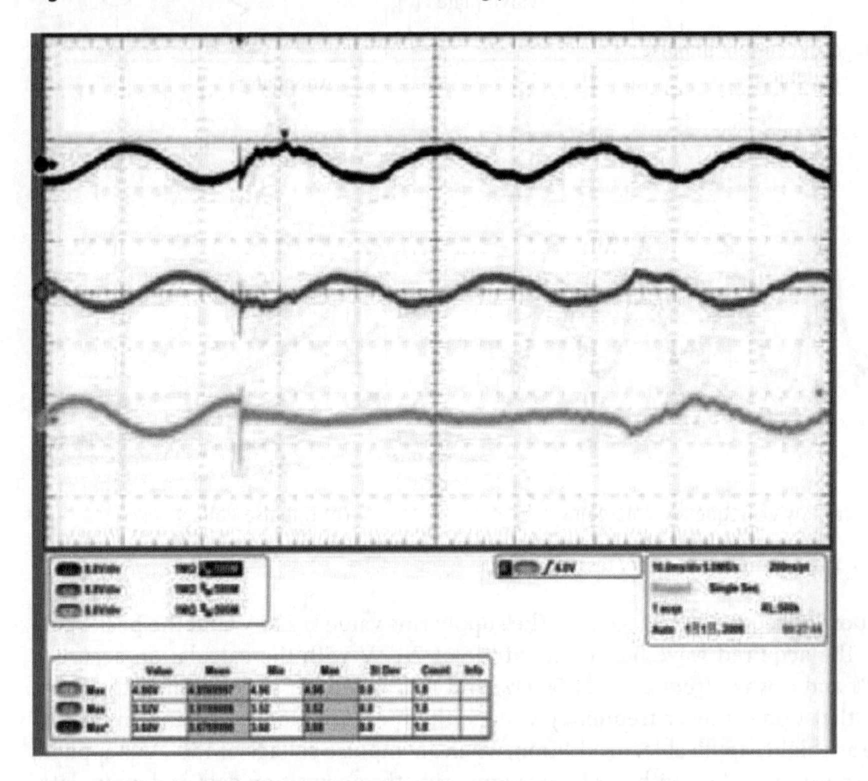

Figure 2.18 The waveforms acquired from the arrester voltage transducer.

The waveforms acquired by the voltage transducer, however, display high-frequency oscillation accompanying operation state changes. Although the trends of the two waveforms are similar, the waveforms on the recording panel lack calculation details. The arrester voltage transducer is a better option when accurate fault analysis needs to be conducted by extracting information such as the high-frequency waveform details.

When the equivalent circuit attenuation is 3 dB, the operating frequency can be obtained by using Equation (2.2.11) below.

$$\frac{1/(2\pi fc)}{2\pi fl + 1/(2\pi fc)} = 10^{\frac{3}{10}} \tag{2.2.11}$$

When the transducer operates in the large current section, the resistance is small and the effects of C_1 and C_2 can be ignored, as illustrated in Figure 2.12, thus the operating frequency using Equation (2.2.11) is simplified to

$$\frac{R}{2\pi fl + R} = 10^{\frac{3}{10}} \tag{2.2.12}$$

With the arrester parameters substituted into Equation (2.2.11) and (2.2.12), the 3 dB attenuation frequency in the small current section is 900 MHz, and 6 MHz in the large current section. Both frequencies can satisfy the highest frequency requirement of 2 MHz for transient overvoltages.

2.3 Voltage Transducers for Transmission Lines

A novel non-contact overvoltage transducer for overhead transmission lines will be introduced in this section, with its operating principles illustrated in Figure 2.19. Installed on the tower, such a transducer adopts the stray capacitance between the line and the induction plate C_1 as the high-voltage arm capacitance, and capacitance connected under the induction plate C_2 as the low-voltage arm capacitance. Overvoltage signals are extracted via the matched resistance from the metal induction plate and transferred to the external data acquisition system by the coaxial cable. The entire transducer is installed in a metal shielding enclosure to shield it from interference from other non-measuring phases. The voltage-dividing ratio of the transducer is

$$k = \frac{C_1 + C_2}{C_1} \tag{2.3.1}$$

2.3.1 Structure Design

The voltage to be measured is the voltage of the signal-phase conductor when the transducer is installed beneath the conductor. However, overhead transmission lines in practice are usually systems of polyphase conductors. The overvoltage transducer designed in this section is capable of shielding the interference from other signals, and takes into account the practical atmospheric conditions. The structure of the shielding enclosure is shown in Figure 2.20. The shielding shell is a cube with an opening at the upper end; the metal induction plate is mounted above the insulation guardrail; the insulation guardrail propped by bolts and nuts plays the role in insulation and support. The gap between the insulation and the metal induction plate is sealed off by

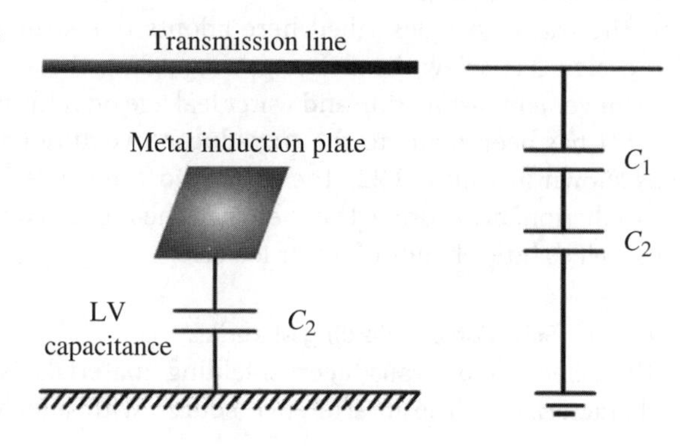

Figure 2.19 Operating principles of the overvoltage transducer.

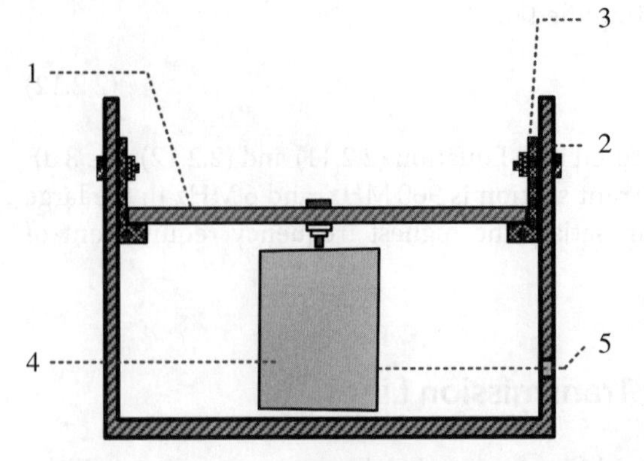

Figure 2.20 Schematic of the shielding enclosure (1 – metal induction plate, 2 – shielding shell, 3 – insulation guardrail, 4 – voltage-dividing unit, 5 – coaxial cable)

Figure 2.21 Photo of the overvoltage transducer.

the sealant to keep water from infiltrating the transducer on rainy and snowy days. The voltage-dividing unit consists of capacitors, matched resistors and discharge gaps. The shell of the voltage-dividing capacitor is fastened to the shielding shell by nuts, and the voltage-dividing unit is closely connected to the metal induction plate. Overvoltage signals are first transmitted from the metal induction plate to the voltage-dividing unit, then guided to the coaxial cable through the matched resistance and ultimately reach the signal acquisition unit. The photo of the actual voltage transducer of this category is as shown in Figure 2.21.

The transducer described here adopts this structure. Experiment shows that such a design has a few disadvantages regarding the structure and installation, including inconvenient installation and water leakage on rainy or snowy days. Therefore, improvement has been made to the transducer structure based on the engineering demands, as shown in Figure 2.22. The improved transducer has the advantages of convenient installation, firm connection between the metal induction plate and the shielding shell, as well as little chance of water leakage.

2.3.1.1 Selection of Shielding Materials

The selection of transducer shielding materials is determined by the field-source characteristics. The interference source with high voltage and small current can be

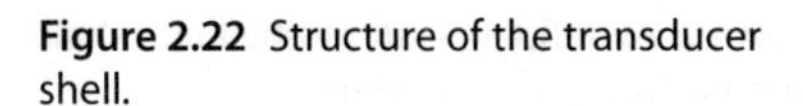

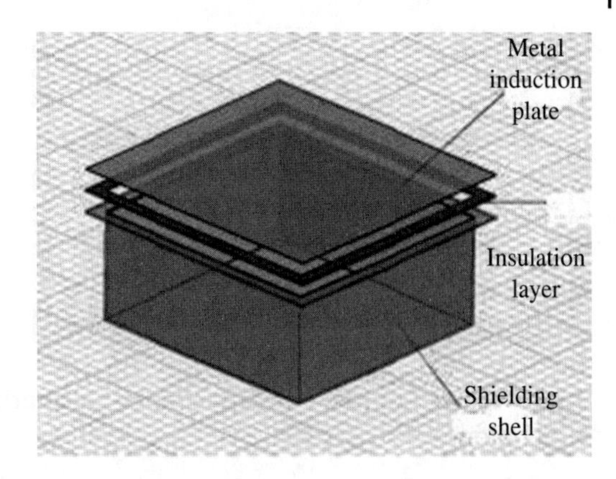

Figure 2.22 Structure of the transducer shell.

regarded as possessing a large characteristic impedance, and appears as the electric field interference. The transducer at practice is mainly subject to electric field interferences, for which the shielding has to take the main form of reflection loss, which depends on the forms and characteristic impedance of the interference source. The lower the wave impedance, the greater the reflection loss. Highly conducting material such as copper, aluminum or steel is usually adopted as the shielding material due to its high reflection loss. The shielding effectiveness is expressed as

$$\text{э} = 60\pi\sigma d^{*}\begin{cases} 1, when\dfrac{d}{\delta} < 0.1 \\ \dfrac{\delta}{2\sqrt{2}d}e^{d/b}, when\dfrac{d}{\delta} > 1 \end{cases} \tag{2.3.2}$$

where d is the shielding shell thickness; δ is the penetration depth; b is the gap length of the shielding enclosure meshes. According to Equation (2.3.2), the shielding effectiveness is proportional to the conductivity and independent of the magnetic permeability when $d/\delta < 0.1$. Hence, the shielding effect of copper is better than that of aluminum of the same thickness, and the shielding effect of steel is better than that of copper of the same thickness. The situation, however, changes with the increase of d and frequency; when $d > 0.2$ mm and $f > 100$ kHz, the shielding effect of steel is better than that of copper. The frequency of impulse transients caused by lightning impulses can be as high as several MHz. In this context, steel, which has the best shielding effect, is selected as the shielding material for the transducer.

2.3.1.2 Influence of the Shielding Shell on the Measurement

The influence of the shielding shell on transducers is shown in Figure 2.23. The capacitance C_3 between the metal induction plate and the bottom of the shielding shell as well as the capacitance C_4 between the plate and the lateral side of the shielding shell can increase the low voltage arm capacitance.

The capacitance between the metal induction plate and the shielding shell is measured as 83.11 pF in the experiment while the capacitance of the low-voltage arm is generally in the order of hundreds of μF. Considering the fact that (a) the capacitance between the metal induction plate and the shielding shell has little influence on voltage division and (b) C_3 and C_4 are taken into account when calibrating the voltage-dividing ratio of the transducer, the shielding shell would not impact upon transducer measurements.

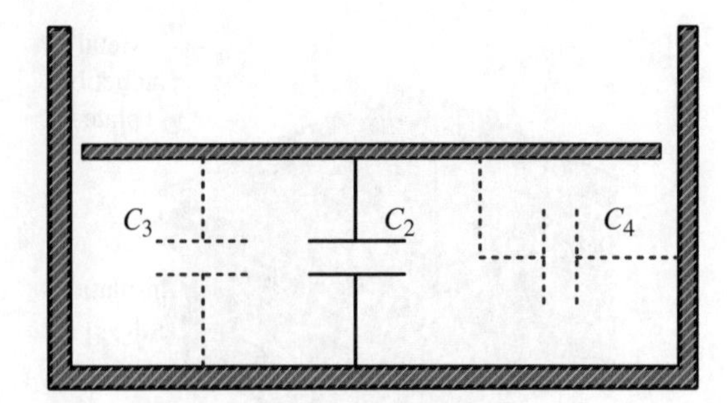

Figure 2.23 Influence of the shielding shell on transducer measurements.

2.3.2 Series-Connection Capacitance Calculation and ANSOFT Field Simulation

According to the analysis of overvoltage transducer fundamentals in Section 2.3.1, the capacitance of the low-voltage arm is a lumped capacitance with a fixed value, and the stray capacitance C_1 has a direct influence on the voltage-dividing ratio. Therefore, the calculation of the stray capacitance C_1 and the accurate calibration of the voltage-dividing ratio are of great concern for transducer design.

2.3.2.1 Calculation of the Stray Capacitance C_1

The capacitance relates to the shape, size and relative position of the conductor as well as the medium between the conductors. The capacitance, usually calculated by the equation $C = Q/V$, may be difficult to obtain due to complexity in electrode shapes in practical engineering. The stray capacitance of high-voltage equipment can be calculated by various methods, such as the surface charge simulation method, finite element method, charge simulation method or moment method. The line-plate model is a three-dimensional model with finite length and width at the floating potential. Therefore calculating the capacitance of this model is scarcely possible using the aforementioned methods.

The shapes of the two poles of the irregular capacitor are simplified in this chapter, so the stray capacitance between the transmission line and the transducer can be calculated by the series method for regular capacitors. Overhead transmission lines can be simplified to facilitate the calculation as well. The terminal effect and the arc sag impact can be neglected, and the transmission line can be regarded as an infinitely long straight conductor parallel to the ground. When calculating the capacitance, a transmission line with the radius r can be seen as the equivalent of a long thin belt with a width of $2r$ and a length of L. For bundled conductors, r refers to the equivalent radius. The length and width of the metal induction plates are a and b, respectively. Assume that the plane with an area of s between the two plates is parallel to the plate and at a distance of dz from the plate. Due to the calculation equation for plate capacitors, the capacitance of the unit plate capacitor with a spacing of dz is

$$\mathrm{d}C = \varepsilon S/\mathrm{d}z \tag{2.3.3}$$

The capacitance between the overhead transmission line and the metal induction plate can be replaced by a number of small plate capacitors in series, represented by

$$\frac{1}{C} = \int \frac{\mathrm{d}z}{\varepsilon S} \tag{2.3.4}$$

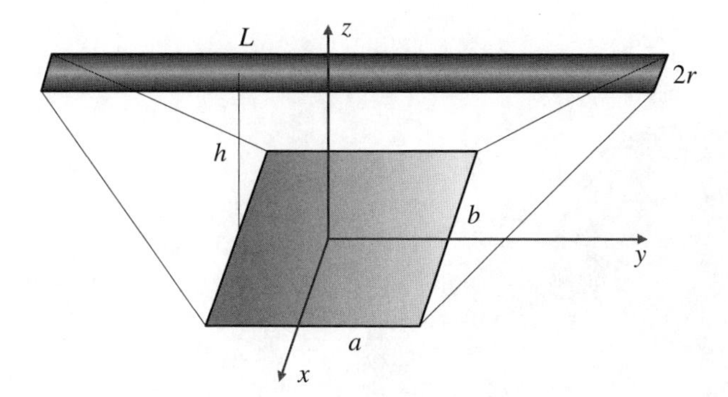

Figure 2.24 Model for capacitance calculation.

where, z changes from 0 to h; the integral region is defined as the area within the connecting lines of all points regarding the two plates, as shown in Figure 2.24.

$$C = \frac{1}{\int_0^h \frac{dz}{\varepsilon\left(2r + \frac{(h-z)(b-2r)}{h}\right)\left(a + \frac{z(l-a)}{h}\right)}} \tag{2.3.5}$$

2.3.2.2 3D Capacitance Simulation Using Ansoft Maxwell

Capacitance calculation is performed on the prerequisite that the electric field between the line-plate capacitor virtually parallels that of a plate capacitor, and that electric field distribution is simulated with the aid of Ansoft Maxwell.

Featured by powerful functions, easy applications and accurate calculation, the Maxwell software is a finite element package developed for electromagnetic calculation of electromechanical systems by the Ansoft Company. Other features of this software include top-down executive user interface, adaptive grid division technology, user-defined material database and multiple equation solvers for static field, eddy current field, static electric field, constant magnetic field and transient field. A top-down executive command menu is used in the software: equation solver, defined model, installation material, boundary/source setting, executive parameter setting, solution setting, solution execution, post-processing etc.

Based on the finite element analysis theory, Ansoft Maxwell 3D is adopted to solve thermal stress and electromagnetic issues often encountered in practical engineering. The 3D equation solver is used in this chapter to calculate capacitor parameters by establishing finite element models under static field.

The capacitance between the line and the plate is simulated and solved using the electrostatic solver in Ansoft Maxwell. The parameter settings of the model are as follows. The conductor length is 316 cm; the diameter is 2.5 cm; the length and the width of the metal induction plate of the transducer are both 40 cm; the region to be solved is a semi-cylinder area; the conductor material is aluminum, while the metal induction plate is made of steel; the voltage across the conductor is 110 kV, and the voltage across the metal induction plate is 0; the bottom voltage of the boundary is 0, and the default boundary balloon is used.

The simulation model is shown in Figure 2.25, and Figure 2.26 shows the details of the line-plate model.

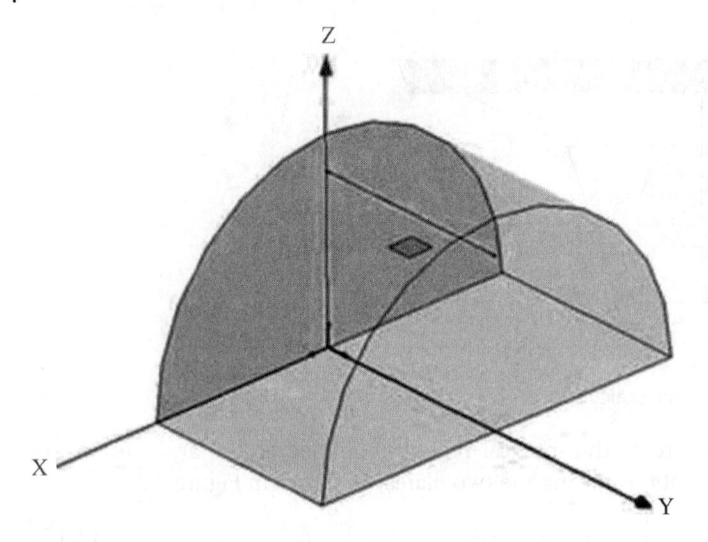

Figure 2.25 Ansoft simulation model.

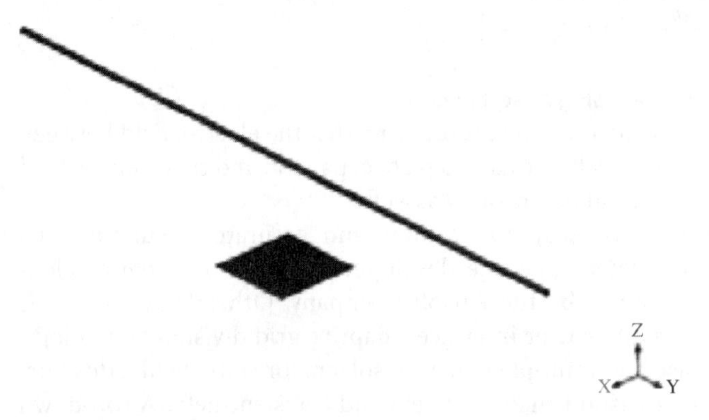

Figure 2.26 Details of the Ansoft model.

The schematic of the voltage distribution of the model is shown in Figure 2.27, where the voltage distribution between the line and the plate is approximately parallel, which agrees with the prerequisite for capacitance approximate calculation.

Comparison between the calculation, tests and simulation results – The engineering interest of the series calculation approach can be verified by comparing Ansoft Maxwell simulations and practical capacitance measurement (determined by the measured voltage of the standard voltage divider, the transducer voltage and the capacitance of the low-voltage arm under the power frequency voltage test). The test parameters are set as follows: for the conductor, the length is 316 cm, and the width is 2.5 cm; for the transducer metal induction plate, the length is 40 cm, and the width is 40 cm. The capacitances acquired by the three methods are presented in Table 2.1.

It can be seen from Table 2.1 that the results of the series approach and Ansoft simulation are quite close, and the practical measurement is comparatively smaller. A reasonable explanation is that the series approach as well as Ansoft simulation uses the interference-free line-plate model, while in practical measurement, the transducer

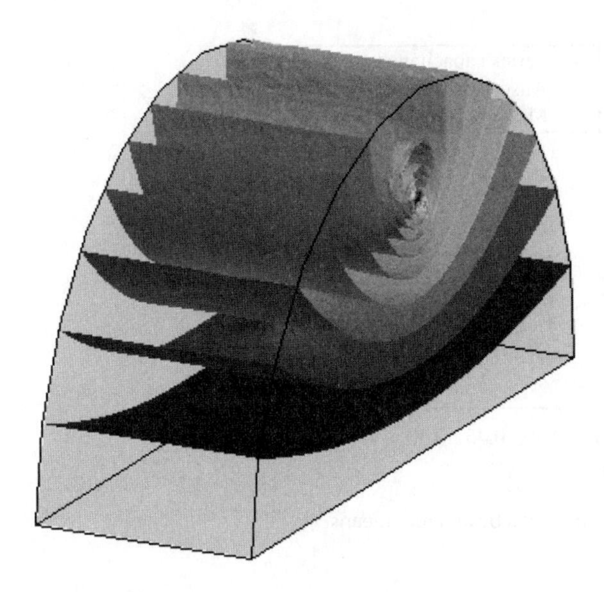

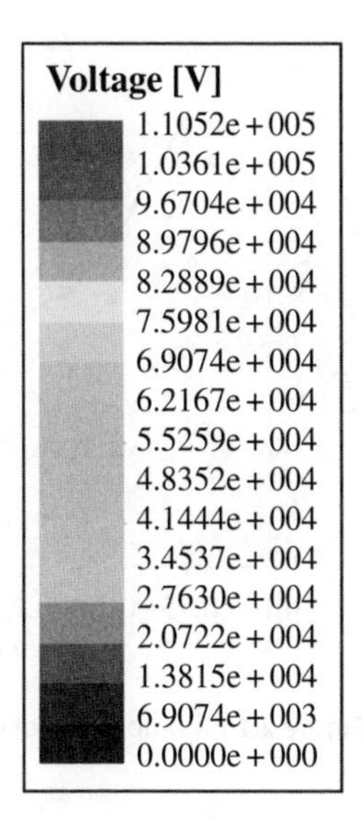

Figure 2.27 Schematic of voltage distribution.

Table 2.1 Comparison of capacitances acquired by different methods.

line-plate distance (m)	0.135	0.195	0.222	0.39	0.47
series approach (pF)	11.731	8.75	7.9	5.2	4.5
simulation result (pF)	11.507	8.914	8.214	5.432	4.6906
practical measurement (pF)	10.22	7.779	6.9578	4.79	3.973

capacitance would be affected by outside objects such as shielding shells and high-voltage generators, making the results lower than the calculated and simulated values.

The formula for capacitance calculation adopted in the series approach is used for instructing transducer design, and not for calibrating the voltage-dividing ratios of transducers. The actual voltage-dividing ratio of a transducer is calibrated after field installation with considerations of external metal conducting bodies (e.g. towers). The following conclusions are based on the experimental results: (a) capacitances obtained through different means are basically the same, and demonstrate identical declining trends: decrease with the increasing distance as shown in Figure 2.28; (b) the capacitance calculation formula used in the series approach can be used to instruct practical engineering.

The influence of the conductor length on transducer capacitance calculation based on Ansoft simulation is illustrated in Figure 2.29, where the simulated capacitance has an identical trend to the capacitance calculated by the series approach.

Preliminary selection of transducer parameters – The requirements for safety distance and line parameters vary with overhead transmission lines of various voltage ratings. The distance regarding transducer installation has to be determined by the minimum safety distance according to relevant specifications. For lines rated 110 kV, 220 kV,

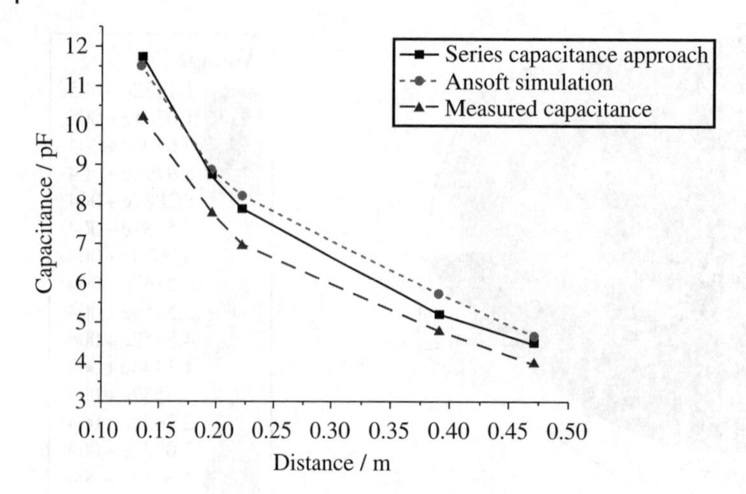

Figure 2.28 Comparison of capacitances acquired by different means.

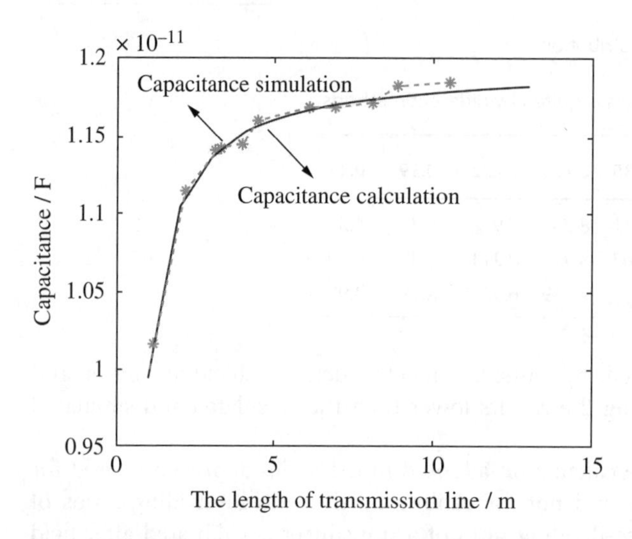

Figure 2.29 Curve of capacitance variation with the conductor length.

330 kV and 500 kV, the installation distances have to be 1.5 m, 3 m, 4 m and 5 m, respectively. A polypropylene capacitor with a withstand voltage of 500 V is chosen for this transducer. The transducer structure is preliminarily designed according to the conductor diameter, installation distance, conductor voltage and capacitance of the low-voltage arm under different voltage ratings. As lines rated 330 kV and 500 kV are equipped with bundled conductors, the equivalent radius is used to calculate the capacitance. According to the previous section, the capacitance approaches a stable value when the conductor length is 15~20 times that of the transducer, and the conductor length of 15 m is selected for calculation.

The parameters of the capacitance and the transducer under different voltage classes are given in Table 2.2. As the voltage-dividing ratio of the transducer may be affected by

Table 2.2 Selection of parameters of the transducer and capacitance.

Voltage class (kV)	110	220	330	500
Type of conductor	LGJ-240/40	LGJ-400/35	2×LGJ-300/40	4×LGJ–400/35
Conductor diameter (mm)	21.66	26.82	23.94	26.82
Safety distance (m)	1.5	3	4	5
Capacitance (μF)	0.15	0.22	0.33	0.47
Transducer dimension (m^2)	0.8×0.8	0.8×0.8	1×1	1×1
Capacitance of the high-voltage arm (pF)	12.1	7.48	9.89	11.5

towers, Table 2.2 can merely be used as a preliminary design guide. Further adjustment has to be made according to the field situation in practical engineering.

2.3.3 Experimental Verification of the Transducer Model

A power frequency voltage experiment and a lightning impulse voltage experiment have been conducted for the simulated measuring system to verify the measuring accuracy, stability and response characteristics of the transducer. Figure 2.30 is the schematic of the experiment's wiring, where a metal induction plate is placed under the simulated transmission line with a length of 3.18 m and a radius of 1.25 mm. The metal induction plate is an aluminum plate with both a length and a width of 0.4 m, below which a voltage-dividing unit is mounted with a capacitance of 0.047 μF. Voltage signals are extracted from the metal induction plate, and transferred to the oscilloscope output by a coaxial cable SYV-50-5 matching a 50 ohm resistor at the sending end. The type of the oscilloscope is LeCroy 44mxi, and its analog frequency band is 400 MHz.

2.3.3.1 Result Analysis of the Power Frequency Experiment

In the power frequency experiment, the transformer type is YDW-50/50 with a rated operating voltage of 0–100 kV. The ac voltage applied on the line ranges from 0–20 kV,

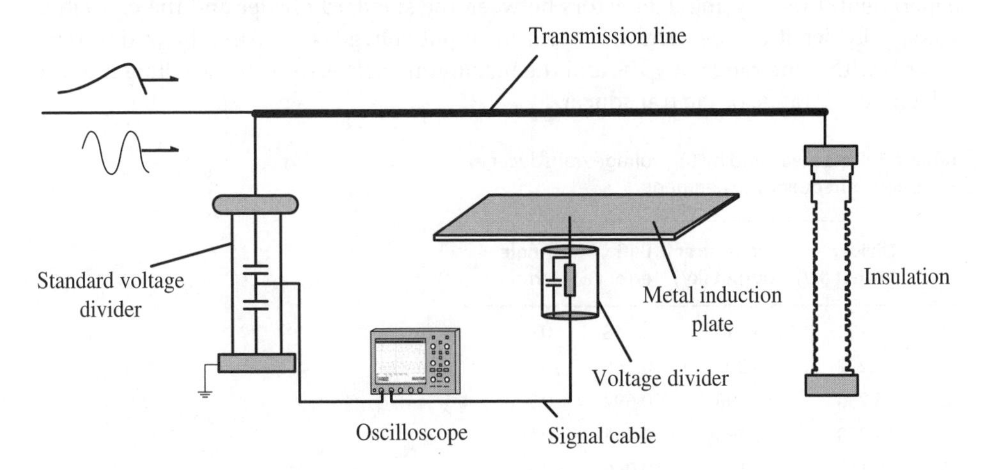

Figure 2.30 Schematic of experiment wiring.

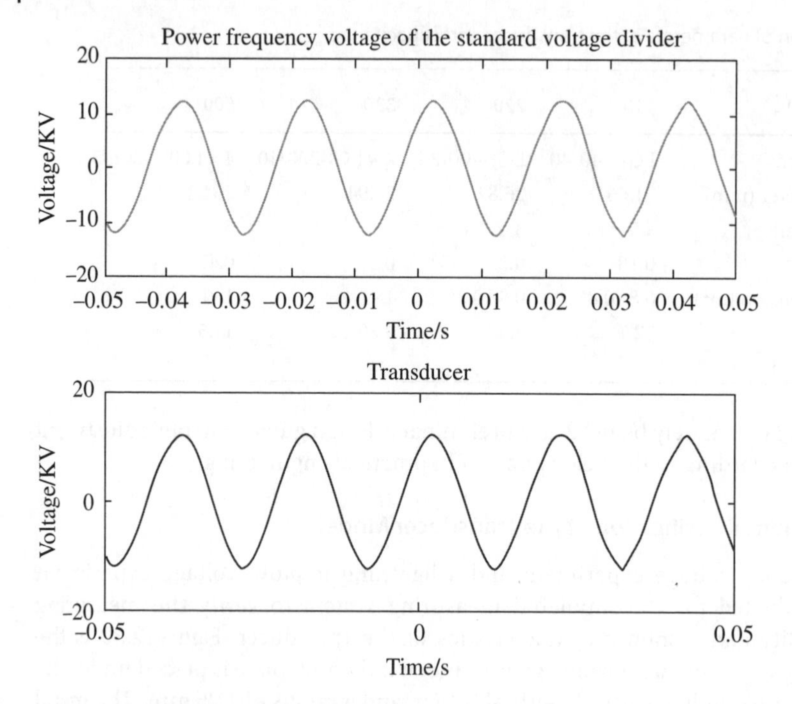

Figure 2.31 Waveforms in the power frequency voltage experiment.

and a standard capacitive voltage divider with a ratio of 1000:1 is connected to channel 1 of the oscilloscope, while the voltage signal output of the transducer is connected to channel 2.

In Figure 2.31, the displayed waveforms are measurements from the standard voltage divider and the transducer in the ac power frequency voltage experiment. Obviously, the transducer can monitor the real-time power frequency voltage of the line, and the two waveforms are quite alike.

Table 2.3 shows the data measured by the voltage transducer in an ac power frequency experiment. By analyzing data errors between the standard divider and the capacitive voltage divider, it can be seen that when the input voltage is 3–20 kV, the output ratio error is within the range of ±1%, and the maximum angle error is 0.8%, falling into the tolerant error range of the transducer.

Table 2.3 Data measured by the voltage transducer in an ac power frequency experiment.

No.	Divider output (kV)	Transducer output (kV)	Ratio error (%)	Angle error (%)
1	5.124	5.148	0.468	0.674
2	8.292	8.305	0.156	0.619
3	12.280	12.195	0.692	0.713
4	14.353	14.295	0.404	0.706
5	18.118	18.130	0.066	0.801

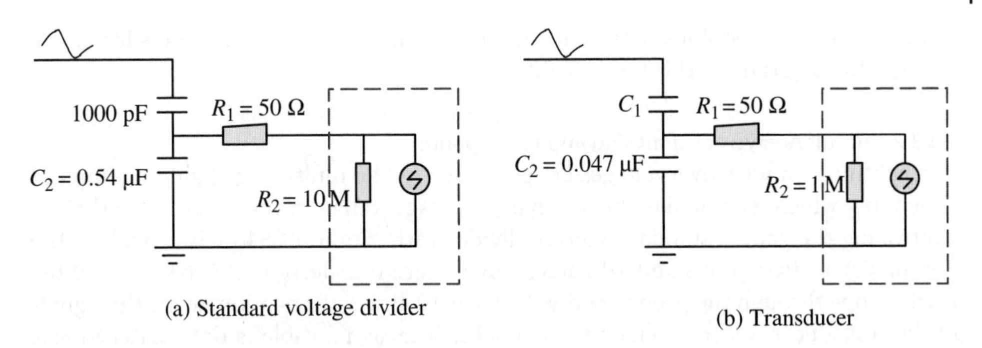

Figure 2.32 Schematic of load impedances in the measuring systems.

Theoretically, when capacitive voltage dividers are used for the measurement, if load impedance is not taken into account, there are only amplitude errors, and phase errors do not exist. However, the load impedance is inevitably introduced as the standard voltage divider and the transducer are both connected to an oscilloscope. The schematic of the load impedances in the two measuring systems are illustrated in Figure 2.32.

The phase of the measured wave will lead $\tan^{-1}(1/(\omega C_1 R_1 + \omega C_2 R_2))$ considering the impact of the load impedance. The load impedance of the standard voltage divider is $10\ M\Omega$, and that of the transducer is $1\ M\Omega$. The phase of the standard voltage divider leads the phase of transmission line by 0.0006, and the phase of the transducer leads the phase of the transmission line by 0.067, hence the measured wave phase of the transducer leads that of the standard voltage divider by 0.064.

The relation curve, consisting of 13 groups of measurements of the transducer and the standard voltage divider under applied ac power frequency upto 25 kV, is plotted in Figure 2.33. The curve appears linear with a correlation coefficient of 0.99 by

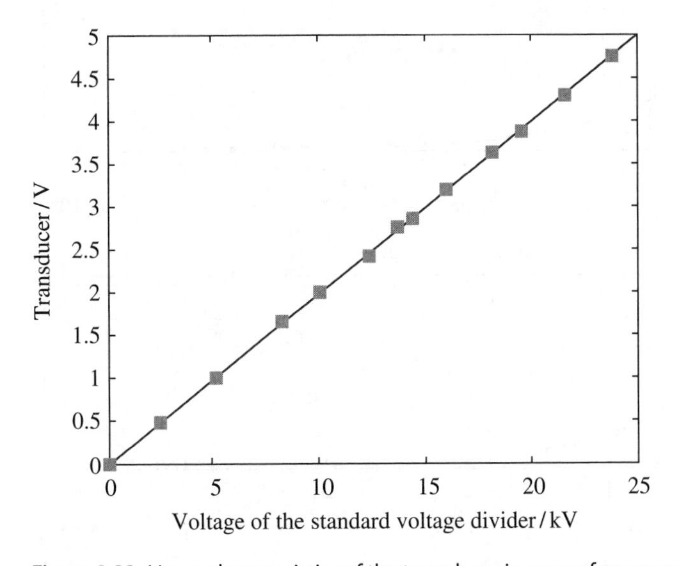

Figure 2.33 Linear characteristics of the transducer in power frequency voltage test.

Matlab, and the curve slope represents the voltage-dividing ratio of the transducer. The voltage-dividing ratio in this test is 5000.

2.3.3.2 Result Analysis of Lightning Impulse Experiment

The lightning impulse overvoltages are generated by the multi-stage lightning impulse generator, which can make the discharge voltage range from 40 kV to 50 kV by controlling the gap. A standard voltage divider with a ratio of 540:1 is provided. The output signals from the standard voltage divider are transferred into channel 1 of the oscilloscope through the probe Lecroy PP009 at 10-times attenuation, while the signals of the transducer are transferred to channel 2. The signal cable is only matched at a single end, and travelling waves are completely reflected at the other end. In the lightning overvoltage experiment, there are no changes in the relative location between the line and the transducer. The voltage-dividing ratio of 5000 is adopted for recovering the lightning overvoltage data measured by the transducer. In Figure 2.34, data1 represents the overvoltage waveform measured by the standard voltage divider with a rise time of 1.511 μs, a half-amplitude decay time of 50.231 μs, and a peak value of 43.198 kV. Data2 refers to the measured voltage waveform of the transducer with a rise time of 1.551 μs, a half-amplitude decay time of 50.75 μs, and a peak value of 43.688 kV. It can be seen that the relative error of the peak value is within ±2%, and that the total uncertainty of the measurement time is within ±3%, indicating that the voltage transducer performs well enough to satisfy the engineering requirements in measuring switching overvoltages.

The detailed views of the wave fronts in the lightning overvoltage experiment are shown in Figure 2.35, where the two waves are similar in fluctuation details.

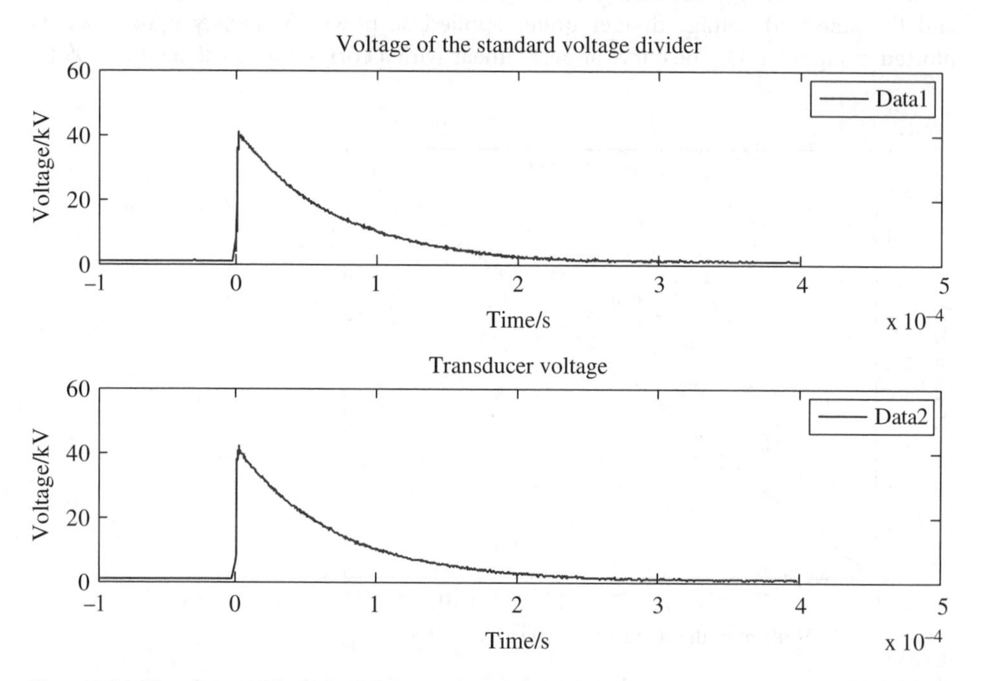

Figure 2.34 Waveforms of the lightning overvoltage experiment.

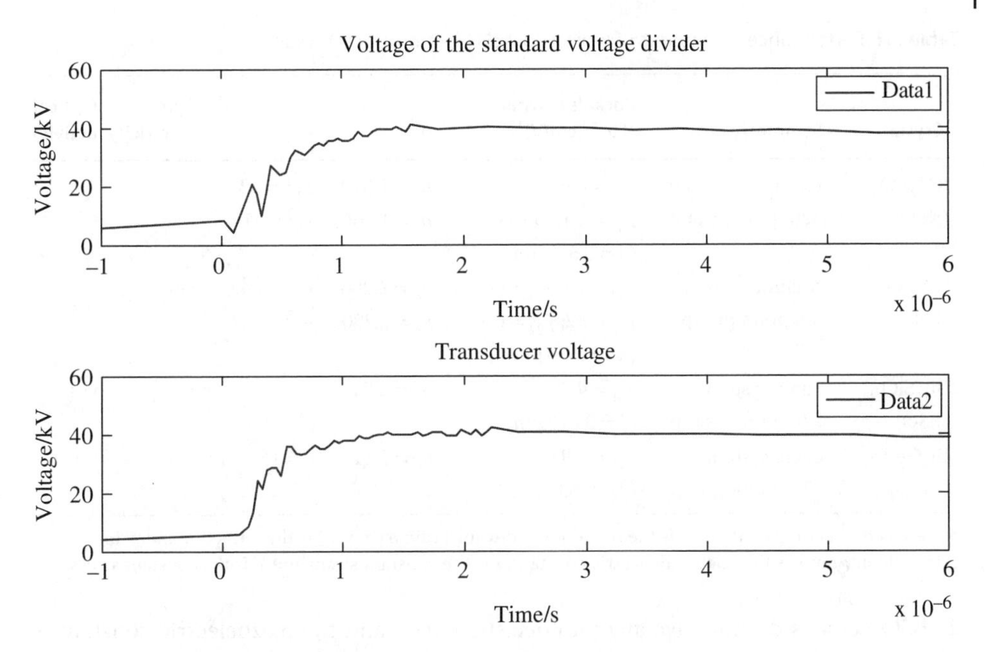

Figure 2.35 Detailed views of the wave fronts in lightning impulse tests.

2.4 Full-Waveform Optical Online Monitoring Technology

2.4.1 Pockels Sensing Material Selection and Crystal Design

2.4.1.1 Selection of Pockels Sensing Material

Optical voltage transducers have been developed based on three effects of special materials: the Pockels effect, the inverse piezoelectric effect and the electro-optical Kerr effect. After comparing the three effects and combining the field conditions, we use the Pockels effect to develop full-waveform optical sensors. There are a variety of crystals with linear electro-optic properties, but very few can meet the practical requirements. Some typical Pockels electro-optical crystals are presented in Table 2.4.

The Pockels effect generally occurs in non-centrosymmetric crystals (such as uniaxial and biaxial crystals). Biaxial crystals, however, are not considered as options for Pockels sensitive materials because the optical axis direction tends to be affected by light wavelength and crystal temperature. Therefore, Pockels sensitive materials are selected from uniaxial crystals.

$LiNbO_3$ (LN for short) and $Bi_{12}SiO_{20}$ (BSO crystal for short) are commonly used as materials for optical voltage sensors. LN has large natural binary refraction and is susceptible to temperature; BSO is affected by optical activity, which may result in nonlinearity of the measurement.

According to the above principles, $Bi_4Ge_3O_{12}$ (BGO for short) crystals from the 43m point group of the cubic system are used as Pockels sensitive elements, which outperform LN and BSO crystals. The advantages of BGO crystals are:

1. Ideal BSO crystals do not have natural binary refraction, optical rotation or thermo-electric effect, so electro-optical modulation is almost independent of temperature.

Table 2.4 Performance parameters of typical Pockels electro-optical crystals.

Crystal	Symmetry	Pockels constant (10^{-10} cm/V)	n	ε	Pyro-electricity	Optical activity
KH_2PO_4 (KDP)	tetragonal system 42m point group	$\gamma_{41} = 8.6$ $\gamma_{63} = -10.5$ ($\lambda = 0.550$ μm)	$n_o = 1.512$ $n_c = 1.466$	$\varepsilon_1{}^T = 42$ $\varepsilon_2{}^S = 21$	–	–
$LiNbO_3$ (LN)	triclinic system 3m point group	$\gamma_{13} = 10$; $\gamma_{33} = -32.2$ $\gamma_{51} = 32$; $\gamma_{22} = 6.7$ ($\lambda = 0.630$ μm)	$n_o = 2.286$ $n_c = 2.220$	$\varepsilon_1{}^T = 84.6$ $\varepsilon_2{}^S = 28.6$	yes	–
$Bi_{12}SiO_{20}$ (BSO)	cubic system 23m point group	$\gamma_{41} = 4.35$ ($\lambda = 0.870$ μm)	$n_o = 2.54$	—	no	yes
$Bi_4Ge_3O_{12}$ (BGO)	cubic system 43m point group	$\gamma_{41} = 0.95$ ($\lambda = 0.630$ μm)	$n_o = 2.11$	$\varepsilon = 16$	yes	no

Note: n is the refractive index, n_o is the refraction index for ordinary ray, n_c is the refraction index for extraordinary ray, ε is the relative dielectric constant, s is the constant strain and T is the constant stress.

2. BSO crystals do not have any photoelastic effect, and the piezoelectric constant is almost zero.
3. Because there is no optical activity, the crystal thickness can be designed larger, to increase the withstand voltage of the crystal.
4. The relative dielectric constant of BGO is very small ($\varepsilon = 16$), and the resistivity is large (greater than 10^{15} Ω·cm), which has very little influence on the electric field distribution of the voltage to be measured.
5. BGO also boasts a broad photic zone, excellent light transmittance, and high half-wave voltage.

When no electric field (or voltage) is applied, BGO is an isotropic optical crystal; when an external electric field (or voltage) is aplied, BGO transforms from an isotropic crystal to an anisotropic crystal, so changes take place in both its refractive index and the polarization state of light passing through the crystal. Therefore, BGO is an ideal sensing material. Table 2.5 shows its general properties.

2.4.1.2 Technical Requirements for BGO Crystals

The requirements for the crystal material properties are:

1. good optical homogeneity, free of impurities and bubbles;
2. Czochralski processing conducted at least three times, and proper annealing treatment conducted after crystal formation to ensure sufficiently small interior residual stress;
3. exceedingly small additive effects, e.g. optical activity, thermo-optic effect, pyroelectric effect and natural birefringence.

Requirements for crystal processing and other technical specifications are:

1. crystal size: 10 × 10 × 10 mm, tolerance −0.05 mm, edge angle 0.1 mm × 45°;
2. cut along the crystal faces labelled (001), (100), (010), and the deviation of crystallographic axis orientation angle ≤ 1′;

Table 2.5 General Properties of BGO.

Categories	Nature
Color	Colorless
Density (g/cm^2)	7.12
Melting point (°C)	1050
Coefficient of thermal expansion (°C/cm/ m)	7.15×10^{-6}
Dielectric constant	16
Refractive index at 0.48 μm	2.149
at 0.63 μm	2.098
at 0.85 μm	2.070
Photic zone (μm)	0.33~5.5
Electro-optic coefficient γ_{41}(cm/V)	1.03×10^{-10}

3. two faces labelled (001) must be light pass surfaces precisely polished, with smooth finish degree of IV, and the flatness $\leq \lambda/4$@850 nm;
4. other surfaces must have fine grinding, and the faces labelled (100), (010) must be marked (Figure 2.37 is the schematic of a BGO crystal);
5. parallel degree of each pair of parallel end planes must be <3′.

2.4.1.3 Technical Requirements for BGO Transparent Conductive Films

As shown in Figure 2.36:

1. Crystal size: 10 × 10 × 10 mm.
2. Coat a layer of ITO film on the two finely polished surfaces (001), with a size of 10 × 10 mm.
3. Conductive film transmittance must exceed 90% at 850 nm.
4. The film must be highly firm, strongly adhesive and uniformly coated.
5. As crystalline materials are soft, brittle and expensive, care has to be taken to avoid damage during coating.

2.4.2 Longitudinal and Transverse Electro-Optical Modulation

According to the way in which optical signals are modulated by the applied electric field, optical voltage transducers (OVT) based on the Pockels effect use two kinds

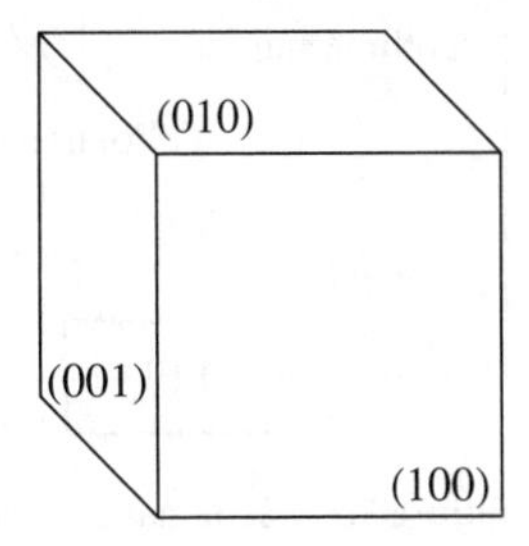

Figure 2.36 Schematic of a BGO Crystal.

of modulation approaches: longitudinal electro-optical modulation and transverse electro-optical modulation. Figure 2.37 is a schematic diagram of a longitudinal modulated OVT. The optical path of this OVT consists of optical elements such as polarizer, $\lambda/4$ wave plate, BGO crystal, analyzer and collimating lens.

It is impossible to directly measure photophase changes due to lack of precision; intensity modulation is therefore usually adopted to reflect the phase change. The BGO material from the 43m point group of the cubic crystal system is adopted in this section. For longitudinal modulated OVTs, the phase difference δ between two outgoing beams is

$$\delta = \frac{2\pi}{\lambda} n_0^3 \gamma_{41} El = \frac{\pi E}{E_\pi} \tag{2.4.1}$$

When expressed in the form of the applied voltage V,

$$\delta = \frac{2\pi}{\lambda} n_0^3 \gamma_{41} V = \frac{\pi V}{V_\pi} \tag{2.4.2}$$

$$V_\pi = \frac{\lambda}{2n_0^3 \gamma_{41}} \tag{2.4.3}$$

where V_π is the half-wave voltage; V is the voltage applied on the two ends of the crystal; λ is the wavelength of the incoming light wave, which takes the value of 820nm; E is the electric field produced by voltage V; n_0 is the the inherent refractive index of the BGO crystal from the 43m point group of the cubic crystal system, which takes the value of 2.11; γ_{41} is the the linear electro-optical coefficient of the BGO crystal from the 43 m point group of the cubic crystal system, which takes the value of 101.0310 cm/V.

It can be seen from Equation (2.4.2) that by calculating the applied voltage V from the phase difference, the light intensity signal detected by the photoelectric detector can be expressed as

$$I = I_0 \sin^2\left(\frac{\pi}{4} + \frac{\pi V}{V_\pi}\right) = \frac{1}{2} I_0 \left(1 + \sin\frac{\pi V}{V_\pi}\right) = \frac{1}{2} I_0 (1 + \sin\delta) \tag{2.4.4}$$

When $\delta \ll 1$ and $\sin\delta \approx \delta$, linear response, as a first approximation, we have

$$I = \frac{1}{2} I_0 \left(1 + \frac{\pi V}{V_\pi}\right) = \frac{1}{2} I_0 (1 + \delta) \tag{2.4.5}$$

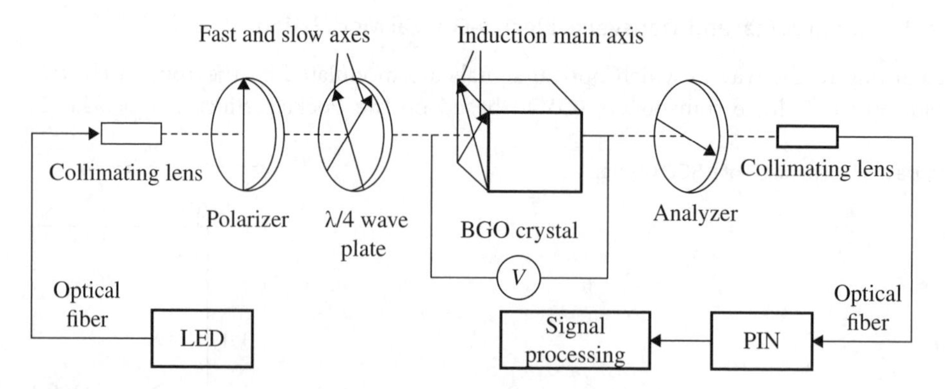

Figure 2.37 Schematic diagram of longitudinal modulated OVT.

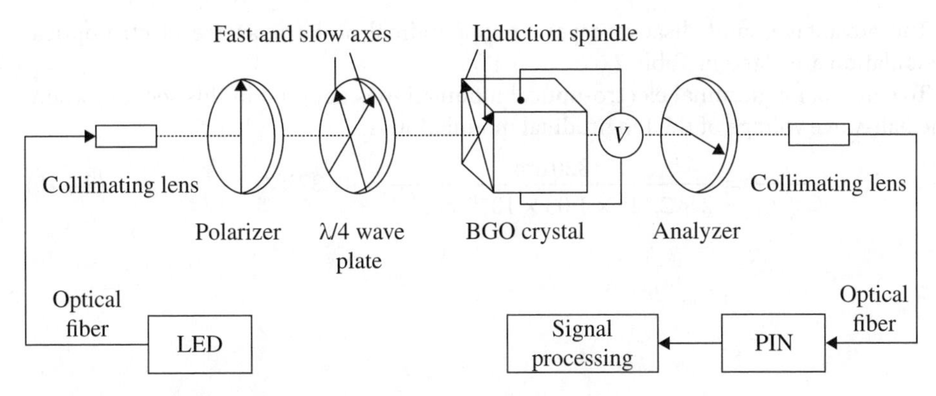

Figure 2.38 Schematic diagram of a transverse modulated OVT.

Figure 2.38 shows the schematic diagram of a transverse modulated OVT.

For transverse modulated OVTs, the phase difference φ between two outgoing beams is

$$\varphi = \frac{2\pi}{\lambda} n_0^3 \gamma_{41} E l \tag{2.4.6}$$

When expressed in the form of the applied voltage V,

$$\varphi = \frac{2\pi}{\lambda} n_0^3 \gamma_{41} \frac{l}{d} V = \frac{\pi V}{V_\pi} \tag{2.4.7}$$

With the half-wave voltage,

$$V_\pi = \frac{\lambda}{2 n_0^3 \gamma_{41}} \left(\frac{d}{l}\right) \tag{2.4.8}$$

At this moment, the light intensity signal detected by the photoelectric detector is

$$I = I_0 \sin^2\left(\frac{\pi}{4} + \frac{\pi V}{V_\pi}\right) = \frac{1}{2} I_0 \left(1 + \sin\frac{\pi V}{V_\pi}\right) = \frac{1}{2} I_0 (1 + \sin\delta) \tag{2.4.9}$$

Table 2.6 Advantages and disadvantages of longitudinal and transverse electro-optical modulation.

Comparison	Longitudinal electro-optical modulation	Transverse electro-optical modulation
Half-wave voltage	Irrelevant to the crystal dimension	Adjustable with the changes of the crystal dimension
Phase delay	Free of phase delay caused by natural double refraction	Phase delays caused by natural double refraction means it is prone to temperature changes
Influence of the applied electric field	Longitudinal effect measures voltage values, which is free of any interference of the external electric field; longitudinal effect is superior in the complex electromagnetic environment near the GIS	Electric field produced by the transverse effect is perpendicular to the light, which makes it handy in use, but this effect acquires the voltage via electric field measurement, which is prone to interferences of external electric field

The advantages and disadvantages of longitudinal and transverse electro-optical modulation are listed in Table 2.6.

To sum up, longitudinal electro-optical modulation is adopted in this section, where the half-wave voltage of the longitudinal modulation is

$$V_{\pi} = \frac{\lambda}{2n_0^3\gamma_{41}} = \frac{820\,\text{nm}}{2 \times 2.11^3 \times 1.03 \times 10^{-10}\,\text{cm/V}} = 42.37\,\text{kV} \tag{2.4.10}$$

3

Online Overvoltage Monitoring System

3.1 Overview

Overvoltages in power systems have different causes and can result in different hazards. External overvoltages, one of the major causes for insulation failures, mainly refer to the atmospheric overvoltages with steep wavefronts and high amplitudes. Even though the internal overvoltages have low amplitudes, their long durations may pose threats to electrical equipment insulation as well. Results have been obtained by researchers regarding the mechanism, amplitude and frequency of overvoltages, but most are based on simulations in laboratories or by computers, and lack sufficient field data for verification. In order to analyze failure causes, to improve insulation coordination and to avoid overvoltage accidents, overvoltages need to be monitored in real time via online overvoltage monitoring systems to meet the requirements from the utility operation department.

Real-time monitoring on operation states of the power grid by the overvoltage monitoring technology is able to discover, in good time, any abnormal operation states that may result in possible failures so that emergency measures can be taken to avoid failure. Intelligent overvoltage monitoring centers need to be built, inside which intelligent analysis of data from overvoltage monitoring terminals can be made and automatic identification and alarm for abnormal operation states could be achieved.

The online overvoltage monitoring system is used to monitor and record transient overvoltage events in the substation, as shown in Figure 3.1.

The recorded data of a bus at a substation is shown in Figure 3.2. At this time no overvoltage occurs.

The closing of other lines on site causes obvious impacts on the line and gives rise to overvoltages, which are well recorded by the equipment, as shown in Figure 3.3.

The transient overvoltage signals can be browsed and studied in detail after localized amplification, as shown in Figures 3.4 and 3.5.

After being recorded by the monitoring instrument at the substation, this data and these states are sent back to the monitoring center and the data center. The operating personnel are able to acquire the monitoring results of any specified substation equipment in real time.

Therefore, research on online overvoltage monitoring is of great concern for the following reasons.

1. Current research mainly focuses on monitoring of overvoltages in distribution networks, whereas overvoltage monitoring for power grids rated 110 kV or above is

Measurement and Analysis of Overvoltages in Power Systems, First Edition. Jianming Li.

Figure 3.1 Online overvoltage monitoring system.

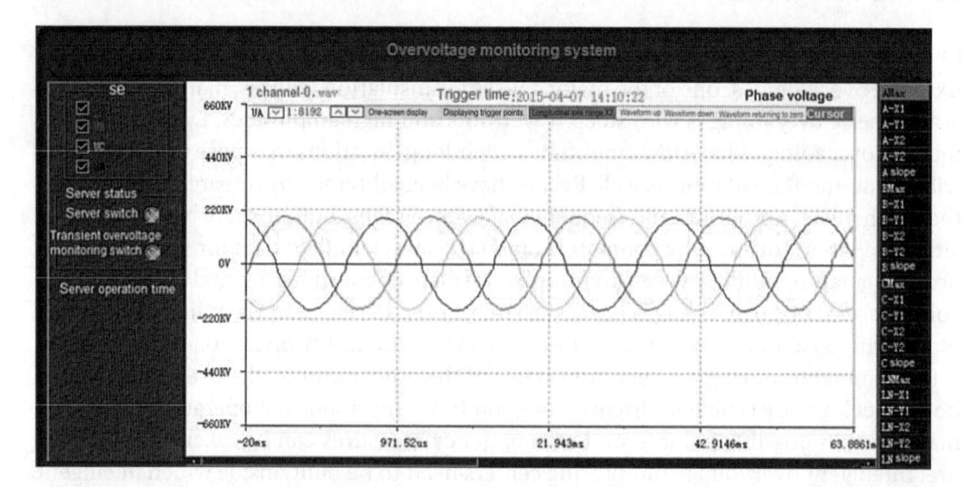

Figure 3.2 Overvoltage monitoring system.

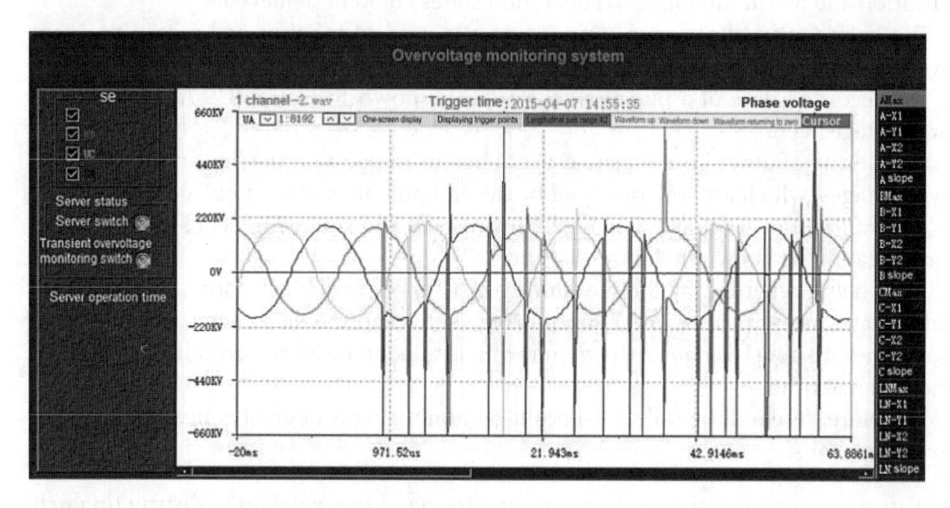

Figure 3.3 Screenshot of a line closing overvoltage.

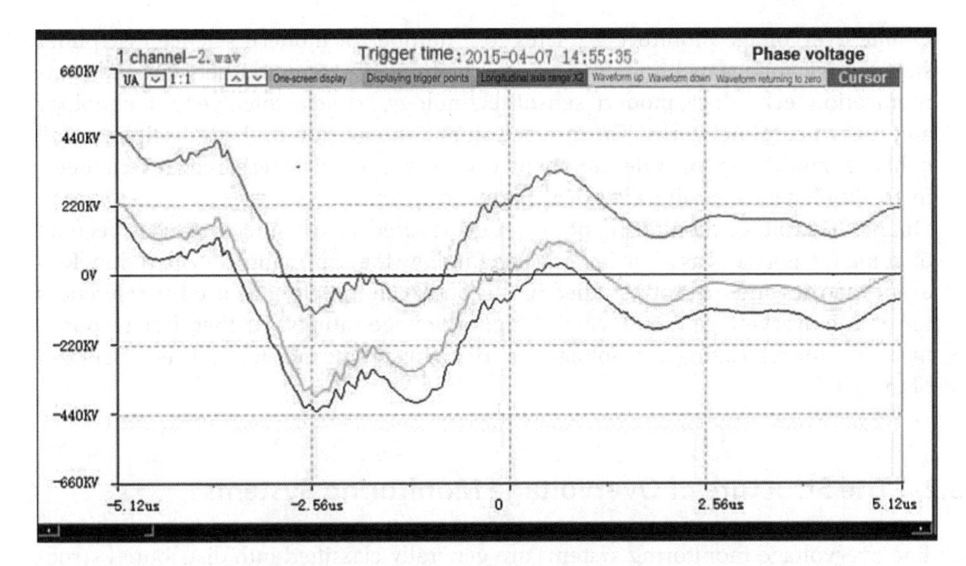

Figure 3.4 Localized amplification of the transient overvoltage signals.

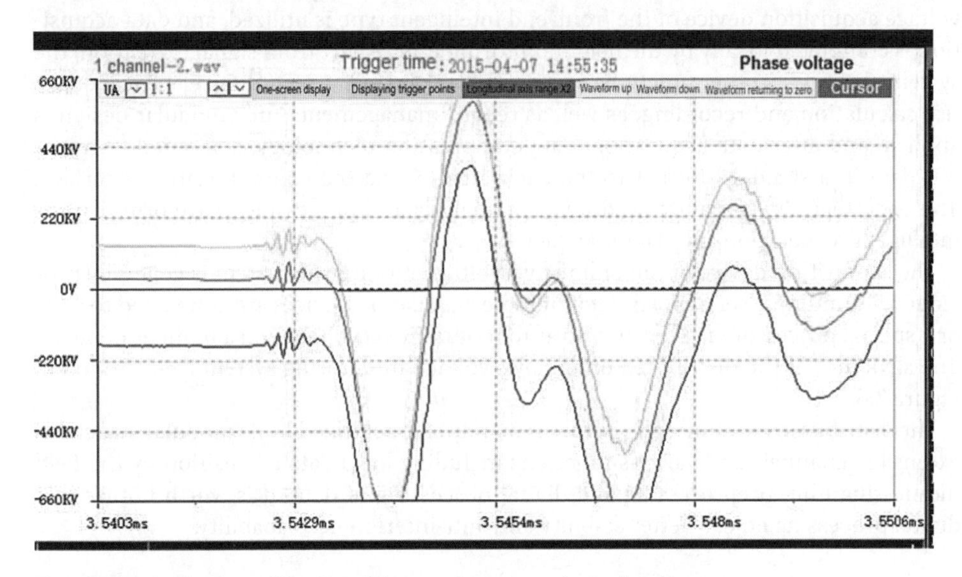

Figure 3.5 Localized amplification of the transient overvoltage signals.

still to be studied. Research on mechanism of lightning overvoltage is often based on numeric simulation. However, lack of onsite measurement data fails to provide necessary verification for the simulation models, thus significantly restricting the advances in overvoltage mechanism research. The fundamental objective of overvoltage monitoring is to study overvoltage mechanisms and to provide a scientific basis for damage prevention and measures for giving a fast response in suppressing overvoltages. Results thereof offer strong data support for foundational theories concerning overvoltage mechanism, physical process of electrical equipment discharge, online monitoring for operating states and condition-based maintenance.

2. Online overvoltage monitoring relates to a number of modern science disciplines such as high voltage technology, electronic technology, computer network and communication technology, modern sensor technology, artificial intelligence technology and fuzzy neural networks. The interaction and integration of these disciplines will enhance multidisciplinary development as well as theoretical research and engineering technologies regarding electrical engineering itself.
3. Online overvoltage monitoring of power grids rated 110 kV and above has been an ongoing focus as well as a cutting-edge technology for improving safe operation. Relevant theories and methods studied in this book can therefore be used as references for online overvoltage monitoring of higher voltage ratings. Further, fast-response measures for restricting overvoltages are of great significance for secured operation of the grid.

3.2 The Structure of Overvoltage Monitoring Systems

Online overvoltage monitoring systems are generally classified into distributed structure and centralized structure. For the distributed monitoring structure, the online overvoltage acquisition device of the front-end intelligent type is utilized, and data acquisition is completed locally by the field monitoring unit. Apart from signal extraction, the system also boasts other functions such as failure and event recording, electrical quantity calculation and recording, as well as record management. Fully modular design is applied, and after data acquisition, data compression technology is adopted to transmit data in a standard format to the backstage PC via the communication interface. The backstage PC is mainly used for data analysis and processing, waveform display, intelligent recognition and database queries.

The acquisition device of the online overvoltage monitoring system is generally provided with multiple serial ports, and multiple acquisition devices are connected to a PC or a special processor via Ethernet, so as to monitor overvoltages on a number of buses. The structure of the distributed online overvoltage monitoring system is illustrated in Figure 3.6.

The distributed monitoring structure is mainly characterized by its modular structure, extensible channels and various functions including local data acquisition by the field monitoring unit, preprocessing and digital processing of data. It is worth noting that digital processing enjoys better stability and anti-interference capability.

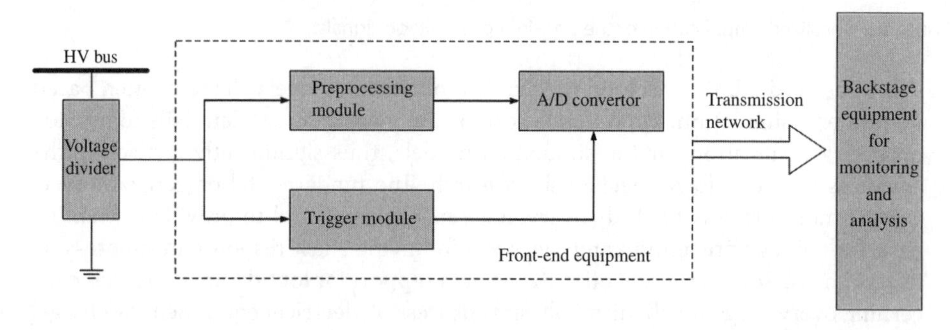

Figure 3.6 Structural diagram of the distributed online overvoltage monitoring system.

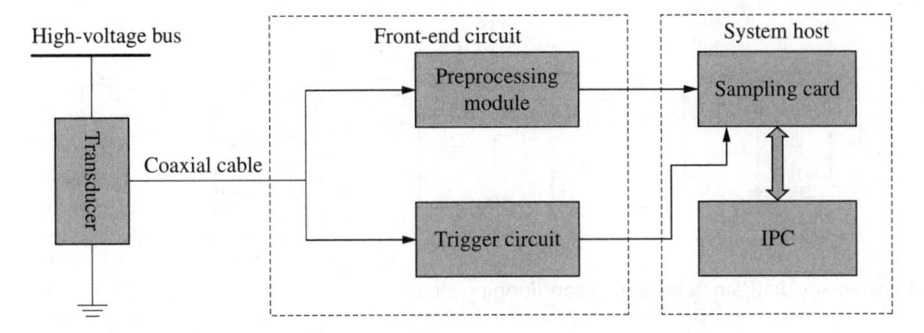

Figure 3.7 Framework of the centralized online overvoltage monitoring system.

The other structure used in online overvoltage monitoring systems is the centralized one, in which the measured signals are introduced into the host via shielded cables and then are subject to centralized circular detection and data processing by the host. The centralized monitoring device usually adopts a multi-level structure, with a management level and a data acquisition level. Industrial personal computers (IPC) are usually employed at the management level for storage, analysis, processing and display of data and at the data acquisition level for timed mutual inspection and clock synchronization. Single chip microcomputers (SCM) are often used as intelligent components at the data acquisition level. A dual-CPU structure composed of SCMs from the MSC196 and MSC51 series is very common, where one is used for data acquisition and A/D conversion, and the other is used for calculation and judgment purposes, for example, whether or not to initiate data storage. In some cases, the slave templates may employ only one industrial control board for data acquisition and startup logic. The management level is connected with the data acquisition level via PCI or ISA buses. However, limited by hardware configuration, this design leads to a low sampling frequency and a 12-digit resolution, and may give rise to a frequent bottleneck effect in the data transmission, which is detrimental to timely transmission and processing of failure messages. The framework of the centralized online overvoltage monitoring system is as shown in Figure 3.7.

The centralized online overvoltage monitoring system is mainly characterized by simple structure, mature techniques, high reliability and easier data processing enabled by the high performance of IPCs. In terms of disadvantages, the measuring system impedance is difficult to accurately match; high-frequency analog overvoltage signals would be attenuated and distorted, with poor protection against interference during long-distance transmission via coaxial cables; a special sampling card is required for extension purposes; the system real-time performance is rather poor in the presence of numerous monitoring objects; a single monitoring system is rather large in size and high in cost.

3.3 Acquisition Devices of Overvoltage Monitoring Systems

3.3.1 Design of the Signal Preprocessing Circuit

The signal conditioning circuit includes a voltage-follower circuit and an active second-order low-pass filter, as shown in Figure 3.8. The circuit is composed of

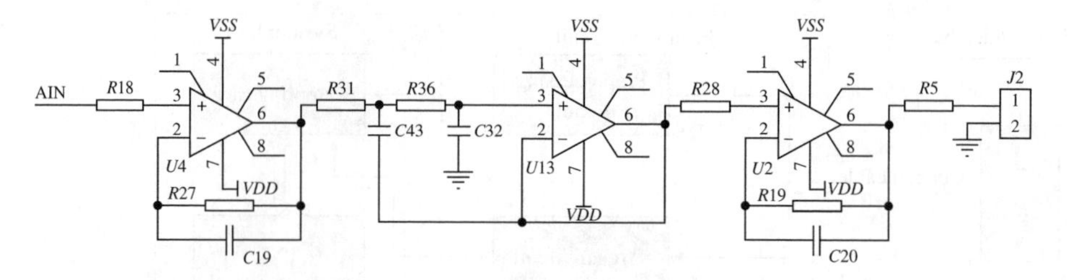

Figure 3.8 Schematic diagram of the signal conditioning circuit.

three-stage operational amplifiers, with the first and third stages used for impedance isolation while the second stage acts as the low-pass filter. A high-speed low-power operational amplifier is selected, which has a unity gain bandwidth of 500 MHz, a conversion rate of 350 V/μs and good load-driving capacity. The signal conditioning circuit composed of such amplifiers can meet the requirements for overvoltage signal preprocessing.

The active low-pass filter has to be a second-order voltage-controlled voltage source low-pass filter with the high cutoff frequency given by

$$f_H = \frac{1}{2\pi R_{36} C_{43}} \tag{3.1}$$

The cutoff frequency of the low-pass filter is about 8 MHz when $R_{36} = 200\,\Omega$ and $C_{43} = 100$ pF.

3.3.2 Design of the Trigger Circuit

The trigger circuit is used to compare the output signal of the conditioning circuit with the preset trigger level, so as to identify whether overvoltages have occurred. If overvoltages do occur, the acquisition unit will be activated in real time to collect the data. It is impossible to identify the overvoltage polarity in advance due to the randomness of overvoltage signals, so the window detection circuit composed of dual comparators is utilized in the trigger circuit to achieve the triggering of overvoltages of different polarities. The schematic diagram of the trigger circuit is shown in Figure 3.9.

The trigger circuit mainly comprises a trigger level circuit, comparators and a photoelectric coupled isolating circuit. The positive trigger level is generated from the precise voltage stabilizing circuit composed of an LM336, while the negative trigger level is provided by reversing the positive trigger level via the phase inverter. In this way, synchronization of the positive and negative reference levels could be regulated. A high-speed dual comparator LM319 is adopted with its response delay shorter than 80 ns, and the three-way voltage comparator outputs are allowed to be directly connected. In order to suppress the interference signals and ensure reliable triggering of the acquisition circuit, the output impulses from the comparator are isolated and transmitted into the external trigger channel of the acquisition module via the high-speed photoelectric coupler 6N137.

The overvoltage monitoring systems applicable to different voltage classes have different trigger levels. For the overvoltage monitoring systems in power grids rated 10 kV and 35 kV, the trigger level is 1.5 times the transducer output signals under the power-frequency steady state; for the monitoring system in power grids rated 110 kV,

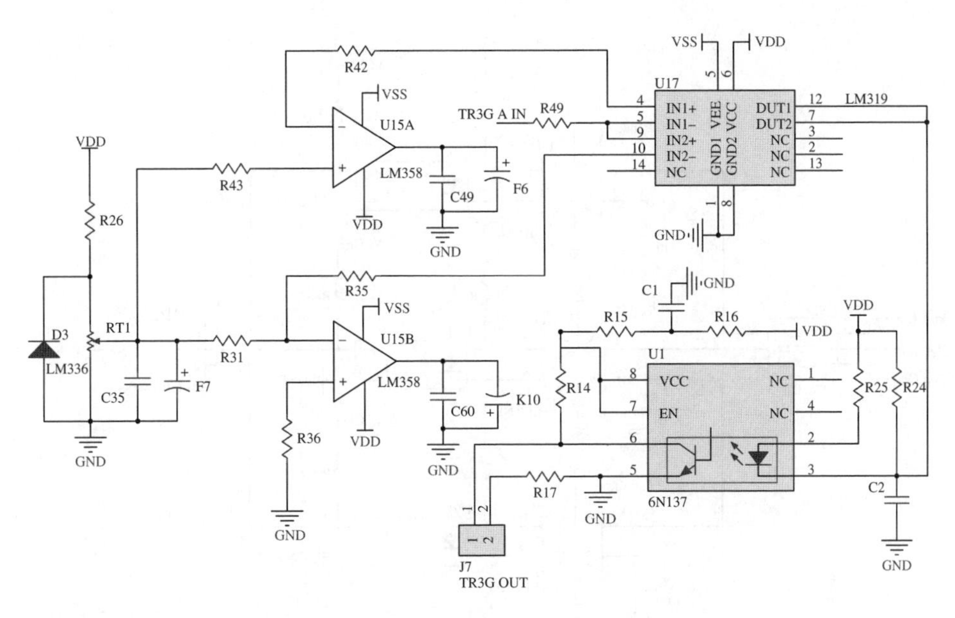

Figure 3.9 Schematic diagram of the trigger circuit.

the trigger level is 1.3 times the transducer output signal under the power-frequency steady state.

3.3.3 Design of the Protection Unit Control Circuit

The protection unit control circuit is mainly used for on-off control of the relays of the transducer protection unit. When the transducer is under normal operation, the relay is in the OFF position; in case of transducer failures or loss of control supply, the relay is in the ON position, thereby avoiding open circuit of bushing taps. The principles of the control circuit in the protection unit are shown in Figure 3.10.

The control circuit consists of three voltage comparators, whose threshold voltages are determined by the maximum withstand safety voltage of the bushing tap plus a reliable margin.

The protection circuit is designed on the basis of the following guidelines: two threshold voltages are set, the lower one for initiating the trigger circuit and the higher one for initiating the protection circuit. When the voltage measured by the transducer is beyond the triggering voltage of the trigger circuit and below the operation voltage of the protection circuit, overvoltages are identified as occurring in the power grid, and the data acquisition card is activated to collect the data. When the voltage measured by the transducer exceeds the operation voltage of the protection circuit, the bushing tap is identified as being open-circuit, and protective relays are immediately activated to short-circuit the tap grounding wire. The unipolar comparator MAX913 is selected for the protection circuit working on similar principles to that of the trigger circuit. A unipolar comparison model is adopted because the voltage polarity issue in the trigger circuit would not occur in the case where the tap voltage rise is caused by open circuit of the bushing tap. The feedback signal from the NAND output 71 is connected to the

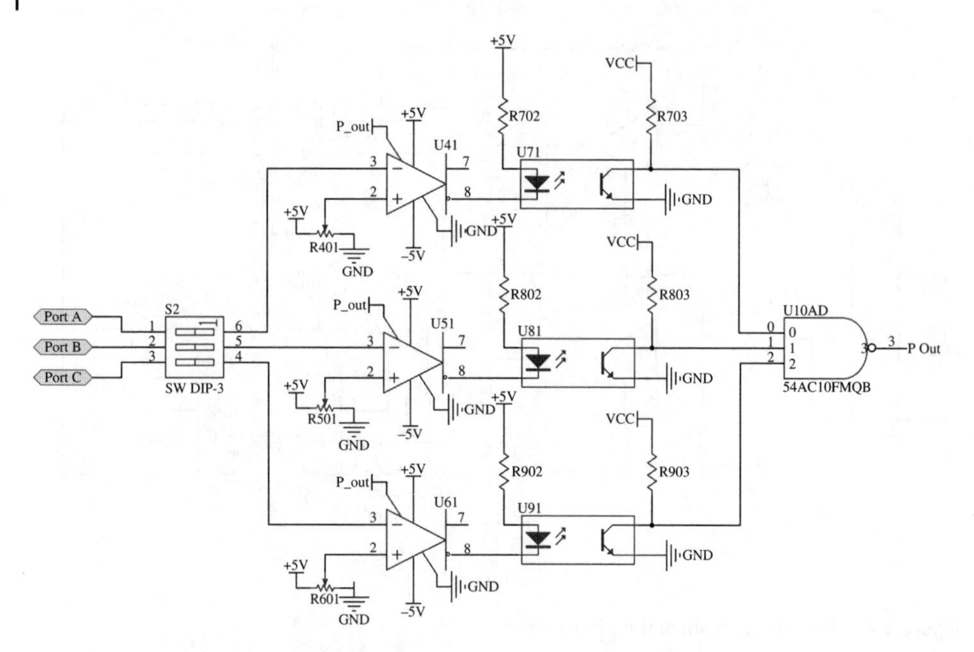

Figure 3.10 Schematic diagram of the control circuit in the protection unit.

latch enable terminal of MAX913 for prevention of relay contact jitter so as to ensure the reliability of the relay actions.

3.3.4 Selection of the Data Acquisition Card

The sampling theorem suggests that for reproducing the original signal, the sampling rate must be more than twice the signal bandwidth, i.e. $f_s > 2BW$, where f_s is the sampling rate, and BW is the signal bandwidth. The above relationship that determines the theoretical minimum sampling rate ensures that the sine wave with a frequency within the bandwidth limit will be sampled more than twice per cycle. In practical applications, a higher sampling rate will be adopted for precise reflection of the original signals.

Overvoltage signals contain abundant frequency components with a frequency band up to tens of MHz, which is very demanding for data acquisition cards.

3.3.4.1 Properties

The acquisition card adopted will be a three-way high-speed parallel A/D converter board that has a parallel conversion speed of 40 MHz for each channel and supports multiple trigger modes. The plug-in card based on the PC bus is located at the 16-bit ISA, and is designed using large-scale gate arrays. In addition, it supports online programming and provides changeable hardware designs. An acquisition card comprises input channels, signal conditioning circuits, ADC circuits, RAM, an internal clock, trigger control logic, memory interfaces, clock/timing control and bus interfaces. The schematic diagram of the system framework is shown in Figure 3.11.

The specific functions and properties of the data acquisition card are:

1. Input signal range: ±1 V, ±5 V.
2. Input signal bandwidth: 0 Hz~10 MHz.

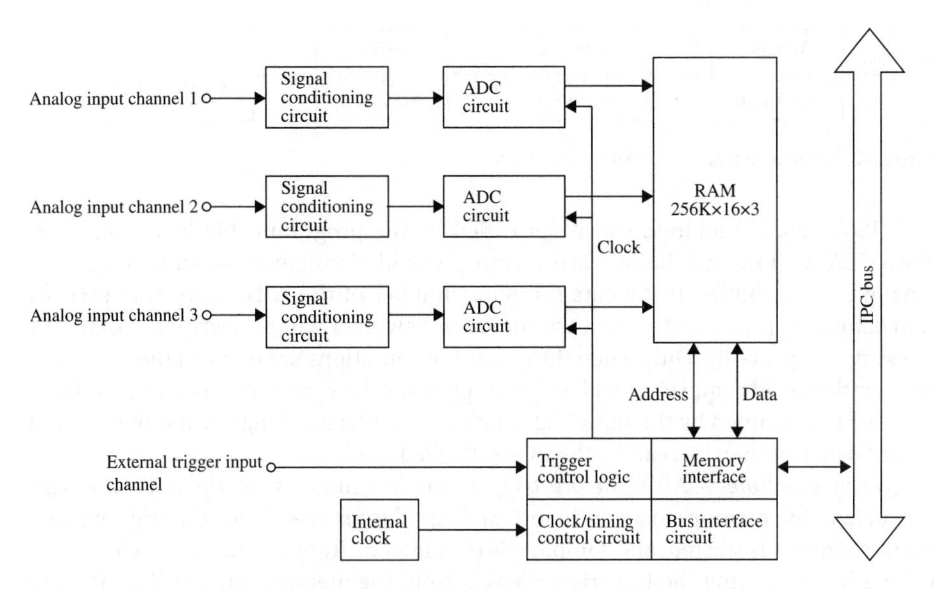

Figure 3.11 The structure framework of the data acquisition card.

3. Input impedance: 50 Ω, 1.25 kΩ, 1 MΩ.
4. Three-way 12-bit A/D parallel sampling, with a maximum conversion speed of 40 MHz/channel.
5. 12-bit resolution, system precision: ±0.1%; peak noise: ±1 LSB (±5 V input).
6. Storage depth: 256K RAM/channel. Three A/D converters store the sampling results to their RAM, according to the sampling sequences.
7. Three sampling modes: normal sampling, pre-trigger sampling, variable frequency pre-trigger sampling.
8. Two A/D trigger modes: software and hardware triggering; among others, the hardware trigger is divided into rising edge triggering and falling edge triggering.
9. Pre-trigger sampling with a pre-trigger length from 1K to 255K.
10. Variable frequency sampling with a variable frequency length from 1K to 255K.
11. Clock mode: internal crystal frequency division with eight division factors of 1, 2, · · · 128.
12. External trigger input level: TTL level; high level: 2.5–5 V; low level: 0–0.6 V. Sinking current input lower than 1 mA.
13. A/D sampling termination: query or interruption. Interruption can be selected with a 4-bit DIP switch, and with the numbers 10, 11, 12 and 15.
14. Base address range: 200H~2F0H, 16 base addresses occupied and selected by a 4-bit DIP switch.
15. Three-way analog signal input and one-way external trigger signal (TTL level) input.

3.3.4.2 Data Sampling Theory of the Acquisition Card

A/D conversion – The input impedance of 50 Ω, 1.25 kΩ or 1 MΩ can be selected for the input channel via the jumper. After passing the voltage follower and the differential amplifier, analog signals are inputted into the ADC for A/D conversion (Figure 3.12).

After initiating data acquisition, data acquired after the A/D conversion is latched and subsequently stored int the independent on-card memory of each channel. The static

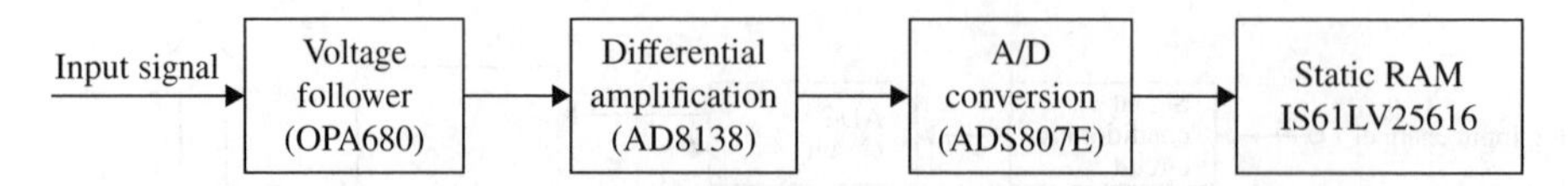

Figure 3.12 Schematic diagram of A/D conversion.

RAM (IS6lLV25616) address can be provided by the programmable logic controller (EPM3256ATC144), and the on-card memory would circulate between 0 and $2^{18}-1$, acting like a loop buffer. If the data sampling number of the A/D conversion exceeds the maximum capacity of the on-card memory, new data will overwrite the old. This process runs repeatedly. Only when the triggering conditions are met can the gate array start counting. And acquisition will not be stopped until the gate array reaches the specified value (determined by the sampling length and pre-trigger length), and the on-card memory data can then be read by the PC-controlled PLC.

Triggering principles – After the pre-trigger sampling mode is set, the A/D converter stores data in the memory in a continuous and circular manner. Once the trigger signal arrives, counter 1 is started, and sampling is terminated after the counter reaches a certain length (subtracting the pre-trigger length from the memory length). The point of sampling termination is deemed as the pre-trigger point. After the data is read into the computer in sequence, pre-trigger data is located at the front, while post-trigger data is located at the back. The schematic of the pre-trigger principles are demonstrated in Figure 3.13.

Theory of variable frequency sampling – The variable frequency sampling technology is developed, based on the requirements for the sampling of online overvoltage monitoring systems. In order to resolve the conflicts between sampling speed and frequency, and to satisfy demands for monitoring internal and atmospheric overvoltages, variable frequency sampling is utilized by Data Acquisition Card INSULAD2053. The basic theory of variable frequency sampling is that after startup, sampling is performed at frequency Fl by the acquisition system; after triggering, counter 2 starts counting, and when counter 2 reaches a certain length (post-trigger length = memory length – pre-trigger length – variable frequency length), A/D sampling and the address generator clock are switched to the set frequency F2 until data acquisition is completed. The schematic of the variable frequency sampling are illustrated in Figure 3.14.

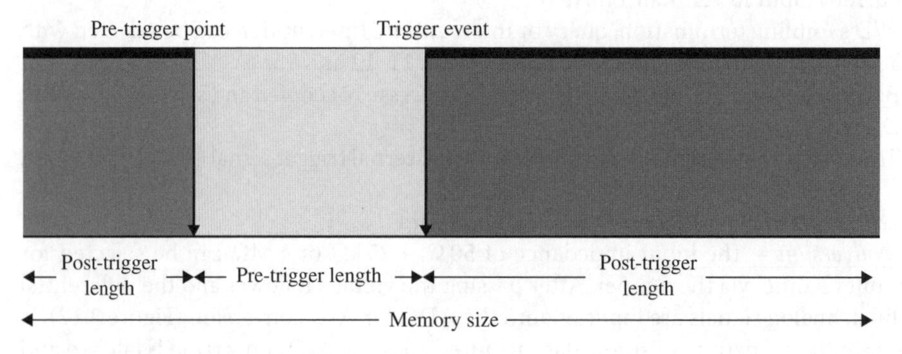

Figure 3.13 Schematic of pre-trigger.

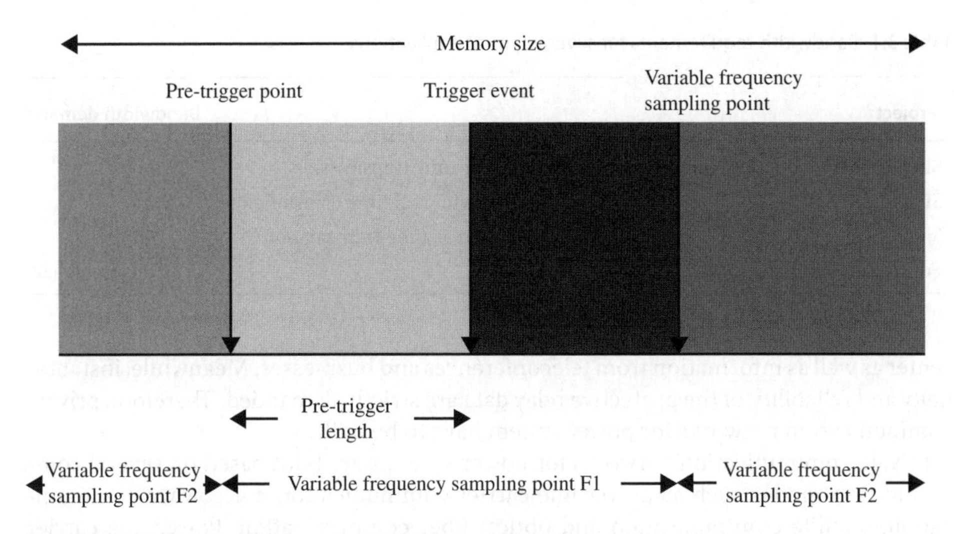

Figure 3.14 Schematic diagram of variable frequency sampling.

Lightning overvoltages and internal overvoltages possess different equivalent frequencies and durations. The front/tail time of the standard lightning wave is 1.2/50 μs, while that of the standard switching overvoltage is 50/2500 μs. Overvoltages generated due to other causes may last from 1 second to several dozens of seconds (with a lower equivalent frequency). If continuous sampling is performed according to the sampling frequency for lightning overvoltages (10 Msps), i.e. with a sampling time of 100 ms, other overvoltage waveforms may be omitted. In a variable frequency sampling scheme, data may, for instance, be first sampled at 10 MHz for 5 ms and then at 100 kHz for 10s. In this way, the development of different types of internal overvoltages can be recorded over a long time.

3.4 Overvoltage Signal Transmission System

In order to transmit the collected onsite overvoltage signals to the monitoring center for remote monitoring, a remote signal transmission system is essential. Depending on network properties, overvoltage signal transmission systems can be divided into private network (intranet) and public network (extranet) transmission system.

Overvoltage signal transmission systems are characterized by large amounts of data and high bandwidth demands and high probability of data concurrency.

Table 3.1 shows the bandwidth requirements for several typical applications.

Once generated, lightning overvoltages are very likely to be observed at many stations and on multiple buses in the same region, calling for high-quality network infrastructure.

3.4.1 Overvoltage Signal Transmission and Monitoring in Internal Networks

3.4.1.1 Private Communication Networks

To ensure safe, stable and economical operation of the power system, it is essential to transmit and share the data of each station, and to dispatch orders of each scheduling

Table 3.1 Bandwidth requirements for several typical applications.

Project	Bandwidth demand
Single bus (A + B + C + N) monitoring with non-real-time transmission	10 Mbps
Single bus (A + B + C + N) compressed monitoring with real time transmission	50 Mbps
Single bus (A + B + C + N) lossless monitoring with real-time transmission	800 Mbps
Future development needs	3.2 Gbps or higher

center as well as information from teleconferences and businesses. Meanwhile, instantaneity and reliability of the protective relay data are strictly demanded. Therefore, private communication networks for power system have to be built.

Private communication networks for power systems are built based on several communication modes such as power line carrier communication, microwave communication, satellite communication and optical fiber communication. Power line carrier communication, using transmission and distribution lines as the transmission medium is highly economical but not suitable for backbone communication networks due to its low transmission rate and low capacity. Microwave and satellite communication stand out in districts with wiring difficulties, as they both belong to wireless communication and have the advantages of high bandwidth and large capacity. As a result, they have become one of the key communication modes.

The widely accepted optical fiber communication enjoys excellent properties such as high bandwidth, large capacity, low attenuation and strong anti-interference capability. Also, optical cables such as OPGW cables and ADSS cables together with transmission lines can be erected on towers. At present, optical fiber communication has become the primary communication mode in dedicated communication networks for power systems.

Optical fiber communication has virtually covered every substation, and the dispatching center is thus able to remotely monitor and control the substations via the private communication networks to achieve substation automation. Overvoltage signals of the substation can also be transmitted back to the monitoring center via the same network for remote monitoring.

3.4.1.2 General Structure

The transmission and monitoring system is a network system that makes use of modern communication and computer technology for remote monitoring and control of operating substation equipment. As there are substantial amounts of monitoring equipment in the substation, in many cases it is impossible to connect all the monitoring equipment terminals directly to the substation control center, especially when the wired communication mode is selected. Hence, the transmission and monitoring network system is generally composed of three tiers: remote monitoring terminals, communication networks and substation monitoring terminals, each of which implements communication and control through the communication medium.

Shown in Figure 3.15 is an overvoltage monitoring network system that includes remote monitoring terminals, communication networks and at least one substation monitoring terminal. The remote monitoring terminal, which contains the main server

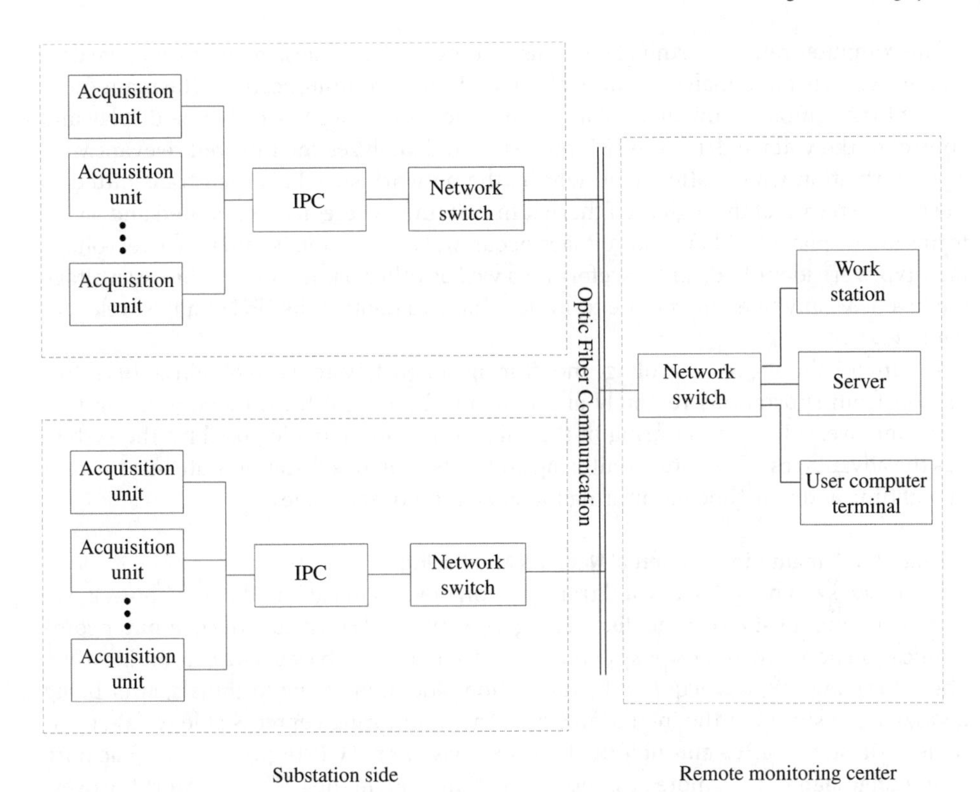

Figure 3.15 Overvoltage monitoring system based on internal networks.

and at least one user computer terminal, is connected to the substation monitoring terminal through the communication network for which the Ethernet is chosen. The substation monitoring terminal includes network switches, IPCs and at least one acquisition unit; the IPCs connect to the Ethernet through the network switches, and to the acquisition units through the communication cables.

Among other things, the IPC acquires real-time voltage messages from each acquisition unit, and comprehensively monitors and controls the substation. With the help of the IPC, three-phase voltage signals can be monitored and analyzed in real time by analyzing the operation states of the substation. Furthermore, fault voltages can be recorded and faults can be identified of various types including voltage flicker, voltage variation, lightning overvoltages, switching overvoltages, three-phase unbalance, zero-sequence overvoltages and temporary voltage rise or drop. By these means, the entire substation can be effectively managed.

The acquisition unit is responsible for acquiring and processing voltage information in a given region of the substation. Data acquired from each phase is output to the IPC via the communication cable, and the IPC is equipped with network communication interfaces (Ethernet cards with a rate over 100M), to guarantee that the three acquisition units are completely isolated from each other. The clock synchronization channel, synchronization trigger channel and data output channel of the IPC all use optical fiber communication. For details of acquisition and sampling methods, refer to Chapter 2 of this book.

The computer remote terminal contains a number of modules, including voltage data storage, voltage fault analysis, failure alarm and serial communication. The main functions of the remote terminal are data communication as well as real-time display and control of the voltage data. The IPC processes and analyzes the received relevant voltage information which, after treatment by the network switch, is then transmitted to the remote terminal through the Ethernet in real time where data is restored and waveforms are displayed. When overvoltages occur, wave recording is initiated, overvoltage fault types are identified, and waveforms as well as other parameters of the overvoltage are saved. Meanwhile, the remote terminal is able to control the IPC to adjust relevant parameters.

To sum up, the above overvoltage monitoring network system can obtain status information from substations remotely and dynamically in real time, demonstrating that real-time overvoltage monitoring of the substation can be made possible. The system has the advantages of remote monitoring and collection of substation state information in real time and real-time monitoring for substation overvoltages.

3.4.1.3 Implementation of Internal Network Monitoring

Computers, switching devices and transmission devices in each node of the power system (including substations and dispatching departments) constitute the computer communication network for power system, which can realize IP-based data communication. The overvoltage signals acquired by the station-side monitoring terminals, after being digitized, are stored in the in-station IPCs. The monitoring center is able to (a) access remote substation IPCs through the Ethernet, based on TCP/IP protocols, (b) acquire overvoltage signals by remote desktop connection to login the IPCs, (c) export the overvoltage data to the database servers or work stations in the monitoring center for later query and analysis.

Remote desktop connection – This method needs to set the IP address of the in-station IPCs and the permission of the remote desktop connection. Firstly, the IPC requires a fixed IP address (generally a private IP address communicating only within the internal network of the power system) so that the computers in the monitoring center are able to establish remote desktop connection. Meanwhile, the permission of remote desktop connection has to be open to the IPC with a set username and a password so that the computers of the monitoring center can successfully log onto the IPC.

The following discussion will detail how remote desktop connection is implemented. Acquired by the monitoring terminal and subsequently digitized, overvoltage data is sent through the coaxial cable to the in-station IPC that receives and stores the data and generates corresponding data files. The work station of the remote monitoring center runs the remote desktop connection program where the IP address or the substation IPC name has to be entered for access application in accordance with a corresponding mapping table between IPCs and IP addresses, and the username and password has to be entered for verification. If login is successful, the monitoring center can remotely operate the IPC as if the center were in the substation. By utilizing the overvoltage data processing software, operations including real-time waveform display, historic data review, spectrum analysis and overvoltage type identification can be performed on the voltage signals received by the IPC from the monitoring terminal. Also, the staff of the monitoring center who log onto the IPC can also remotely export the IPC data to the database server or the work station of the monitoring center.

Remote data access can be enabled by simple settings by using existing programs for remote desktop connection. Monitoring sites, however, generally come in large numbers. If every IPC in the substation adopts this method, it will be inconvenient because the monitoring center has to build the desktop remote connection one by one.

Real-time C/S remote mode – Besides remote desktop connection, by adopting the client/server (C/S) mode and by compiling remote communication programs, communication between the IPCs and the monitoring center can be realized, and the IPCs of each substation can automatically transmit the acquired overvoltage data to the monitoring center.

The working of the C/S remote mode is described below. The IPCs in each monitored substation run the client programs of remote communication and receive and package the voltage data from the monitoring devices. This data is subsequently uploaded onto the power communication networks through the network switch, and finally transmitted via the routers to the servers in the remote monitoring center. The way the IPCs send the data is optional, sending either real-time voltage messages or overvoltage data only in the presence of overvoltages. The former choice can monitor voltages in real time at the expense of large data amount. The remote monitoring center runs the server program; the servers connecting to the network switches receive voltage messages from the IPCs in the monitored substations, unpack the data based on the communication protocol and record and store the overvoltage data. The work stations or PCs in the monitoring center are thus able to access the server data through the switch so that real-time waveform display, overvoltage spectrum analysis and overvoltage category recognition can be achieved.

The C/S remote mode can enable the monitoring center to monitor multiple substations simultaneously in real time, which solves the one-by-one access deficiency of remote desktop connection.

3.4.2 Transmission and Monitoring Based on Wireless Public Networks

Remote transmission of overvoltage data, when utilizing the existing private optical fiber communication networks, has advantages such as high transmission speed and no requirements for networking, and is thus suitable for real-time remote overvoltage monitoring in substations. However, the private communication network may not suit certain cases such as monitoring overvoltages on transmission lines distant from the substation. In such a case, public wireless networks can be used, and can significantly save the cost of line construction.

3.4.2.1 GSM Networks – Introduction

The so-called GSM (global system for mobile communications) is a second-generation mobile communication system. Unlike the first-generation analog communication system, GSM is a digital system that provides remote data services, meaning that GSM can be used for remote transmission of overvoltage data, mainly for low data rate services due to its low transmission rate of 9.6 kbps. For overvoltage signals with high data rates, GSM cannot guarantee real-time performances of online monitoring. However, GSM can provide information on overvoltage faults and alarms by using SMS (short messaging service). When the IPC receives overvoltage data or fault signals, the GSM module installed on the IPC would be triggered and alarm information would be transmitted through the GSM network to the working staff, in the form of short messages.

GPRS (general packet radio service) is a newer packet data transmission service aiming to improve the data transmission rate of the GSM networks. The organic combination of mobile communication techniques and IP Over PPP Technique featuring high-speed remote access of data terminals makes GPRS a mobile IP network seamlessly connecting the rapidly developing fixed IP network. GPRS provides users with multimedia services involving data, voice, images and so on. GPRS can reach a relatively high data transmission rate of 114 kbps, so it can be used as the transmission mode for real-time remote overvoltage monitoring.

Apart from the networks described above, public wireless networks include 3G and 4G networks, respectively referring to third and fourth generation mobile communication. In comparison to GPRS, these two both have higher transmission rates, with 4G even exceeding 100 Mbps. Besides, due to their better performance in real-time response for remote overvoltage monitoring, 3G and 4G networks will no doubt become the direction of development of remote overvoltage monitoring on transmission lines.

3.4.2.2 General Structure of Systems Based on GSM Networks

The described system is an application based on the SMS of GSM with overvoltage signals as the monitored object. The function of the system is to remotely monitor overvoltage signals and to achieve automatic alarm, which is made possible by the overvoltage signal acquisition circuit, the IPC and its peripheral interface circuit as well as the GSM communication module and its connection with the host computer. When they occur, the overvoltage signals will be processed by the IPC and then transmitted to the monitoring center via the GSM communication network, while at the same time, information is edited in the form of short messages and sent to the personnel concerned.

As shown in Figure 3.16, the system is divided into two parts: the overvoltage signal monitoring center with host computers and the wireless monitoring branches consisting of IPCs and peripheral interface circuits (hereinafter referred to as the branch). The hardware of the host computer is connected to the GSM wireless communication module via the RS232 serial port.

The basic functions of the monitoring center servers include the following functions:

1. receive the voltage signals processed by the branch, display them on the, host computer interface as well as classify and store them;
2. monitor the branch and handling emergency alarms;
3. database management, real-time display of branch situations and printing backup. Note that the branch consists of the IPC and its peripheral interface circuit, the overvoltage acquisition circuit, and the GSM wireless communication module.

The basic functions of the branch include:

1. acquire, filter and amplify the overvoltage signals through the overvoltage signal acquisition system;
2. convert and process the data from the peripheral sensor through the data acquisition card;
3. display data on the LCD and process the alarm signals;
4. transmit real-time data to the monitoring center or the managerial personnel;
5. control the GSM wireless communication module to receive and send texts.

3.4.2.3 General Structure of the GPRS-Based Transmission Systems

There are two structures for the GPRS-based system for remote overvoltage data transmission. One is similar to that of the GSM mode: slave computers (the IPCs) and host computers (the monitoring center) both need GPRS communication modules, and the entire data flow is transmitted on the GPRS network, as illustrated in Figure 3.16.

The Internet is accessible through the GPRS networks because GPRS supports the TCP/IP protocol. Hence, the GPRS communication module at the host computers (the monitoring center) can be omitted; host computers can access the Internet directly through switches or routers. However, to make sure that the slave computer can communicate with the host computer, the latter must have its own fixed public

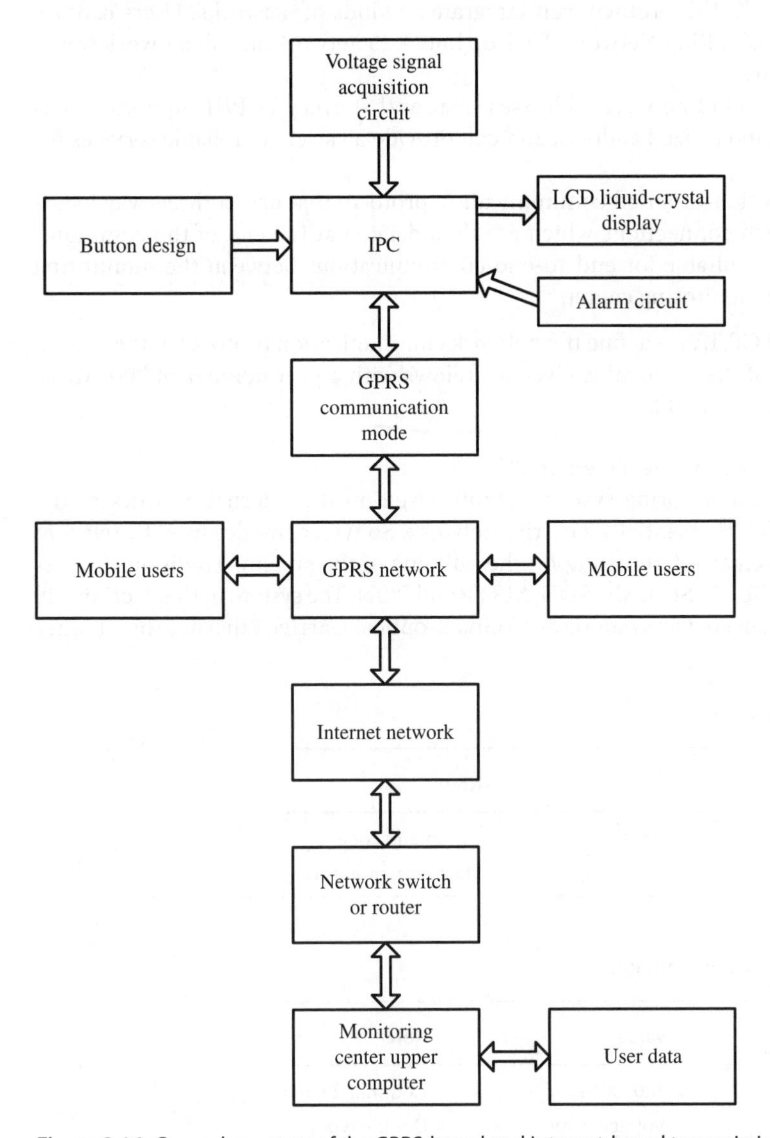

Figure 3.16 General structure of the GPRS-based and Internet-based transmission systems.

IP address. If the host computer is connected to the intranet switch and only has a private IP address, port mapping will be needed on the router, which enables extranet clients (GPRS modules of the slave computers) to access intranet servers (servers in the monitoring center). The general structure of the system is shown in Figure 3.16.

The TCP/IP protocol has the following characteristics:

1. The TCP/IP protocol does not depend on any specific computer hardware or operation system, and it provides open standards. It receives extensive support, irrespective of its application in the Internet. So the TCP/IP protocol has become a practical system that combines all kinds of hardware and software.
2. The TCP/IP protocol does not rely on specific network transmission hardware, and that is why the TCP/IP protocol can integrate all kinds of networks. Users can use the Ethernet, Token Ring Network, Dial-up line, X25 network and all network transmission hardware.
3. Unified assignment of network addresses ensures that every TCP/IP equipment has a unique and standardized address, and can provide a variety of reliable services for users.
4. TCP is a connection-oriented communication protocol that uses a three-way handshake to establish connections which are then deleted at the end of the communication. It is very suitable for end-to-end communications between the monitoring devices and the monitoring center.

On the basis of TCP/IP, we define the network communication protocol of the system in this book. Part of the protocol is given as follows with a port number of 900. These are shown in Tables 3.2 to 3.6.

3.4.2.4 10 Gb All-Optical Carrier Ethernet (CE)

Online overvoltage monitoring system is built based on the internal network of the power system, which is essentially a carrier network. So typical modes used by telecom carriers can be taken as references upon the building of the system. Traditional access modes include ADSL/VDSL, LAN, SDH/MSTP and PON. The system in this section will utilize a more advanced access mode – 10 Gb all-optical Carrier Ethernet (hereinafter referred to as CE).

Table 3.2 Start\stop monitoring.

Byte serial number	Value	Note
0–3	8d00000A	Command code
4–7	1/0	Start/stop monitoring

Table 3.3 Voltage monitoring settings.

Byte serial number	Value	Note
0–3	8d000014	Command code
4–11	Voltage value	Double type

Table 3.4 Working directory settings.

Byte serial number	Value	Note
0–3	8d00000B	Command code
4–73	Server directory	Limited to 70 bytes

Table 3.5 File list access.

Byte serial number	Value	Note
0–3	8d000007	Command code
4–67	Server directory	Limited to 64 bytes
68–71	0	retained
72–75	0	retained
76–79	0	retained

Table 3.6 File downloads.

Byte serial number	Value	Note
0–3	8d000009	Command code
4–67	Filename	Limited to 64 bytes
68–71	Increment number	Packet index
72–75	File starting point	Initial location of the file data
76–79	Data length	Data length

Table 3.7 Comparison between CE and other access modes.

Characteristics	SDH/MSTP	PON	LAN	xDSL	CE
Branch bandwidth	2–10M	100M	100M	1–10M	1000M
Protection	<50 ms	TypeB level (s)	no	no	<50 ms
L3 capability	none	poor	none	none	good
Maintainability	good	good	poor	poor	good

Advantages and disadvantages between CE and other access modes are described in Table 3.7.

CE provides high-speed and reliable integrated services for access within metropolitan areas, which has become the best technique for metropolitan area high-speed access. The following contents will give a detailed discussion on its properties.

Ten gigabyte – xDSL, SDH/MSTP and PON are mostly utilized by the traditional IP-based MAN (metropolitan area network) for access purposes. In the case of

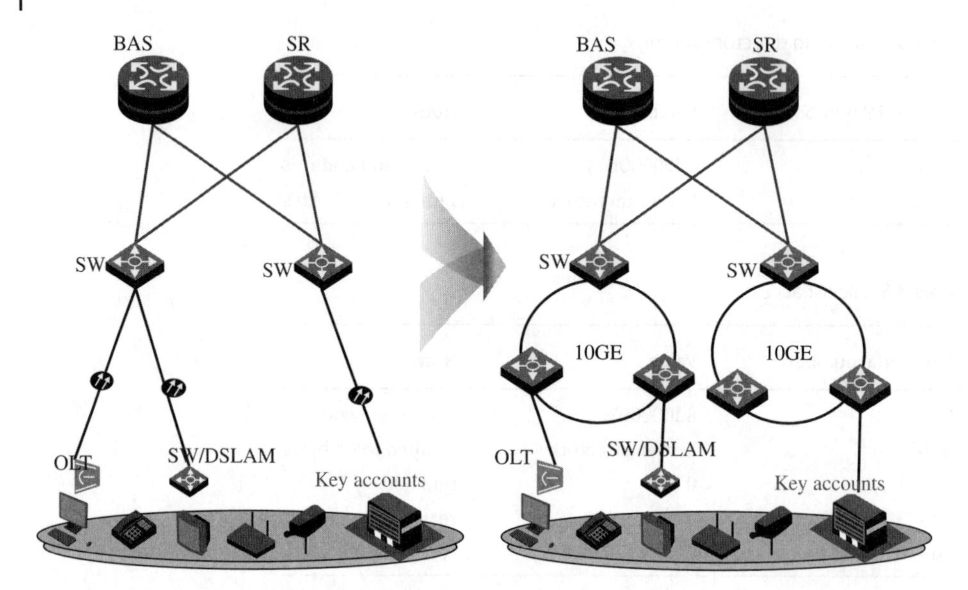

Figure 3.17 Comparison between the traditional IP access modes and CE.

overvoltage monitoring, xDSL and SDH/MSTP have difficulty in satisfying bandwidth requirements; PON, which can be used for high-speed access in small areas, lacks conditions for metropolitan area transmission where a high-speed metropolitan area transmission and access platform is required. CE, characterized by coverage of metropolitan areas and high access speed of ten gigabyte, exactly caters to this requirement. CE access saves upstream fiber optics of OLTs and SWs, guarantees the safety of optical networks and improves the link utilization ratio (Figure 3.17).

All optical networks – The traditional low-speed EoP access of SDH (2–8M) will inevitably introduce protocol converters and fiber optic transceivers (FOTs), and may cause stack issues at the central office. Too many protocol converters and FOTs may bring about both management issues and more faults. Due to lack of network management, faults cannot be found and handled until user complaints are received (Figure 3.18).

Although the EoS access (2–50M) of MSTP does not use protocol converters, it still needs FOTs to extend to users from the central office. Presently, MSAP is another option used by the industry, which tries to avoid stack issues and improve network management by installing card-based SDH modules in central office terminals (COTs), but user-side FOTs are still potential fault sites, and its reliance on low-speed SDH access is a fatal defect.

The "all optical network" of CE means that users can connect with the optical line terminal (OLT), which solves the issues such as protocol handling and photoelectric conversion and provides an access capacity of 2–1000M (Figure 3.19).

Ethernet ring – At present, ring networks are largely adopted by SDH to facilitate service protection. PTN, not yet fully standardized, also proposes the shared protection ring technology. The advantages of the Ethernet ring include the following

1. Low network investment – When deploying RRPP, special interface boards are not needed as all its interfaces are ordinary Ethernet optical ports. Therefore the

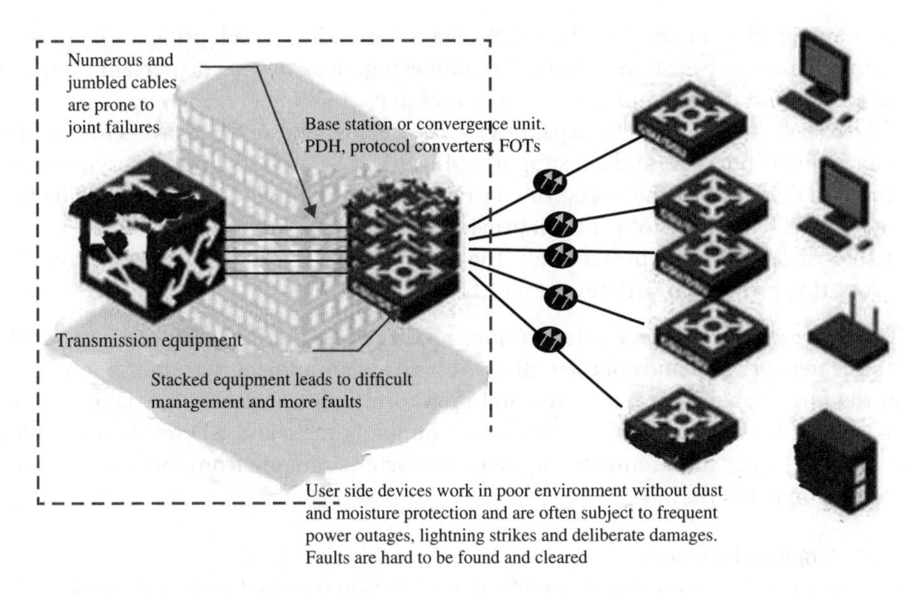

Figure 3.18 Issues of photoelectric mixed access.

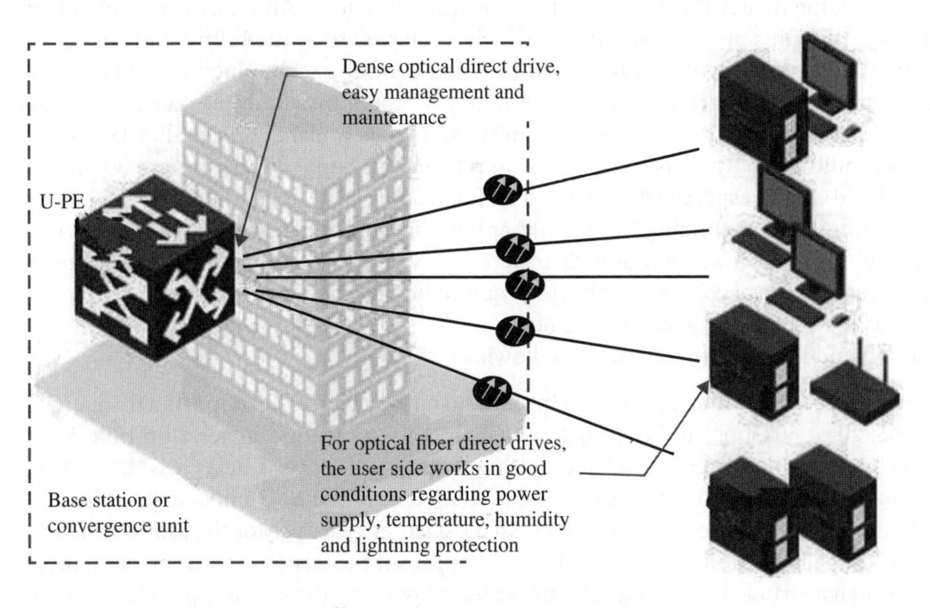

Figure 3.19 Direct-driven optical fiber mode.

Ethernet ring is characterized by rapid deployment and low investment. The ring may well replace the dual-homing deployment of traditional packet switching equipment, which greatly reduces link consumption, saves valuable optical fiber resources and reduces the complexity of implementing the project.

2. High link utilization rate – Whatever ring protection method is adopted in the traditional SDH ring, 50% of the resources are used for redundant backup, resulting

in a link utilization rate less than 50%. The carrier Ethernet adopting a ring protection technology based on statistical multiplexing, does not need special redundant resources and its link utilization rate can reach over 50%.

3. Three layers to the edge (L3 capability) – The deployment of three layers to the edge can be made possible, and CE supports MPLS/VPLS and L2/L3 multi-service transmission demands of the users. In contrast, the L3 function cannot be provided by SDH/MSTP, and has not yet been provided by the PTN standard to date.
4. Cross-ring end-to-end protection – The traditional SDH is unable to realize complex cross-ring protection without means of SNCP and DNI.

To sum up, as the best technology among all the access methods, CE can fully satisfy the basic network demands of the online overvoltage monitoring system.

In the foreseeable future, research and monitoring on transient overvoltage signals will accomplish data acquisition with higher sampling rates and AD resolution, which put forward higher requirements for the entire system, ranging from the devices to the transmission network.

3.4.2.5 InfiniBand Network

Currently, gigabit Ethernet is adopted by the substation terminal networks, while 10 Gb Ethernet is adopted by the backbone networks of monitoring centers and data centers. Both are going to face the problem of bandwidth bottleneck. An innovative technology with regard to this problem is utilized, which is referred to as the InfiniBand network.

InfiniBand is a high-bandwidth, high-speed networking technology with open standards, and also a "conversion cable" technology that supports multiple concurrent connections; it supports three kinds of connections (1x, 4x and 12x). Note that 1x, 4x and 12x are multiples of the basic transmission rate, i.e. the supporting rates are (QDR) 10, 40 and 120 Gbps, respectively, as shown in Table 3.8.

The InfiniBand technology is mainly applied in communications between servers (such as replication servers and distributed servers), between servers and terminal equipment (like overvoltage monitoring equipment) and between servers and networks (such as LANs, WANs, and the Internet).

InfiniBand is characterized by the following.

1. High speed – The first-generation InfiniBand technology DDR supports a throughput of 5, 20 or 60 Gbps, and a delay less than 1.3 μs. The second-generation InfiniBand technology QDR supports a bandwidth as high as 120 Gbps, a delay less than 100 ns and a transmission bandwidth far higher than that of the 10 Gb network (10 Gbps).
2. Remote direct memory access – This is a most suitable feature for the monitoring system as it allows the server to acquire and use part of the terminal equipment memory through a virtual addressing scheme without involving the operating system kernel.

Table 3.8 Supporting rates (Gbps) under different modes and channel numbers.

	SDR mode	DDR mode	QDR mode
Channel number 1x	2.5	5	10
Channel number 4x	10	20	40
Channel number 12x	30	60	120

3. Transmission uninstallation – Remote direct memory access can benefit transmission uninstallation which switches the packet router from the operating system to the chip level. In this way, the burdens of the processor are considerably lessened.

To date, the system cost of the InfiniBand network is still relatively high, but it will be widely supported and applied with the development of electronic technology. When its cost reaches a reasonable level, InfiniBand would undoubtedly play a key role in overvoltage online monitoring.

3.5 Overvoltage Waveform Analysis System

3.5.1 Introduction to the Waveform Analysis Software

The offline overvoltage waveform analysis software is able to load the recorded overvoltage data for further analysis so that users can analyze the acquired overvoltage data in a specific format on any computer instead of on a given computer.

The basic main functions (Figure 3.20) of the software include:

1. overvoltage waveform display;
2. collecting sampling information, such as the overvoltage impulse number, sampling frequency, sampling points and sampling period;
3. displaying three-phase voltage waveforms simultaneously or the voltage waveform of any single phase;
4. measuring the maximum overvoltage magnitude, the maximum multiple, front time and duration as well as calculating and counting front gradient;
5. analyzing the spectrum and spectral power density of the acquired overvoltage data as well as the main frequency range of overvoltages;
6. basic operations such as waveform scaling and stretching as well as print services.

The offline analysis program for overvoltage waveforms is characterized by high operability and a simple but highly functional user interface in accordance with the user

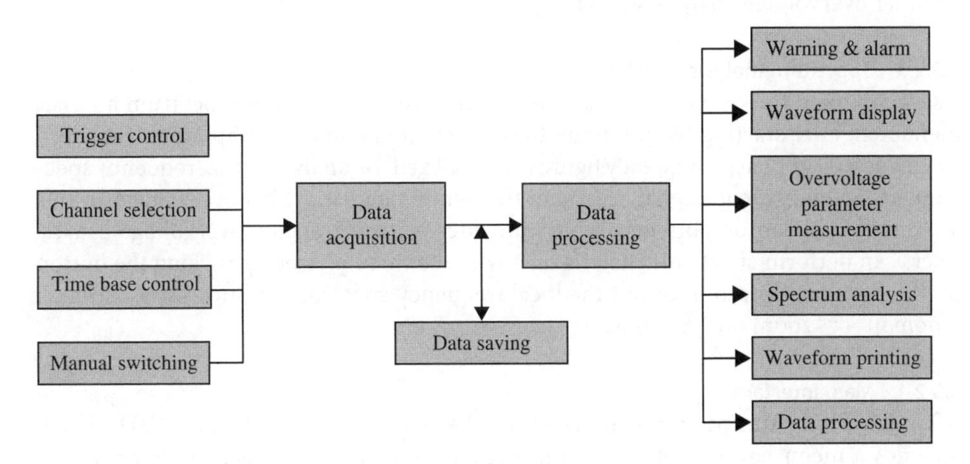

Figure 3.20 Descriptions of the analysis software functions.

requirements. Users are provided with a platform for analyzing overvoltage events. The offline analysis program in this book is able to process all overvoltage data files based on the data exchange standard for transient overvoltages.

3.5.2 Functions and Interface Design

In the following contents, the main functions of the designed offline analysis program for overvoltage waveforms will be detailed.

3.5.2.1 Serial Communication Interface

Overvoltage data is sent through the serial ports of the host computer to the serial communication program of the backstage computer that receives the data in time and generates saved files in the corresponding format. This serial communication program has such functions as data reading and writing, file management and hard drive operation.

Serial communication will be started after executing the program. The received data is named after the overvoltage instant, and the corresponding data files are generated and saved on a given disk.

3.5.2.2 Data-Loading Interface

After initiating the program, users can open the dialog box for loading overvoltage data files by the file menu on the menu bar. This dialog box is similar to the many interfaces compatible with the Windows system, thus making the operation simple and clear.

The folder where overvoltage data is stored can be figured out by specifying the local disk path, and overvoltage data files can be figured out by file names. A double click can then load the overvoltage data into the analysis program.

The "current file path" displays the location and path of the selected data files, and lists all overvoltage data files in the relevant folder. Users can choose the data files according to the overvoltage occurrence time and load them into the analysis program. In order to be highly compatible with other software, "save as Excel" is included in the pull-down menu. Saved in the Excel format, these data files are able to be processed and analyzed by other overvoltage analysis software.

3.5.2.3 Spectrum Analysis Interface

Users can open the overvoltage spectrum analysis interface via the spectrum analysis menu item after starting the program. In this functional module, amplitude-frequency figures as well as phase-frequency figures are analyzed. By analyzing the frequency spectrum of the overvoltage signal, the main frequency range in which overvoltage signals are distributed can be obtained, which provides evidence on the overvoltage causes. Users can perform spectrum analysis on any of the three phases by clicking the button on the right, and zoom in or out the local frequency spectrum by successive clicks on "zoom in" or "zoom out" buttons.

3.5.2.4 Main Interface

After launching the program, users enter the main interface (Figure 3.21), which includes a menu bar, operation buttons, waveform display interface, basic operation settings, sampling information and overvoltage parameters.

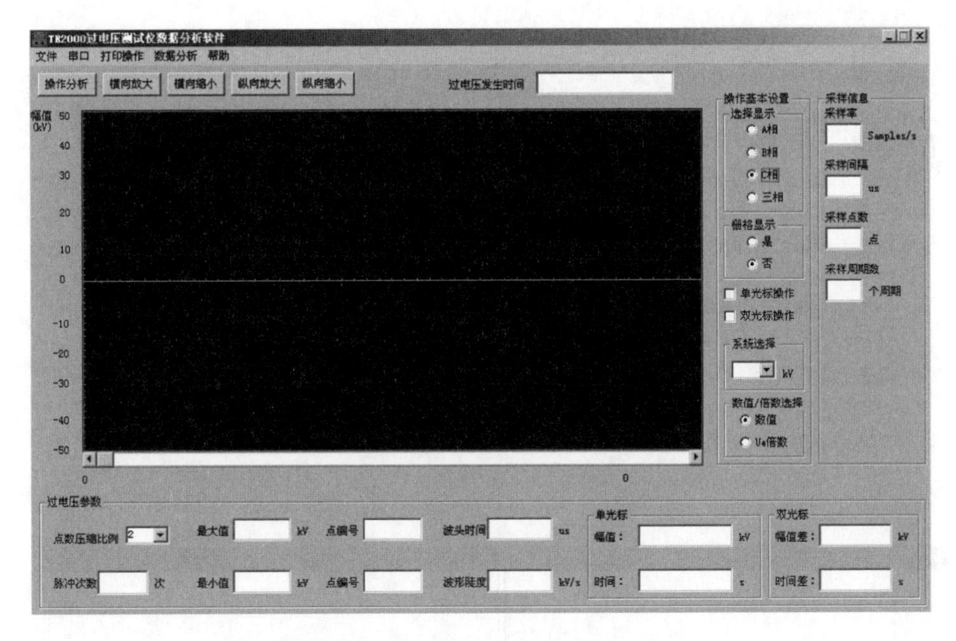

Figure 3.21 Main interface of the overvoltage waveform analysis software.

The menu bar includes items such as opening data files, overvoltage data analysis and overvoltage waveform printing. The waveforms can be locally zoomed in or out by the operation buttons. The menu items of basic operation settings can choose the phase of the waveform to be displayed and decide whether to display the waveform numerically or in the form of multiples. The sampling information menu includes sampling rate, sampling points and sampling periods, and is clear to users. The overvoltage menu involves calculation of the impulse frequency, waveform steepness and the maximum value of the overvoltage data.

3.6 Remote Terminal Analysis System for Overvoltage Waveforms

The basic principle of this system is shown in Figure 3.22. When overvoltage faults occur on the line, the signals will propagate to both ends of the line. When signals reach the substation, the current signals are received and processed by the current monitoring unit connected to the ground wire, while the power frequency components and overvoltage signals are received and processed by the voltage monitoring unit. The signal data is stored in the mass storage array of the system in real time. The warning messages are sent to the monitoring center via GSM and to the mobile phones of the staff in charge of line maintenance in the form of text messages. Monitoring personnel can remotely review the fault information via the GPRS system. quickly view the fault parameters such as overvoltage types and fault locations a on the large screen after the processing of the mass data storage system, and eventually plan for emergency repairs and instruct on power recovery as soon as possible. The system can collect and record electrical parameters of the lines in real time in the absence of overvoltage events.

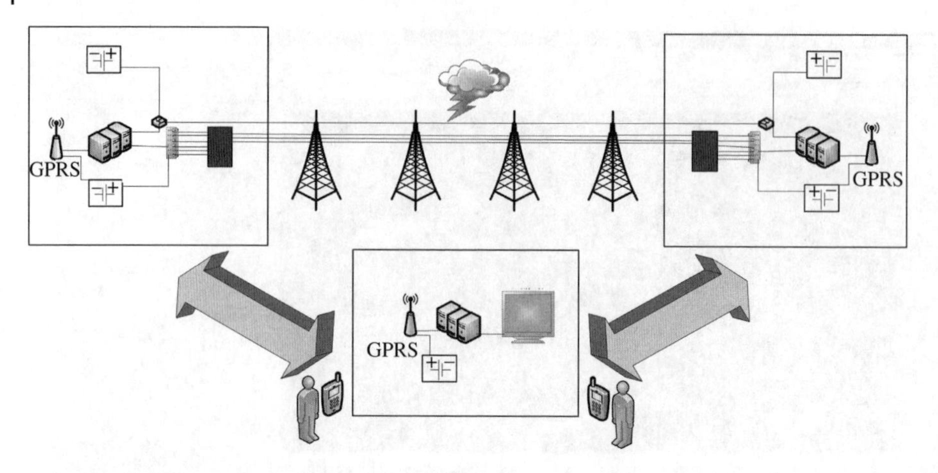

Figure 3.22 Condition monitoring and fault analysis system for transmission lines.

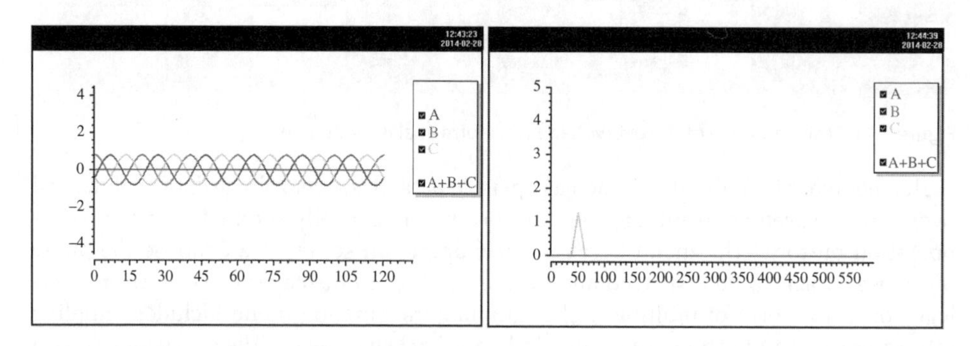

Figure 3.23 Waveforms in real-time monitoring.

The real-time monitoring window is shown in Figure 3.23 This interface can achieve the following functions: data reading from the voltage acquisition system (A), direct data display on the curves (B), switching display (C) between line voltages and phase voltages as well as spectrum and unbalanced degree analysis.

Fault loggings and fault waveform recordings are illustrated in Figure 3.24. The waveform recording system will be launched to record corresponding data files when voltage faults are detected by the system.

This system can analyze signal properties in time and frequency domains and calculate a number of time-domain and frequency-domain parameters as characteristic parameters for identifying signals and for preliminary overvoltage pattern recognition. Due to low sampling rates, the existing fault recorders cannot record transient overvoltages accurately and quickly. The high-speed recording and storage devices for transient waveforms can be used in this case with a sampling frequency of 40M, an automatic overvoltage trigger mode and a dual memory to store power frequency and high-frequency overvoltages in a variable frequency manner. This device is able to accurately record transient overvoltages and send them via the networks to the remote monitoring center where equipment failure causes are analyzed and corresponding prevention measures are proposed.

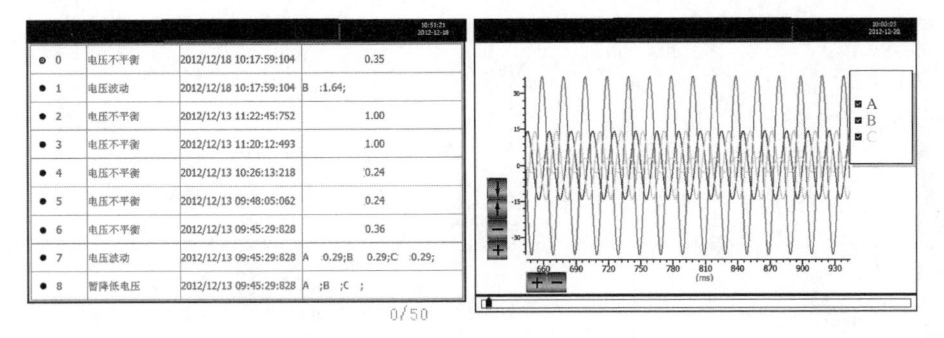

Figure 3.24 Fault logging and fault waveform recording.

3.7 Case Study of Online Overvoltage Measurements

The operation reliability of power systems is mainly judged by power outage numbers and blackout time. There are a variety of outage causes, which to a large extent depend upon the insulation levels and operating conditions of the equipment. No damage will be caused to the insulation of equipment under normal operating voltages. For various reasons, equipment insulation may still withstand overvoltages of different shapes, amplitudes and durations, and the presence of overvoltages may result in accidents. Once an accident happens, it is very difficult to figure out the causes, as there are too many of them. Therefore, overvoltage analysis in power system is exceedingly important; it can provide reliable and accurate information for the impact of overvoltage development on power grids and provide references for better accident treatment and countermeasures.

With the help of online monitoring on transient overvoltages and analysis of field data, overvoltage waveforms can be displayed, and overvoltage parameters including amplitude, duration, rising steepness and so on. can be obtained. Hence, the type of the overvoltages that the substation is subject to can be roughly judged, which provides a scientific monitoring basis for secured operation of the power system. The various overvoltages are grouped into external overvoltages and internal overvoltages based on general classification. In the following, analysis will be performed on typical waveforms acquired by the "TR2000 transient overvoltage monitoring and recording system". Overvoltage amplitude (or multiple), steepness and duration will be analyzed to recognize overvoltage types and find out the causes. In this way, a reliable basis for overvoltage prevention, performance prediction of equipment insulation and power system improvement will be provided, and essential conditions for further study on overvoltages will also be created.

3.7.1 Lightning Overvoltages

3.7.1.1 Direct Lightning Overvoltages

Figure 3.25 shows a set of lightning overvoltage waveforms acquired in some substation. It is demonstrated that the voltages of phases A, B and C rise suddenly at the time instant of point a and later stabilize quickly. After about five power frequency cycles, the three-phase voltages undergo sudden changes again at point b, which is subsequently followed by a voltage rise in phase A and simultaneous voltage drops of the same amplitude in phases B and C.

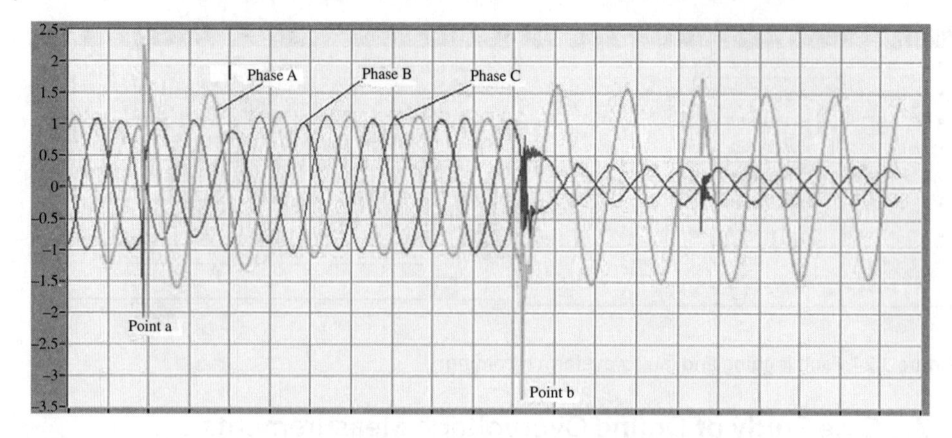

Figure 3.25 Three-phase waveform display of lightning overvoltages.

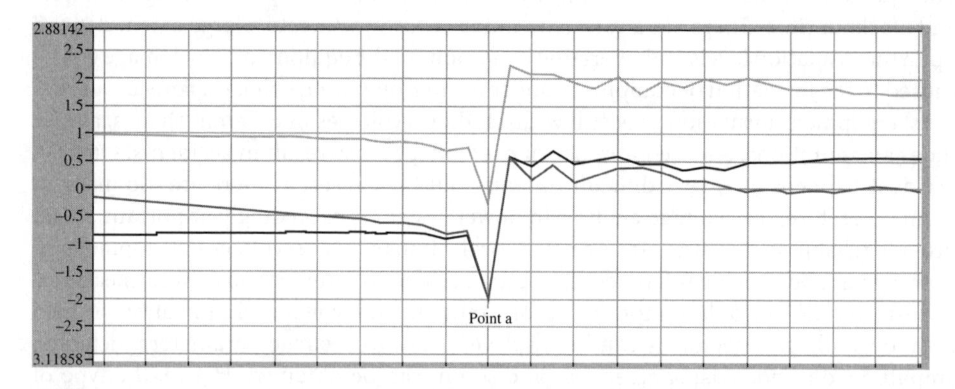

Figure 3.26 An enlarged view of the three-phase waveforms of lightning overvoltages at point a.

In order to have a clear understanding, the overvoltage waveforms at point a are amplified for further analysis. As shown in Figure 3.26, there are downward spikes in the fronts of the three-phase waveforms with the front steepness of phase A being the largest (387643.204 kV/s). It can be inferred that the A-phase line is hit by lightning strokes and the lightning wave propagates to the substation along the line. Overvoltages of similar wave shapes are induced on the other two phases. The time lags of the three phase fronts are very small, meaning that the lightning occurs at the line (bus) quite close to the substation. Since the rising time is large, it can also be considered as a switching overvoltage. The phase sequence has not changed on the whole.

After slight oscillations, the normal waveforms last for about five cycles until the second overvoltage occurs, as shown in Figure 3.27. During the high-frequency oscillations, phase A reaches 1.7 p.u., which is close to the line voltage; the other two phase voltages are less than 0.4 p.u.; the phase sequence of the three phases changes slightly (abnormal) with the voltages of phase B and C lagging about 30°. This situation lasts for 5.5 power frequency cycles. It can be inferred from the above observations that the lightning overvoltage occurs again at point b and is soon followed by a short-circuit of phases B and C.

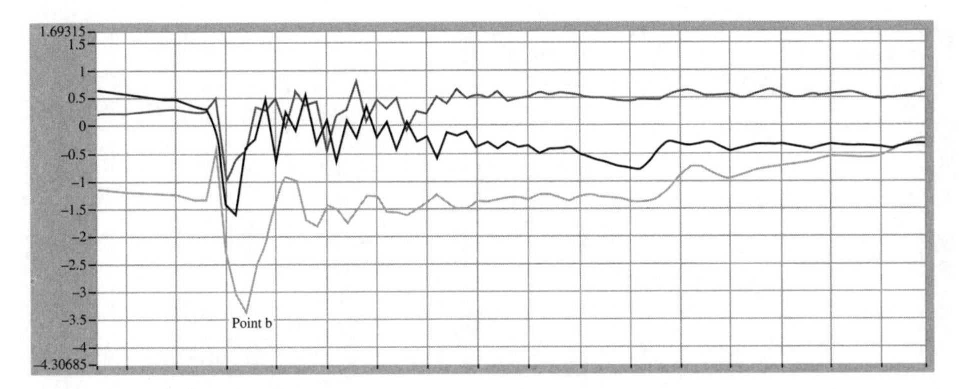

Figure 3.27 An enlarged view of three-phase waveforms of lightning overvoltages at point b.

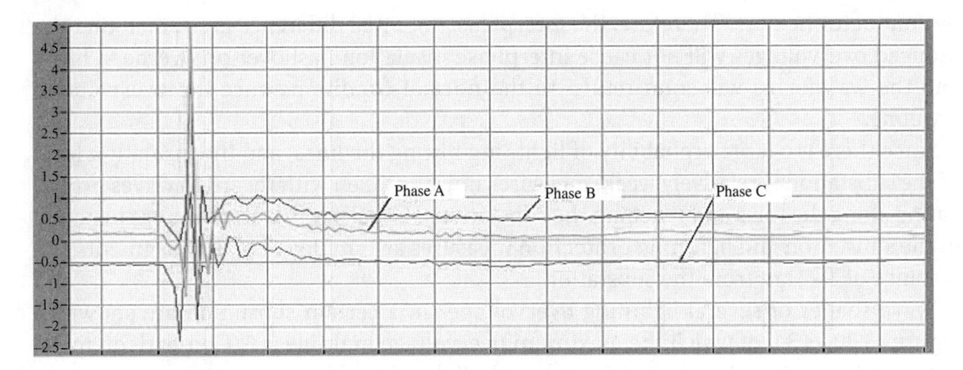

Figure 3.28 Waveforms of the direct lightning overvoltage.

Figure 3.28 shows a set of direct lightning overvoltage waveforms acquired in some substation. The three phase voltages are virtually stable before the overvoltage occurs. After the lightning strike, there are obvious upward spikes in the fronts of the three-phase waveforms; the rise time is about 6 μs, and phase A has the largest front steepness of 1.5605E+08 kV/s. The overvoltage multiples of phase A, B and C are 5, 4 and 2.3, respectively. The accident is inferred to be caused by lightning strokes to the A-phase line, while similar waveforms are induced on the other two phase lines. After the lightning strike, small oscillations appear, and the voltages recover quickly. The surge arresters protecting the substation operate in normal conditions so that lightning overvoltages stabilize to the power frequency voltage rapidly.

3.7.1.2 Induced Lightning Overvoltages

Another set of lightning overvoltage waveforms acquired in the substation are shown in Figure 3.29. It can be seen that the overvoltages in the figure are transient lightning overvoltages with the maximum value of 1.5 p.u. Overvoltage waveforms of the three phases are similar, and the oscillations are quite different from those of the previous group. The overvoltage in this case is probably an induced lightning overvoltage in the near zone of the substation so the figure does not show any sign of incoming surges. As induced lightning overvoltages are simultaneously induced on three-phase conductors, the differences between the overvoltage values of the conductors are merely caused

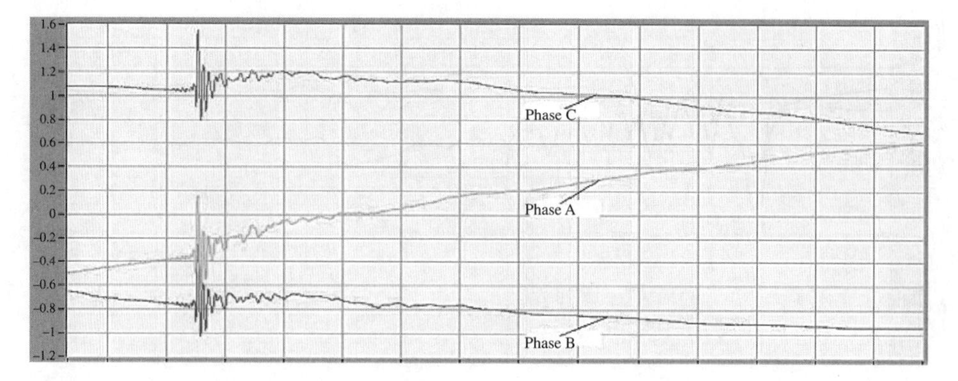

Figure 3.29 Waveforms of induced lightning overvoltages.

by conductor heights. Therefore, the interphase potential difference is so small that the induced overvoltages will not cause interphase insulation flashover on overhead lines, and the waveforms will soon return to the normal conditions once the overvoltages are gone.

Since there is a large probability of lightning hitting the lines, and the insulation level of the substation is relatively weak, measures must be taken with regard to waves propagating along the lines and reaching the substations. Usually, valve arresters are installed in the substation and lightning protection measures are employed on line segments at a distance of 1–2 km from the substation.

Wave shapes of several lightning overvoltages in a certain substation are shown in Figures 3.30–3.34, of which the maximum overvoltage multiple is 5. Comparison made between all measured wave shapes and the LLS data has shown consistency.

3.7.2 Power Frequency Overvoltages

Figure 3.35 shows a set of power frequency overvoltage waveforms acquired in a certain substation. In this figure, the phase sequence remains basically unchanged; the envelope curve of the three-phase voltages tends to be normal; the maximum overvoltage multiple is 1.5 p.u.

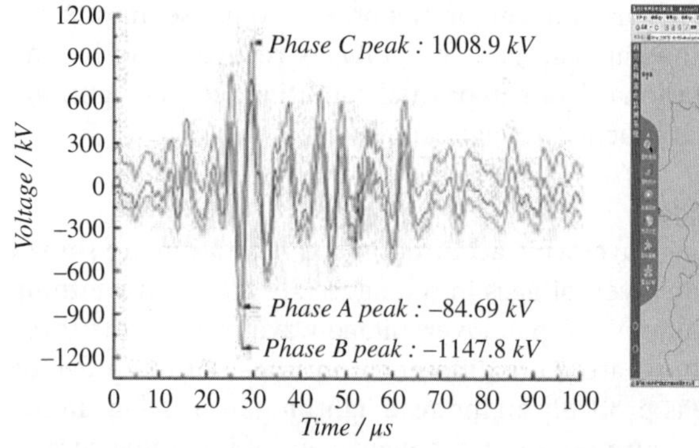

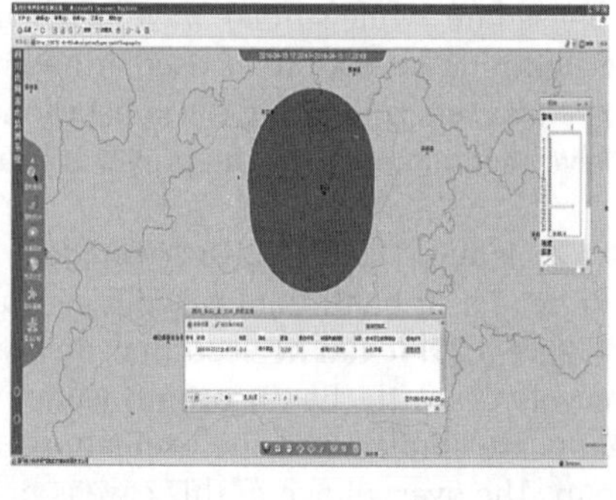

Figure 3.30 Waveforms of overvoltages on some line at instant 1. Peak: Phase A −847.69 kV, Phase B, −1147.8 kV, Phase C, 1008.9 kV.

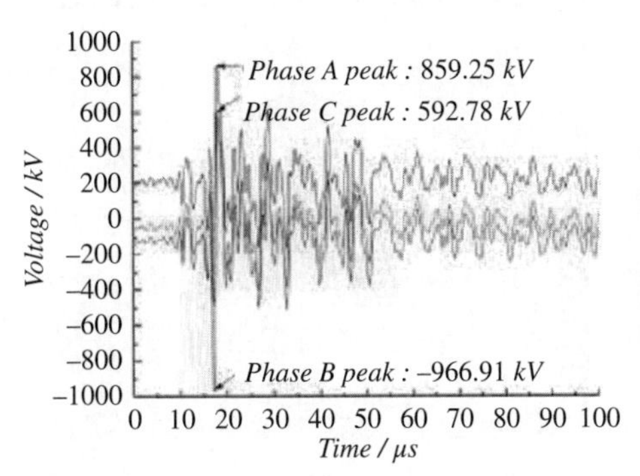

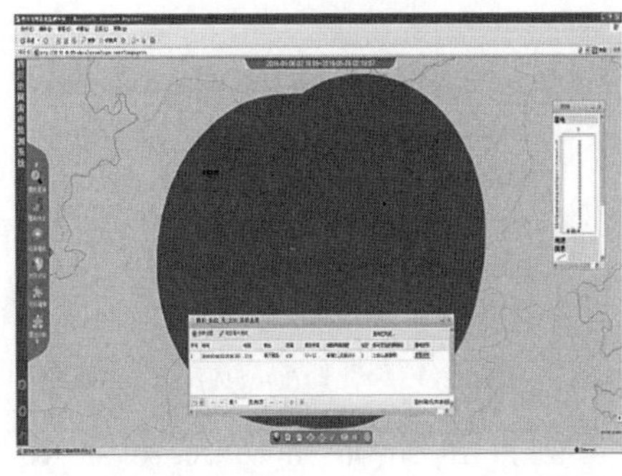

Figure 3.31 Waveforms of overvoltages on some line at instant 2. Peak: Phase A 859.25 kV, Phase B, –966.91 kV, Phase C, 592.78 kV.

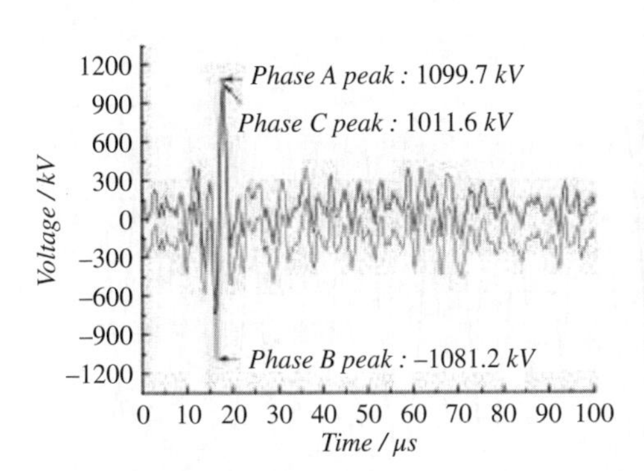

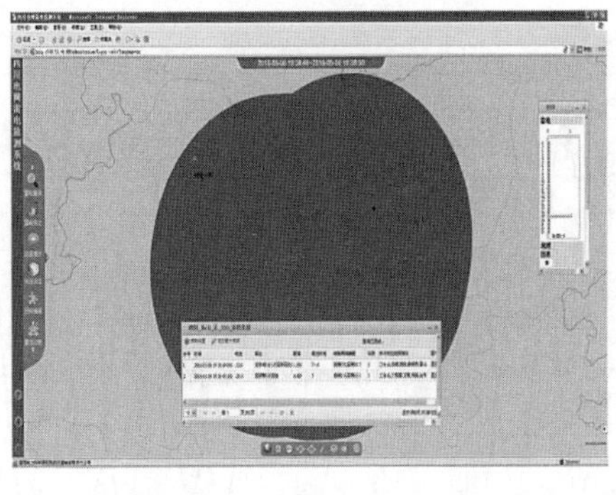

Figure 3.32 Waveforms of overvoltages on some line at instant 3. Peak: Phase A 1099.7 kV, Phase B, –1081.2 kV, Phase C, 1011.6 kV.

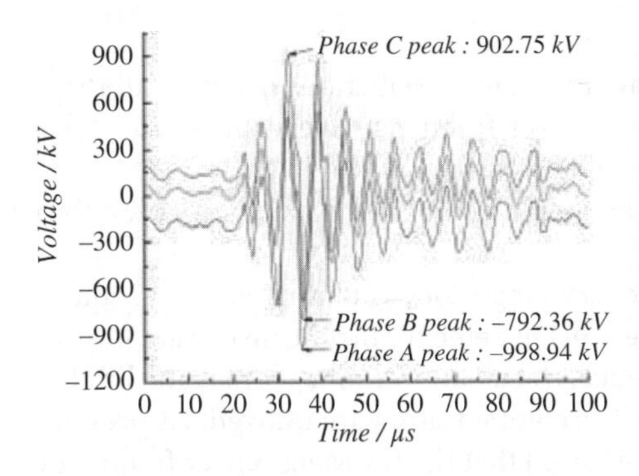

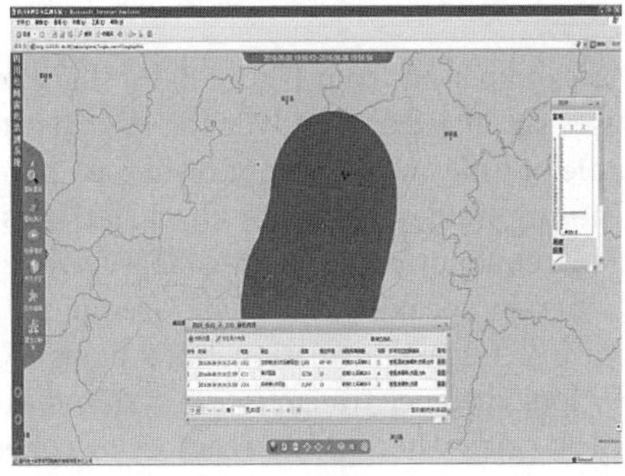

Figure 3.33 Waveforms of overvoltages on some line at instant 4. Peak: Phase A –998.94 kV, Phase B, –792.36 kV, Phase C,902.75 kV.

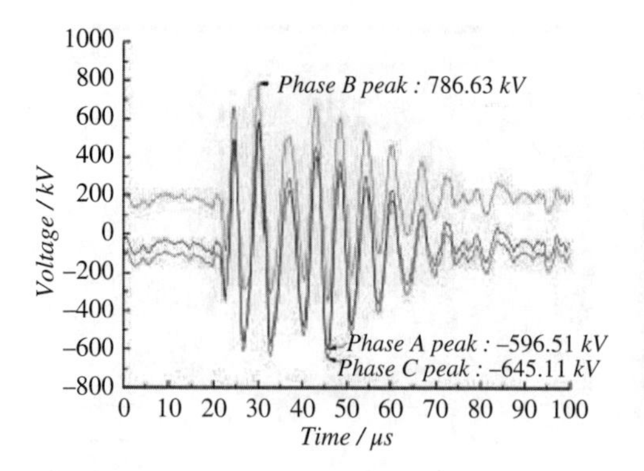

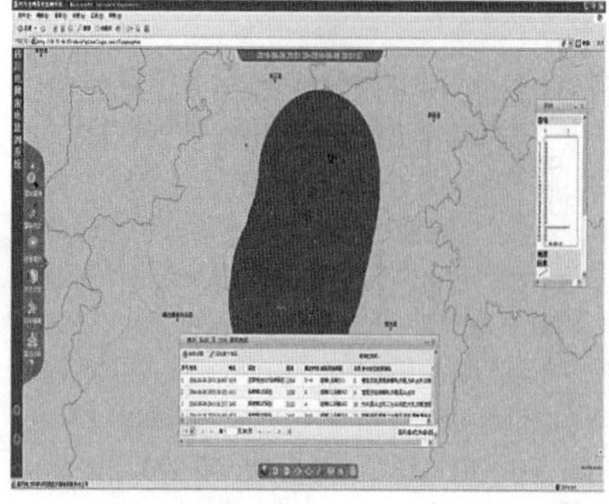

Figure 3.34 Waveforms of overvoltages on some line at instant 4. Peak: Phase A −596.51 kV, Phase B, 786.63 kV, Phase C −645.11 kV.

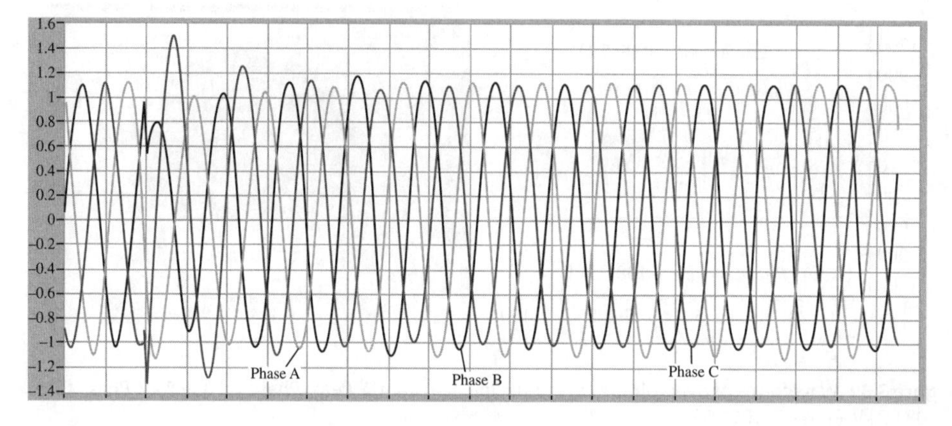

Figure 3.35 Power frequency overvoltage waveforms.

As can be seen from Figure 3.36, slight low-frequency oscillations occur to all three phases simultaneously. Over the next period, power frequency overvoltage rises for a couple of times, which is not discussed here in details.

The amplitude–frequency characteristics are depicted in Figure 3.37, in which only a 50 Hz frequency component is observed.

Generally speaking, the power frequency overvoltage does not harm electrical equipment with normal insulation, but it plays a key role in determining the insulation level of EHV long distance transmission systems. Hence, transient overvoltages must be limited in EHV grids where shunt reactors or other var compensators are utilized. At present, for power grids rated 500 kV, it is generally required that the transient power frequency voltage rise of the bus and of the lines must not exceed 1.3 times and 1.4 times the power frequency voltage, respectively. For 500 kV unloaded transformers, 1.3 p.u. power frequency voltage is allowed to last for 1 min, while for shunt capacitors, 1.4 p.u. power frequency voltage is allowed to last for 1 min.

The main causes of power frequency overvoltages include Ferranti effect of unloaded lines, unbalanced grounding faults and sudden load rejection of generators.

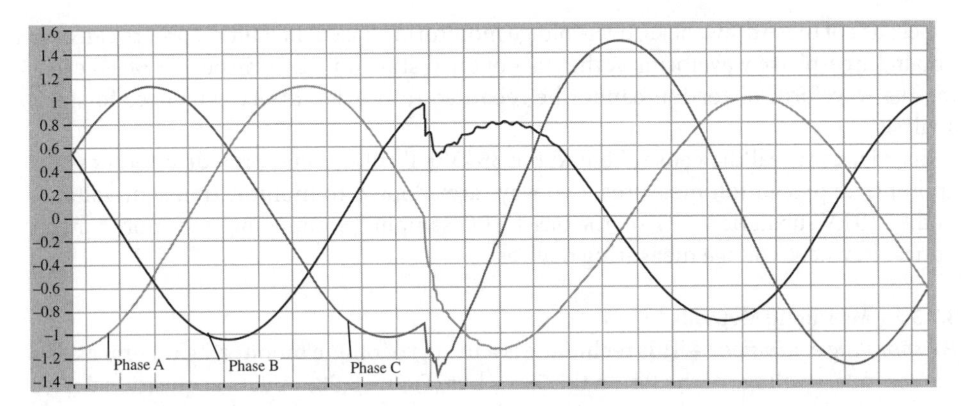

Figure 3.36 An enlarged view of the power frequency overvoltage waveforms.

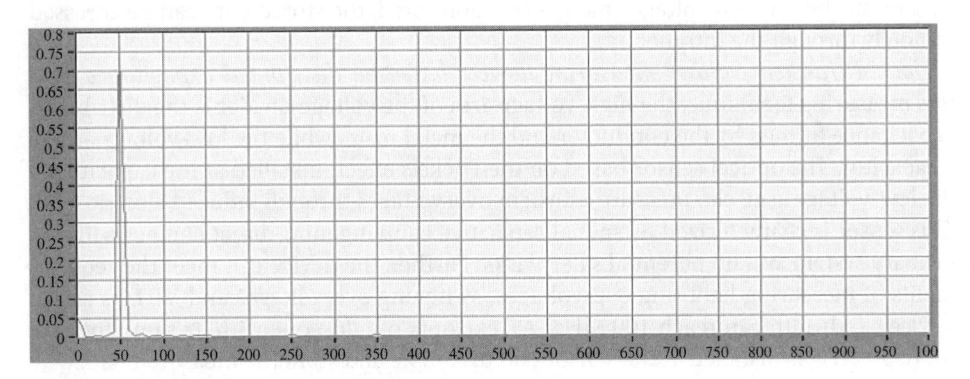

Figure 3.37 Amplitude–frequency characteristics.

3.7.3 Resonance Overvoltages

3.7.3.1 Case One

A 35 kV system of a substation is operating in the following conditions: the MV side of the main transformer is connected to a 35 kV bus via two 3 × 400 mm HV cables; the 35 kV system has only one outgoing line supplying a photovoltaic plant, and is composed of a 170 m overhead line and a 3 × 300 mm cable of 50 m; the PT primary winding is wye-connected and the secondary winding is wye-connected with an open delta; the high-voltage fuse model adopted is RW9-35/1A. When put into operation, the PTs of phases B and C on the 35 kV bus exploded, and the phase-B HV fuse was melted after replacing the PTs.

Previously, transient voltage monitoring in the 110 kV substation has merely relied on in-station fault recorders, which have the following two drawbacks:

1. The sampling rate of the fault recorder is only 5 kHz, namely recording five points per millisecond, not nearly enough to record microsecond transient voltages.
2. The fault recorder is mainly triggered by either switching operations or faults. The fault recorder is thus hardly able to record the transient voltages that do not result in faults.

Because of these drawbacks, overvoltage monitoring based on fault recorders can only obtain three-phase waveforms at the start of the distortion. The transient process causing the waveform distortion cannot be recorded to provide direct evidences for fault analysis.

In order to reveal the causes behind the 35 kV PT failures and provide references for appropriate precaution measures, two ways are adopted to monitor transient voltages in the 110 kV substation: "PT on the bus + EPO sampling equipment" and "non-contact optical transient voltage measuring devices".

3.7.3.2 Measuring Methods

PT sampling equipment – This technique is a means of online monitoring that measures the secondary voltage of the PT on the 35 kV bus (Figure 3.38). A data acquisition device having a high sampling rate of 40 MHz and multiple triggering modes such as "voltage transients", "three-phase voltage imbalance" is adopted to monitor the transient voltages on the bus. When voltage changes are monitored, the stored data can be accessed remotely through the intranet.

Optical transient voltage measuring devices based on distributed capacitance – In this measuring technique, the high-voltage arm of the voltage divider is the distributed capacitance formed by the conductor and the metal plate, while the low-voltage arm is a capacitor. The optical sensor based on the Pockels effect, installed at the capacitor of the low-voltage arm, can measure transient overvoltages through induced electric field. This sensor is characterized by spatial capacitive coupling, non-direct contact with the primary system and no potential safety risks. Further, the device can meet the requirements of measuring lightning overvoltages, switching overvoltages and VFTOs as the device bandwidth can reach 100 MHz. A Tektronix oscilloscope with its sampling rate set to 50 MHz is installed at the sensor end to sample and store the measured data.

Figure 3.38 Monitoring the secondary voltage of a PT on the 35 kV bus.

3.7.3.3 Measurement Results

Before the PT was damaged, ferro-resonance of the PT fundamental frequency took place on the two 70 m cables connecting the 35 kV outgoing line of the transformer and the bus, as shown in Figure 3.39.

Before the PT was damaged, no resonance eliminator was installed at the open delta of the secondary side. The ferro-resonance of the fundamental frequency lasted for 25 s until the PT fuse was melted, as shown in Figure 3.40.

In order to prevent PT faults, the two cables connecting the 35 kV outgoing line and the bus has to be replaced by a single cable, and resonance eliminators have to be installed at the secondary open-delta of the 35 kV PT. After changing the cable quantity (i.e. reducing the resonant circuit capacitance), ferro-resonance of fractional frequency still happens frequently while closing the PT for several times, but the current of PT and the overvoltage level are limited within the margin range (as shown in Figure 3.41), and PT faults and fuse melting would not be caused. After installing the secondary resonance eliminator, ferro-resonance of fractional frequency lasts for 3 seconds and then the system returns to normal (as shown in Figure 3.42).

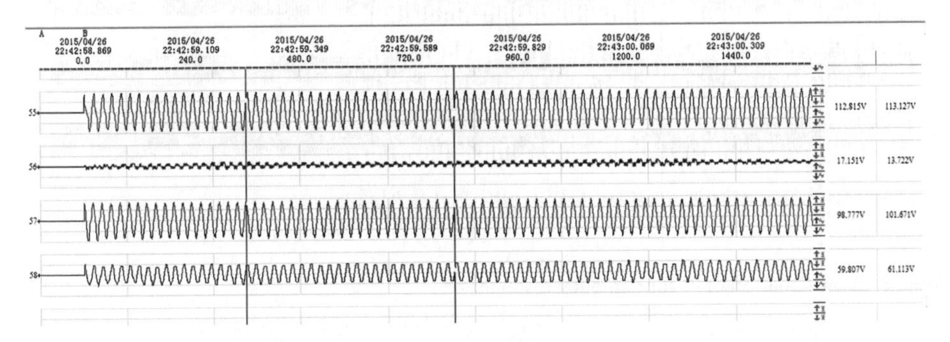

Figure 3.39 Waveforms of PT fundamental resonance.

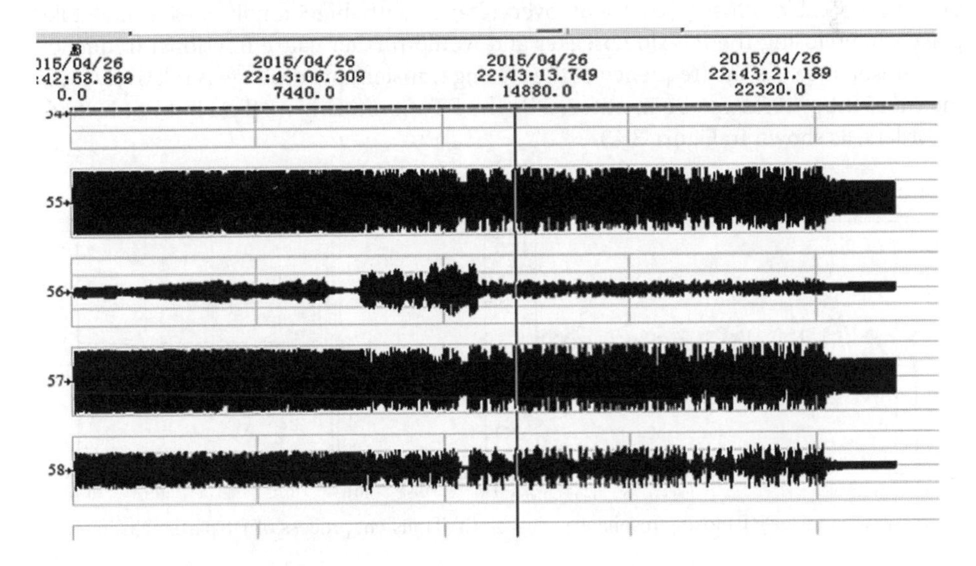

Figure 3.40 Attenuation of PT fundamental resonance.

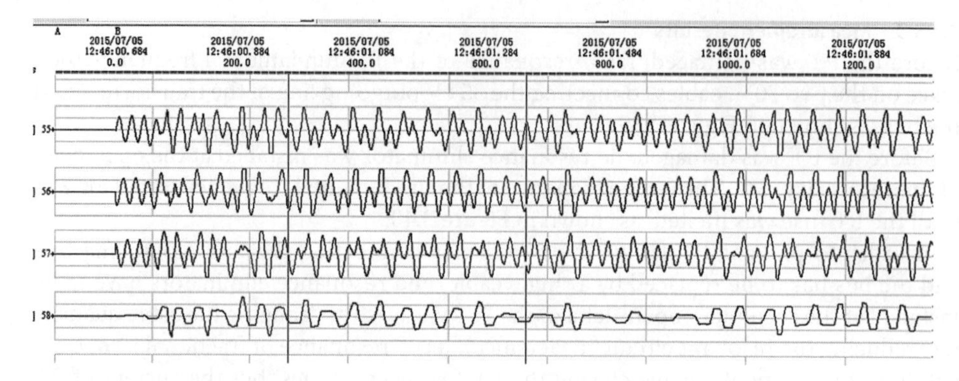

Figure 3.41 Waveforms of PT resonance of fractional frequency.

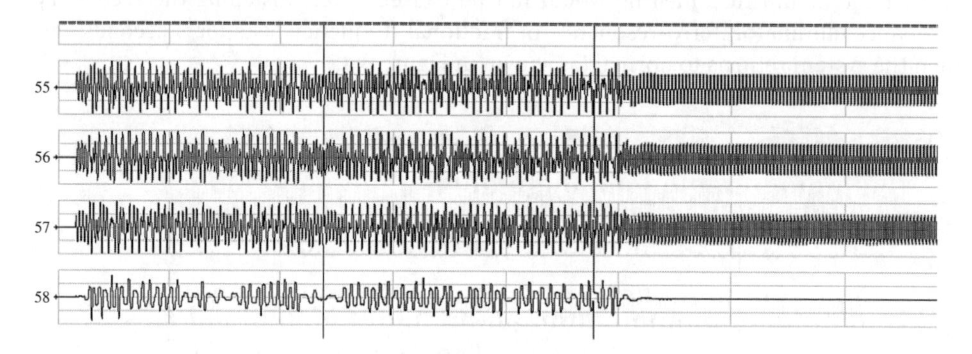

Figure 3.42 Attenuation of PT fundamental resonance.

After replacing the two cables connecting the 35 kV outgoing line and the bus by a single cable, closing transient overvoltages with high amplitudes would take place when closing the PT-side breaker and would further cause fractional frequency ferro-resonance. As the frequency of the closing transient overvoltage is relatively high, the transient process can be monitored by the PT monitoring method but not by fault recorders, as shown in Figure 3.43.

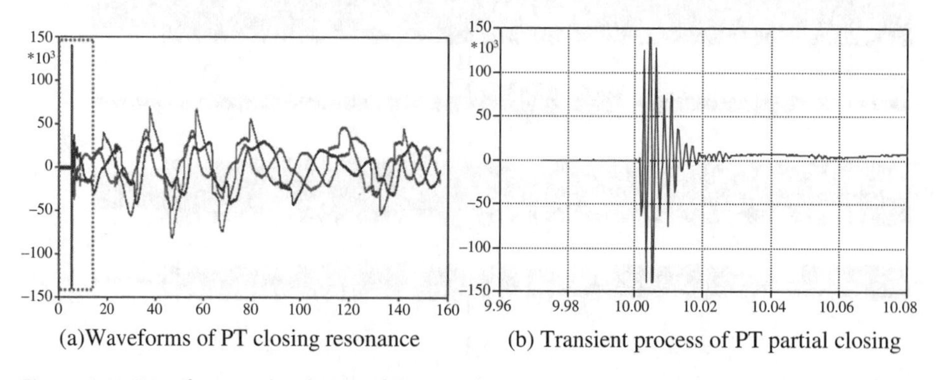

Figure 3.43 Waveforms at the closing of PT in a substation.

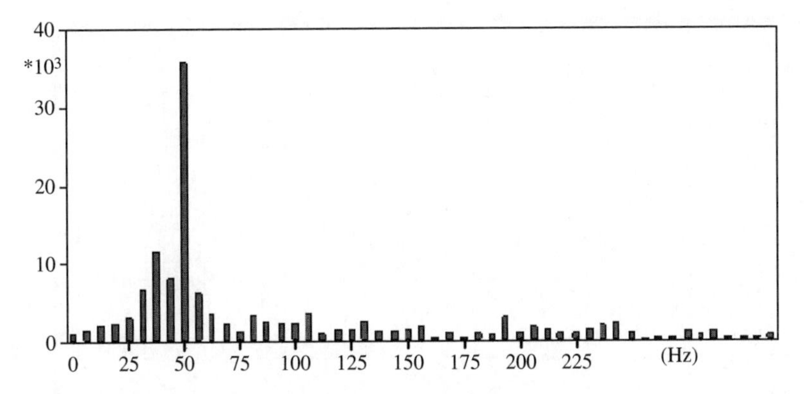

Figure 3.44 Spectrum curve of fractional frequency resonance.

Figure 3.44 shows the spectrum curve of fractional frequency resonance. The main distribution range of fractional frequency resonance is 25–50 Hz, which is different from that of fundamental frequency resonance with a single phase virtually grounded (which mainly consists of power frequency components).

3.7.3.4 Analysis

1. For the same PT circuit, ferro-resonance of multiple frequency requires the maximum electromagnetic transient disturbance, while ferro-resonance of fundamental frequency comes second, and ferro-resonance of fractional frequency requires the minimum amplitude. Correspondingly, this is also the case for the current and the overvoltage of PT. PT resonance in the substation is mainly caused by improper parameter coordination. The electromagnetic transient disturbance falls within the range of the fundamental frequency resonance, and the PT current and the overvoltage usually have large magnitudes. By changing the parameters, the magnitude of electromagnetic transient disturbance falls within the range of fractional frequency. Although smaller electromagnetic transient disturbance is more likely to cause resonance, the magnitudes of the PT current and the overvoltage are both within the margin.
2. The resonance eliminator installed at the secondary side works well for eliminating the resonance within the PT internal circuit, but it does not play a significant role in eliminating resonance of fundamental or fractional frequency.
3. A necessary condition for the PT resonance is the presence of transient electromagnetic disturbances that would cause iron core saturation. Fault recorders cannot record transient electromagnetic disturbances, so online devices for monitoring these transient processes are needed which can monitor the disturbance sources and magnitudes so as to predict the types of resonance likely to occur, to configure circuit parameters reasonably and to restrict PT currents and overvoltages to ensure normal equipment operation.

3.7.3.5 Case Two

Figure 3.45 shows a set of overvoltage waveforms acquired in a certain substation. According to historic data, these very typical resonance overvoltages last around a

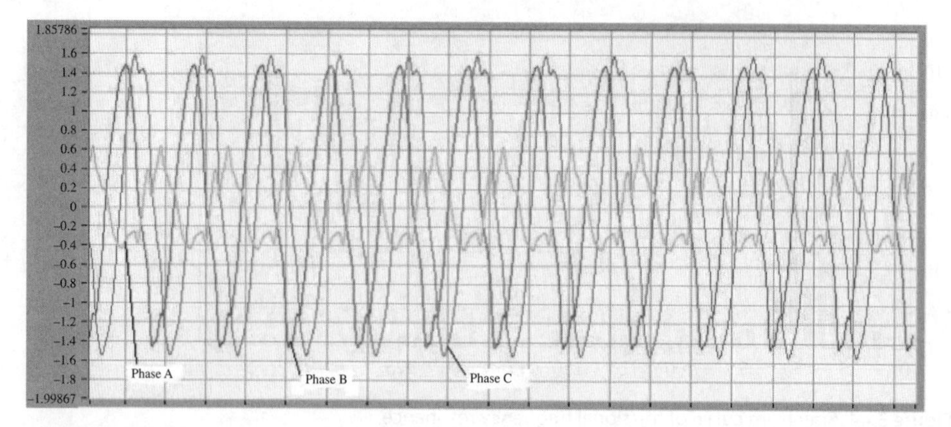

Figure 3.45 Resonance overvoltage waveforms.

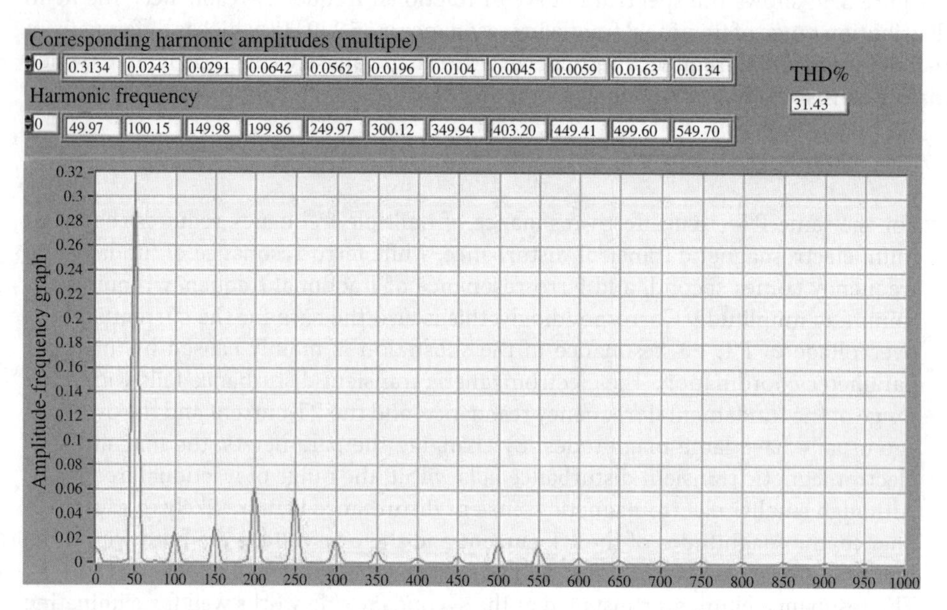

Figure 3.46 Harmonic analysis diagram (phase A).

total of 50 minutes. As shown in Figure 3.46, in the frequency spectrum of phase A, there are obvious spikes in the waveforms of 2, 3, 4, 5, 6, 7, 9 and 10 times frequency, and among them, the magnitude of 4 and 5 times frequency component is larger than that of others. This is also true for phases B and C. As can be seen from the figure, the time-domain waveforms are deformed; the A-phase magnitude is 0.625 times p.u. and the magnitudes of the other two phases are close to the line voltage, about 1.6 times p.u. Before these overvoltages occur, the recording device has not been triggered.

Resonance is a steady-state phenomenon, and the hazard level of resonance overvoltages depends on both the magnitude and duration. Because of the long duration, resonance overvoltages can cause great harm. They would not only endanger the insulation of electrical equipment, but also produce sustained overcurrents which can cause equipment burnout. Also, they may affect the working conditions of overvoltage protective devices. Operating experience has demonstrated that resonance overvoltages can be found in power grids of a variety of voltage levels, especially in the grids rated 35 kV and below where resonance is a concern due to many resulting accidents. Before grid design and operation, assessment and arrangement are essential to prevent resonance or to eliminate the conditions under which resonance can exist.

Internal ferro-resonance may easily be caused by external overvoltages when resonance overvoltages occur inside PTs. This kind of overvoltage can be eliminated by employing resistance-type resonance eliminators at the PT secondary winding. When resonance overvoltages are caused by inappropriate coordination of external parameters, installing resistive resonance eliminators at the PT secondary winding will be of no use. In this case, parameters of the external system have to be re-coordinated to eliminate the resonance.

3.7.4 Statistical Analysis of Overvoltages in a 110 kV Substation

Online monitoring is performed to a 110 kV substation to address equipment failures on the 35 kV side. Overvoltage occurrence is recorded to provide a basis for analyzing accident causes and for developing preventive measures. The monitored data over a period of time will be analyzed next.

Switching overvoltages and resonance overvoltages occur frequently in this substation according to the monitored data. Combined with the typical waveforms, statistical analysis is done on several aspects, including the multiple, frequency and duration of the overvoltages.

3.7.4.1 Typical waveforms

The typical waveforms of the switching overvoltages and resonance overvoltages monitored in this substation are shown in Figures 3.47 and 3.48, respectively. The vertical axis represents the overvoltage multiple with the power frequency voltage peak during normal operation taken as the reference value, and the horizontal axis represents the time.

Waveforms of switching overvoltages – This waveform belongs to a switching overvoltage which happened at 14:46:28 on July 25th, 2015. All three phases oscillated because of the switching operations. Among others, overvoltages happened on phases A and B with the multiple being 1.6 times and 1.49 times, respectively.

Waveforms of resonance overvoltages – The monitored waveforms of resonance overvoltages are shown in Figure 3.48, which are of two types, ones with transient impulses and ones without. For resonance overvoltages with transient impulses, the magnitudes of the phase voltage and the line voltage can reach 2.37 times the normal operating voltage, and the oscillation time of the transient impulse is 3–5 power frequency cycles, as shown in Figure 3.48(b). Regardless of the presence of transient impulse, the resonance would last a total of 1 s or so. It is demonstrated by using spectrum analysis that the low-frequency resonance components center around 10 Hz, i.e. a one-fifth frequency resonance. Moreover, the line voltage magnitude remains essentially unchanged throughout the low-frequency resonance.

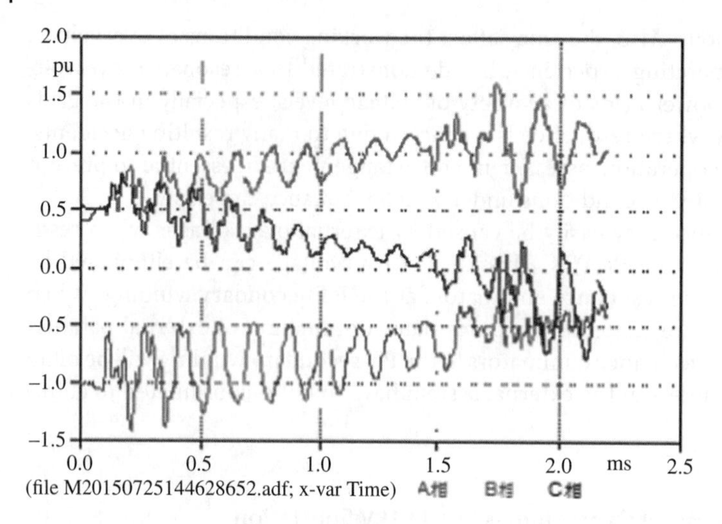

Figure 3.47 Waveforms of switching overvoltages.

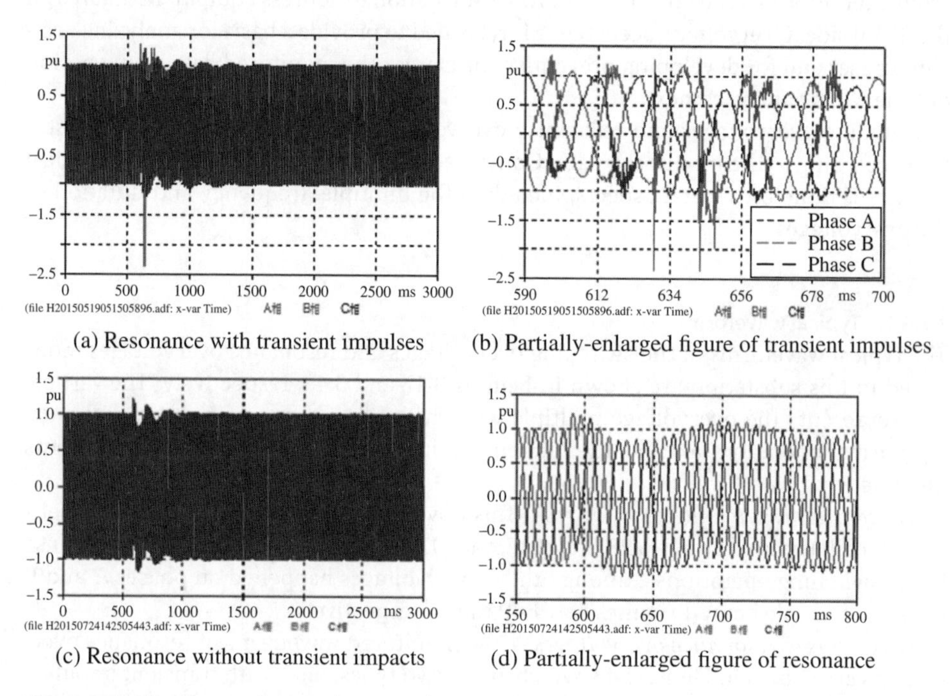

(a) Resonance with transient impulses
(b) Partially-enlarged figure of transient impulses
(c) Resonance without transient impacts
(d) Partially-enlarged figure of resonance

Figure 3.48 Waveforms of resonance overvoltages.

3.7.4.2 Data Statistics and Analysis

From the statistical data, it is known that switching overvoltages occur most frequently with a record of 317 times over 12 days, i.e. about 27 times daily. In terms of the overvoltage multiple, about 70% of the switching overvoltage magnitudes center on the range of 1.1–1.49 times the power frequency peak, and the percentage of the overvoltages whose magnitudes fall within 1.5–1.99 times the power frequency peak is 27.7%. The

Table 3.9 Statistics of switching overvoltages from May 2015 to July 2015.

Time	Overvoltage type	Occurrence times in each multiple range			Total number of times	Maximum overvoltage multiple
		1.10–1.49	1.50–1.99	2-		
2015/5/18		13	10	1	24	2.06
2015/5/19		13	7	2	22	2.37
2015/5/21		1	2	0	3	1.58
2015/7/19		10	3	0	13	1.89
2015/7/20		8	3	0	11	1.61
2015/7/21	Switching overvoltages	7	6	0	13	1.92
2015/7/22		28	10	0	38	1.82
2015/7/23		55	16	4	75	2.37
2015/7/24		41	18	0	59	1.99
2015/7/25		27	8	1	36	2.06
2015/7/26		15	4	0	19	1.53
2015/7/27		3	1	0	4	1.46
Aggregate		221	88	8	317	2.37

number of overvoltages with magnitudes exceeding double the power frequency peak is 8, according to the fault recorder, of which the maximum overvoltage magnitude is 2.37 times the power frequency peak. Table 3.9 shows the frequency distribution of overvoltages in each period and multiple range.

In addition to switching overvoltages, another kind of overvoltage frequently monitored is the resonance overvoltage, which happened 252 times during the nine counting days, i.e. 28 times daily. Distribution of the resonance overvoltage magnitude is nearly the same as that of the switching overvoltage. Still, it centers mostly within 1.1–1.49 times the power frequency peak and next in the range of 1.5–1.99 times the power frequency peak, while the number of the overvoltages with the magnitudes of double the power frequency peak is relatively small. The proportion of the overvoltages in each multiple range is 80%, 17.8% and 2.2%, respectively, and the maximum magnitude is 2.37 times. According to the resonance overvoltage waveforms, most are with transient impulses. Therefore, the statistical characteristics of the switching overvoltage and the resonance overvoltage are largely consistent. Table 3.10 shows the frequency distribution of overvoltages in each period and multiple range.

3.7.4.3 Conclusions

1. The monitored overvoltages are found of two main types, switching overvoltages and resonance overvoltages.
2. Switching overvoltages and resonance overvoltages occur frequently, and are consistent statistically. The resonance overvoltages are mainly caused by overvoltages arising from switching operations and system parameter changes.
3. The monitored resonance overvoltage mainly comes from 1/5 frequency resonance with a duration of nearly 1 s.

Table 3.10 Statistics of the resonance overvoltages from May 2015 to July 2015.

Time	Overvoltage type	Occurrence times in each multiple range		Total number of times	Maximum overvoltage multiple
		1.10–1.49	1.50–1.99	—	
2015/5/18	Resonance overvoltages	22	10	33	2.06
2015/5/19		31	7	40	2.37
2015/5/21		3	2	5	1.58
2015/7/22		24	3	27	1.64
2015/7/23		48	11	60	2.03
2015/7/24		41	6	47	1.83
2015/7/25		22	2	24	1.66
2015/7/26		12	3	15	1.78
2015/7/27		0	1	1	1.66
Aggregate		203	45	252	2.37

4. Although the multiples of the overvoltages of these two types are less than 1.5 times, they occur at a high frequency, and the resonance overvoltage even lasts for up to 1 s. Therefore the device insulation may be damaged if subject to these overvoltages for a long time.

3.7.5 Single-Phase Grounding Overvoltages

Figure 3.49 shows a set of overvoltage waveforms in a single-phase grounding accident in some substation. Obviously, phase B is grounded and the voltages of phases A and C rise to 1.85 times the line voltage. The overvoltage lasts about 20 minutes.

Single-phase grounding is the major form of fault in the power grid. In the grid with ungrounded neutrals, the single-phase grounding fault will not alter the symmetry of the three-phase voltages of the transformer winding, and the grounding current is generally not large. So there is no need to immediately de-energize the lines and cut off the power supply to users. With the help of the grounding indicators, operation personnel can locate and handle the faults in time, which greatly improves the reliability of the power supply.

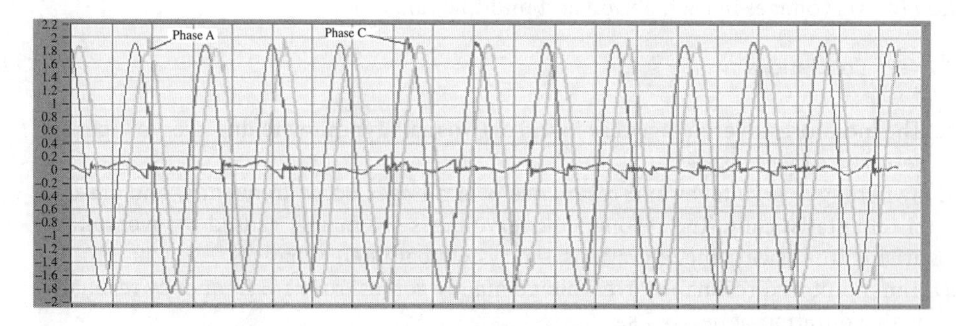

Figure 3.49 Waveforms of the single-phase grounding overvoltage.

3.7.6 Two-Phase Grounding Overvoltages

Figure 3.50 shows a set of overvoltage waveforms in a two-phase grounding accident in some substation. Obviously, the voltages of phase A and phase B are grounded and the voltage of phase C rises to 1.18 times the line voltage.

3.7.7 Intermittent Arc Grounding Overvoltages

Figure 3.51 shows a set of waveforms monitored in some substation with oscillation in the wavefront, which greatly resembles that of the typical arc grounding overvoltage. Figures 3.52 and 3.53 are enlarged views of phase A and phase B, respectively. As can be seen from the figures, the waveforms of these two phases are very similar. The maximum overvoltage multiples are both around 3.5 times and grounding both occurs at phase C (Figure 3.54).

As the magnitude of the intermittent arc grounding overvoltage is not too large, normal equipment in modern grids with ungrounded neutrals is able to withstand it because of the large insulation margin. However, major threats will be presented to aging equipment with poor insulation as well as transmission lines with insulation weakness due to long duration of this overvoltage. This means that the protection of arc grounding overvoltages relies on excellent insulation of electrical equipment. Hence, it is necessary to perform regular preventive tests and overhauls, and to monitor and maintain the equipment in operation.

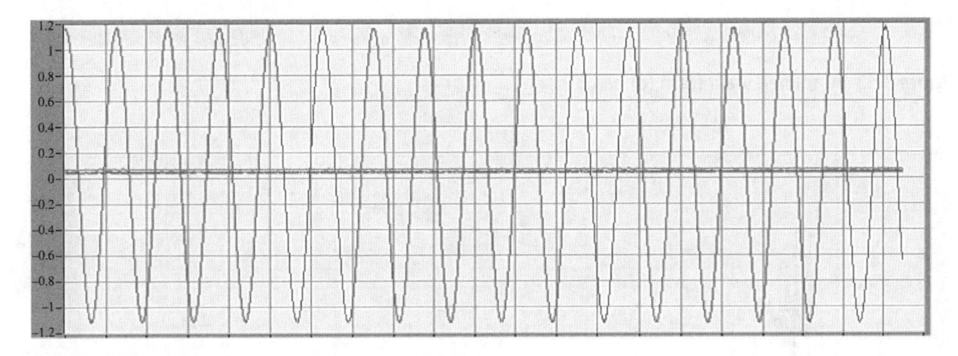

Figure 3.50 Waveforms of the two-phase grounding overvoltage.

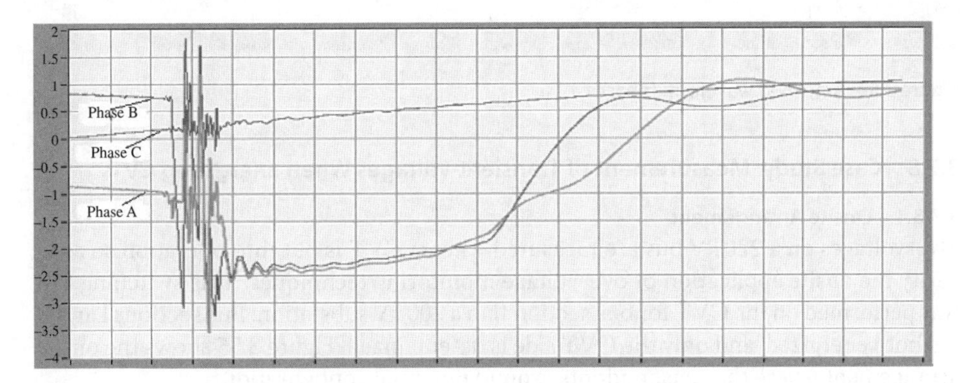

Figure 3.51 Intermittent arc grounding overvoltages.

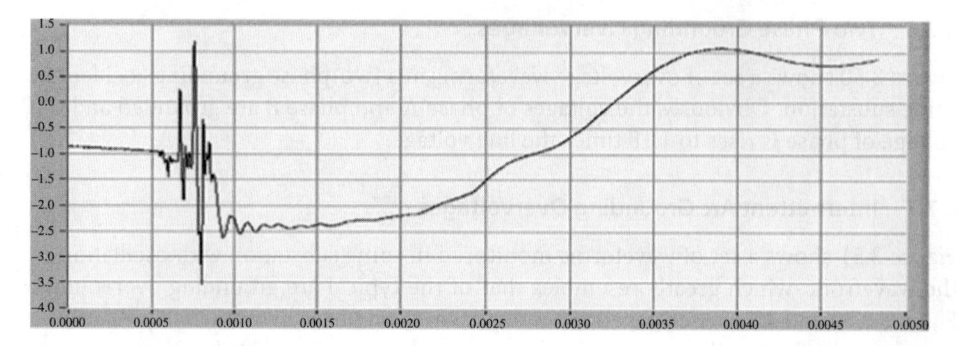

Figure 3.52 Amplified waveform (phase A).

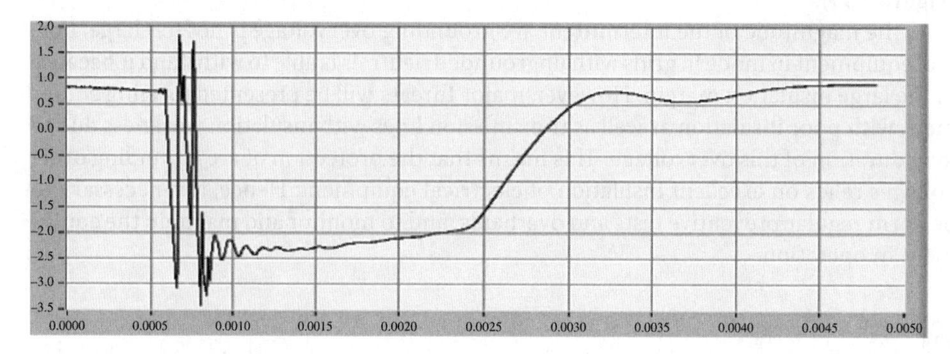

Figure 3.53 Amplified waveform (phase B).

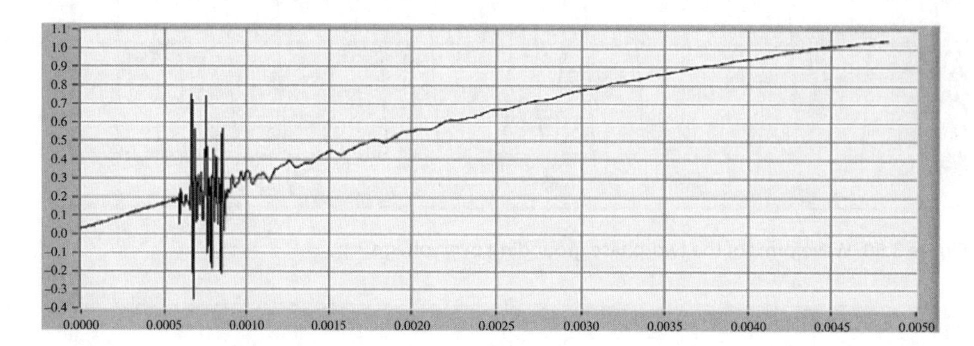

Figure 3.54 Amplified waveform (phase C).

3.7.8 Case Study: Measurement of Transient Voltages When Energizing CVTs

3.7.8.1 Onsite Arrangement

Overvoltages on a 220 kV bus are measured when a CVT is put into operation so as to verify the onsite application of overvoltage monitoring techniques. This switching test was performed on the CVT for bus section II in a 200 kV substation. Bus sections I and II are both energized, and only the CVT side is de-energized. Figure 3.55 shows the onsite arrangement where the sensor adopts ground potential configuration.

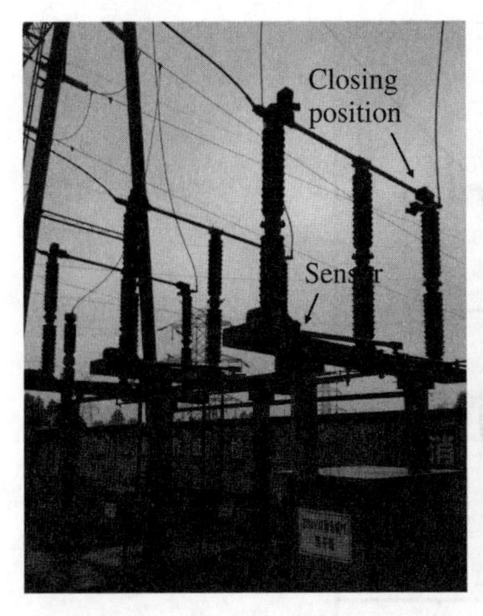

Figure 3.55 Onsite arrangement of the sensor.

3.7.8.2 Measurement Results

Transient overvoltages caused by switching CVTs – The measurement results acquired by onsite field monitor during the switching process of the CVT on the 220 kV section II bus are given in Figure 3.56. It is demonstrated by the measurement that before the energization of the CVT via the isolating switch, the three field sensors all record power frequency fields of small magnitudes as bus section I and section II are energized. Even if the isolating switch is opened, fields would be generated by other energized conductors. Fields of this kind are characterized by constant low amplitudes and power frequency, and can be deemed as background noise. The waveforms after background noise treatment are as shown in Figure 3.57.

Figure 3.57 shows the concave parts on the field wave shapes, and this is inconsistent with the physical fact that the voltage has to change continuously and slowly when the capacitor is charged. This is caused by the effect of the voltage saltation of the other phase voltages on a phase voltage. Figure 3.58 shows the three phases in the one picture. It can be seen that when overvoltage rising or descending edges appear in the wave shape of phase B, the other two phases are impacted and exhibit apparent jumps. On the other hand, when wave shapes of phase A and phase C encounter rising or descending edges, phase B produces a jump as well.

The above analysis has demonstrated that the measured field is the coupling field of the three phase conductors as there is distance between the field sensor and the conductor to be measured. The relative three-phase coupling matrix M can be calculated using the saltation in the waveforms, and the measurements are able to be decoupled.

$$\begin{bmatrix} U_{\text{S-a}} \\ U_{\text{S-b}} \\ U_{\text{S-c}} \end{bmatrix} = \begin{bmatrix} 1 & \dfrac{k_{\text{Ba}}}{k_{\text{Bb}}} & \dfrac{k_{\text{Ca}}}{k_{\text{Cc}}} \\ \dfrac{k_{\text{Ab}}}{k_{\text{Aa}}} & 1 & \dfrac{k_{\text{Cb}}}{k_{\text{Cc}}} \\ \dfrac{k_{\text{Ac}}}{k_{\text{Aa}}} & \dfrac{k_{\text{Bc}}}{k_{\text{Bb}}} & 1 \end{bmatrix} \begin{bmatrix} U_{\text{S-Aa}} \\ U_{\text{S-Bb}} \\ U_{\text{S-Cc}} \end{bmatrix} = \mathbf{M} \begin{bmatrix} U_{\text{S-Aa}} \\ U_{\text{S-Bb}} \\ U_{\text{S-Cc}} \end{bmatrix} \tag{3.2}$$

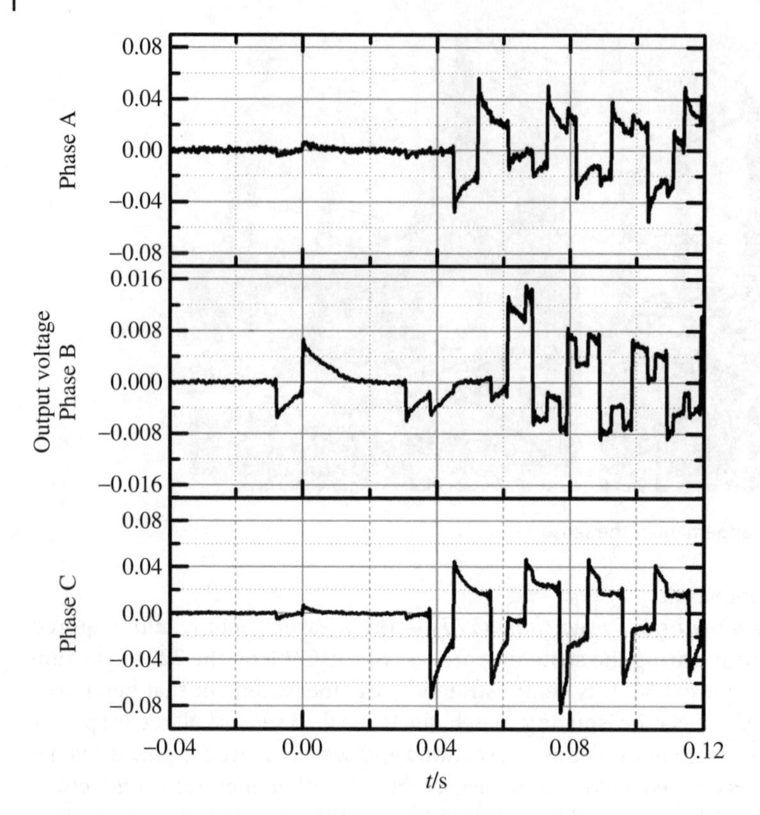

Figure 3.56 Sensor outputs upon closing the switch.

where M is the matrix of relative three-phase coupling coefficients; $U_{S\text{-}Aa}$, $U_{S\text{-}Bb}$, $U_{S\text{-}Cc}$ are the outputs of the transient monitoring device under the individual effects of U_A, U_B and U_C with their wave shapes showing the same changing trends as U_A, U_B and U_C. Once M is solved, the wave shape of the conductor voltage of each phase can be acquired by decoupling the three-phase measurements. For this reason, the solution of the coupling coefficient matrix is the key to decoupling the complex fields of three-phase conductors.

Taking Figure 3.59 as an example, the waveform of phase B shows an apparent drop at instant T_1 which leads to downward saltation of waveforms of phases A and C. The saltation of phase A and phase C is entirely caused by waveform changes of phase B, as this is the instant when the overvoltage occurs on phase B. Assuming that the waveform change of phase B is unit 1, the waveform changes of phase A and phase C are 0.98 and 0.52, respectively according to the relative changes in waveform amplitudes. By adopting this principle, the relative coupling coefficients between phases can be worked out.

The matrix of relative coupling coefficients M could be solved via the above method:

$$M = \begin{bmatrix} 1 & 0.98 & 0.06 \\ 0.07 & 1 & 0.09 \\ 0.05 & 0.52 & 1 \end{bmatrix} \tag{3.3}$$

The waveform of the decoupled field is shown in Figure 3.60.

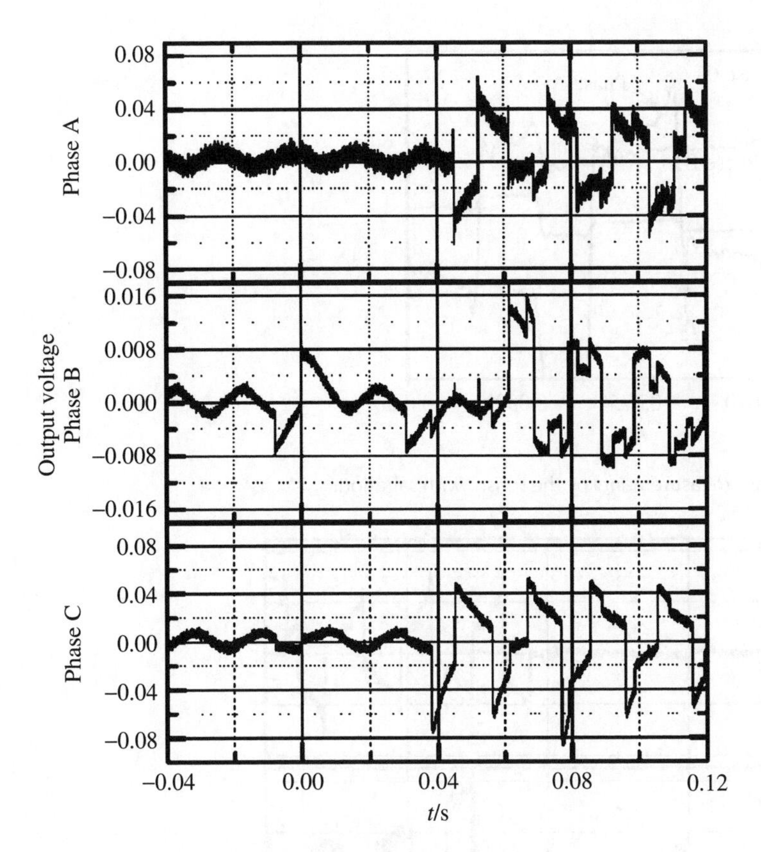

Figure 3.57 Waveforms after background noise treatment.

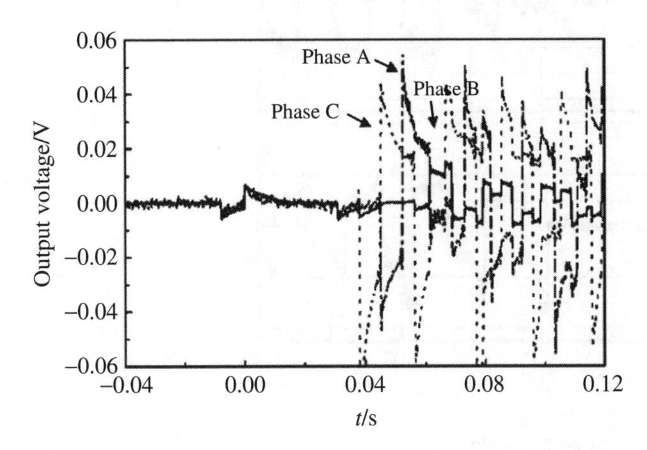

Figure 3.58 Three-phase sensor outputs after subtracting the background noise.

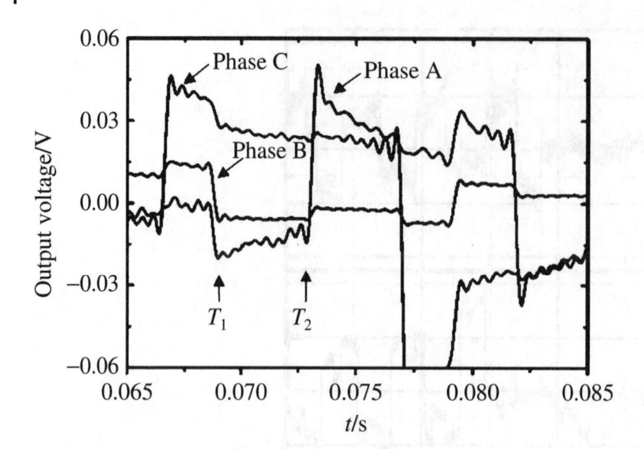

Figure 3.59 Three phase measurements by the sensor with saltation.

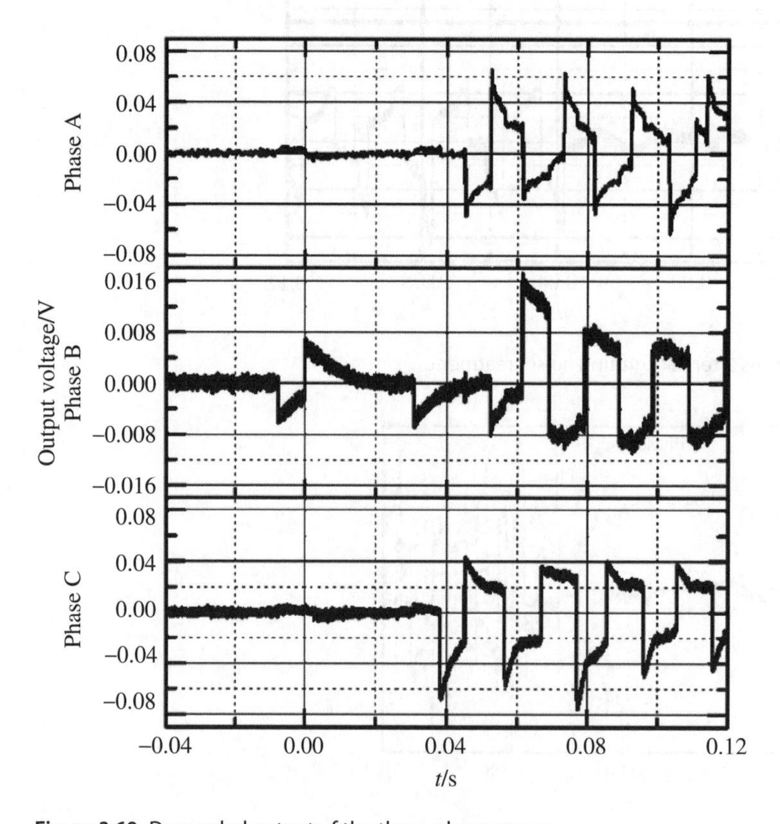

Figure 3.60 Decoupled output of the three-phase sensor.

From the perspective of the operating process, a CVT can be deemed as a capacitor. The equivalent circuit during switching is as illustrated in Figure 3.61 where C, the equivalent capacitance of the CVT, is approximately 1 nF; R, the equivalent bus impedance, is less than 1 ohm; S_{AC} is the equivalent bus supply with a 220 kV phase voltage rms value and a 180 kV single-phase voltage peak; S_w, the equivalent switch,

Figure 3.61 Equivalent circuit of the switching process.

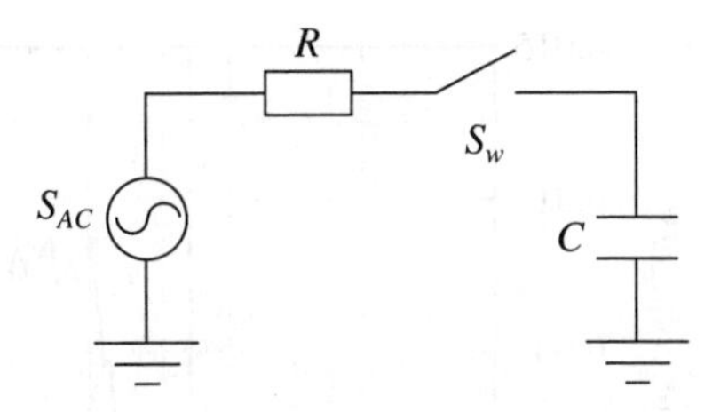

is closed when the gap breaks down due to the voltage between contacts, and opened when the arc current is unable to sustain the arc. According to the equivalent bus impedance and CVT capacitance, the time constant for charging the CVT after contact breakdown is estimated to be in the order of 1 ns. Compared to the power frequency cycle, this charging time can be ignored. Besides, after the breakdown of the contacts, the arc current, estimated to be less than 0.06A based on the calculation of the CVT's equivalent capacitance and the supply voltage, cannot sustain the continuous burning of the arc.

Based on the above discussions, the switching of CVTs can be described by the following stages:

1. The switch contacts gradually become closer, and the voltage across the contacts is the bus alternating voltage U_s.
2. When the contact gap is not large enough to withstand U_s, the gap breaks down, and the bus voltage charges the CVT; the terminal voltage of the CVT rises to U_s instantaneously. At this instant, the potential of the bus-side contact is U_s as well.
3. As the sustained current is not large enough, the arc is quenched, and the bus-side potential alternates with the bus voltage. The CVT gradually releases the charge via stray capacitance, and the potential at the CVT side thus decreases from U_s.
4. The CVT-side potential continues to decrease, and the bus-side potential alternates to the opposite polarity; the potential difference between the contacts will result in gap breakdown when the bus-side potential reaches the gap breakdown voltage.
5. The arc is quenched again; the CVT-side potential decreases, and the bus-side potential alternates with the bus voltage. Stages (4) and stage (5) will be repeated until the switch is closed.

The above analysis agrees with the measurement results in Figure 3.59.

The overvoltage of phase B is as shown in Figure 3.62 where there are two rising edges. The rise time of the two edges are 14.5 μs and 18 μs, respectively. The rise time and decay time for the other overvoltages, by statistics, range between 10 and 20 μs.

Field waveforms under power frequency – The waveforms under power frequency, after the CVT is put into operation, are collected in order to analyze the overvoltage magnitudes and multiples. Similarly, the three phases would exert influences on one another. Decoupling can be performed using the method described above. The decoupled waveforms are shown in Figure 3.63.

Calculating overvoltage multiples are as follows:

Channel 1, phase B: background power frequency voltage: peak-to-peak value, 0.00365 V; in-service power frequency voltage peak-to-peak value, 0.027 V; the maximum overvoltage value: 0.017 V.

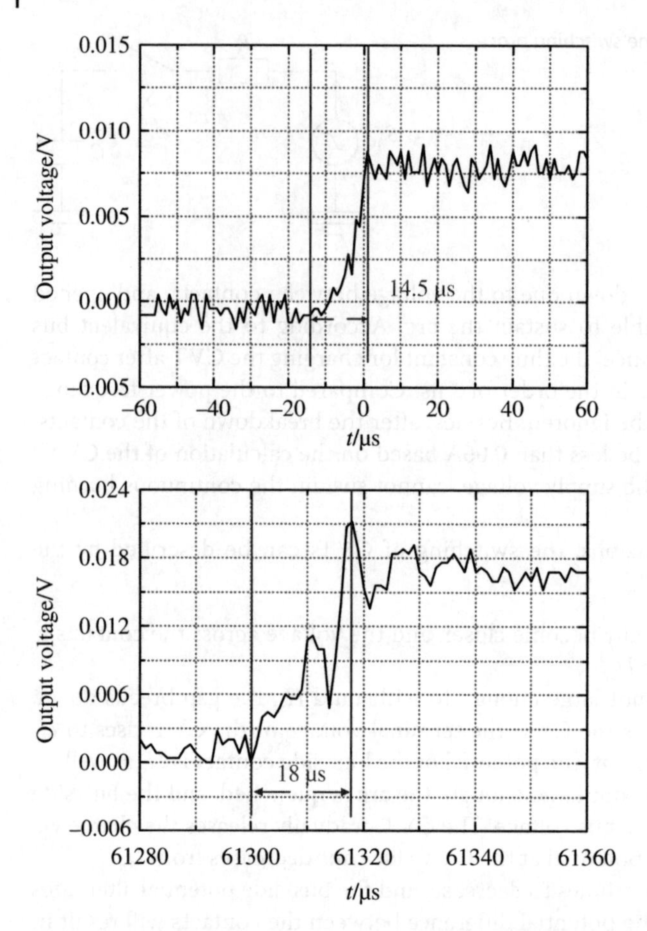

Figure 3.62 Rising edges of the sensor output waveform.

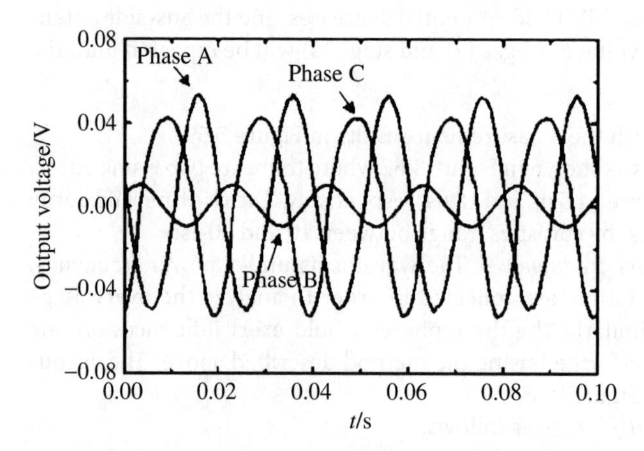

Figure 3.63 Decoupled output waveforms of the three-phase sensor.

Channel 2, phase C: background power frequency voltage: peak-to-peak value, 0.012 V; in-service power frequency voltage peak-to-peak value, 0.0946 V; the maximum overvoltage value: 0.075 V.

Channel 3, phase A: background power frequency voltage: peak-to-peak value, 0.012 V; in-service power frequency voltage peak-to-peak value, 0.103 V; the maximum overvoltage value: 0.08 V.

The response scale factors within the frequency range are the same due to the large frequency response range of the sensor. Therefore, the overvoltage multiples can be worked out by comparing the overvoltage and the output of the measuring system when inputting power frequency voltage.

$$\begin{aligned} k_A &= \frac{0.08}{(0.103-0.012)/2} \times 100\% = 1.76 \\ k_B &= \frac{0.017}{(0.027-0.00365)/2} \times 100\% = 1.46 \\ k_C &= \frac{0.075}{(0.0946-0.012)/2} \times 100\% = 1.82 \end{aligned} \tag{3.4}$$

Of the three phases, the largest overvoltage multiple is 1.82.

3.7.9 Case Study: Transient Voltage Measurement When Energizing Unloaded Lines

Transient voltages in the case of energizing unloaded lines via breakers are measured using the high-potential measuring method. The onsite arrangement of the measuring device is as shown in Figure 3.64. This device is configured at Song-Fan east line of 220 kV Songlin substation. Song-Fan east line and Song-Fan west line are connected to Fanba substation 11 km away.

The measurement of transient voltages during the energization of the unloaded line (Song-Fan east line) via the breaker is shown in Figure 3.65. The line voltage becomes

Figure 3.64 Onsite arrangement of transient monitoring devices installed at high potentials.

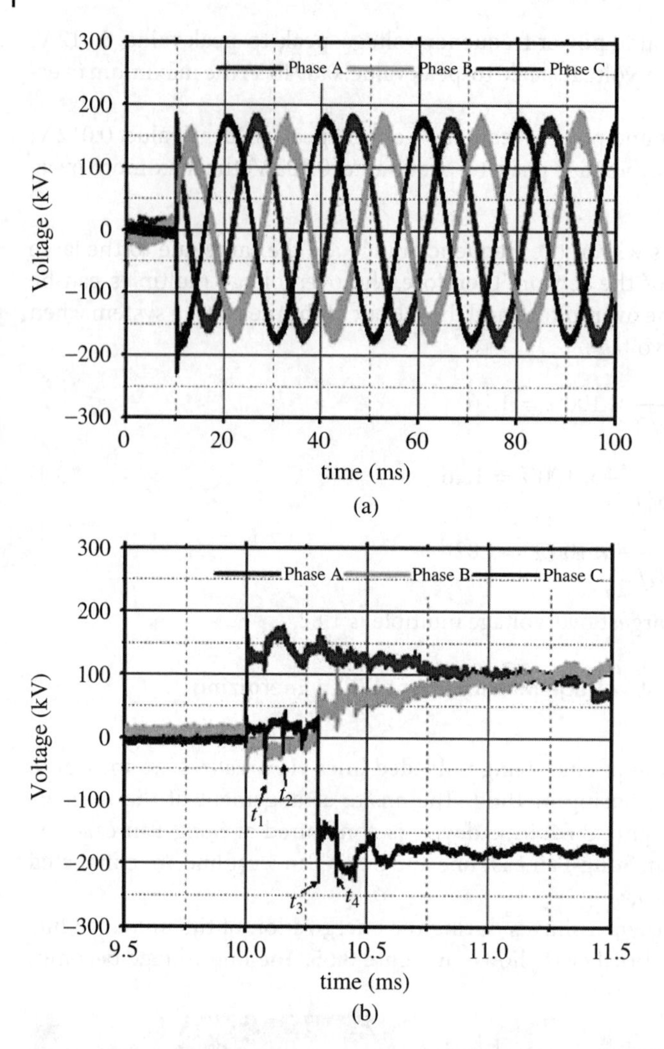

Figure 3.65 Transient voltage waveforms during the connecting of an unloaded line with a 220 kV bus via a breaker.

stable after several transient processes in the order of ms, due to the short length of the unloaded line (only 11 km long) and the short charging time. The maximum transient voltage occurs during the closing of phase C; the maximum amplitude reaches 228 kV, and the voltage multiple reaches 1.27.

The partial enlarged view of waveforms during the closing process is as shown in Figure 3.65(b). For each phase, the time to complete the closing operation may have a difference of several hundred μs with the other phases, because of the non-synchronous closing of the breaker. In this case, phase A is closed about 300 μs earlier than phases B and C. Song-Fan east line was put into service in Songlin substation while Fanba substation side stays opened; thus the transient voltage would be entirely reflected at the Fanba side. In Figure 3.65(b), t_3 is the instant when the closing of phase C is completed, and t_4 is the instant when the transient voltage wave returns back to Songlin substation from Fanba substation. The time difference between t_3 and t_4 is about 76 μs. The length of Song-Fan east line calculated based on the single-end positioning principle for travelling waves is 11.4 km, which matches the actual length. Hence, the measurement results from the transient voltage monitor can also be used for positioning the faults on the transmission lines. The rising edges of transient switching voltages are several μs, much faster than that of the standard switching impulse wave with a 250 μs front time.

3.8 Statistical Analysis of Overvoltages

3.8.1 Statistical Analysis of Phase-to-ground Overvoltages

Statistical data of the recent overvoltages in some substation is listed in Table 3.11.

3.8.2 Calculation of Overvoltage Characteristic Values

Since the overvoltages of the substation obey normal distribution, $U_{2S} = U_{50\%} + 2.05\sigma$, where $U_{50\%}$ is calculated as the overvoltage average value and σ is calculated as the standard deviation (Table 3.12).

Table 3.11 Impact coefficients of phase-to-ground overvoltages in some substation.

A		B		C	
1.19096	1.31774	1.24747	1.14282	1.56887	1.57654
1.92772	1.92144	1.36631	1.51095	1.59492	1.47351
1.45112	1.21142	1.28305	1.71841	1.35442	1.28933
1.65166	1.46306	1.65584	1.46119	1.36468	1.77702
1.54849	1.18049	1.19541	2.35472	1.35422	2.28145
1.51305	1.79167	1.15328	1.29305	1.40236	1.57165
1.18258	1.93425	1.48602	1.68238	1.47983	1.24538
1.93412	1.32918	1.45631	1.71841	1.36125	1.62213
1.25375	1.51561	1.34235	1.22863	1.17842	1.78702
1.20561	1.81596	1.46562	2.22636	1.61166	2.01
1.49625	1.70853	2.12237	2.34587	2.05598	1.488
2.17856	2.2456	2.34	1.88487	1.6657	1.737
1.86325	2.436	1.34589	1.71836	2.31756	1.681

Table 3.12 Characteristic values.

	A	B	C	Combining three phases
Maximum value	3.436	2.35472	2.31756	3.436
Minimum value	1.18049	1.15328	1.17842	1.15328
Average value	1.63	1.61	1.61	1.617
Standard deviation	0.13	0.15	0.39	0.34

3.8.3 Determination of Insulation Levels of Substation Electrical Equipment

Determining the insulation levels of substation electrical equipment (represented by the maximum system operating voltage U_m) are:

Insulation coordination for lightning overvoltages – The insulation level of electrical equipment under lightning overvoltages is usually represented by its basic impulse insulation level (*BIL*), $BIL = K_1 U_{P(i)}$, where $U_{P(i)}$ is the arrester protection level under lightning overvoltages (kV), and is normally simplified to the residual voltage U_r under the coordination current. K_1 is the coordination coefficient of lightning overvoltages, and is in the range 1.2–1.4. It is regulated by experience that K_1 takes the value of 1.25 when the equipment is close to the arrester and 1.4 when the equipment is far from the arrester, i.e. $BIL = (1.25–1.4)\ U_r$.

Insulation coordination for switching overvoltages – When determining insulation coordination based on internal overvoltages, resonance overvoltages are usually not considered because the attempt is made to avoid them by design and selection of the operation state. Besides, power frequency voltage rises are not separately considered either, because its impact is included in terms of the maximum long-term operating voltage.

Two kinds of conditions will be discussed as follows:

1. For all types of electrical equipment in the substation, the switching impulse insulation level (*SIL*) is expressed as $SIL = K_s K_0 U_\phi$, where K_s is the coordination coefficient for switching overvoltages; $K_s = 1.15–1.25$; the calculation multiple of the switching overvoltage $K_0 = 4.0$.
2. Zinc oxide arresters are widely used in the power system to limit switching and lightning overvoltages. The switching impulse insulation level is represented by $SIL = K_s U_{P(s)}$. In this equation, the coordination coefficient for switching overvoltages $K_s = 1.15–1.25$. It is smaller than K_1 (the coordination coefficient for lightning impulse overvoltages) due to smaller wavefront steepness compared to that of the lightning overvoltage. Besides, the voltage difference caused by the distance between the protected equipment and the arrester is small enough to be neglected. The protection level of the arrester under switching overvoltages is represented as $U_{P(s)}$, which equals the value of residual voltage under the specified switching impulse current: (a) discharge voltage under the standard 250/2500 μs switching impulse voltage; (b) residual voltage under the specified switching impulse voltage.

3.8.3.1 Case Study of the Overvoltages in the 35 kV System of a Substation

The analysis of insulation coordination in this case is based on real-time data in the year of 2008 (Table 3.13). The most severe switching overvoltage and lightning overvoltage over the specific period are taken for analysis.

The basic impulse insulation level of electrical equipment under lightning overvoltages is given by $BIL = (1.25–1.4)U_r$. Considering the minimum insulation level, the coefficient is selected as 1.25, namely $BIL = 1.25 \times U_r$.

The rated rms voltage of the arrester is 35 kV, and the rms value of the arc-quenching voltage is 73 kV. Therefore, the basic impulse insulation level of the electrical equipment under lightning overvoltages is given by $BIL = 1.25 \times 73 = 91.25$ kV.

The absolute value of the maximum amplitude of the lightning overvoltage which the transformer withstands is 49.15 kV, less than the calculated *BIL*.

Table 3.13 List of the overvoltage waveform parameters.

Date	2008-8-11	Time 0:40:6	Amplitude (kV) −49.15	Multiple −1.720
		Maximum duration (μs)	Rising time of wavefront (μs)	Steepness (kV/s)
A		107.5	5	8.474×10^5
B		122.5	5	10.777×10^5
C		112.5	5	11.429×10^5
Overvoltage type			Lightning overvoltage	

Table 3.14 List of the switching overvoltage waveform parameters.

Date	2008-8-8	Time 16:45:6	Amplitude (kV) 50.302	multiple 1.760
		Maximum duration (μs)	Rise time (μs)	Steepness (kV/s)
A		237.5	7.5	-5.215×10^5
B		250	7.5	-6.721×10^5
C		260	7.5	-6.605×10^5
Overvoltage type			Switching overvoltage	

As mentioned above, the switching impulse insulation level (*SIL*) can be represented by $SIL = K_s K_0 U_\varphi$ Here, considering the minimum insulation level, $K_s = 1.15$, $K_0 = 3.2$ (assume that the neutral is grounded effectively via small resistance), $SIL = 1.15 \times 3.2 \times 28.57 = 105.13$. According to Table 3.14, the maximum amplitude of the switching overvoltage in this substation is 50.302 kV, less than the calculated *SIL*.

4

Wave Process of Incoming Surges and Transient Response Characteristics

4.1 Current State of Incoming Surge Research

East Coast, Southern China, West China and Central China are all areas having frequent lightning activities. Lightning current produced by lightning discharge can be as high as dozens or even hundreds of kilo-ampere, and may give rise to severe electromagnetic, mechanical and thermal effects. In the vicinity of the strike site, lightning striking the transmission lines will not only endanger insulation safety, but also cause strong electromagnetic interferences on the nearby communication lines, and high potential differences may even pose serious threats to the safety of personnel near the fault point. On the other hand, lightning currents propagate along transmission lines in the form of lightning waves, and waves of large amplitudes would seriously impact insulation coordination as well as electromagnetic compatibility and stability of adjacent substations. Statistics of operating transmission lines rated 220 kV and above in the past two decades (1990–2010) has suggested that the lightning overvoltage, as one of the major causes for insulation failures of electrical equipment, has become a major threat to secured operation of the grid and cannot be neglected.

Upon determining the lightning withstand level of high-voltage overhead lines, the frequency-dependent properties of the line parameters, the impulse corona and the actual lightning wave process will significantly impact the electromagnetic transient process caused by lightning strokes. Therefore, when determining the insulation levels of electrical equipment, precisely acquiring the actual incoming surge levels and waveforms is critical for appropriate insulation coordination between equipment.

The 1.2/50 μs standard lightning wave specified by IEC is a typical waveform of the transient overvoltage caused by lightning without considering the impact of the substation and electrical equipment on the wave process. On account that there are a variety of wave shapes under practical conditions, many problems arise if the non-standard lightning wave is substituted with the standard lightning wave to assess and test the insulation properties of electrical equipment. Therefore, experts began to study the parameters of non-standard lightning waves at the end of the last century. In his statistical lightning wave research, S. Okabe suggested that the overvoltage wave of the substation transformer has many high-frequency oscillating components overlaid, owing to wave reflection and refraction in the substation, i.e. oscillating voltage waves with natural winding frequency would be obtained even when standard lightning waves are applied on the transformer. Four typical non-standard lightning waveforms are defined on the basis of waveform statistics: (a) single impulse wave, (b) front impulse wave,

(c) oscillating decay wave and (d) oscillating rising wave. I.D. Couper and K.J. Cornick et al. studied the polarity effect of the oscillating impulse voltage, and concluded that under the impact of stray inductance and capacitance, overvoltage waveforms on the electrical equipment are normally oscillating decay waveforms, and the damping coefficient is dependent on the resistance and loss in the system.

The $(1.2 \pm 30\%)/(50 \pm 20\%)$ μs standard waveform is mostly adopted in China for current insulation design and withstand voltage tests under lightning impulses. In practical situations, the power system and electrical equipment today are quite different from those in the past. For instance, the wide application of GIS substations, overhead ground wires and high-performance arresters will both lead to variations between the past and present waveforms of direct lightning strokes to transmission lines and also cause great differences between the present incoming surge waveforms and the past waveforms of lightning on transmission lines.

Due to protection for incoming lines and limitation by in-station surge arresters, the magnitude and steepness of incoming surges will be attenuated to some degree, and the shapes of the waves applied on transformers will become very complex due to a number of factors such as arrester locations, station topology, operation state changes and refraction and reflection inside the system. As a consequence, the lightning wave applied on the transformer is no longer the 1.2/50 μs standard one. Therefore, lightning waveforms used in traditional design and tests are incapable of fully reflecting the parameters of lightning waves which the equipment insulation has to withstand in practical conditions, which may affect the reliability of equipment insulation design and withstand voltage tests. In recent years, online overvoltage monitoring has been developing rapidly, and research on lightning waveform measurement in substations has been reported from time to time. Nevertheless, only simple qualitative summaries – not quantitative analysis – are provided for lightning waveforms from actual measurements. Let alone carrying out tests to reveal the impact of actual lightning overvoltages on transformer insulation and comparing the test results with that under the standard lightning wave.

The actual lightning wave process needs to be studied based on statistics of long-term measurement of incoming surges in the substation by the online overvoltage monitoring system. This research is of great reference value for equipment insulation design, and can guarantee safe and economic operation of the power grid.

4.2 Wave Process under Complex Conditions

4.2.1 Wave Process in Lossless Parallel Multi-Conductor Systems

When a transmission line is not long, the impact of the line resistance R_0 on the wave process is negligible; also, the line-to-ground conductance G_0 is small enough to be neglected. Such a line is thus known as a single lossless line. In practical conditions, transmission lines consist of three-phase conductors and overhead ground wires; the number of conductors in dual-circuit lines is even larger. When ground losses are ignored, the wave process of multi-conductor systems would approximate the propagation of one-dimensional parallel electromagnetic waves, and by introducing the wave velocity v, the Maxwell's equations in the electrostatic field are applicable to the parallel multi-conductor system.

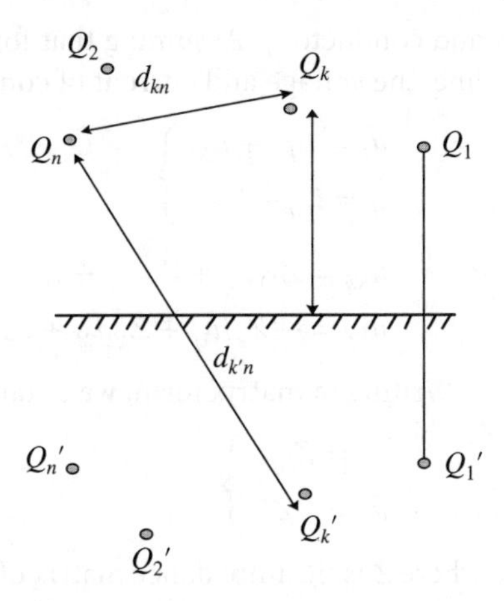

Figure 4.1 Parallel multi-conductor (number n) system.

For parallel multi-conductor systems as shown in Figure 4.1, the charges of the conductors per unit length are $Q_1, Q_2, \ldots, Q_n$, respectively, and the conductor-to-ground voltages are $u_1, u_2, \ldots, u_n$, respectively. Using Maxwell's equations, we obtain

$$\left.\begin{aligned} u_1 &= \alpha_{11}Q_1 + \alpha_{12}Q_2 + \ldots + \alpha_{1n}Q_n \\ u_2 &= \alpha_{21}Q_1 + \alpha_{22}Q_2 + \ldots + \alpha_{2n}Q_n \\ u_n &= \alpha_{n1}Q_1 + \alpha_{n2}Q_2 + \ldots + \alpha_{nn}Q_n \end{aligned}\right\} \tag{4.2.1}$$

where α_{kk} with the same subscript is the self-potential coefficient of conductor k; α_{kj} with different subscripts is the mutual potential coefficient between conductor k and conductor j. The potential coefficients are determined by geometric dimensions between conductors and between conductors and their mirror images. The calculation equation is expressed as

$$\left.\begin{aligned} \alpha_{kk} &= \frac{1}{2\pi\varepsilon_0}\ln\frac{2h_k}{r_k} \\ \alpha_{kj} &= \frac{1}{2\pi\varepsilon_0}\ln\frac{d_{j'k}}{d_{jk}} \end{aligned}\right\} \tag{4.2.2}$$

where $\alpha_{kk} = \alpha_{jk}$; r_k is the radius of conductor k; h_k the height above the ground of conductor k; d_{jk} the distance between conductor j and conductor k; $d_{j'k}$ the distance between the mirror image of conductor j and conductor k; and ε_0 the dielectric constant of air. Multiplying each item on the right side of Equation (4.2.1) by v, with $v = \frac{1}{\sqrt{\varepsilon_0\mu_0}}$ (propagation velocity of lightning waves along overhead lines) and applying $Q_k v = i_k$ (the current of k conductor), Equation (4.2.1) is rewritten as

$$\left.\begin{aligned} u_1 &= Z_{11}i_1 + Z_{12}i_2 + \ldots + Z_{1n}i_n \\ u_2 &= Z_{21}i_1 + Z_{22}i_2 + \ldots + Z_{2n}i_n \\ u_n &= Z_{n1}i_1 + Z_{n2}i_2 + \ldots + Z_{nn}i_n \end{aligned}\right\} \tag{4.2.3}$$

where Z_{kk} with the same subscript refers to the self wave impedance of the conductor k, Z_{kj} with different subscripts refers to the mutual wave impedance between conductor k

and conductor j. Assuming that forward and backward travelling waves coexist on the line, the voltage and current of conductor k would be respectively given by

$$\left.\begin{aligned} u_k &= u_{Kq} + u_{Kf} \\ i_k &= i_{Kq} + i_{Kf} \end{aligned}\right\} \tag{4.2.4}$$

$$\left.\begin{aligned} u_{Kq} &= Z_{k1}i_{1q} + Z_{k2}i_{2q} + \ldots + Z_{kn}i_{nq} \\ u_{Kf} &= -(Z_{k1}i_{1f} + Z_{k2}i_{2f} + \ldots + Z_{kn}i_{nf}) \end{aligned}\right\} \tag{4.2.5}$$

Writing in matrix form, we obtain

$$\left.\begin{aligned} u_q &= Zi_q \\ u_f &= -Zi_f \end{aligned}\right\} \tag{4.2.6}$$

where Z is the impedance matrix of the multi-conductor line; u and i refer to the voltage and current vector of the multi-conductor line, respectively; and the subscripts q and f refer to the forward and backward travelling wave, respectively. In practical use, if specific boundary conditions are given, analysis could be provided for the wave process of parallel multi-conductor system.

In Figure 4.2, the voltage wave u_1 propagates along Conductor 1, and Conductor 2 is insulated at both sides (the current is approximately zero due to the low induction current). The equations of travelling waves in the multi-conductor system are expressed as

$$\left.\begin{aligned} u_1 &= Z_{11}i_1 + Z_{12}i_2 \\ u_2 &= Z_{21}i_1 + Z_{22}i_2 \end{aligned}\right\} \tag{4.2.7}$$

When Conductor 2 is insulated and the line is not long, $i_2 = 0$ (conductors too long would otherwise induce current waves), we obtain

$$u_2 = \frac{Z_{12}}{Z_{11}}u_1 = Ku_1 \tag{4.2.8}$$

where K refers to the coupling coefficient between Conductor 1 and Conductor 2, and is determined by spatial geometric dimensions of the conductors. As can be inferred from Equation (4.2.8), when voltage waves propagate along Conductor 1, voltage waves of similar shapes and polarities would be induced on Conductor 2. According to the wave impedance formula, $Z_{12} \le Z_{11}, K \le 1$. The smaller the conductor spacing, the larger

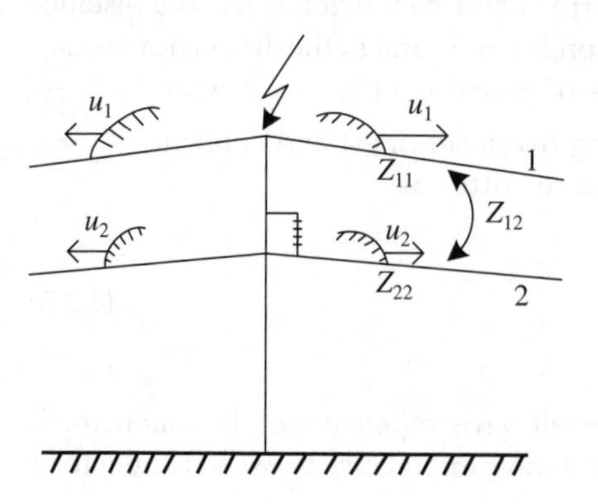

Figure 4.2 A parallel two-conductor system. Note: Conductor 1 is hit by lightning and Conductor 2 is insulated against the ground.

the coupling coefficient, and the voltage difference between the conductors could be calculated by $u_1 - u_2 = (1 - K)u_1$. From this equation, it is clear that the larger the value of K, the smaller the voltage difference between the conductors.

4.2.2 Wave Propagation along Lossy Lines

4.2.2.1 Line Loss

When travelling waves propagate along ideal lossless lines, energy would not disappear (stored in the electromagnetic field), and waves would experience neither attenuation nor distortion. In practical conditions, all lines are lossy due to:

1. conductor resistance (including the impact of skin and proximity effects);
2. ground resistance (including the impact of waveforms on ground current distribution);
3. leakage conductivity and dielectric loss of the insulation (the latter exists solely in cables);
4. radiation loss of high-frequency or steep waves;
5. impulse corona.

These loss factors can lead to changes in travelling waves:

1. decrease in amplitudes (accompanied by attenuation);
2. decline in front steepness and flattened fronts;
3. extended waveforms;
4. smoothed waveforms;
5. wave shapes of the voltage waves not being similar to those of current waves.

4.2.2.2 Impact of Line Resistance and Line-to-Ground Conductance

Taking into account the line resistance R_0 and the line-to-ground conductance G_0 per unit length, the equivalent circuit of the distributed parameter transmission line is shown in Figure 4.3, where R_0 contains conductor resistance and ground resistance, and G_0 contains insulation leakage and dielectric loss. When travelling waves are propagating along lossy lines, part of the energy is dissipated into heat energy due to the existence of R_0 and G_0, and the waves would be attenuated and distorted.

Provided that the transmission line parameters satisfy the condition of zero distortion, i.e. $\frac{R_0}{L_0} = \frac{G_0}{C_0}$, the attenuation law of overvoltage waves would be obtained in accordance

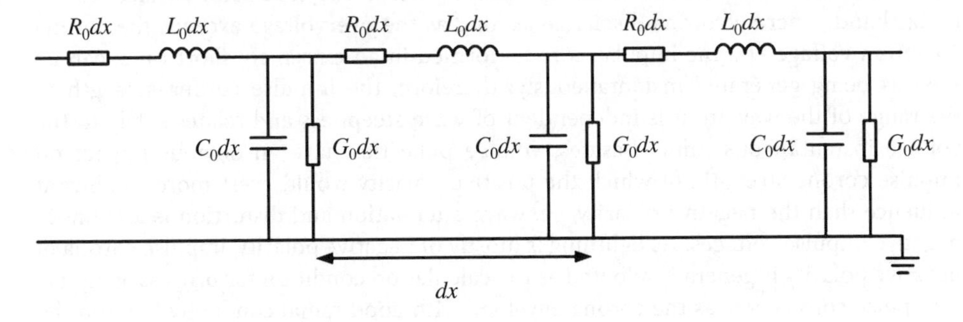

Figure 4.3 Equivalent circuit of distributed parameter lossy lines.

with the equation of long uniform lines:

$$U_x = U_0 e^{-\frac{1}{2}\left(\frac{R_0}{L_0}+\frac{G_0}{C_0}\right)t} = U_0 e^{-\frac{1}{2}\left(\frac{R_0}{Z}+G_0 Z\right)x} \tag{4.2.9}$$

where U_0, U_x – the original amplitude of the voltage wave and the amplitude after the wave propagates for a distance x; t, x – the propagation time and distance of the travelling wave; Z – characteristic impedance of the line.

From the above equation, it is clear that the voltage wave decays exponentially but is not distorted. In engineering practice, zero distortion is hard to realize. In most cases $\frac{R_0}{L_0} \neq \frac{G_0}{C_0}$, meaning that distortion would accompany attenuation. Generally speaking, the leakage conductance and dielectric loss of overhead lines are so low that G_0 could be ignored; hence, attenuation during propagation along the transmission line can be calculated by

$$U_x = U_0 e^{-\frac{1}{2}\frac{R_0}{Z}x} \tag{4.2.10}$$

which shows that the larger the travelling distance x and the ratio $\frac{R_0}{Z}$, the greater the attenuation. Because the ratio $\frac{R_0}{Z}$ of cables is much greater than that of overhead lines, waves propagating along cables undergo greater attenuation. R_0 is related to the equivalent wave frequency. The faster the waveform changes, the more significant the skin effect, and the greater is R_0. Hence, short waves are attenuated more significantly when propagating along the lines.

4.2.2.3 Impact of Impulse Corona

Electromagnetic waves will not be attenuated and distorted when propagating along single-phase lossless lines. In practice, line resistance and line-to-ground conductance will generate energy loss and attenuate the waveforms. When $R_0/L_0 \neq G_0/C_0$, the waveforms would be distorted. Additionally, the earth is a non-ideal conductor with a non-zero conductivity, therefore currents are distributed both over the ground surface and into the ground. The presence of the skin effect would result in frequency-dependent parameters of lines that cause waveform distortion. It is also indicated by theory and actual measurement that waves would be distorted when propagating along multi-conductor lines. The reason in this case is that the differential equations for multi-conductor lines are different from those for single conductor lines due to coupling between conductors.

For high-voltage lines, impulse corona is the major cause for wave attenuation and distortion. When the line is hit by lightning, air at the conductor surface will be ionized and generate corona discharge as long as the overvoltage exceeds the corona inception voltage U_c. The impulse corona, formed in an extremely short time, can be seen as being generated instantaneously; therefore, the impulse corona strength in the range of the wavefront is independent of wave steepness and relates solely to the voltage instantaneous value. Besides, voltage polarities have an obvious impact on impulse corona strength, of which the positive polarity would exert more significant influence than the negative polarity, i.e. wave attenuation and distortion is less under negative impulse voltages. As lightning is mostly of negative polarity, impulse corona of negative polarity is generally adopted as the calculation condition for process analysis.

Impulse corona acts as the corona envelope with good radial conductivity upon the conductor surface, which has the same effect as increasing the conductor radius and thus increasing the conductor capacitance. And the line inductance remains unchanged due to the poor axial conductivity of the corona envelope. On the whole, impulse corona could influence the wave process in many aspects such as the decrease in energy loss and the wave impedance as well as the increase in conductor coupling coefficients.

Decrease in the conductor characteristic impedance

$$Z' = \sqrt{\frac{L_0}{C_0'}} = \sqrt{\frac{L_0}{C_0 + \Delta C_0}} < Z \left(= \sqrt{\frac{L_0}{C_0}} \right) \tag{4.2.11}$$

There is generally a decrease of 20% ~ 30%. In the presence of impulse corona, the wave impedances of the ground wire and the single conductor shall both take the value of 400 Ω, and the wave impedance for dual ground wires in parallel shall take the value of 250 Ω.

Decrease in the wave velocity

$$v' = \frac{1}{\sqrt{L_0 C_0'}} = \frac{1}{\sqrt{L_0(C_0 + \Delta C_0)}} < v \left(= \frac{1}{\sqrt{L_0 C_0}} \right) \tag{4.2.12}$$

In the presence of intensive impulse corona, v' can be reduced to $0.75c$ (c is the light velocity).

Increase in coupling coefficients – In the presence of impulse corona, the coupling coefficient between the lines is increased due to the larger conductor effective radius, the smaller conductor self-wave impedance and the slightly larger mutual wave impedance with the adjacent conductor. When impulse corona is taken into account, the coupling coefficient between the ground wire and the transmission line is given by

$$k = k_1 k_0 \tag{4.2.13}$$

where k_0 and k_1 are the geometric coupling coefficient and the corona correction coefficient, respectively, as shown in Table 4.1.

Wave attenuation and distortion – The wave velocity becomes smaller and smaller with the increase in the wavefront voltage until reaching a value related to the instantaneous voltage value. The voltage at each point of the wavefront corresponds to a different wave velocity, and the higher the voltage, the smaller the wave velocity, which results in severe distortion of the wavefront.

As already shown in Figure 1.9, curve 1 represents the original waveform, and curve 2 represents the waveform of the voltage wave after propagating a distance l. When the voltage is higher than the corona inception voltage u_k, the waveform will be subject to dramatic attenuation and distortion; the line capacitance will be increased, and the line inductance will have a few changes due to the corona envelope; the wave velocity

Table 4.1 Correction table of the coupling coefficient k_1.

Voltage class (kV)	20–35	66–110	154–330	500
Dual ground wire	1.1	1.2	1.26	1.29
Single ground wire	1.15	1.24	1.3	—

Note: When lightning strikes the central span of the ground wire, k_1 takes the value of 1.5.

will become smaller. After propagating a distance l, point A on the curve 1 lags to A' with a time of $\Delta\tau$. The wave velocity corresponding to the voltage u_k is called the phase velocity; the higher the voltage, the lower the phase velocity. Different voltages correspond to different line capacitance parameters, which can be determined by the empirical V–C characteristic formula. The lag time $\Delta\tau$, related to the wave propagation distance and the voltage value, is specified by the recommended practices regarding overvoltage protection as

$$\Delta\tau = l\left(0.5 + \frac{0.008u}{h}\right) \tag{4.2.14}$$

where l is the wave propagation distance (km); u is the amplitude of the travelling wave voltage; h is the conductor average height (m); and the unit of $\Delta\tau$ is μs.

As impulse corona can attenuate and distort travelling waves, incoming line protection is of great concern with respect to lightning protection at the substation.

4.2.3 Wave Process on Transformer Windings

When power transformers are subject to external incoming surges or internal switching overvoltage waves, complex electromagnetic oscillations occur inside the winding, causing high potential differences between points on the coil and the ground or between different points on the coil. In view of the complexity of the winding structure and nonlinearity of the iron inductance, it is difficult to solve the voltage between each winding point and the ground as well as its impulse voltage or the current response under impulse waves of various shapes.

In order to analyze the wave process in transformers, a simple equivalent circuit must be built first. In the following contents, analysis will be performed from the simple and ideal representation to the complex and practical conditions.

4.2.3.1 Wave Process in Single-Phase Windings

Only when windings are connected to all three phases are both single-phase and three-phase transformers able to operate. However, the wave process in only the single-phase winding is required to be studied when (1) the neutral point of the high-voltage wye-connected winding is directly grounded without considering the coupling between the three-phase windings; (2) the neutral point of the high voltage winding is ungrounded, and the incoming waves appear simultaneously at the three phases. In this case, research on the single-phase winding with an open-circuit end only is required because the three phases are completely symmetric.

The wave impedance of the transformer winding is far higher than that of the transmission line. The simplified representation of the winding (see Figure 4.4) is obtained based on these assumptions: parameters of all windings are the same without considering the impact of other windings; resistance, conductance and mutual inductances are ignored; K_0, C_0 and L_0 are the longitudinal capacitance, the capacitance and the inductance to ground per unit length of the winding respectively, and U_0 is a dc voltage source.

At the instant $t = 0$, the power supply is closed suddenly, and the currents of all the inductors are zero as they cannot change instantaneously. The equivalent circuit at this particular instant is illustrated in Figure 4.5, where the voltage at a distance x from the

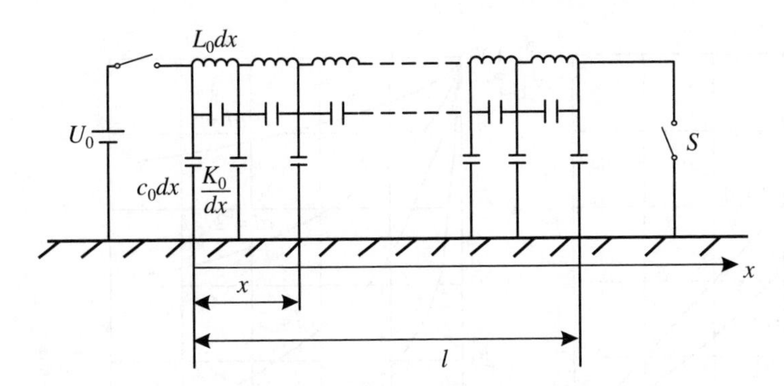

Figure 4.4 Simplified representation of the single-phase winding.

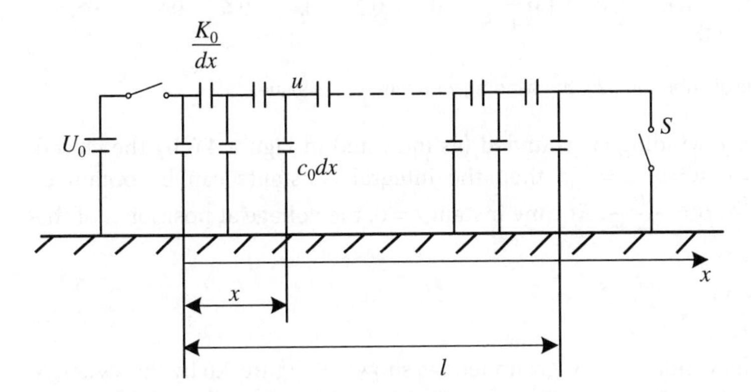

Figure 4.5 Equivalent circuit of the winding at time instant $t = 0$.

winding end is u, the charges on the longitudinal capacitance K_0/dx is Q, and the charges on the capacitance to ground C_0dx is dQ. Thus the equation is given by

$$Q = \frac{K_0}{d_x} d_u \tag{4.2.15}$$

K_0 is the longitudinal capacitance per unit length of the winding; the shorter dx, the larger the capacitance. The charges on the capacitance to ground are given by

$$dQ = (C_0 d_x)u \tag{4.2.16}$$

Since the circuit is infinitely small, the charges can be deemed as spatial point charges. Differentiating Equation (4.2.15) with respect to x and substituting it into Equation (4.2.16) gives

$$\frac{d^2u}{dx^2} - \frac{C_0}{K_0}u = 0 \tag{4.2.17}$$

with its solution given by

$$u = Ae^{\alpha x} + Be^{-\alpha x} \tag{4.2.18}$$

with $\alpha = \sqrt{C_0/K_0}$, where A and B are integration constants determined on the basis of the boundary condition.

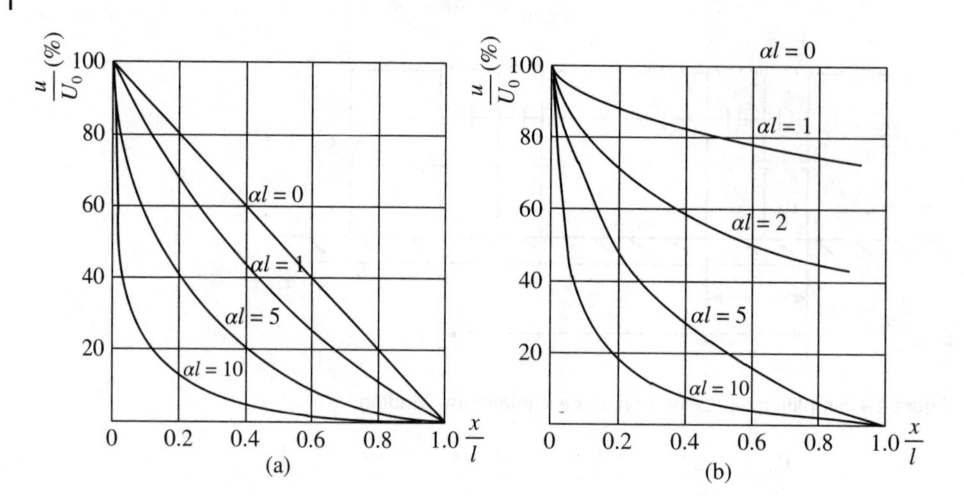

Figure 4.6 Initial voltage distribution over the winding with different αl values.

When the end of the winding is grounded (as indicated in Figure 4.6 by the closed switch S), $x = 0$, $x = l$, $u = 0$, $u = U_0$, then the integral constants can be obtained: $\left|\frac{du}{dx}\right|_{x=0}\right| = \alpha U_0 = \alpha l \frac{U_0}{l}$, $B = \frac{U_0}{1-e^{-2\alpha l}}$. At time instant $t = 0$, the voltage at position x of the winding is expressed as

$$u = U_0 \frac{sh\alpha(l-x)}{sh\alpha l} \tag{4.2.19}$$

When the end of the winding is not grounded (as shown in Figure 4.6 by the switch S that is not switched on), $x = 0$, $x = l$, $i = 0$, $u = U_0$. Due to the shunting effect of the capacitance to ground, the voltage across a differential segment of the winding line terminal is virtually zero, i.e. $K_0 \frac{du}{dx} \approx 0$; hence the voltage distributed over the winding is given by

$$u = U_0 \frac{ch\alpha(l-x)}{ch\alpha l} \tag{4.2.20}$$

The above two voltage equations reflect the distribution of the voltage to ground for each point on the winding at the closing instant of the transformer winding, which is referred to as the initial voltage distribution. For continuous windings taking special measures, αl is a value between 5 and 15, and is approximately 10 on average; when $\alpha l > 5$, $sh\alpha l \approx ch\alpha l \approx \frac{1}{2}e^{\alpha l}$; hence, when $x < 0.8l$, $sh\alpha(l-x) \approx ch\alpha(l-x) \approx \frac{1}{2}e^{\alpha(l-x)}$. According to the above two equations, no matter whether the winding end is grounded, the initial voltage distribution for most windings is similar at the closing instant, which can be given by

$$u \approx U_0 e^{-\alpha x} \tag{4.2.21}$$

It is worth noting that the initial voltage is distributed unevenly and is related to α. The larger α, the more uneven the distribution will be (as illustrated in Figure 4.7). The voltage drops dramatically near the line terminal with a large potential gradient (du/dx) which can be calculated by

$$\left|\frac{du}{dx}\right|_{x=0}\right| = \alpha U_0 = \alpha l \frac{U_0}{l} \tag{4.2.22}$$

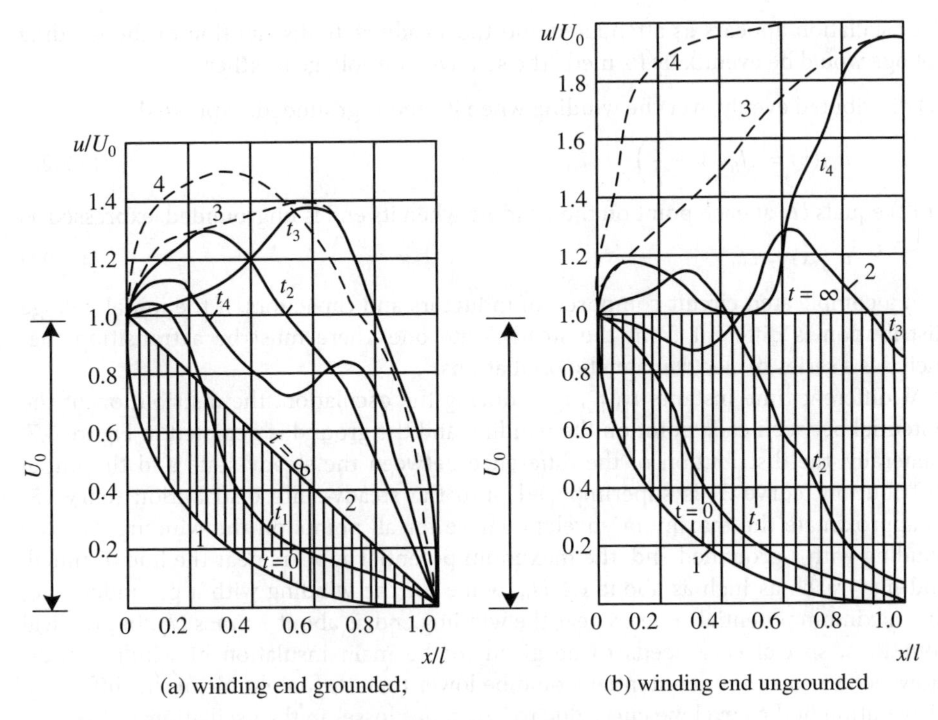

Figure 4.7 Voltage distribution in the initial, steady-state and oscillating stages.

where U_0/l is the average potential gradient of the winding. At the instant when $t=0$, the potential at the line terminal of the winding can reach up to 10 times the average value. When $\alpha l=10$, an extremely high potential gradient would be generated at the line terminal by incoming surges, which will severely threaten turn-to-turn insulation. Therefore, measures to prevent longitudinal insulation from breakdown are generally taken at the line terminal of the transformer winding.

The inductance of the transformer winding is fairly large, and when the winding is subject to steep waves, the inductive current of the winding within 10 μs is small enough to be ignored. The winding is then equivalent to a capacitance chain (i.e. a lumped parameter capacitance C_T if seen from the outside), and is known as the transformer entrance capacitance. Table 4.2 lists the entrance capacitances of transformers of different voltage ratings.

In the previous analysis, the voltage distribution of the transformer winding is described when dc voltage is applied. In practice, a complicated oscillating circuit will be generated under the influence of lightning waves as there are resistance and inductance within the transformer winding. Since the resistance consumes energy,

Table 4.2 Transformer entrance capacitances.

Voltage ratings (kV)	35	110	220	330	500
Entrance capacitance (pF)	500–1000	1000–2000	1500–3000	2000–5000	4000–6000

the oscillation appears as attenuated and the steady-state distribution of the winding voltage would be eventually formed. The steady-state voltage is either

(1) distributed evenly over the winding when its end is grounded, expressed as

$$u_{\infty}(x) = U_0\left(1 - \frac{x}{l}\right) \tag{4.2.23}$$

or (2) equals U_0 at each point on the winding when its end is ungrounded, expressed as

$$u_{\infty}(x) = U_0 \tag{4.2.24}$$

In a complicated circuit composed of inductors and capacitors, if the initial voltage distribution is different from the steady-state one, there must be a transition that includes a series of electromagnetic oscillations.

At different time instants t_1, t_2, t_3... during the oscillation, the distribution of the potential between each point on the winding and the ground is depicted in Figure 4.7. Generally, the distribution of the difference between the steady-state and the initial distribution (curve 3) is superimposed onto the steady-state distribution (curve 5) to approximate the maximum envelope curve for all points on the winding. For the winding with a grounded end, the maximum potential appears near the line terminal, and may reach as high as about 1.4 U_0, while for the winding with a grounded end, the maximum potential appears near the winding end, is about 2 times U_0. In practical situations, special care needs to be given to the main insulation of winding ends. However, the maximum potential would be lower than the amplitude of the difference distribution of the envelope curve due to resistance losses in the oscillation.

Regardless of the grounding state of the winding end, the maximum potential gradient αU_0 appears at the line terminal when $t = 0$, while as the oscillation develops, the maximum potential gradients of other points appear at different time instants, and may result in excessively high inter-turn voltages, which is a concern regarding the minor insulation of the winding. Inside the winding, oscillation is affected by the shape of the incoming surge; the longer the front time, the more slowly the voltage rises, i.e. a longer front time means a lower equivalent wave frequency, more influence of resistors and inductors on the initial voltage distribution and thus a gentler oscillation with both maximum and minimum potential gradients of lower values. By contrast, if transformer windings are subject to steep waves, the oscillation inside the winding would be greater. Besides, the impulse wave would also result in greater oscillation as its long wave tail provides more energy.

The greatest threat to winding insulation (especially minor insulation) is undoubtedly presented by right-angle short waves. The right-angle wave with an amplitude of "$+U_0$" that enters the winding is followed by another right-angle wave with an amplitude of "$-U_0$" at the winding line terminal. The two waves will be superposed, and result in greater oscillation and higher overvoltages than the insulation withstand. Hence, in addition to the full impulse wave test, the chopped wave test has to be conducted as well.

4.2.3.2 Wave Process in Three-Phase Windings

Wye connection with grounded neutral points – In this case, the transformer can be seen as having three independent windings with grounded ends as the interaction between the three phases is small. Regardless of the number of the phases that the wave reaches, the process can be treated as incoming surges reaching a single phase.

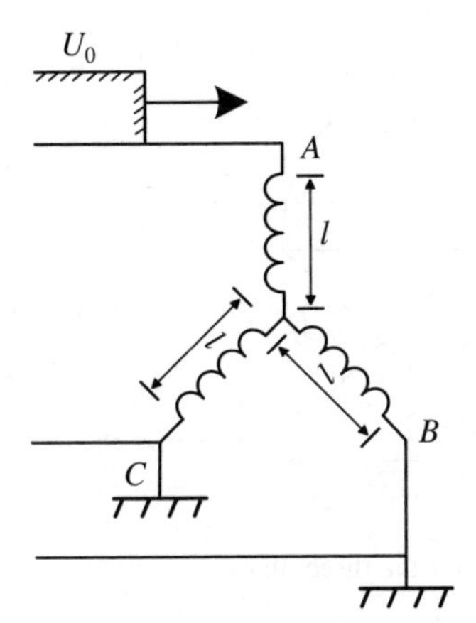

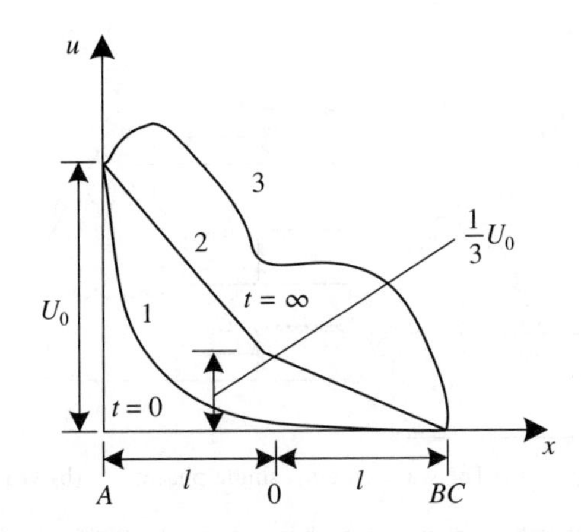

Figure 4.8 Voltage distribution of the wave entering a single phase of wye-connected windings (1: Initial voltage distribution; 2: Steady-state voltage distribution; 3: Envelope curve of the maximum voltage).

Wye connection with ungrounded neutral points – A wave entering a single phase (phase A) is shown in Figure 4.8. The mutual inductances between windings are ignored, and the winding length of each phase is l. Because the characteristic impedance of the winding is much higher than that of the line, points B and C can be regarded as grounded. In Figure 4.8, point A is assumed as the starting point and points B and C the terminal points; curve 1 represents the initial voltage distribution; curve 2 shows the steady-state voltage distribution, and curve 3 is the envelope curve of the maximum voltage between each point on the winding and the ground. The steady-state voltage of the neutral point is $U_0/3$, and the maximum neutral-to-ground voltage will not exceed $2U_0/3$ throughout the oscillation.

When the waves enter two phases simultaneously with the amplitudes of $+U_0$, the superposition theorem can be adopted, i.e. the maximum voltage of the neutral point is $2U_0/3$ when the wave enters phase A or phase B solely, hence the maximum voltage at the neutral point would reach $4U_0/3$ when the waves enter phase A and phase B simultaneously.

The case where the waves simultaneously enter all three phases is the same as that of the single-phase winding with an ungrounded neutral. The maximum neutral voltage could be two times the line terminal voltage, but such a situation rarely happens.

Delta connection – For delta-connected transformer windings, when the wave enters a single phase, as the wave impedance of the winding is much higher than that of the line, the ends of the other windings can be treated as grounded, similar to the case of single-phase windings with grounded ends.

The superposition theorem can be adopted again in the case of waves reaching two and three phases. Figure 4.9 shows the waves entering a single phase and three phases, respectively, where curves 1 and 2 are the initial and steady-state voltage distribution of the winding with the wave entering only one phase; curve 3 is the envelope curve of the maximum voltage between each point on the winding and the ground; curve 4 is the steady-state voltage distribution of the winding with the waves entering two terminals simultaneously. It can be inferred from the figure that the maximum voltage at the central part of the windings may reach as high as $2U_0$.

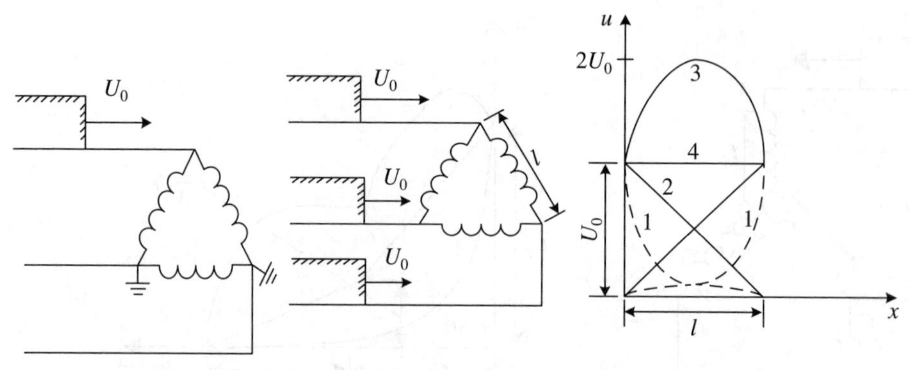

Figure 4.9 Voltage distribution when waves enter the single phase and three phases of delta-connected windings.

4.2.3.3 Transfer of Impulse Voltage Between Windings

When a transformer winding is subject to impulse voltage waves, overvoltages would be present on other windings due to electromagnetic coupling, i.e. the so-called overvoltage transfer between windings. This kind of overvoltage includes two components: the electrostatic component and the electromagnetic coupling component.

Electrostatic component – As shown in Figure 4.10, when the impulse voltage wave reaches the transformer winding, the inductor current cannot change instantaneously, owing to the steep wavefront. The equivalent circuits of winding I and II are both capacitance chains, and capacitance coupling exists between the windings, generating their own initial voltage distribution. The capacitance to ground of winding II is C_2, the capacitance between winding I and winding II is C_{12}, and the electrostatic component present on winding II is given by

$$U_2 = \frac{C_{12}}{C_{12} + C_2} U_0 \tag{4.2.25}$$

In the above equation, as C_2 contains the capacitances to ground of electric apparatus, lines and cables connecting to winding II, it is certain that $C_2 > C_{12}$. Electrostatic components generally do not pose dangers to the secondary side, but in the case of open secondary sides – for instance a three-winding transformer with its high-voltage

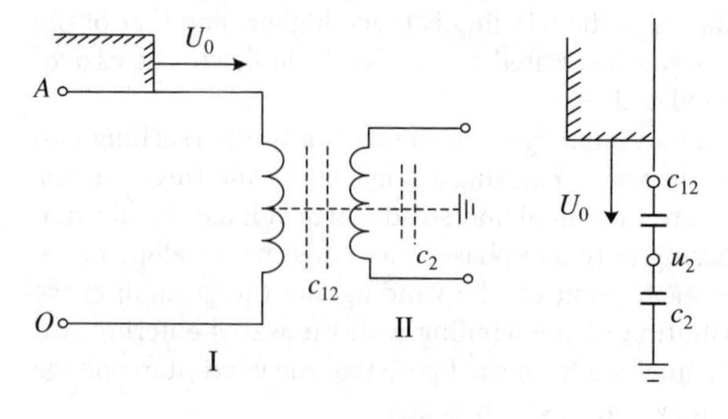

Figure 4.10 Electrostatic coupling between windings.

and medium-voltage sides in service and low-voltage side out of service – C_2 is just the capacitance to ground of the low-voltage winding and has a small value. Therefore, the electrostatic component on the low-voltage winding might be very high and protective measures have to be taken on this occasion.

Electromagnetic component – When impulse waves enter the transformer winding, the inductor current is small over a period because the inductance is large, and the overvoltages in the secondary winding are mainly static coupling components. When the inductor current increases, the secondary winding would be influenced by the overvoltage component induced by magnetic field changes $M\frac{di}{dt}$ (i.e. the electromagnetic component). This component would transfer between windings as per the transformation ratio.

Since the relative impulse strength of the low-voltage windings (a ratio of the impulse withstand voltage and the rated phase voltage) is much higher than that of the high-voltage winding, the overvoltage that the high-voltage winding insulation can survive, after transferring to the low-voltage side, will not pose threats to the low-voltage winding. The induced voltage component may only harm the high-voltage windings when waves first enter the low-voltage windings. For example, it may cause insulation breakdown of the high-voltage winding when lightning strikes the lines on the low-voltage side of the distribution transformer. Usually, three-phase arrester banks are installed close to the high-voltage winding terminals to protect against this kind of overvoltage.

4.3 Generation of Lightning Overvoltages on Electrical Equipment

The lightning impulse waves that electrical equipment withstands in practical situations are generated by the three steps (Figure 4.11):

1. Lightning strikes the transmission line, and generates the initial impulse wave at the strike site.
2. The initial impulse wave gets distorted when propagating along the line towards the transformer.
3. The travelling voltage wave is further distorted under the effect of arresters after reaching the bus.

There are four circumstances under which the lightning strikes the transmission lines and generates the initial impulse waves

1. Lightning bypasses the ground wire and strikes the conductor (shielding failure) without the presence of insulator flashover; the initial impulse wave is a relatively complete double exponential wave.
2. Lightning bypasses the ground wire and strikes the conductor (shielding failure) with the presence of insulator flashover; the initial impulse waveform is similar to that in (1) prior to the flashover, and appears as a chopped wave after the flashover.
3. Lightning strikes the shielding device without the presence of insulator flashover; the initial impulse wave on the conductor is mainly induced from the ground wire.

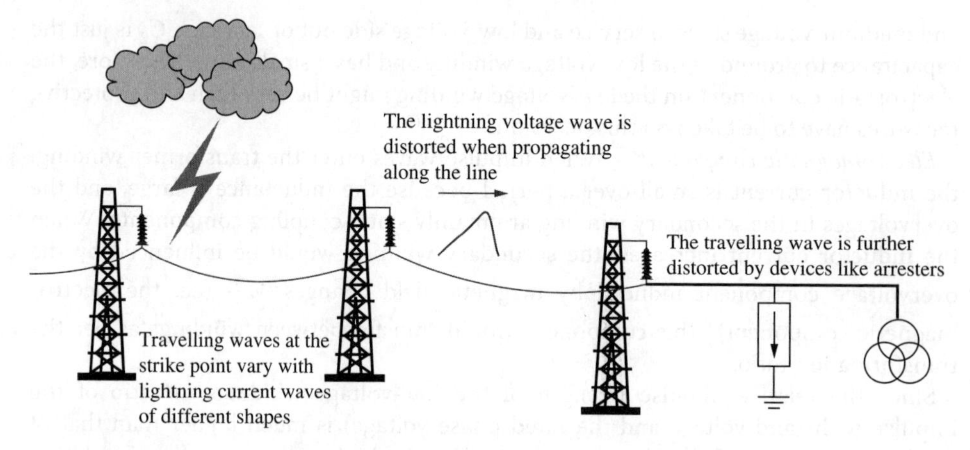

Figure 4.11 Formation of lightning overvoltage waves on electrical equipment.

4. Lightning strikes the shielding device with the presence of insulator flashover; the front part of the initial impulse wave is induced while the back part is a travelling wave caused by the injection of lightning current.

The wave shapes of the above four initial impulse waves on the conductor are illustrated in Figure 4.12.

In addition, the features of the initial impulse waves may be influenced by the dispersive lightning waveforms. Travelling wave signals generated by lightning are a mixture of signals of different propagation modes. Each frequency component has

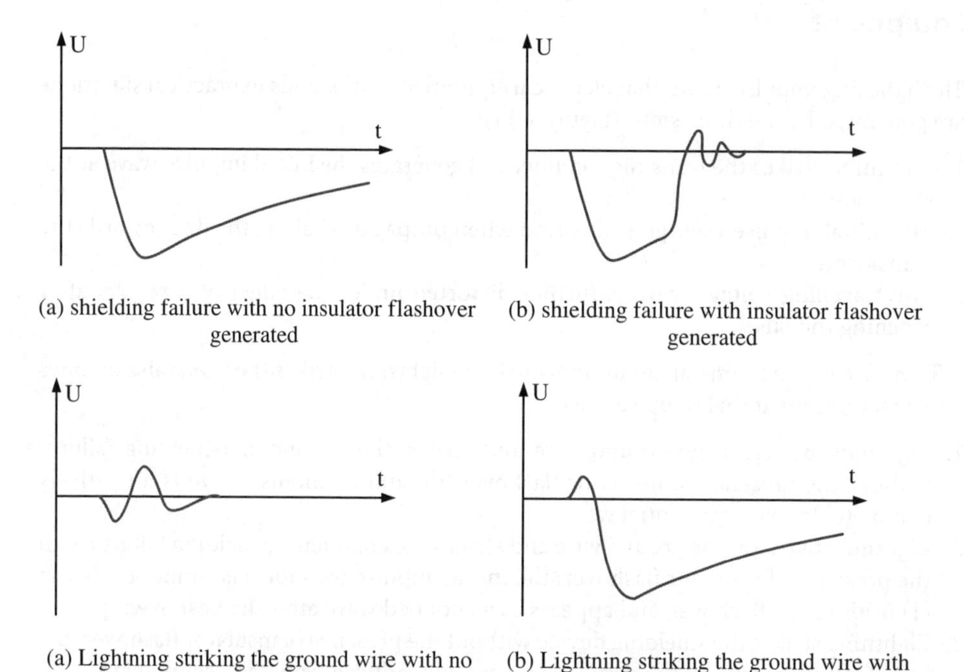

Figure 4.12 Four types of initial lightning impulse waves on conductors.

a different speed and attenuation speed, generating the dispersion that distorts the travelling waves during the propagation. For instance, the shape of the voltage wave tends to be gentler as high-frequency components attenuate at a higher speed.

The attenuation and distortion of travelling voltage waves would be influenced by transmission line parameters, such as earth resistivity, conductor type, sectional grounding wire and line spacing.

As the essential component of the transmission system, substation equipment have to be protected from incoming surges. Arresters must therefore be installed where the overhead transmission line meets the substation to suppress overvoltages. It is generally suggested that for lightning voltage waves, the amplitude of the refracted wave is much lower than that of the incident wave at the substation entrance, i.e. the equivalent wave impedance of the substation Z_2 is much smaller than that of the overhead transmission line Z_1.

Figure 4.13 is taken as an example, where by definition, the direction in which positive charges flow towards the substation is the positive direction of current, and the travelling wave moving towards the substation is the forward travelling wave. After lightning strikes the transmission line, the forward travelling voltage wave u_q of negative polarity and the forward travelling current wave i_q of negative polarity will be generated on the line. On arriving the arrester, the travelling waves will be refracted and reflected. For the voltage travelling wave, the amplitude of the forward refracted wave u_z will be considerably reduced after passing through the arrester used to suppress overvoltages. In order to ensure the voltage uniqueness of arrester configuration, a reflected wave of the opposite polarity will be generated. For the travelling current wave, the reflected wave i_f is a backward travelling wave of a polarity opposite to that of the backward travelling wave u_f, hence i_f is negative.

To sum up, at the connection between the overhead transmission line and the substation, the travelling voltage wave would be negatively reflected and the travelling current wave positively reflected.

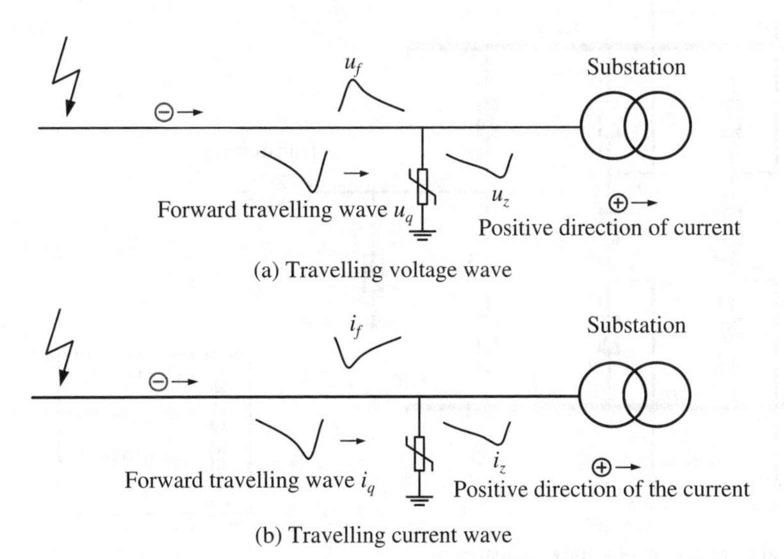

Figure 4.13 Refraction and reflection of overvoltage travelling waves in substations.

4.4 Simulation of Incoming Surges in Substations

In this book, the simulation model for a 500 kV ac substation subject to incoming surges is established based on EMTP-ATP. The basic structure of the ac substation is shown in Figure 4.14. Of the 5 outgoing lines, Line 1 and Line 2 are both 46 km long. PTs are arranged at the substation entrance, and ZnO arresters are arranged at the transformer entrance. Provided that lightning strikes Line 1, given the in-station breaker states, the main factors that may impact the propagation of transients are Line 1, Line 2, the in-station buses, arresters and the transformer entrance capacitance. The ATP-based substation model is illustrated in Figure 4.15.

In this model, a capacitor of 0.5 nF is used to simulate the effect of PT on transient travelling waves, while a capacitor of 3 nF is used to simulate the transformer entrance capacitance. Buses and other connecting lines within the station are simulated using the distributed parameter model with the length of each line shown in Figure 4.14. The parameters of the V–I characteristic curve of the arrester is listed in Table 4.3.

The response of lightning striking the transmission lines is a transient process in the order of μs, and involves the propagation issue of travelling wave components of different frequencies. Therefore, a frequency-dependent model (JMarti model) in ATP is used to simulate overhead transmission lines.

The multi-wave impedance model is adopted by the transmission line tower, and can reflect the practical process of travelling waves reflecting and refracting within the tower and at the tower ground point. The multi-wave impedance model of the 500 kV ac tower of type 5D1X1-ZH1 is shown in Figure 4.16.

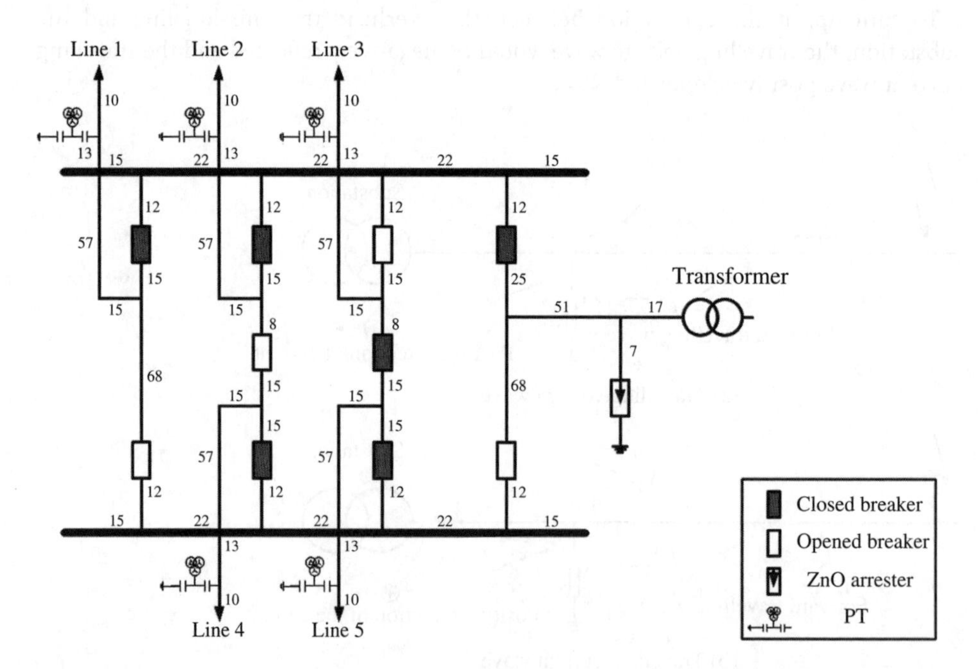

Figure 4.14 Electrical composition of a 500 kV substation.

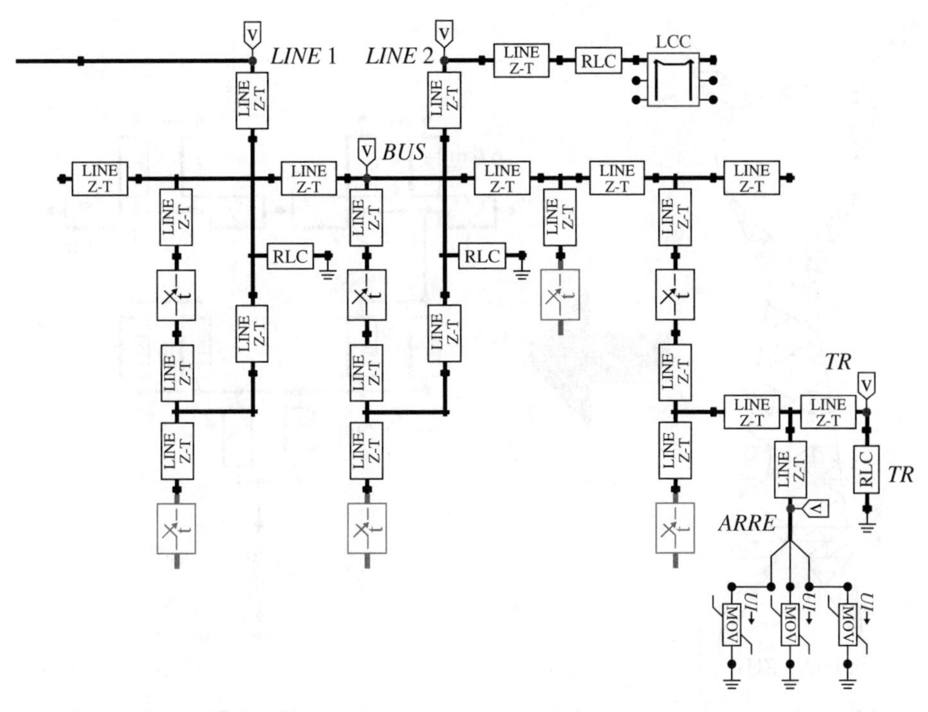

Figure 4.15 ATP-based simulation model of a 500 kV substation.

Table 4.3 Parameters of the arrester V–I characteristic curve.

Current (A)	Voltage (kV)
0.001	565
1000	744
2000	788
5000	863
10,000	903
20,000	960

Insulator string flashover has to be modeled to simulate the string breakdown after lightning strikes the transmission line. Current insulator string flashover models primarily include the definition method, the intersection method and the leader method. In both definition and intersection methods, the occurrence of insulator string flashover is judged based on the comparison between the overvoltage across the string terminals and the V–I characteristic curve acquired by the impulse test under the standard lightning wave (1.2/50 μs). In practice, the voltage wave across the insulator string caused by actual lightning strikes differs greatly from the standard lightning wave. The leader method, rich in physical significance, is based on discharge mechanism research of long air gaps. This method is used in this chapter for identifying

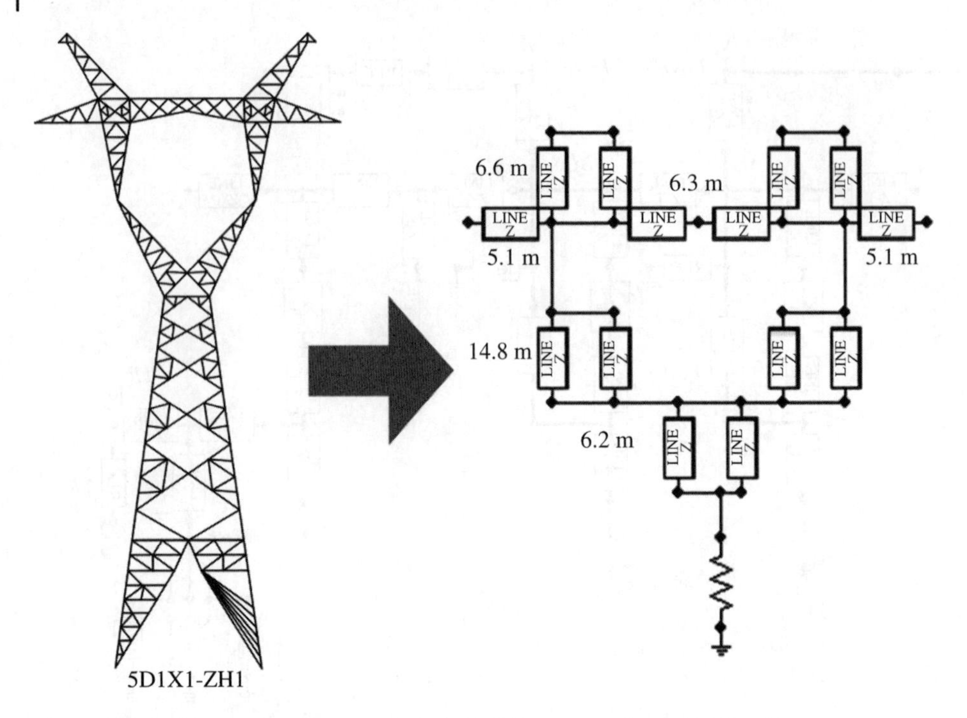

Figure 4.16 Multi-wave impedance model of a 500 kV ac tower of type 5D1X1-ZH1.

insulator string flashover under lightning strikes. The flashover model is built using Models language and user-defined components in ATP.

Moreover, in practical conditions, the overvoltage across the insulator string contains induced lightning of the return stroke channel, whose effect is reflected in the insulator string flashover model using Models language as well. The 500 kV substation subject to incoming surges is modeled in ATP, as illustrated in Figure 4.17.

4.5 Influencing Factors of Substation Incoming Surges

4.5.1 Influences of Lightning Stroke Types on Incoming Surges

Lightning generates travelling overvoltage waves on transmission lines by various means and can be categorized into direct lightning and indirect lightning, based on whether lightning strikes the transmission line directly or strikes the ground in the vicinity of the transmission line. Among others, direct lightning strokes can be subdivided into four groups depending on the strike locations and the presence of insulator flashover, i.e. lightning strokes bypassing ground wires (hitting conductors directly) with or without insulator flashover and lightning strokes on ground wires (or transmission towers) with or without insulator flashover. The first two kinds are also referred to as shielding failures. For different types of lightning, the physical processes of the resulting incoming surges differ from each other; therefore the lightning overvoltages near the strike site are varied as well. In this section, analysis will be made on overvoltages caused by different types of lightning strokes in the vicinity of the strike site (Figure 4.18).

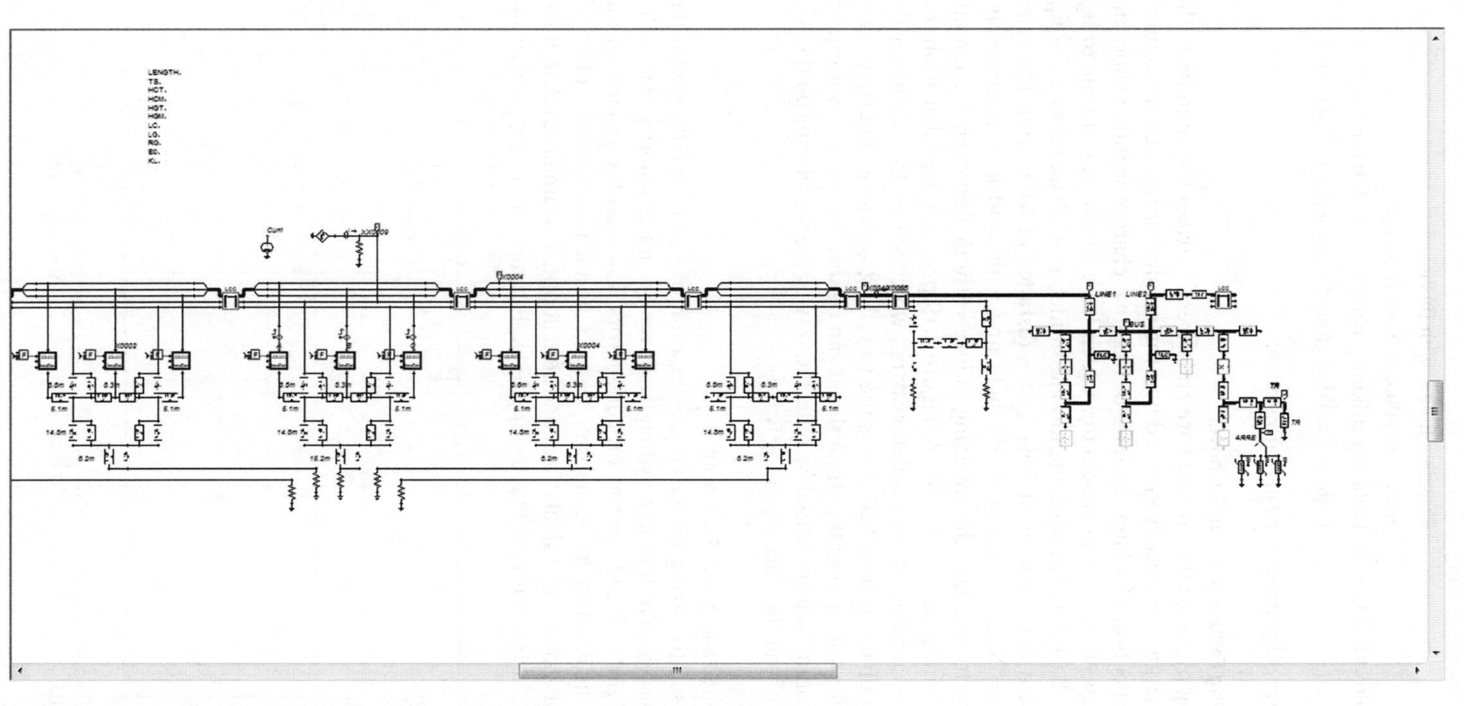

Figure 4.17 ATP-based simulation model of a substation subject to incoming surges.

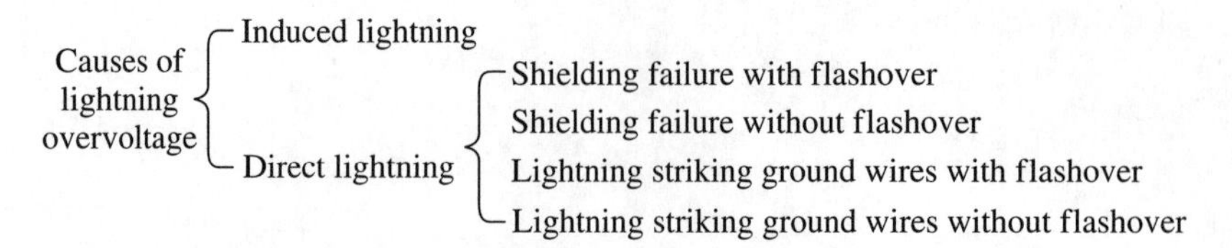

Figure 4.18 Causes of lightning overvoltage generation.

4.5.1.1 Shielding Failure Without Flashover

When lightning bypasses the ground wire and directly strikes the conductor, thunderstorm cloud charges are directly injected into the conductor through the return stroke channel. The injection of a large number of negative charges within a short time will generate voltage waves on the conductor, which have similar waveforms to lightning current waves. Based on the aforementioned model, it is assumed that the lightning strikes the phase A conductor of Line 1 at a distance of 6 km from the substation. For transmission lines rated 500 kV, the lightning withstand level is generally above 20 kA. In order to simulate the shielding failure without flashover, the amplitude of the applied lightning current is −10 kA. Figure 4.19 presents the calculation results of the three-phase currents of the insulator string, which shows that flashover does not happen on any phase. Figure 4.20 shows the travelling voltage and current waveforms on the conductor 400 m from the strike site. It can be seen from the waveforms that the jitter at time instant 50 μs is caused by reflected waves; the waveforms on the whole are similar to that of the standard lightning wave.

4.5.1.2 Shielding Failure with Flashover

Prior to the insulator string flashover, the situation of shielding failure with flashover is much like that without flashover: when flashover occurs, a majority of thunderstorm cloud charges and induced charges on conductors flow into the ground through the channel of the insulator string. As a result, the voltage and current waves on conductors are all chopped waves. The calculation condition adopted is similar to that of shielding failure without flashover except that the applied lightning current amplitude is modified

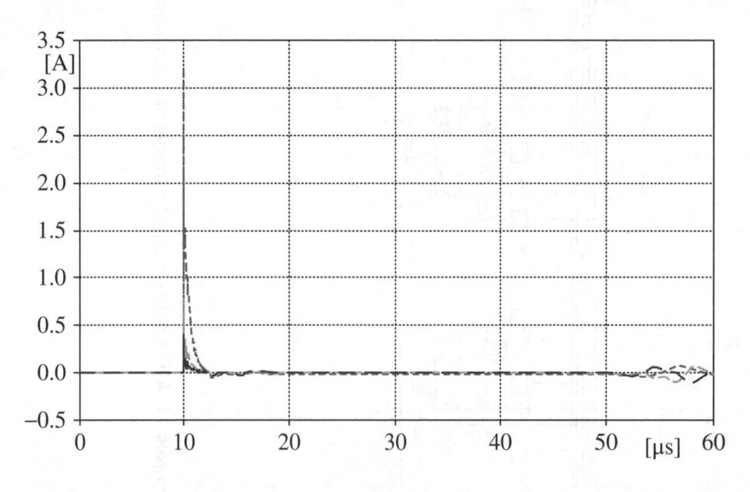

Figure 4.19 Currents flowing through the tower insulator string.

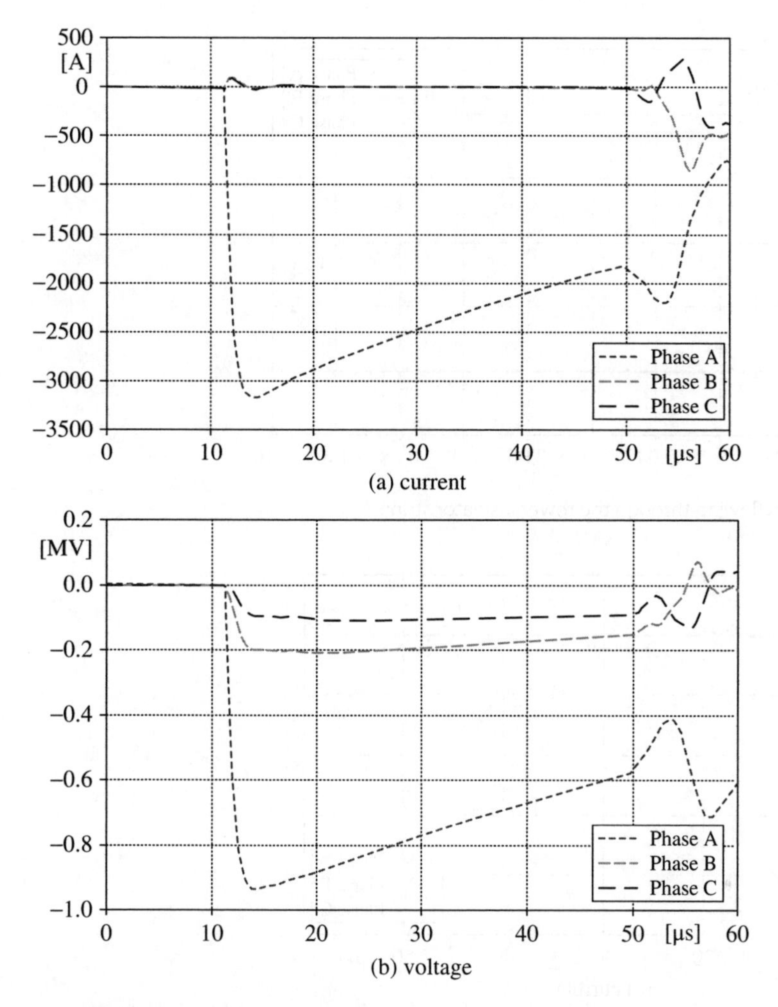

Figure 4.20 Shapes of travelling voltage and current waves on conductors 400 m from the strike site (shielding failure without flashover).

to −30 kA to cause the insulator string flashover. Figure 4.21 displays the calculation results of the currents flowing through the insulator string, which indicates that at time instant 15 μs, flashover happens on phase A.

Figure 4.22 shows the travelling voltage and current waveforms on the conductor 400 m from the strike site. It is indicated by the calculation results that before the flashover, travelling waves are featured by standard fronts, the same as that of shielding failure without flashover, while after the flashover, the amplitudes of travelling voltage and current waves drop rapidly, resulting from the flow of charges into the ground through the arc channel.

4.5.1.3 Lightning Striking Ground Wires (or Transmission Towers) Without Flashover

Under this condition, the voltage and current waves on the conductor are mainly caused by electrostatic and electromagnetic induction by the charges flowing through

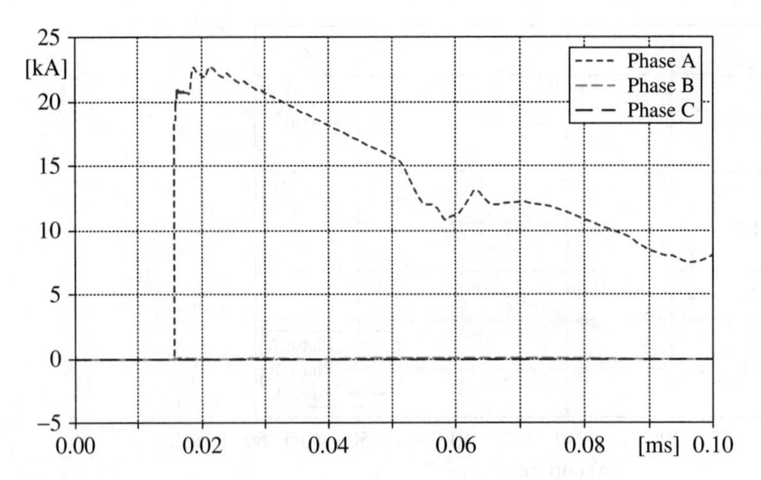

Figure 4.21 Currents flowing through the tower insulator string.

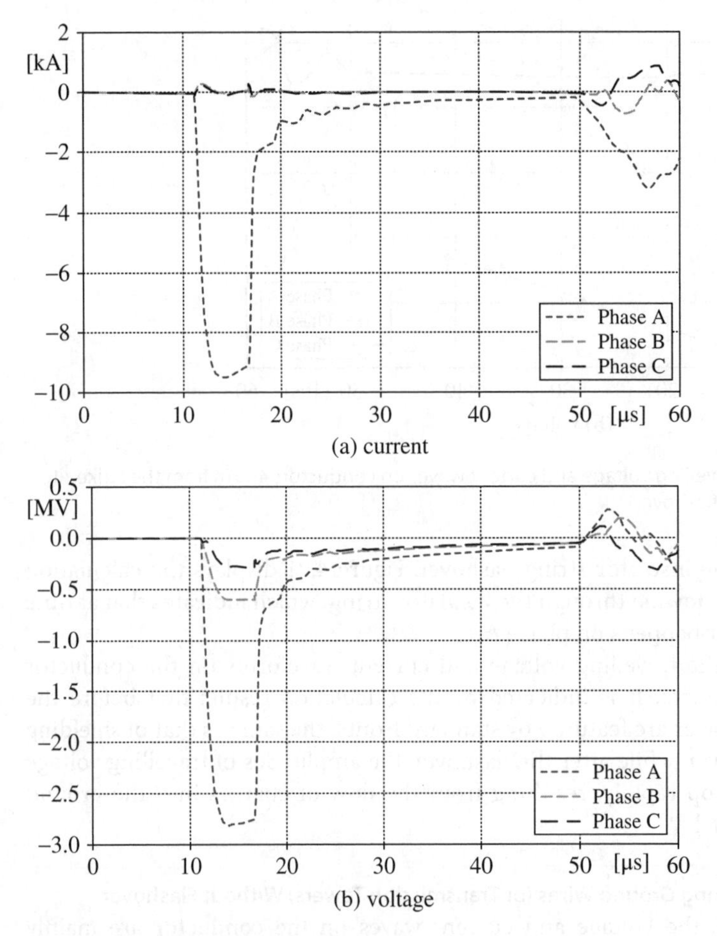

Figure 4.22 Shapes of the travelling voltage and current waves on conductors 400 m from the strike site (shielding failure with flashover).

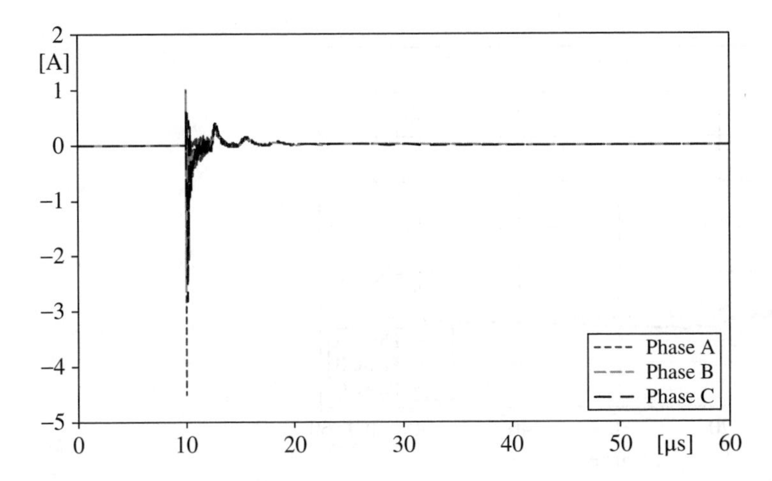

Figure 4.23 Currents flowing through the tower insulator string.

the ground wire and tower. The voltage waves travelling along the conductor originate from electrostatic induction by the ground wire potential while the current waves originate from electromagnetic induction by the ground wire current. Therefore, the voltage waves share similar features with the ground wire potential while the shapes of the current waves resemble that of the reciprocal of the ground wire current. The calculation conditions used are similar to that in Section 4.5.1.1, except that the applied lightning current amplitude is modified to −30 kA and the tower top is set as the striking location. Figure 4.23 shows the currents of the insulator string, and there are no signs of flashover. Figure 4.24 shows the voltage and current waveforms on the conductor 400 m from the strike point. For the applied standard lightning waves, both the front and tail time of the voltage waves show an increase; the waveforms on the whole, however, still highly resemble that of double exponential waves.

4.5.1.4 Lightning Striking Ground Wires (or Transmission Towers) with Flashover

Before the flashover of the insulator string, the features of the voltage and current waves on the conductor are much like that described in Section 4.5.1.3 without the presence of flashover; after the flashover of the insulator string, a portion of thunderstorm cloud charges, originally injected into the ground through ground wires and towers, is injected into the conductor, leading to the rise of the potential and current of the flashover phase. The calculation condition is similar to that described in Section 4.5.1.3 except that the lightning current amplitude is modified to 150 kA. Figure 4.25 shows the currents flowing through the insulator string, which indicates the breakdown of the phase A insulator string at a time instant around 18 μs. Figure 4.26 shows the shapes of the voltage and current waves on the conductor 400 m from the strike site. It is demonstrated by the calculation results that before the flashover, the voltage and current waves are both induced from the ground wire with rather low amplitudes while after the flashover, the injection of the lightning current significantly improves the voltage and current amplitudes, and the waveforms resemble that of double exponential waves with shortened front and tail time compared to the standard lightning wave.

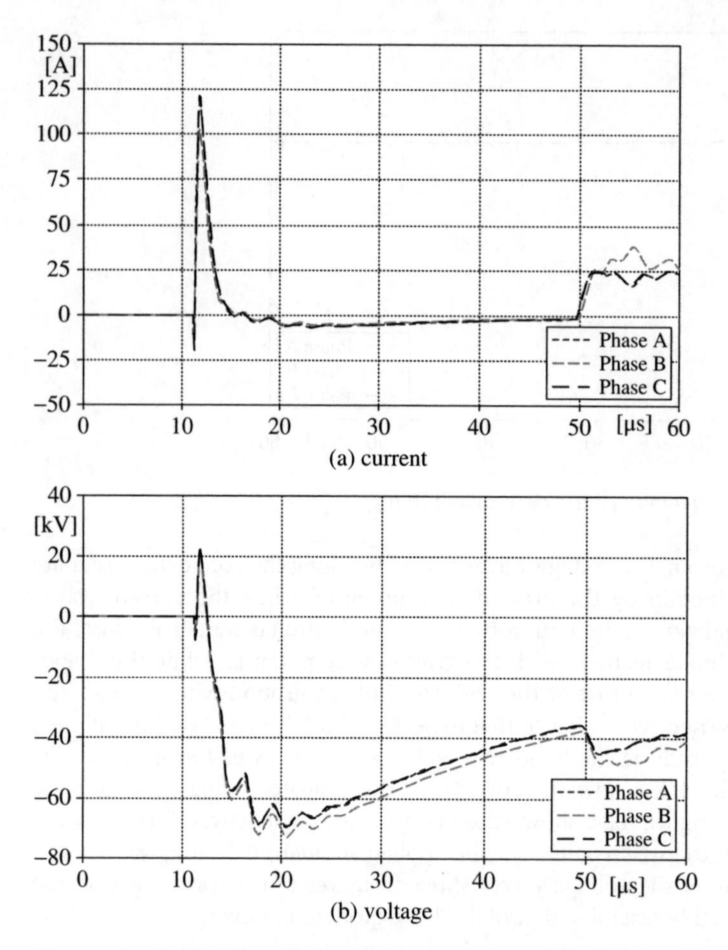

Figure 4.24 Shapes of travelling voltage and current waves on the conductor 400 m from the strike site (back flashover without flashover of the insulator string).

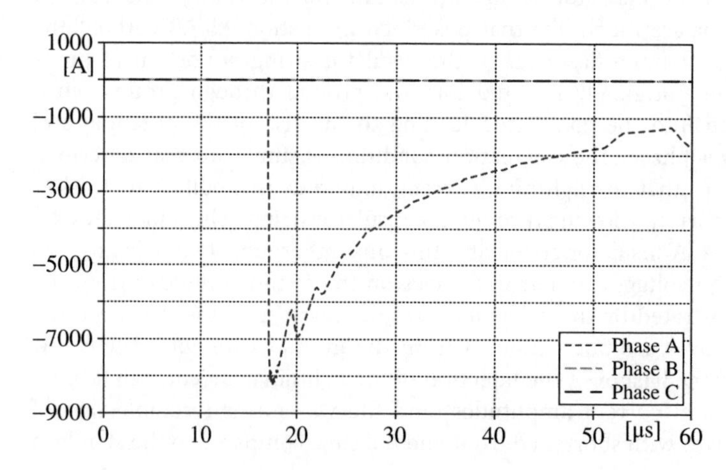

Figure 4.25 Currents flowing through the tower insulator string.

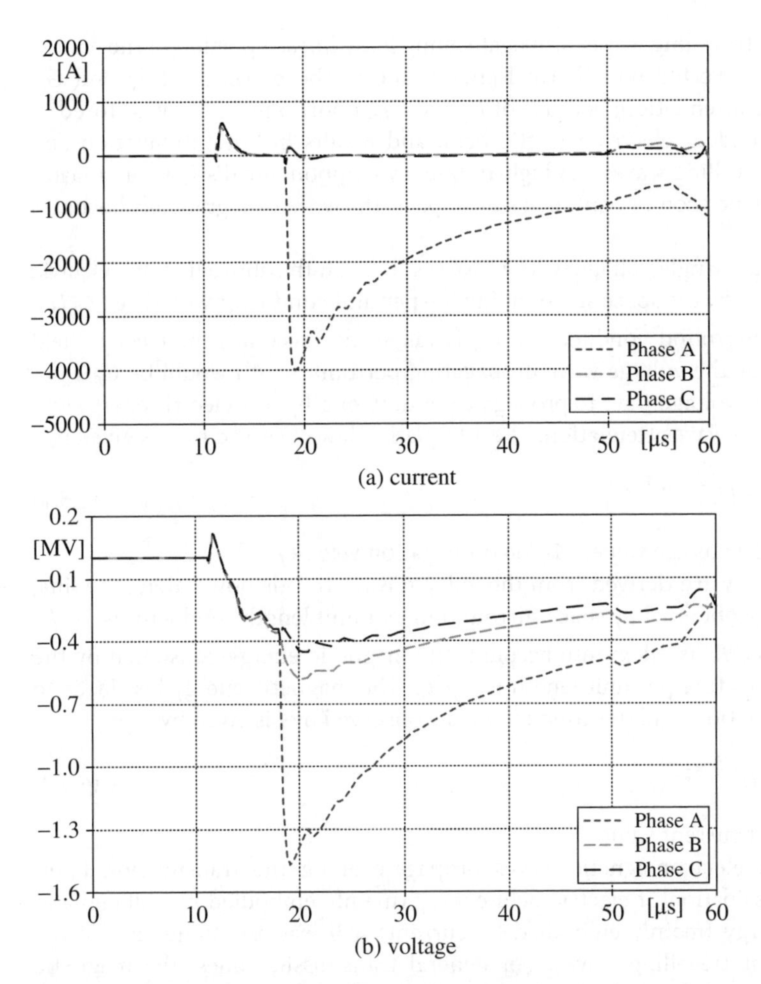

Figure 4.26 Shapes of travelling voltage and current waves on the conductor 400 m from the strike site (back flashover with insulator string flashover generated).

4.5.2 Influences of Transmission Lines on Overvoltage Wave Propagation

When voltage and current waves are propagating on ideal lossless transmission lines, hardly any energy is lost (stored in the electromagnetic field) and waveforms are hardly distorted. In practical conditions, energy losses always happen when waves travel along transmission lines, and are mainly caused by:

1. conductor resistance;
2. conductor-to-ground conductance;
3. conductor corona;
4. ground losses.

The distributed parameter circuit of a transmission line is depicted in Figure 4.3, where R_0 contains conductor resistance and ground resistance; G_0 contains leakage and insertion loss; L_0 contains conductor self-inductance and interphase mutual inductance; and C_0 contains conductor-to-ground capacitance and capacitance between conductors. When travelling waves propagate along lossy lines, a portion of the energy is dissipated and converted into thermal energy due to the presence of R_0 and G_0, and results in wave attenuation; because of L_0 and C_0, a portion of energy is stored in the frequency-dependent electromagnetic field, and results in both attenuation and distortion of the travelling waves. As high-frequency components dissipate at a faster pace and low-frequency components at a slower pace, the waveform gradually becomes gentle.

Assume that a right-angle voltage wave U passes through the above line, the electric field energy obtained by the space surrounding the per unit conductor would be $\frac{1}{2}C_0U^2$. If the conductor-to-ground conductance G_0 is taken into account, the electric field energy consumed by the voltage wave propagating per unit length would be $G_0{U_0}^2t_0$ (t_0 is the time for the voltage wave to propagate per unit length). The electric energy loss would lead to voltage wave attenuation. The attenuation law observed by u is given by

$$u = Ue^{-\frac{G_0}{C_0}t} = Ue^{-\frac{G_0}{C_0}\times\frac{x}{v}} \tag{4.5.1}$$

where U is the initial voltage value; v is the propagation velocity.

When the current wave derived from the voltage wave travels down along the line, the magnetic energy obtained by space surrounding per unit length conductor is $\frac{1}{2}L_0I^2$. If the line resistance R_0 is taken into account, the magnetic energy consumed by the current wave propagating per unit length is $R_0i^2t_0$. The magnetic energy loss leads to current wave attenuation, and the attenuation law observed by i is given by

$$i = Ie^{-\frac{R_0}{L_0}t} = Ie^{-\frac{R_0}{L_0}\times\frac{x}{v}} \tag{4.5.2}$$

where I is the initial current value.

Therefore, when electromagnetic waves propagate along the transmission lines, the storage and dissipation of electric field energy (mainly embodied by voltage) and magnetic field energy (mainly embodied by current) will lead to attenuation of the voltage and current travelling waves. For general transmission lines, the magnetic field energy dissipates faster than the electric field energy, enabling energy conversion from electric field energy to magnetic field energy. As a result, the voltage wave will experience a number of negative reflections during the propagation with its fronts gradually chopped and tails gradually elongated. Also, because of the skin effect, conductor resistance increases with the frequency, and the travelling wave with a steep front is attenuated more significantly when propagating along the line.

The above analysis is based on the condition where the voltage wave is a right-angle one. Under practical conditions, the voltage waves have various complex shapes and tend to contain multiple frequency components. When the voltage components of different frequencies pass through the same line segment, the attenuation coefficients vary greatly. High-frequency components generally attenuate at a faster speed while low-frequency components attenuate at a slower speed. The differences in the attenuation degree between frequency components will also result in distortion.

In order to further explain the attenuation and distortion of the voltage wave propagating along the transmission line, related simulation is performed. Because the research is limited solely to the travelling wave propagation along the transmission line, the influence of substations is not taken into account. The transmission line under study is 45 km in length, with a 40 km line of the same parameters matched at the terminal to avoid the impact of reflected waves.

When phase A is directly hit by lightning of −10 kA in the case of shielding failures, simulations of the voltage waves on the phase A conductor 1, 5, 15, 25, 35 and 45 km from the strike site are presented in Figure 4.27. The calculation has demonstrated that after propagating a distance of 45 km, the amplitudes of the voltage waves attenuate to approximately half that of the original wave, the front time is increased from 2–3 μs to 12 μs, and the tail becomes significantly gentler. Figure 4.28 shows the simulation results of the voltage waves under the shielding failure with insulator string flashover with the amplitude of the applied lightning current being −30 kA. The simulation results come to the same conclusion that after travelling a distance of 45 km, voltage waves exhibit attenuated amplitudes and gentler shapes.

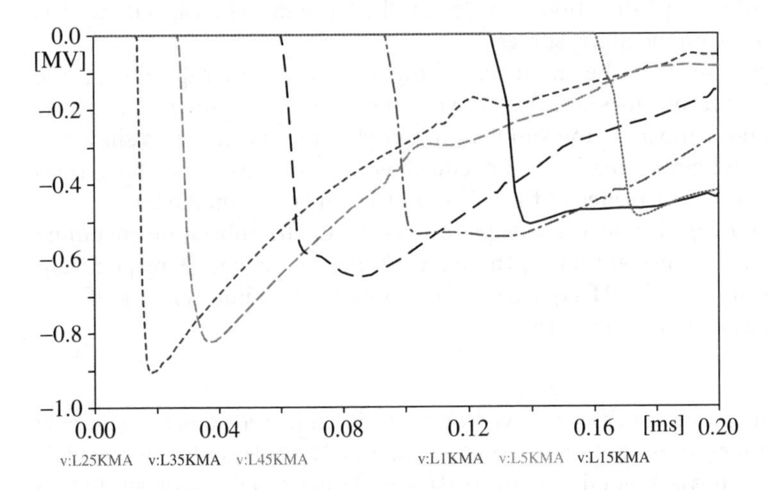

Figure 4.27 Along-the-line voltages under shielding failures without flashover.

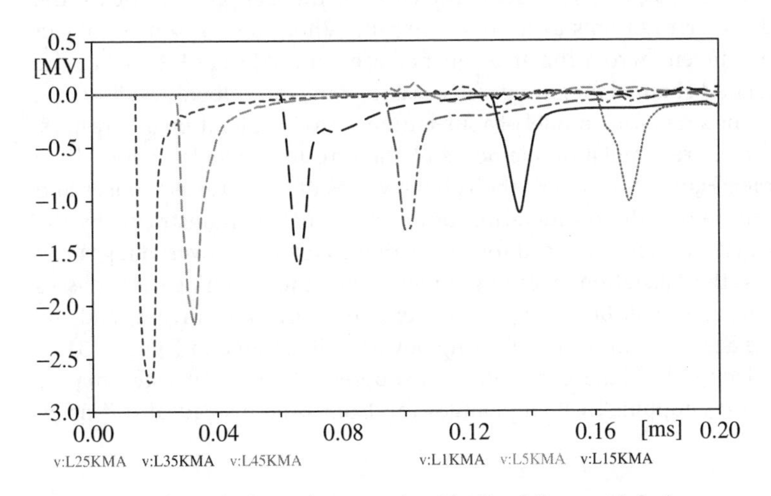

Figure 4.28 Along-the-line voltages under shielding failures with flashover.

4.5.3 Influences of In-Station Equipment on Overvoltage Wave Propagation

4.5.3.1 Potential Transformers (PT)

Potential transformers are usually employed on the outgoing lines of substations. A PT is equivalent to a capacitance when performing transient analysis, and may affect the shape of an incoming surge. If the wavefront time is about 20 μs, the period of the high-frequency components of the incoming surge can be treated as 80 μs, i.e. the frequencies can be seen as 1.25×10^4 Hz, and the equivalent capacitance of the PT is 0.5 nF. Based on the calculation formula of the capacitance given by

$$Z_c = \frac{1}{j\omega C} = \frac{1}{j2\pi fC} \tag{4.5.3}$$

the equivalent impedance of the high-frequency components is estimated to be tens of kΩ, much larger than the characteristic impedance of buses and transmission lines. Consequently, a preliminary estimation is made that the PT equivalent capacitance has no significant influence on incoming surges.

In order to further analyze the influence of the PT equivalent capacitance, the travelling voltage waves on lines and buses with and without consideration of the PT are simulated and compared. The simulation model of the substation is shown in Figure 4.29, where the circles mark the PT equivalent capacitances and the arrows illustrate the monitoring site located at the substation entrance and on the bus.

Figures 4.30(a) and (b) present the calculation results of the substation incoming surge with and without consideration of the PT equivalent capacitance, respectively. The outcomes indicate that the PT equivalent capacitance would not exert significant influences on the shape of incoming surges.

4.5.3.2 Arresters

As the major equipment for transformer overvoltage protection, arresters are generally arranged on substation buses or at transformer entrances. Nowadays, the most widely used ZnO arresters possess good nonlinear *U*–*I* characteristics, as illustrated in Figure 4.31. When the arrester is subject to the voltage with a low amplitude, the current across the arrester is very low, generally close to the leakage current (of the order of mA), and the arrester acts as an open circuit. When the arrester is subject to overvoltages, the current across the arrester increases to the level of A or kA; the arrester becomes a good channel for energy release and limits the voltage amplitude.

Figure 4.32 shows the simulation model of a substation with consideration of arresters, where inside the circle are simulation elements of the arresters installed about 17 m from the transformer. Figure 4.33 shows the voltage waves of the arresters, buses and transformers without and with consideration of arresters. It is demonstrated by the calculation results that the arresters used for protecting transformer overvoltages can significantly suppress the substation incoming surges. When the incoming surge passes through the arrester, there will be an apparent increase of arrester current, and the arrester will become a good channel for releasing power, as illustrated in Figure 4.34.

As illustrated in Figure 4.35, the overvoltage is suppressed by the arrester, mainly through the means of "chopping", which changes the front time instead of the rising

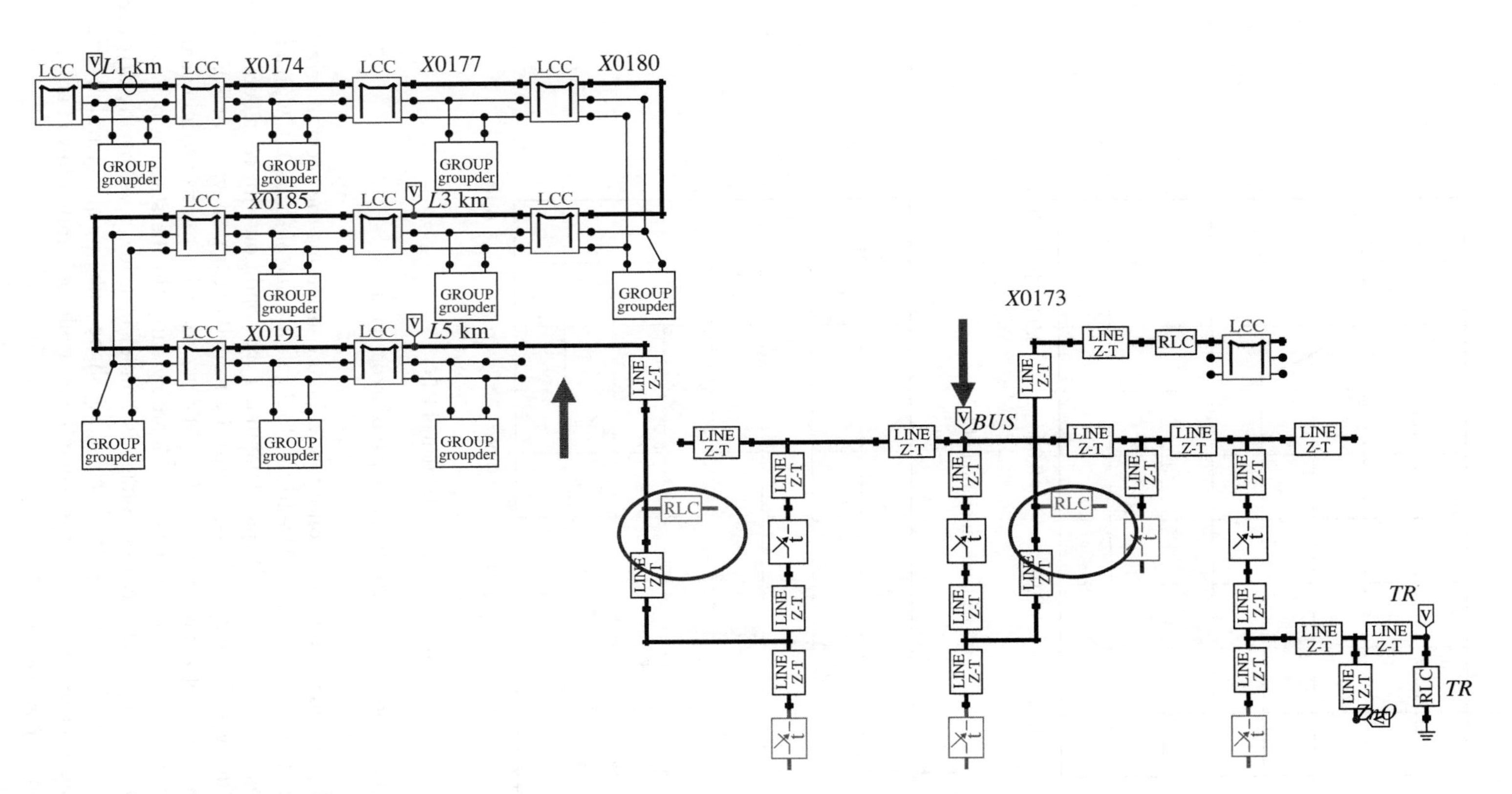

Figure 4.29 Simulation model of a substation.

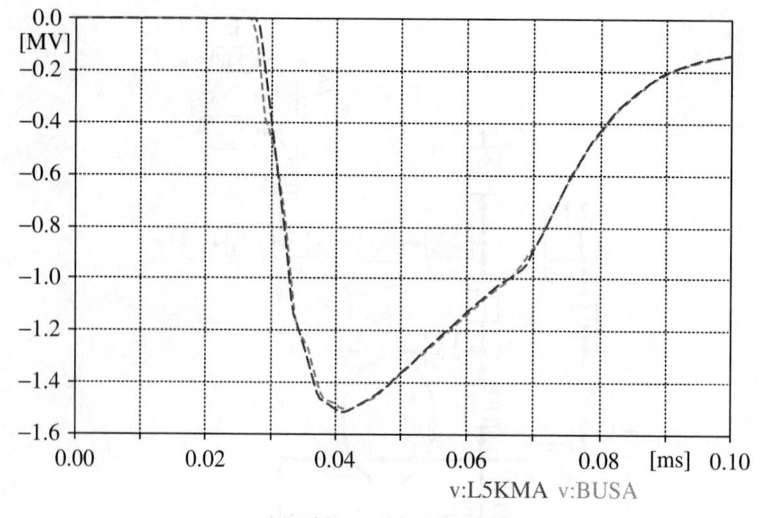

(a) with consideration of PTs

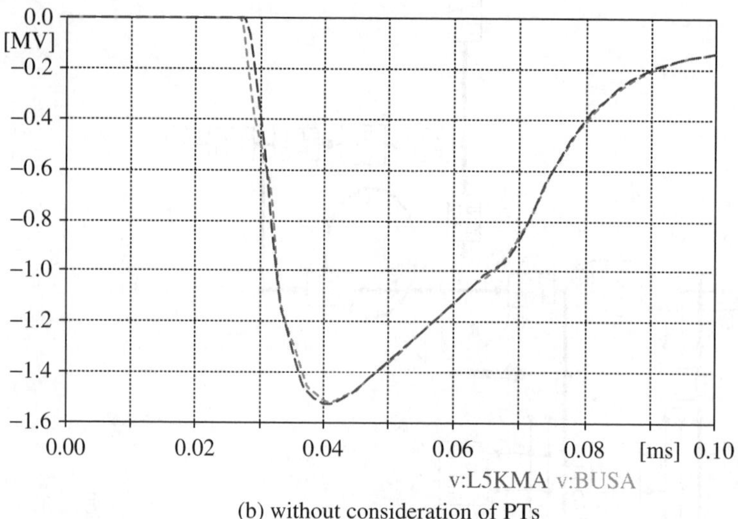

(b) without consideration of PTs

Figure 4.30 Calculation results of the PT's influences on incoming surges.

gradient. For instance, under the condition without considering the effect of arresters, the incoming surge takes 12 μs to rise from 0 to the peak value of −1.52 MV with a front slope of approximately −0.12 MV/μs, while under the condition with consideration of arresters, the incoming surge takes 4.5 μs to rise from 0 to the peak value of −0.78 MV with a front slope of approximately −0.17 MV/μs, and with the front time reduced from 12 μs to 4.5 μs. For the tail of the incoming surge, in the stage of overvoltage suppression (T_2–T_4), the bus overvoltage is limited to a low level by the arrester; the peak is chopped, forming a slowly changing shape with an equivalent tail time of approximately 200 μs. In the tail stage when the overvoltage is rather low in amplitude ($t > T_4$), the voltage drops rapidly.

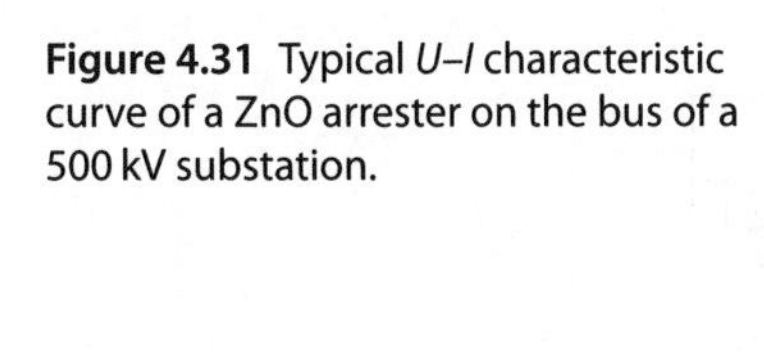

Figure 4.31 Typical U–I characteristic curve of a ZnO arrester on the bus of a 500 kV substation.

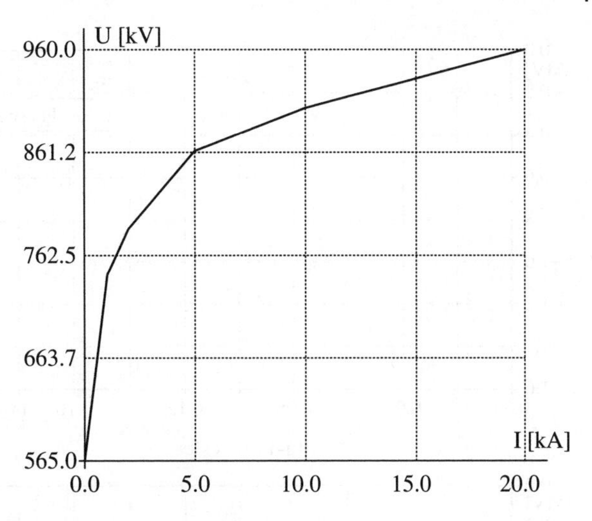

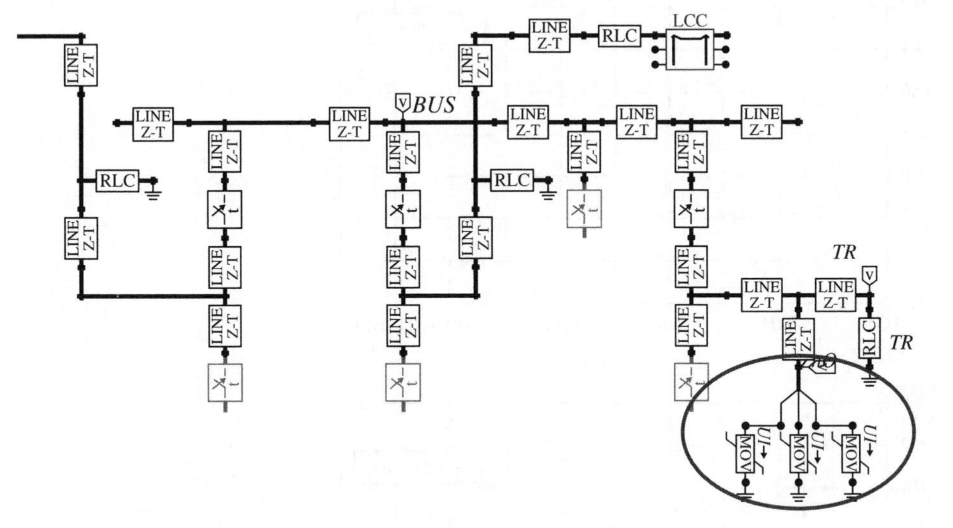

Figure 4.32 The simulation model of a substation with consideration of arresters.

4.6 Typical Waveforms of Substation Incoming Surges

According to the analysis in Sections 4.4 and 4.5, forms of lightning strike, propagation of travelling overvoltage waves along transmission lines as well as through in-station arresters can all distort the lightning overvoltage. Therefore the incoming surges that key equipment (e.g. transformers) withstands differ greatly from the standard lightning wave in terms of waveforms. In this section, several typical waveforms of the substation incoming surges will be simulated and analyzed.

4.6.1 Short-Front-Short-Tail Surges

According to the calculation results in Section 4.5, the front time of the incoming surge primarily depends upon the shapes of the lightning source and the propagation

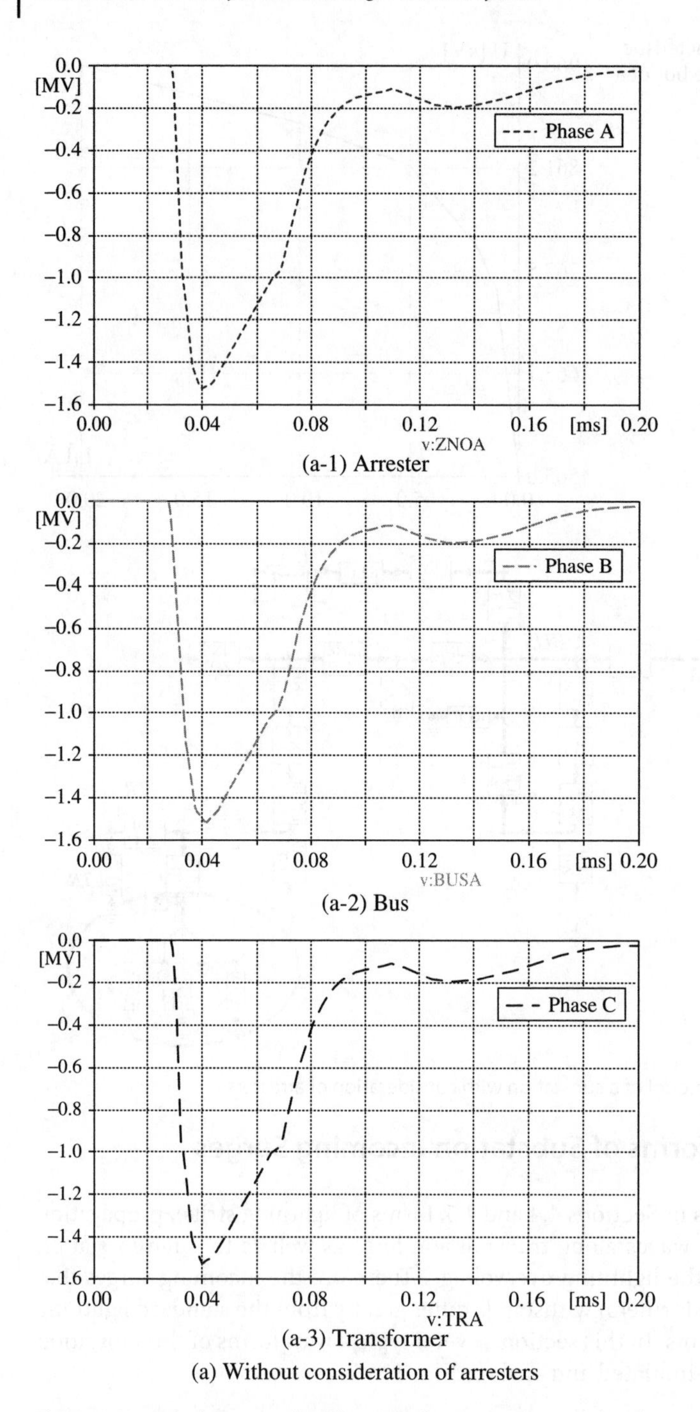

(a-1) Arrester

(a-2) Bus

(a-3) Transformer

(a) Without consideration of arresters

Figure 4.33 Calculation results (Phase A) of waves without and with consideration of arresters.

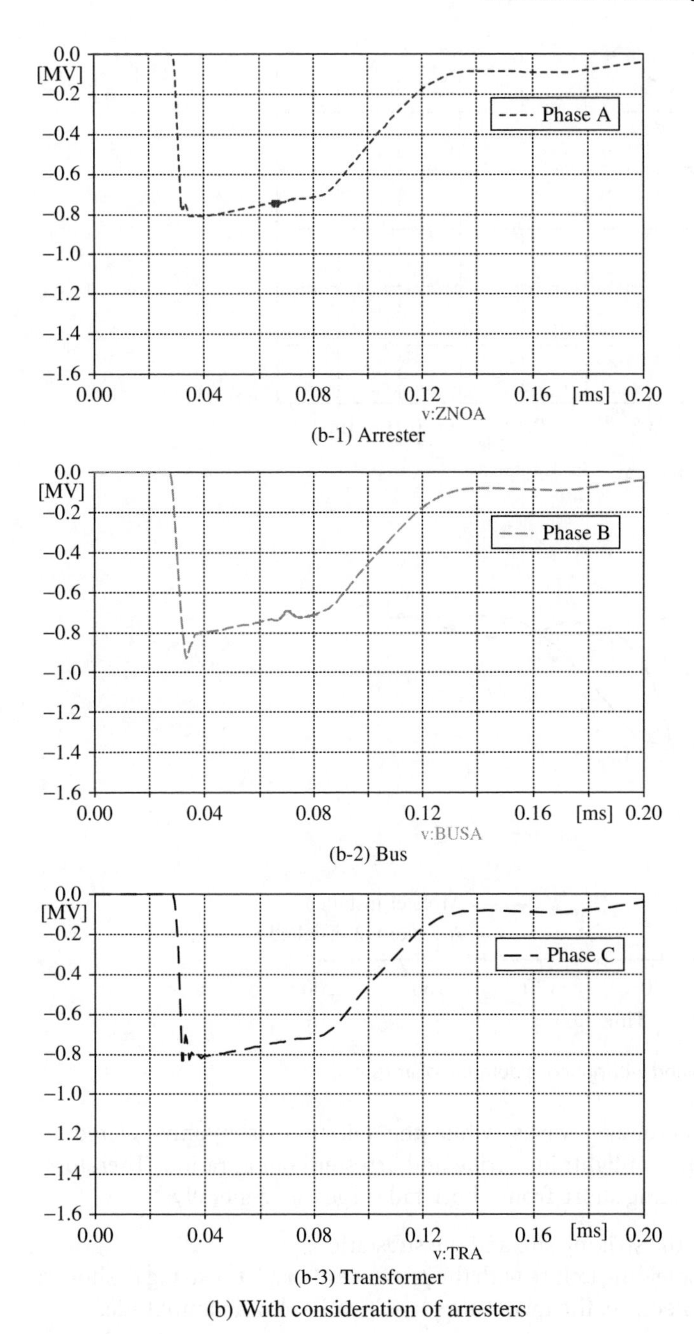

(b) With consideration of arresters

Figure 4.33 (*Continued*)

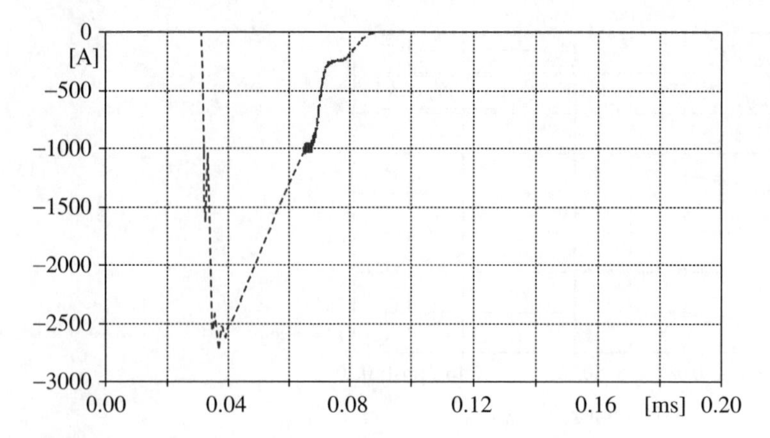

Figure 4.34 Arrester current (phase A).

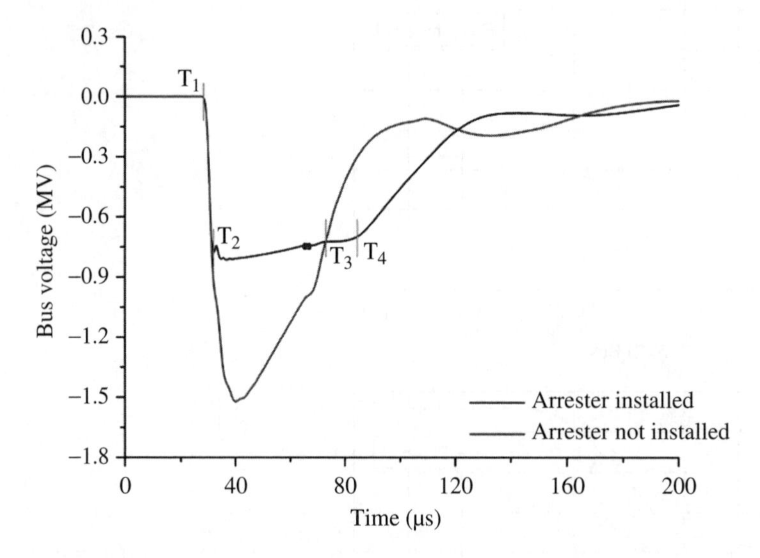

Figure 4.35 Bus voltages with and without consideration of arresters.

distance of travelling overvoltage waves, while the tail time is mainly related to propagation distances, types of lightning strike and arrester configuration. Therefore, the requirements for generating short-front-short-tail surges are generally:

1. close distance between the striking site and the substation;
2. lightning strike type is shielding failure with the presence of insulator string flashover;
3. short time of arrester effect, i.e. the incoming surge is rather low in amplitude.

As protection for incoming lines is generally designed for substations, shielding failures with flashover are assumed not to be present on the incoming lines in the study with the following calculation conditions: (a) a shielding failure occurs at a distance of 2 km from the substation (except for the incoming line section) and leads to insulator string flashover; (b) lightning current amplitude is the lightning withstand level (around −28 kA). The calculation results are shown in Figure 4.36, where the front

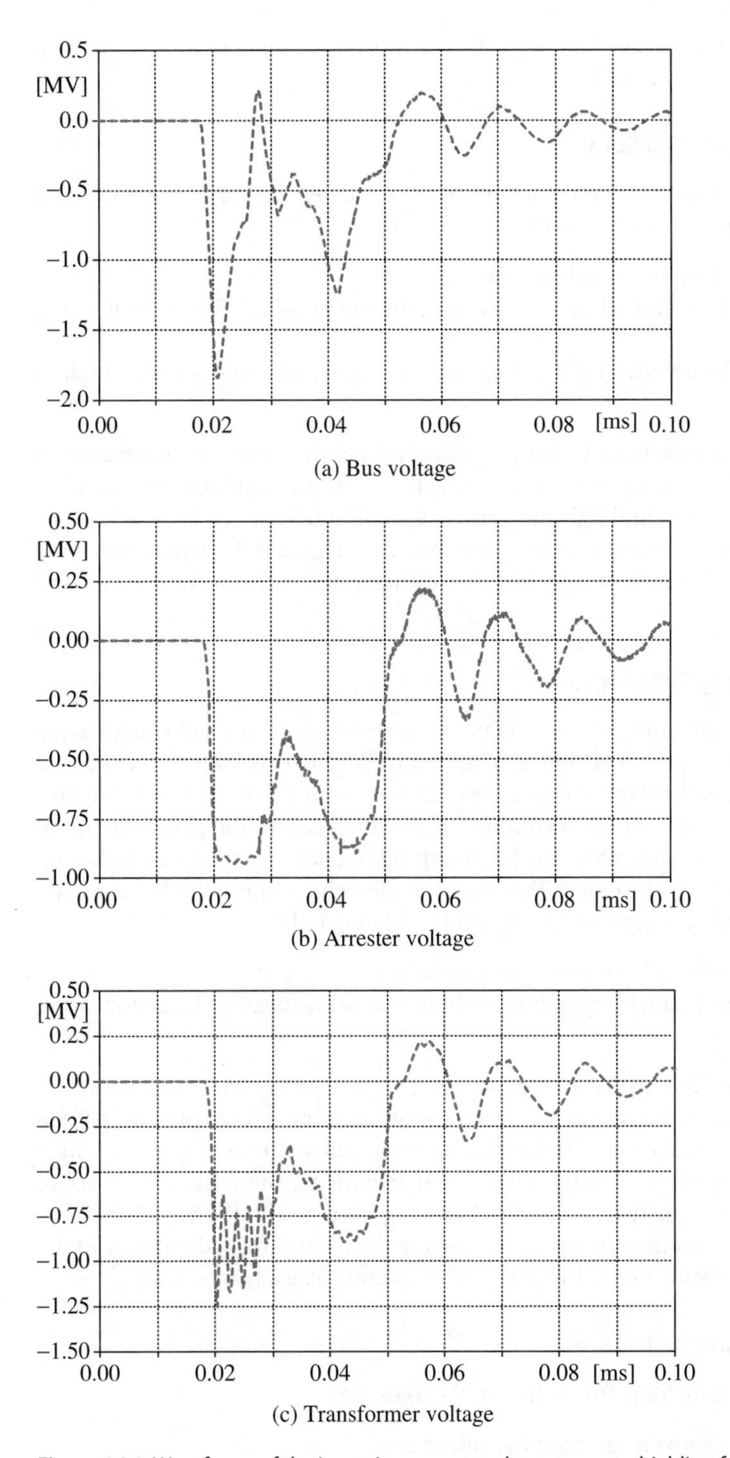

Figure 4.36 Waveforms of the incoming surges under near-zone shielding failures with flashover.

time and tail time of the overvoltage that the transformer is subject to are 2 μs and 12 μs, respectively.

4.6.2 Short-Front-Long-Tail Surges

Adopting similar analytical methods, the requirements for generating waves with short fronts and long tails are generally as follows:

1. striking location is close to the substation;
2. type of lightning strike is shielding failure without the presence of insulator string flashover;
3. operation time of the arrester is fairly long, i.e. the incoming surge is rather high in amplitude.

The calculation condition is as follows: a shielding failure occurs at a distance of 2 km from the substation (except for the incoming line section) without the occurrence of insulator string flashover, and lightning current amplitude is set to be −26 kA (not causing flashover). The calculation results are shown in Figure 4.37, where the front time and tail time of the overvoltage that the transformer withstands are 2 μs and 120 μs, respectively.

4.6.3 Long-Front-Long-Tail Surges

The front time of the incoming surge is mainly concerned with the lightning source waveform and the propagation distance of the travelling overvoltage waves. If the lightning source remains constant, the longer the distance from the strike site to the substation, the gentler the shape of the incoming surge, which is how long-front-long-tail waves are generated. The tail time of the incoming surge mainly depends on lightning strike types and propagation distance. Therefore the following requirements have to be satisfied to form incoming surges with long fronts and long tails:

1. striking location is relatively far from the substation;
2. lightning strike type is shielding failure without the occurrence of insulator string flashover.

Note that distortion as well as attenuation will occur during the propagation of travelling waves. Hence, if the strike site is too distant, the incoming surge amplitude will be lower than the insulation withstand level of the substation equipment, which is of little research interest. Given the overall considerations, the strike site is set to be 50 km away from the substation, and the lightning current amplitude −26 kA. The calculation results are presented in Figure 4.38 where the front time and tail time of the overvoltage that the transformer withstands are 12 μs and 150 μs, respectively.

4.6.4 Long-Front-Short-Tail Surges

The conditions to generate long-front-short-tail surges are:

1. striking location is relatively far from the substation;
2. lightning strike type is shielding failure without the occurrence of insulator string flashover.

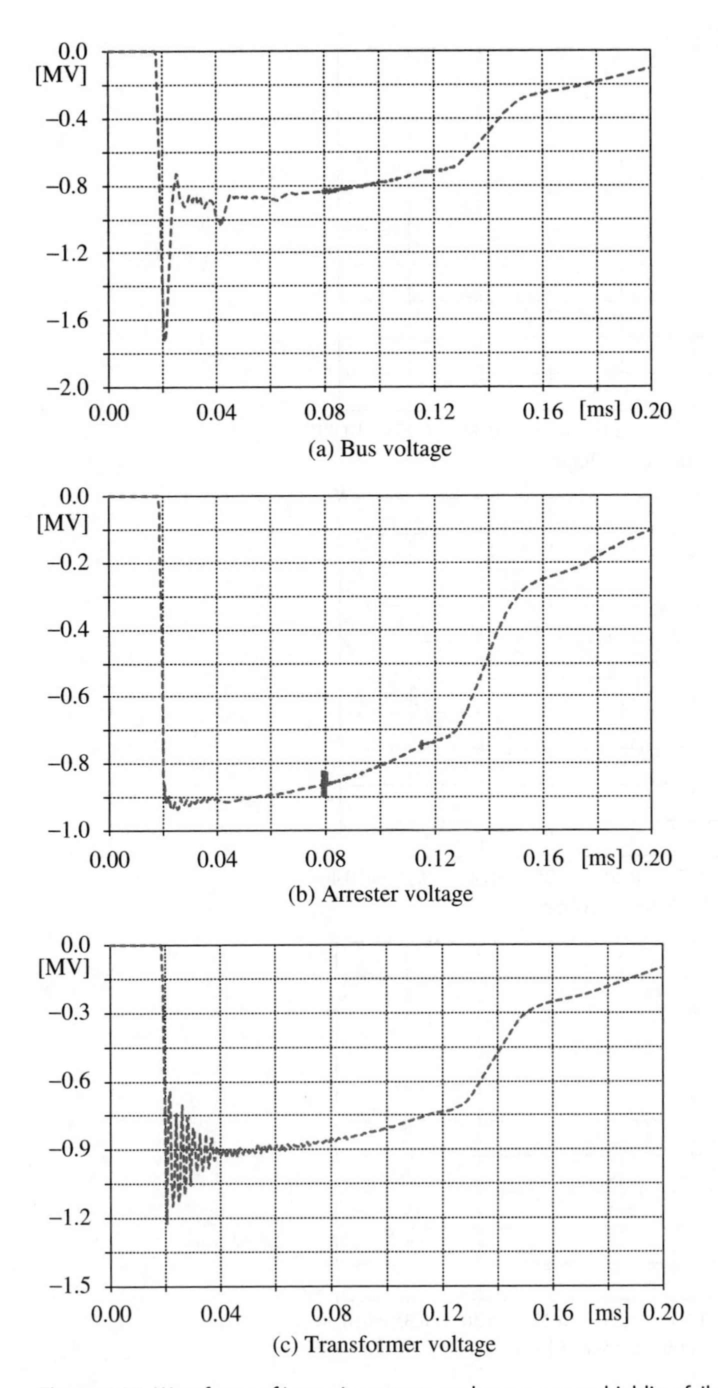

Figure 4.37 Waveforms of incoming surges under near-zone shielding failures without flashover.

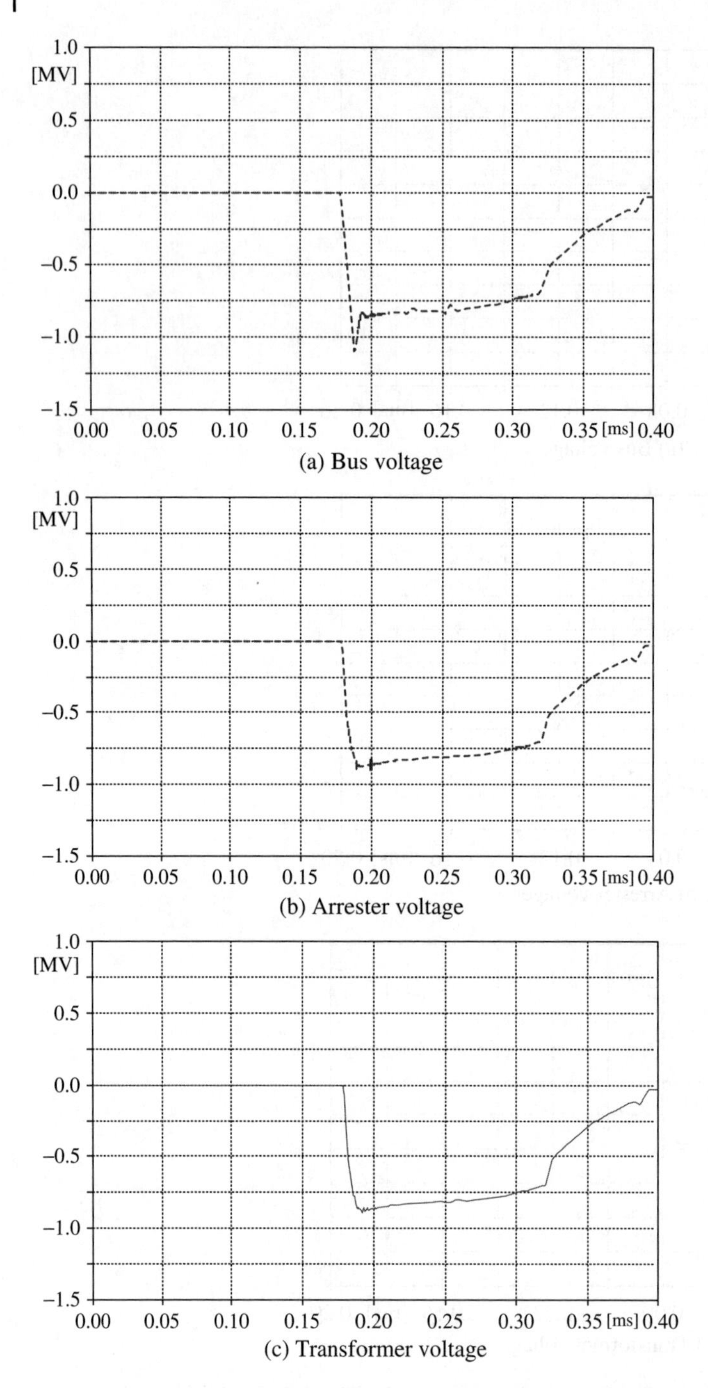

Figure 4.38 Waveforms of incoming surges under far-zone shielding failures without flashover.

According to the above analysis, the strike site is considered to be 50 km away from the substation, and the lightning current amplitude is set to be −28kA. The calculation results are shown in Figure 4.39, where the front time and tail time of the overvoltage that the transformer is subject to are 12 μs and 22 μs, respectively.

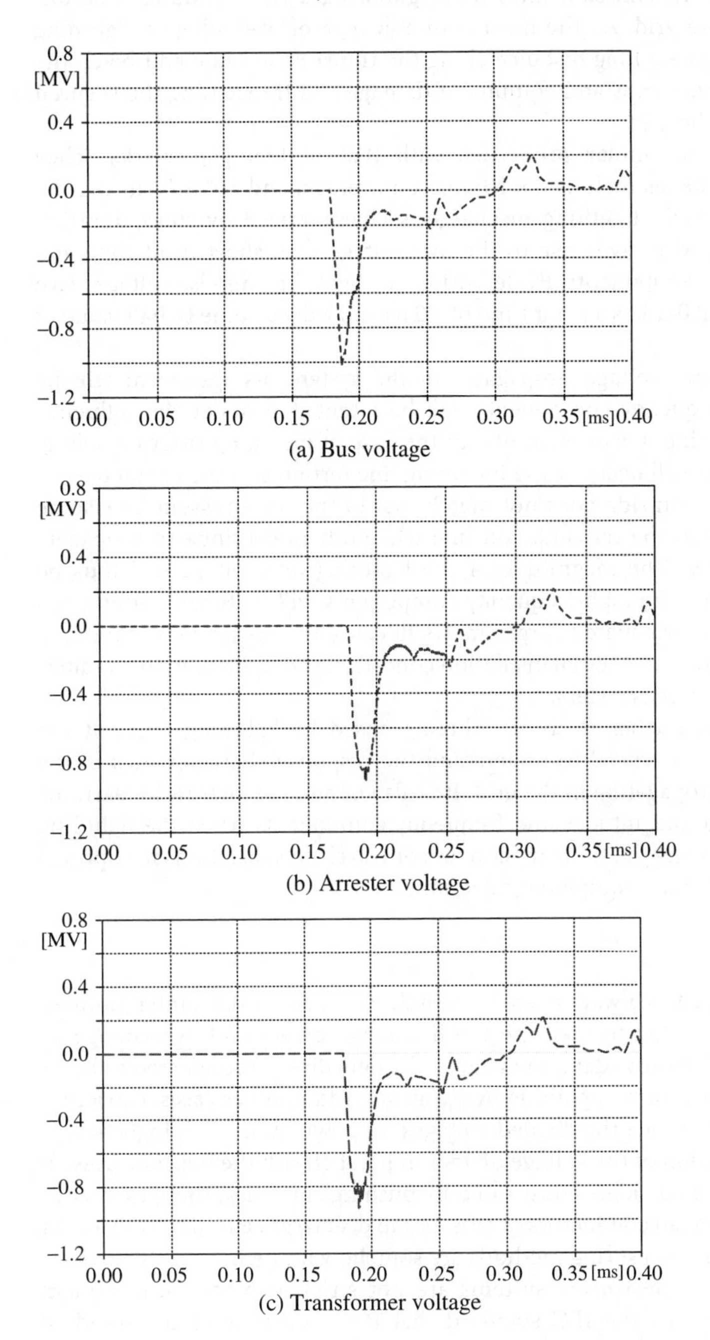

Figure 4.39 Waveforms of incoming surges under far-zone shielding failures with flashover.

4.7 Response Characteristics of Lightning Overvoltages Propagating in the Grid

When lightning strikes transmission lines or neighboring areas, lightning overvoltages are generated in the grid. As the most common type of overvoltages, lightning overvoltages can propagate a long distance along the transmission line and reach the substation, causing breaker trips and equipment damage and threatening the secured and stable operation of the grid.

Lightning leaders share similar properties with that of long-gap sparks. They develop at a very fast rate: as high as 8×10^5 m/s on average, and 10^9 m/s during the main discharge. As a result, lightning discharge is characterized by short duration and steep rising edge, and it gives rise to the overvoltage with short front time and multiple high-frequency components. By definition, the lightning impulse voltage wave standardized by IEC 60060-1 has a front time of 1.2 μs and a decay time to half value of 50 μs.

When the lightning overvoltage propagates in the system, its waveform will be distorted. Also, high-frequency components will be attenuated under the influence of line impedance, resulting in slower fronts. In the case of incoming surges reaching the substation, reflection will occur at the incoming line terminal as the characteristic impedance of the substation side does not match that of the transmission line; some components will return to the transmission line while others continue to propagate along the in-station lines. The magnitude of the lightning overvoltage will thus be reduced considerably, and the high-frequency components will be further attenuated as well. Nevertheless, the attenuated surges may still cause the breakdown of internal insulation and the flashover of external insulation concerning equipment with windings or cause damage to the surge arresters.

Therefore, in order to further study the threats posed by lightning overvoltages to electric equipment, it is crucial to understand the response characteristics of the lightning overvoltages propagating in the grid. It is also of guiding interest to learn the variations of overvoltage magnitudes and frequency components when the lightning overvoltage propagates along the transmission line or travels to substations for optimal grid structure and appropriate equipment selection.

4.7.1 Status Quo

The lightning overvoltage has always been a research focus, and a substantial number of researchers have attempted to measure the lightning current and overvoltage by various means. For instance, low-damping voltage dividers directly connected to buses are specially made for measuring lightning overvoltage. And in other cases, capacitive bushing taps are used to extract the divided voltages. In power grids with high voltage ratings, long-term operation of the voltage divider in parallel with the bus may present potential risks. For voltage dividers constituted by bushings in series, there exist risks such as breakage of grounded conductors of taps, or tap discharge caused by transducer open circuit. Hence, these measuring methods are somehow limited.

In addition, the current measuring systems are not strict with the measurement bandwidth. It is specified by the IEC standard that the front time of the standard lightning impulse voltage is 1.2 μs, and the measurement bandwidth of the measuring system can be lower than 1 MHz, which is because the measuring hardware devices could not meet the requirements half a century ago. For accurate overvoltage measurement, the measurement bandwidth shall be improved to at least 100 MHz, which can be easily achieved by the current techniques.

4.7.2 Research Scheme

In this book, a whole-process scheme for lightning overvoltage measurement is presented, thereby better insights into the response characteristics of the lightning overvoltages propagating in the grid can be acquired. The scheme framework is illustrated in Figure 4.40.

Lightning overvoltages are usually generated by lightning strokes to conductors or towers or by induction, and propagate from the transmission lines towards the substations. The transducers are arranged as illustrated in Figure 4.40. At least two measuring sites are arranged on the line towers with one arranged on the closest tower to the substation, and more measuring sites can be chosen if the condition permits. At least two measuring sites are arranged in the substation, one installed in the vicinity of the incoming line bushing and the other on the incoming line side of the transformer; more measuring sites can be arranged between the two sites if appropriate.

Photoelectric sensors are adopted to measure the overvoltages on transmission lines. The entire measuring system contains the sensor probe, laser source, optical receiver, polarizer and polarization beam splitter. The laser source uses a linearly polarized light source with the specific wavelength. This light is modulated by the applied electric field when passing through the optical waveguide inside the sensor; modulated signals are converted into voltage signals through the optical receiver and ultimately displayed on the oscilloscope. The electric field signals to be measured can thus be calculated according to the voltage signals; the line voltage can thus be obtained through the inversion method based on the electric field signals. The optical electric field sensor adopted by the measuring system is characterized by high integrated level, small

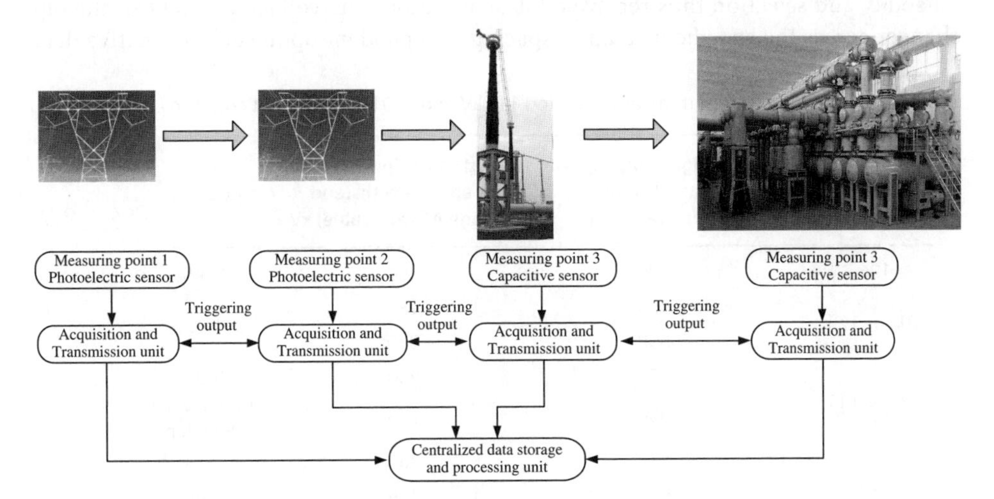

Figure 4.40 Scheme of measuring the response characteristics of the lightning overvoltage propagating in the grid.

size, small impact on electric field distribution and measurement bandwidth of over 300 MHz, and thus suitable for high-field measurement in the high voltage area.

In the case of a GIS station, capacitive voltage dividers are installed in accordance with the GIS structure; impedance conversion units are added and the cable structure of the acquisition system is upgraded to ensure stable voltage division ratios; measurement bandwidth is improved to 300 MHz.

For the acquisition and transmission units of all the sensors, the sampling rate has to be no lower than 200 MHz, the bandwidth no lower than 100 MHz and the sampling duration no less than 20 ms. Self-triggering is adopted by the system, i.e. if the acquisition and transmission unit of any sensor is triggered, all the other measuring sites are triggered by the output trigger signals of this particular sensor. When all the measuring sites complete their data acquisition, the measurement results are transmitted through the Ethernet to the centralized data storage and processing unit for further analysis.

Since the electric strength of major and minor insulation of transformers have a direct impact upon the secured and stable operation of transformers, the reliability of the insulation plays a critical role in transformer design. The acquisition of lightning waveforms can help further specify the transformer insulation level and reduce the cost of components such as transformers and switchgear while at the same time ensuring transformer reliability.

The lightning withstand level of the transformer depends on the residual voltage of the arrester, and insulation coordination is achieved in accordance with the residual voltages of different arresters to determine the insulation levels of various equipment. The insulation levels of the transformers rated 110 kV and 220 kV are shown in Table 4.4 according to IEC 60076-3: 2000 and GB1094.3-2003 *Power transformers. Part 3: Insulation levels, dielectric tests and external clearances in air*.

In Table 4.4, four insulation levels of 110 kV are LI480AC200, LI450AC185, LI550AC230 and LI650AC275; three insulation levels of 220 kV are LI850AC360, LI950AC395 and LI1050AC460. The insulation level remains to be specified.

On-load tap changers are key equipment for transformer voltage regulation, its reliability and selection thus receive a lot of attention. The voltage gradient of the tap changer across the specific insulation spacing is dependent upon such normative data

Table 4.4 Insulation levels of transformers rated 110 kV and 220 kV in different countries.

Equipment maximum voltage U_m (rms value) kV	Rated lightning impulse withstand voltage LI (peak value) kV	Rated short-time induced or applied withstand voltage AC (rms value) kV	Country
126	480	200	China
100 or 123	450	185	USA
	550	230	
123 or 145	550	230	India, Bangladesh, Sweden
	650	275	
245	850	360	USA
	950	395	China
	1050	460	India

of the transformer as the rated voltage, voltage regulation range, voltage regulation methods (linear, coarse-fine and reversing changeover regulation), winding types (pancake, cylindrical, helix and layered type) and winding arrangement.

The tap changer insulation is divided into two categories, the internal and external insulation. The withstand voltage of the external insulation and the corresponding maximum voltage of the equipment U_m have already been included in the national standard. In the case of single-phase and three-phase tap changer at the neutral, the external insulation is the insulation against the ground. In the case of three-phase tap changers for delta-connected windings, the external insulation includes the insulation to ground and interphase insulation, both of which are determined by the maximum voltage of the equipment U_m.

The internal insulation of the tap changer is impossible to normalize, the rated withstand voltages can thus only be defined by grades divided based on practical needs and the experience of voltage gradients in transformer tests. With the rapid development of simulation techniques, it becomes easy to accurately calculate the voltage gradients of different winding types, under different winding arrangement and by different voltage regulation methods, which enables early selection of tap changers in the stage of transformer design.

In the full-wave or chopped-wave lightning impulse tests, as voltage distribution over the winding is nonlinear, the highest voltage gradient appears on the internal insulation of the tap changer. Consequently, the necessary insulation spacing can be determined according to these gradients. In the switching impulse tests and power frequency overvoltage tests, the voltage distribution over the winding is approximately linear and produces rather low gradients, which the insulation spacing determined by lightning impulse test is able to withstand. Under the applied power frequency voltage tests, no gradient will be present on the internal insulation of tap changers.

For 110 kV large-capacity transformers with the insulation levels of LI480AC200 and LI450AC185, completely continuous windings can be utilized for the high-voltage windings. Under this condition, care should be given to the voltage gradients under the chopped-wave impulse tests as they are more severe than those under the full-wave impulse tests. The potential between the HV winding end A_k and the ground as well as the voltage gradients in the range of 1–9 are all equivalent to the 20 kV lightning impulse voltage; and the oscillating coefficient of the potential between the HV winding end A_k and the ground, of the voltage gradients in the range of 1–9 and of the second, fourth and sixth grade voltage $k = 2.6$, is larger than that of the HV windings utilizing the kink-continuous or screen-continuous method ($k = 1.85$). Completely continuous HV windings are characterized by simple manufacture, excellent insulation reliability, small short-circuit mechanical strain and low stray losses.

The impulse properties of transformer windings vary with winding structure. As impulse waves can generate oscillations within the windings, oscillating potentials much higher than the experimental voltage value will be generated within the windings; it is also possible that the impulse voltage may largely fall on the several segments near the line terminal. In order to suppress the oscillating potential, appropriate winding structures are adopted by different products. For instance, the HV windings of the transformers rated 110 kV and 220 kV use the partial parallel capacitance compensation technique (using electrostatic plate at the terminals) and the particular winding types such as completely continuous type, completely kink type and kink-continuous type so as to reduce the overvoltages on major and minor insulation caused by the impulse voltage, to improve the reliability of transformer insulation, as well as to reduce the winding size and lower the cost.

Taking the load-tapchanging and regulating transformer SSZ11-50 MVA/110 kV qualified by the lightning impulse test as an example, the voltage gradients between windings or across segments are calculated in detail using the wave process software.

The transformer model is SSZ11-50 MVA/110 kV, and the other parameters are:

1. rated capacity of the HV, MV and LV windings: 50/50/50 MVA;
2. rated voltage and regulating range: $110 \pm 8 \times 1.25\%/38.5 \pm 2 \times 2.5\%/11$ kV;
3. insulation level:

HV line terminal	LI/AC	480/200
HV neutral terminal	LI/AC	325/140
MV line terminal	LI/AC	200/85
MV neutral terminal	LI/AC	200/85
LV line terminal	LI/AC	75/35;

4. rated frequency: 50Hz;
5. connection symbol: YNyn0d11.

The 50 MVA/110 kV transformer adopts the reversing changeover regulation with its connection diagram shown in Figure 4.41 where voltage tapping 1 is defined as $X \rightarrow 1$, $A_k \rightarrow +$, i.e. the maximum tap voltage; the rated tapping voltages have two states: rated 1 is defined as $X \rightarrow 9, A_k \rightarrow +$; rated 2 is defined as $X \rightarrow 1, A_k \rightarrow -$; voltage tapping 17 is defined as $X \rightarrow 9, A_k \rightarrow -$, which has the minimum tap voltage.

The comparison concerning oil duct height, insulation thickness, the maximum voltage gradient and its duration, permissible voltage and safety factor is given below.

1. Comparison between 1.2/50 μs and 2.4/50 μs (Table 4.5)
2. Transposition coefficients at the zero point of the chopped waves 0.3 & 0.4 (Table 4.6)
3. Preliminary discharge time of the chopped impulse waves 2.5 & 5.0 (Table 4.7)

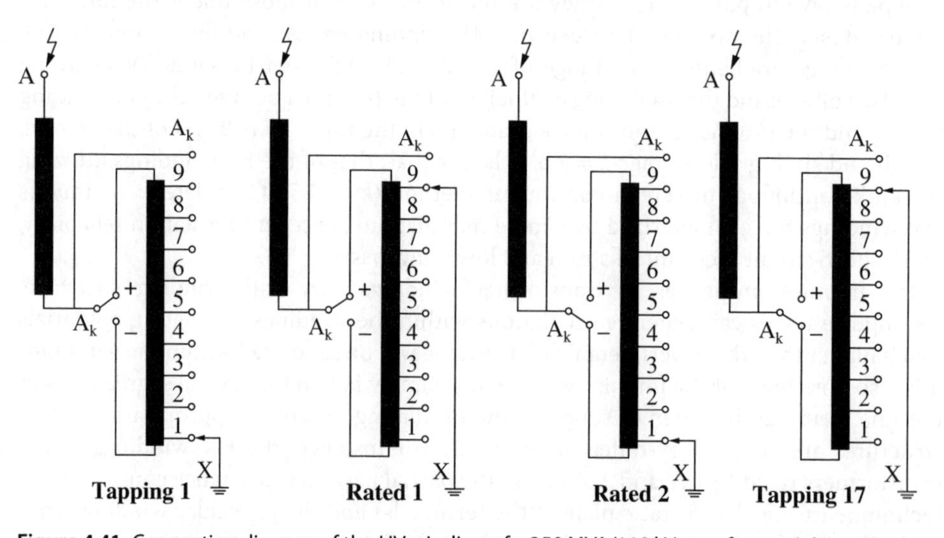

Figure 4.41 Connection diagram of the HV winding of a 350 MVA/110 kV transformer (phase A).

Table 4.5 Comparison between 1.2/50 μs and 2.4/50 μs.

Test voltage 480.0 kV FW 1.20/50.00

Gap No.	Gap designation	Oil duct height, mm	Turn insulation thickness per two sides, mm	Applied voltage: maximum value, %	Applied voltage: duration, μsec	Permissible voltage, kV	Safety factor
1	HV_1_2	3.20	1.52	12.08	4.00	120.49	2.08
2	HV_2_3	3.20	1.52	16.64	3.50	121.43	1.52
3	HV_3_4	3.20	1.52	18.67	3.00	122.37	1.37
4	HV_4_5	4.12	.85	19.14	3.50	123.03	1.34
5	HV_5_6	4.12	.85	17.49	4.00	121.88	1.45
6	HV_6_7	4.12	.85	16.30	4.00	121.88	1.56
7	HV_7_8	4.12	.85	16.05	4.50	120.72	1.57
8	HV_8_9	4.12	.85	15.83	4.50	120.72	1.59
9	HV_9_10	2.74	.85	15.18	5.00	94.16	1.29
10	HV_10_11	2.74	.85	14.74	5.50	93.23	1.32
11	HV_11_12	2.74	.85	14.30	5.50	93.23	1.36
12	HV_12_13	2.74	.85	13.62	5.50	93.23	1.43
13	HV_13_14	2.74	.85	13.25	6.00	92.30	1.45
14	HV_14_15	2.74	.85	13.07	5.50	93.23	1.49
15	HV_15_16	2.74	.85	13.02	5.50	93.23	1.49
16	HV_16_17	2.74	.85	12.84	5.50	93.23	1.51
17	HV_17_18	2.74	.85	12.58	5.50	93.23	1.54
18	HV_18_19	2.74	.85	12.26	5.50	93.23	1.58
19	HV_19_20	2.74	.85	12.07	5.50	93.23	1.61
20	HV_20_21	2.74	.85	12.00	5.50	93.23	1.62
21	HV_21_22	2.74	.85	11.92	5.50	93.23	1.63
22	HV_22_23	2.74	.85	11.85	6.00	92.30	1.62
23	HV_23_24	2.74	.85	11.77	6.00	92.30	1.63
24	HV_24_25	2.74	.85	11.81	6.50	91.37	1.61
25	HV_25_26	2.74	.85	11.98	6.00	92.30	1.61
26	HV_26_27	2.74	.85	12.20	5.50	93.23	1.59
27	HV_27_28	2.74	.85	12.55	5.50	93.23	1.55
28	HV_28_29	2.74	.85	11.84	5.50	93.23	1.64
29	HV_29_30	2.74	.85	11.62	6.00	92.30	1.65

Test voltage 480.0 kV FW 2.40/50.00

Gap No.	Gap designation	Oil duct height, mm	Turn insulation thickness per two sides, mm	Applied voltage: maximum value, %	Applied voltage: duration, μsec	Permissible voltage, kV	Safety factor
1	HV_1_2	3.20	1.52	9.83	5.00	118.61	2.51
2	HV_2_3	3.20	1.52	13.66	5.00	118.61	1.81
3	HV_3_4	3.20	1.52	15.12	4.50	119.55	1.65
4	HV_4_5	4.12	.85	15.88	4.00	121.88	1.60
5	HV_5_6	4.12	.85	15.85	4.50	120.72	1.59
6	HV_6_7	4.12	.85	15.18	4.50	120.72	1.66
7	HV_7_8	4.12	.85	14.65	5.50	118.42	1.68
8	HV_8_9	4.12	.85	14.41	5.50	118.42	1.71
9	HV_9_10	2.74	.85	14.11	5.50	93.23	1.38
10	HV_10_11	2.74	.85	13.91	6.00	92.30	1.38
11	HV_11_12	2.74	.85	13.66	5.50	93.23	1.42
12	HV_12_13	2.74	.85	13.24	6.00	92.30	1.45
13	HV_13_14	2.74	.85	12.84	6.00	92.30	1.50
14	HV_14_15	2.74	.85	12.52	6.00	92.30	1.54
15	HV_15_16	2.74	.85	12.24	6.50	91.37	1.56
16	HV_16_17	2.74	.85	12.02	6.50	91.37	1.58
17	HV_17_18	2.74	.85	11.87	6.50	91.37	1.60
18	HV_18_19	2.74	.85	11.67	6.50	91.37	1.63
19	HV_19_20	2.74	.85	11.53	6.50	91.37	1.65
20	HV_20_21	2.74	.85	11.48	6.50	91.37	1.66
21	HV_21_22	2.74	.85	11.46	6.50	91.37	1.66
22	HV_22_23	2.74	.85	11.47	7.00	90.44	1.64
23	HV_23_24	2.74	.85	11.48	7.00	90.44	1.64
24	HV_24_25	2.74	.85	11.53	6.50	91.37	1.65
25	HV_25_26	2.74	.85	11.69	6.50	91.37	1.63
26	HV_26_27	2.74	.85	11.93	6.00	92.30	1.61
27	HV_27_28	2.74	.85	11.95	6.50	91.37	1.59
28	HV_28_29	2.74	.85	11.60	6.50	91.37	1.64
29	HV_29_30	2.74	.85	11.33	7.00	90.44	1.66

Table 4.6 Transposition coefficients at the zero point of the chopped waves 0.3 and 0.4.

Test voltage 550.0 kV CW 2.50/ .30

Gap No.	Gap designation	Oil duct height, mm	Turn insulation thickness per two sides, mm	Applied voltage: maximum value, %	Applied voltage: duration, μsec	Permissible voltage, kV	Safety factor
1	HV_1_2	3.20	1.52	17.59	.40	156.23	1.62
2	HV_2_3	3.20	1.52	17.71	2.40	130.18	1.34
3	HV_3_4	3.20	1.52	18.78	2.40	130.18	1.26
4	HV_4_5	4.12	.85	19.27	2.40	133.50	1.26
5	HV_5_6	4.12	.85	17.94	2.20	136.60	1.38
6	HV_6_7	4.12	.85	17.16	2.20	136.60	1.45
7	HV_7_8	4.12	.85	15.57	2.40	133.50	1.56
8	HV_8_9	4.12	.85	12.56	2.60	130.39	1.89
9	HV_9_10	2.74	.85	11.65	2.40	106.38	1.66
10	HV_10_11	2.74	.85	10.83	2.40	106.38	1.79
11	HV_11_12	2.74	.85	9.72	2.40	106.38	1.99
12	HV_12_13	2.74	.85	9.10	1.60	117.74	2.35
13	HV_13_14	2.74	.85	8.75	1.60	117.74	2.45
14	HV_14_15	2.74	.85	8.37	1.60	117.74	2.56
15	HV_15_16	2.74	.85	8.43	1.60	117.74	2.54
16	HV_16_17	2.74	.85	8.06	1.60	117.74	2.66
17	HV_17_18	2.74	.85	7.76	1.80	114.90	2.69
18	HV_18_19	2.74	.85	7.41	1.80	114.90	2.82
19	HV_19_20	2.74	.85	7.37	1.60	117.74	2.91
20	HV_20_21	2.74	.85	7.23	1.80	114.90	2.89
21	HV_21_22	2.74	.85	7.07	1.80	114.90	2.95
22	HV_22_23	2.74	.85	6.78	1.80	114.90	3.08
23	HV_23_24	2.74	.85	6.37	1.80	114.90	3.28
24	HV_24_25	2.74	.85	6.01	2.20	109.22	3.31
25	HV_25_26	2.74	.85	6.69	2.20	109.22	2.97
26	HV_26_27	2.74	.85	7.45	2.00	112.06	2.73
27	HV_27_28	2.74	.85	7.55	1.80	114.90	2.77
28	HV_28_29	2.74	.85	6.46	2.00	112.06	3.15
29	HV_29_30	2.74	.85	6.76	2.00	112.06	3.01

Test voltage 550.0 kV CW 2.50/ .40

Gap No.	Gap designation	Oil duct height, mm	Turn insulation thickness per two sides, mm	Applied voltage: maximum value, %	Applied voltage: duration, μsec	Permissible voltage, kV	Safety factor
1	HV_1_2	3.20	1.52	20.08	.40	156.23	1.41
2	HV_2_3	3.20	1.52	17.71	2.40	130.18	1.34
3	HV_3_4	3.20	1.52	18.79	2.40	130.18	1.26
4	HV_4_5	4.12	.85	19.27	2.20	136.60	1.29
5	HV_5_6	4.12	.85	17.95	2.20	136.60	1.38
6	HV_6_7	4.12	.85	17.15	2.20	136.60	1.45
7	HV_7_8	4.12	.85	15.54	2.40	133.50	1.56
8	HV_8_9	4.12	.85	12.55	2.60	130.39	1.89
9	HV_9_10	2.74	.85	11.64	2.40	106.38	1.66
10	HV_10_11	2.74	.85	10.82	2.40	106.38	1.79
11	HV_11_12	2.74	.85	9.71	2.40	106.38	1.99
12	HV_12_13	2.74	.85	9.19	1.60	117.74	2.33
13	HV_13_14	2.74	.85	8.78	1.60	117.74	2.44
14	HV_14_15	2.74	.85	8.38	1.60	117.74	2.55
15	HV_15_16	2.74	.85	8.48	1.60	117.74	2.52
16	HV_16_17	2.74	.85	8.09	1.60	117.74	2.65
17	HV_17_18	2.74	.85	7.81	1.80	114.90	2.68
18	HV_18_19	2.74	.85	7.41	1.80	114.90	2.82
19	HV_19_20	2.74	.85	7.38	1.60	117.74	2.90
20	HV_20_21	2.74	.85	7.25	1.80	114.90	2.88
21	HV_21_22	2.74	.85	7.10	1.80	114.90	2.94
22	HV_22_23	2.74	.85	6.79	1.80	114.90	3.08
23	HV_23_24	2.74	.85	6.37	1.80	114.90	3.28
24	HV_24_25	2.74	.85	6.02	2.20	109.22	3.30
25	HV_25_26	2.74	.85	6.69	2.20	109.22	2.97
26	HV_26_27	2.74	.85	7.45	2.00	112.06	2.73
27	HV_27_28	2.74	.85	7.56	1.80	114.90	2.76
28	HV_28_29	2.74	.85	6.46	2.00	112.06	3.15
29	HV_29_30	2.74	.85	6.77	2.00	112.06	3.01

Table 4.7 Preliminary discharge time of the chopped impulse waves 2.5 and 5.0.

Test voltage 550.0 kV CW 2.50/ .30

Gap No.	Gap designation	Oil duct height, mm	Turn insulation thickness per two sides, mm	Applied voltage: maximum value, %	Applied voltage: duration, μsec	Permissible voltage, kV	Safety factor
1	HV_1_2	3.20	1.52	17.59	.40	156.23	1.62
2	HV_2_3	3.20	1.52	17.71	2.40	130.18	1.34
3	HV_3_4	3.20	1.52	18.78	2.40	130.18	1.26
4	HV_4_5	4.12	.85	19.27	2.40	133.50	1.26
5	HV_5_6	4.12	.85	17.94	2.20	136.60	1.38
6	HV_6_7	4.12	.85	17.16	2.20	136.60	1.45
7	HV_7_8	4.12	.85	15.57	2.40	133.50	1.56
8	HV_8_9	4.12	.85	12.56	2.60	130.39	1.89
9	HV_9_10	2.74	.85	11.65	2.40	106.38	1.66
10	HV_10_11	2.74	.85	10.83	2.40	106.38	1.79
11	HV_11_12	2.74	.85	9.72	2.40	106.38	1.99
12	HV_12_13	2.74	.85	9.10	1.60	117.74	2.35
13	HV_13_14	2.74	.85	8.75	1.60	117.74	2.45
14	HV_14_15	2.74	.85	8.37	1.60	117.74	2.56
15	HV_15_16	2.74	.85	8.43	1.60	117.74	2.54
16	HV_16_17	2.74	.85	8.06	1.60	117.74	2.66
17	HV_17_18	2.74	.85	7.76	1.80	114.90	2.69
18	HV_18_19	2.74	.85	7.41	1.80	114.90	2.82
19	HV_19_20	2.74	.85	7.37	1.60	117.74	2.91
20	HV_20_21	2.74	.85	7.23	1.80	114.90	2.89
21	HV_21_22	2.74	.85	7.07	1.80	114.90	2.95
22	HV_22_23	2.74	.85	6.78	1.80	114.90	3.08
23	HV_23_24	2.74	.85	6.37	1.80	114.90	3.28
24	HV_24_25	2.74	.85	6.01	2.20	109.22	3.31
25	HV_25_26	2.74	.85	6.69	2.20	109.22	2.97
26	HV_26_27	2.74	.85	7.45	2.00	112.06	2.73
27	HV_27_28	2.74	.85	7.55	1.80	114.90	2.77
28	HV_28_29	2.74	.85	6.46	2.00	112.06	3.15
29	HV_29_30	2.74	.85	6.76	2.00	112.06	3.01

Test voltage 550.0 kV CW 5.00/ .30

Gap No.	Gap designation	Oil duct height, mm	Turn insulation thickness per two sides, mm	Applied voltage: maximum value, %	Applied voltage: duration, μsec	Permissible voltage, kV	Safety factor
1	HV_1_2	3.20	1.52	23.19	.40	156.23	1.23
2	HV_2_3	3.20	1.52	20.09	1.00	148.41	1.34
3	HV_3_4	3.20	1.52	18.78	2.80	124.98	1.21
4	HV_4_5	4.12	.85	19.27	3.00	124.18	1.17
5	HV_5_6	4.12	.85	17.94	3.20	123.72	1.25
6	HV_6_7	4.12	.85	17.16	3.60	122.80	1.30
7	HV_7_8	4.12	.85	16.33	3.80	122.34	1.36
8	HV_8_9	4.12	.85	15.83	4.40	120.95	1.39
9	HV_9_10	2.74	.85	15.11	4.60	94.90	1.14
10	HV_10_11	2.74	.85	14.80	4.40	95.27	1.17
11	HV_11_12	2.74	.85	14.51	4.00	96.01	1.20
12	HV_12_13	2.74	.85	14.07	3.60	96.76	1.25
13	HV_13_14	2.74	.85	13.25	4.20	95.64	1.31
14	HV_14_15	2.74	.85	10.99	5.00	94.16	1.56
15	HV_15_16	2.74	.85	10.38	5.20	93.79	1.64
16	HV_16_17	2.74	.85	9.90	5.20	93.79	1.72
17	HV_17_18	2.74	.85	8.99	2.80	100.71	2.04
18	HV_18_19	2.74	.85	8.29	2.80	100.71	2.21
19	HV_19_20	2.74	.85	7.88	3.00	97.87	2.26
20	HV_20_21	2.74	.85	7.93	4.20	95.64	2.19
21	HV_21_22	2.74	.85	8.03	4.60	94.90	2.15
22	HV_22_23	2.74	.85	8.23	4.60	94.90	2.10
23	HV_23_24	2.74	.85	8.39	4.60	94.90	2.06
24	HV_24_25	2.74	.85	8.37	5.00	94.16	2.04
25	HV_25_26	2.74	.85	8.36	4.60	94.90	2.06
26	HV_26_27	2.74	.85	8.45	4.80	94.53	2.03
27	HV_27_28	2.74	.85	8.78	4.60	94.90	1.97
28	HV_28_29	2.74	.85	8.19	5.00	94.16	2.09
29	HV_29_30	2.74	.85	8.71	3.80	96.38	2.01

For lightning chopped waves with different transposition coefficients or preliminary discharge time, only the safety factors of the line terminal are greatly varied, whereas for the waves with different front time (1.2/2.4 μs), the safety factors of the turn insulation vary greatly (Table 4.7).

Acquiring the response characteristics of lightning overvoltages and utilizing the software to calculate the impulse voltage within the winding are very instructive for better design of the insulation dimensions of transformers rated 110 kV and 220 kV and thus for improved reliability. The voltage gradient, duration and safety factor of each position of the winding have to be accurately calculated so as to adjust the gap of the major and minor insulation, thereby making the design more reliable and reasonable.

4.8 Lightning Location System (LLS)

4.8.1 Overview and Current Situation

Lightning is a process in which the thundercloud that carries a large quantity of charges releases enormous amounts of energy in an exceedingly short time, and the large current and high voltage thus generated have destructive powers. Owing to its massive influence, lightning has drawn wide attention from such industries as meteorology, aeronautics, aviation, power and petroleum. Among others, the power grid, characterized by wide-range distribution and long geometric dimensions (several hundred or even several thousand kilometers), is extremely prone to lightning strikes.

Lightning detection is the foundation for lightning protection. Modern lightning telemetery and location technology was initially proposed by American scientists at the end of the 1970s, and the US National Lightning Detection Network (NLDN), unified and appropriately arranged, was established in the 1990s. Over the past 20 years, the Lightning Location System (LLS) has been widely applied in the fields of aeronautics and aviation, disaster reduction and prevention as well as for the power industry. The application of LLS in power systems around the globe has in turn greatly improved the locating accuracy and detection efficiency. In China, LLS was first introduced from America in the 1980s, and independent research has been carried out ever since. In 1993, the first national LLS equipment was developed and put into operation in the country. So far, a lightning detection network covering 25 provinces is out there for network-based detection and information sharing in the grid.

LLS is a fully automatic lightning detection system characterized by large coverage area, high accuracy and real-time detection. It can telemeter and display the time and location of cloud-to-ground discharge (ground flash), the peak value and polarity of lightning current, and the return stroke parameters. The time-sharing color image of a strike site can clearly reflect the thunderstorm movement. LLS can provide real-time monitoring of the lightning activities in the grid, collect information regarding lightning movement trends, and send lightning alarms, so that the power scheduling department can develop contingency plans and provide dispatching schemes. Also, LLS is used for fast query of fault sites and recognition of fault properties especially in the thunderstorm season. The LLS data upon which lightning protection has long relied provides essential resources needed by lightning engineering. Over the few past years, the LLS data has become a vital basis for operations personnel at the front line to carry out troubleshooting for transmission lines as well as for assessment and design of lightning prevention.

The technologies of acquiring and processing lightning signals are of great significance in improving lightning locating accuracy as well as in monitoring and predicting lightning parameters. The theory and methods concerning lightning detection and location will be discussed in this book by taking the example of a specific research project where a highly accurate system of lightning signal detection and location is established based on the TOA (time of arrival) method. This system consists of such modules as lightning signal data acquisition, GPS frequency calibration and data transmission, and is applied in the actual grid.

4.8.2 Detection Principles

In the presence of lightning, strong light, sound and electromagnetic radiation are generated, among which the most suitable signal for wide-range detection is the electromagnetic radiation field which transmits along the earth's surface with low frequency or very low frequency (LF/VLF). The transmitting range, which depends on discharge capacity, can reach hundreds of meters or above. Cloud-to-ground lightning (ground flashes) and cloud-to-cloud lightning (cloud flashes) are the main forms of lightning discharge. The former consists of the main discharge and subsequent discharge, and can endanger ground objects. According to modern optical observations, over 50% of the subsequent discharge will break out of the main discharge channel when approaching the ground and generate new cloud-to-ground discharge sites.

LLS is a comprehensive system that uses multiple detection stations to measure the LF/VLF electromagnetic radiation fields of lightning simultaneously, take out cloud flash signals and locate the ground flashes. The broadband antenna system and the specially designed electronic circuit in the detection station can recognize ground flash signals, collect the peak values of return strokes and align the measured values with the initial parts of the return strokes, i.e. the lower parts of the vertical return stroke channels. At this moment, the influence of refraction and reflection in the ionosphere and of the longitudinal branch of the channel is the minimum, thereby ensuring the accuracy of measuring lightning strike spots and the peak values of lightning current both theoretically and technically.

Lightning duration is primarily determined by the number of return strokes. Generally, the lightning duration is less than 1 s, and is 0.2 s on average while the duration of a return stroke is less than 0.1 ms. The interval between return strokes is 20–200 ms, and the average value is 50–70 ms. The discharge time acquired by the LLS is the instant when the first peak of the electromagnetic pulse generated by the return stroke arrives at the monitoring station; it equals the occurrence time of the return stroke plus the propagation delay.

4.8.2.1 Typical LLS Locating Methods

Orientation method – When the ground-to-cloud magnetic radiation wave passes through the orthogonal antenna frame of the detection station, the magnetic field strength generated in the north-to-south and west-to-east directions of the antenna (with respect to the x-y axes) are H_{NS} and H_{WE}, respectively (Figure 4.42). By measuring $\tan\alpha = H_{NS}/H_{WE}$, the azimuth angle between the lightning strike point A and the detection station could be obtained. The coordinate of point A can be calculated using the coordinates and azimuth angles collected by two detection stations. More data can be used to calculate the optimum value and to estimate the error.

Time-difference method – The locating principle of the time-difference method is shown in Figure 4.43 where A, B and C are the detection stations. P is assumed to be the ground flash location, and the arrival time of the lightning occurring at location P is recorded by each monitoring station. For each pair of the stations, there exists a time difference and a corresponding distance difference, and a hyperbolic curve can be depicted on which the lightning may strike a certain point. If the third detection station meets the location requirements, another hyperbolic curve would be formed. The

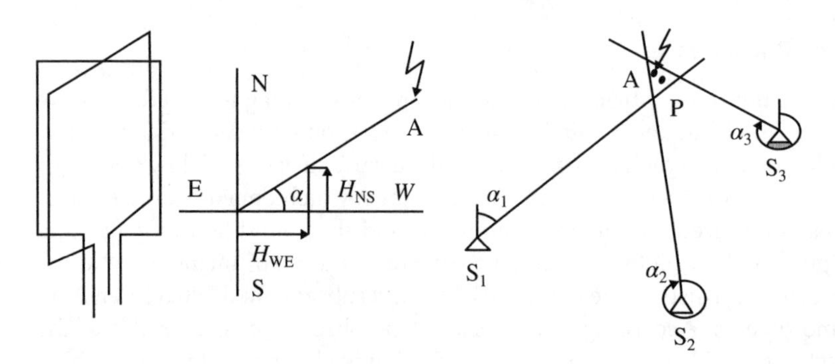

Figure 4.42 The orientation method.

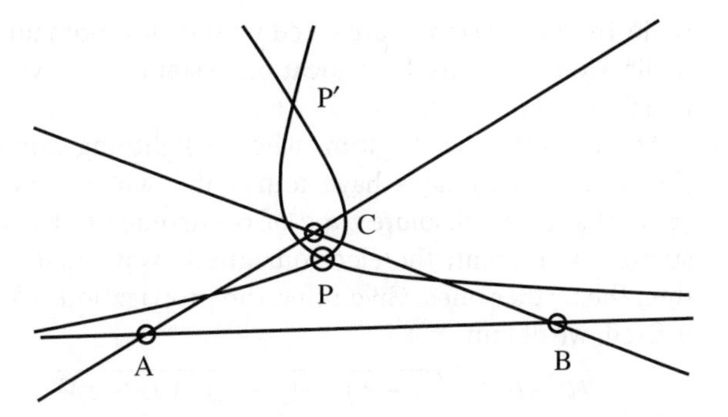

Figure 4.43 Principle of the time-difference method.

intersection point of the two hyperbolic curves is the strike point, while P′, the other mathematical solution, is the fake strike point. A location system with four or more stations is able to eliminate P′, calculate the optimal value and estimate the accuracy.

Comprehensive locating method – The error of the orientation method is large, but only a few items need to be measured; the time-difference method is of high locating accuracy but requires more objects to be measured. With a combination of the two methods, the measuring objects regarding orientation and time can all be acquired in one station. With more measuring objects, accuracy can be improved and the particular issue of eliminating fake strike points would be solved. At present, the comprehensive method is the locating method most widely used.

4.8.2.2 Time of Arrival Method

Time of arrival (TOA) is a multi-station lightning location system that uses the time differences of the electromagnetic waves arriving at different lightning detection stations, as illustrated in Figure 4.44. By acquiring the time when the electromagnetic wave arrives at each detection station due to lightning events, the time differences between stations can be calculated to obtain the 3D position and occurrence time of the lightning event. Hence, the locating accuracy of the system depends on the standard time differences between stations, i.e. station synchronization. In addition, the acquired

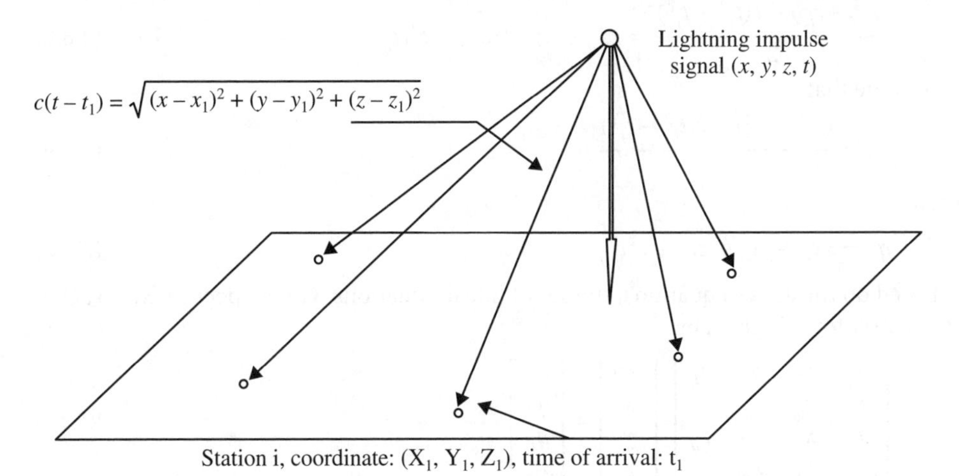

Figure 4.44 Schematic diagram of the TOA method.

radio-frequency signals are faced with major radio interference during the propagation in air. Hence, in this TOA location system, time synchronism is considered to be the most important factor.

According to the diagram, when a lightning impulse occurs, the electromagnetic signals that it radiates have four unknown factors: the three-dimensional location (x, y, z) and the absolute time of occurrence t. The arrival time of the event at each station is different; the electromagnetic wave first arrives at the nearest station and then the farther ones. Given that the propagation velocity of the electromagnetic wave is fixed, we obtain

$$C(t_i - t) = \sqrt{(x - x_i)^2 + (y - y_i)^2 + (z - z_i)^2} \tag{4.8.1}$$

The above equation further gives

$$t_i = t + \frac{1}{c}\sqrt{(x - x_i)^2 + (y - y_i)^2 + (z - z_i)^2} \tag{4.8.2}$$

where C is the propagation velocity of the signal, (x_i, y_i, z_i, t_i) is the three-dimensional position of station i, and t_i is the time when the electromagnetic wave reaches station i.

Define

$$r_i^2 = x_i^2 + y_i^2 + z_i^2 \tag{4.8.3}$$

$$r^2 = x^2 + y^2 + z^2 \tag{4.8.4}$$

we can obtain

$$c^2(t^2 + t_i^2) = r^2 + r_i^2 - 2(xx_i + yy_i + zz_i - c^2tt_i) \tag{4.8.5}$$

Replacing i with j in Equation (4.8.5) and subtracting Equation (4.8.5) gives

$$c^2(t_i^2 - t_j^2) - (r_i^2 - r_j^2) = -2[x(x_i - x_j) + y(y_i - y_j) + z(z_i - z_j) - c^2t(t_i - t_j)] \tag{4.8.6}$$

Define

$$t_{ij} = t_i - t_j; x_{ij} = x_i - x_j; y_{ij} = y_i - y_j; z_{ij} = z_i - z_j \tag{4.8.7}$$

Hence,

$$\frac{(r_i^2 - r_j^2) - c(t_i^2 - t_j^2)}{2} = xx_{ij} + yy_{ij} + zz_{ij} - c^2tt_{ij} \tag{4.8.8}$$

Assume that

$$q_{ij} = \frac{(r_i^2 - r_j^2) - c(t_i^2 - t_j^2)}{2} \tag{4.8.9}$$

then

$$q_{ij} = xx_{ij} + yy_{ij} + zz_{ij} - c^2tt_{ij} \tag{4.8.10}$$

Based on the above equations, the set of linear equations with respect to (x, y, z, t) in the matrix form is given by

$$\begin{bmatrix} ct_{ij} & x_{ij} & y_{ij} & z_{ij} \\ ct_{ik} & x_{ik} & y_{ik} & z_{ik} \\ ct_{il} & x_{il} & y_{il} & z_{il} \\ ct_{im} & x_{im} & y_{im} & z_{im} \end{bmatrix} \cdot \begin{bmatrix} -ct \\ x \\ y \\ z \end{bmatrix} = \begin{bmatrix} q_{ij} \\ q_{ik} \\ q_{il} \\ q_{im} \end{bmatrix} \tag{4.8.11}$$

By solving this matrix, the location of the lightning can be obtained.

4.8.2.3 Calculation Model for the Lightning Current Peak

The calculation of the peak value of lightning current I is shown in Figure 4.45. According to Uman's return stroke current model for ground flashes, assuming that the return stroke channel of a ground flash is a current source that propagates upward at the velocity of v, the radiation field component generated at a distance of D under the transmission line mode would be represented by

$$B(D,t) = (\mu_0 v/2\pi CD)I(t - D/C) \tag{4.8.12}$$

where C is the velocity of light, m/s; v takes the value of 1.3×10^8 m/s; μ_0 is the air permeability, H/m; D is the distance between the strike site and the detection station, m.

After integration and amplification, the signal strength of the induced voltage of the magnetic antenna frame is expressed by

$$M = knS\int \frac{dB}{dt}dt = k'B = knS(\mu_0 v/2\pi CD)I(t - D/C) \tag{4.8.13}$$

where k is the amplification multiple of the integrator; n is the number of turns of the frame antenna coil; S is the area of the coil, m^2, with

$$I = ADM,\ (\text{with} A = 2\pi C/knS\mu_0 v) \tag{4.8.14}$$

Clearly, M is dispersive for different detection stations. Normalization is a method suited to solve the dispersiveness of lightning current calculation under the multi-station situation. The signal strength is normalized by $r = 100$ km, i.e. the normalized value M_{RN} of the signal strength M in the detection station is expressed as

$$M_{RN} = M(r/100)^b e^{\left(\frac{r-100}{\lambda}\right)} \tag{4.8.15}$$

where b and λ are selected after comparison with the measured values of lightning current. It is recommended that b be 1.13 and A be 100,000. If substituting Equation (4.8.15) into Equation (4.8.14), the output current of LLS would be given by $I = AM_{RN}$.

4.8.2.4 Error Analysis

The LLS locating accuracy primarily depends on the clock synchronization and clock calibration technologies for lightning signal measurement by detection stations. At present, clock synchronization is a mature technology. The time-keeping clock, constituted by highly stable thermostatic crystals and a core 20–50 ns GPS timing module, is able to achieve a high precision in the order of 10^{-7} s. Compared with clock synchronization, calibration of the arrival time of lightning waves is even more critical. The lightning wave with numerous frequency components will be attenuated and distorted during its propagation. Hence, for ground flashes reaching different detection

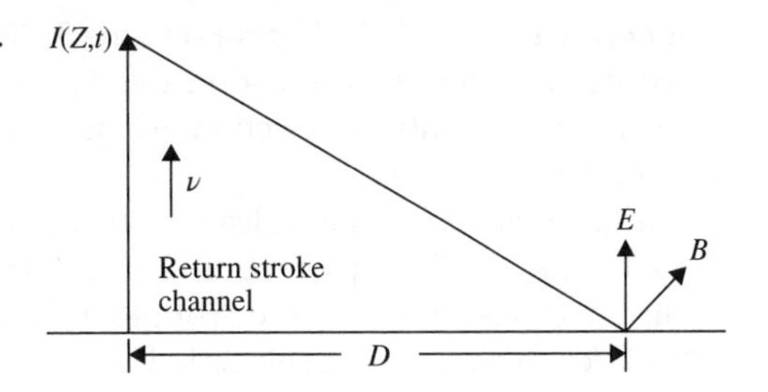

Figure 4.45 Peak value of the lightning current.

stations, the peaks of the characteristic return strokes move backwards with varying degrees, resulting in clock calibration errors in the order of ps. Two methods are currently adopted in China to address this issue: the real-time calibrating technology for characteristic points and waveform inversion method. In the first method, the main characteristic points of the lightning wave are calibrated through narrow-wave hardware filtering as the propagation performances of the main characteristic points are superior to the complete signal characteristics. The advantage of this method is that the detection synchronizes with time calibration, which can reduce peak time errors by 75%. In the second method, the clock is calibrated by inversion of the characteristic wave peaks of the ground flashes based on the given conditions of propagation path and medium. The challenge of this method lies in the development of the inversion models for propagation path and medium.

When the lightning current amplitude I is calculated by measuring the electromagnetic radiation field with the help of LLS, the amplitude error has long been a concern. The Uman's model has kept being improved ever since it was first proposed, in 1975. Although newly built models for lightning current calculation have been emerging, they are never as simple as the Uman's model in regard to engineering application. The verification for the Uman's model by Weidman, Willett et al. in 1988 demonstrated that the measurement result was fairly accurate within the first several milliseconds of the return stroke, but the error would then gradually increased over the time. The simple equations to measure I can still be adopted while performing the calibration tests. So far, it has proved impossible to accurately quantify the factors contributing to the errors of lightning current peaks, and to calculate the error of each current. However, the probability distribution of lightning current amplitudes observes a statistical law, and can therefore be used in engineering applications.

4.8.3 System Structure

Structure of the detection station – High-frequency and low-frequency lightning signals are captured by the signal receiving system through the antenna, and transmitted to the data acquisition system through coaxial cables after being pre-filtered and amplified. The two-channel analog signals are amplified by an analog operational amplifier, and then sampled and digitized by the A/D module. At the CPLD module, treatment such as information extraction and time calibration is performed, and the selected data after processing, together with the time marks are sent to an embedded ARM controller. Finally, the ARM controller saves the data into the USB (Figure 4.46).

LLS system structure – The LLS system structure is shown in Figure 4.47.

4.8.4 Applications

Fault overview – At 18 hr 25 min on June 3, 2013, phase C of a 500 kV line tripped and succeeded in reclosing. Protection 1 and protection 2 in the substation both belong to the current differential protection; when the fault occurred, the load was –85.25 MW and the current was 93.75A.

Fault location – The fault distance measured by protection 1 is 145.4 km; the fault distance measured by protection 2 is 143 km; fault recorder information: the fault occurred on phase C at a distance of 140.077 km; travelling wave distance measurement: 133.103 km away from the substation.

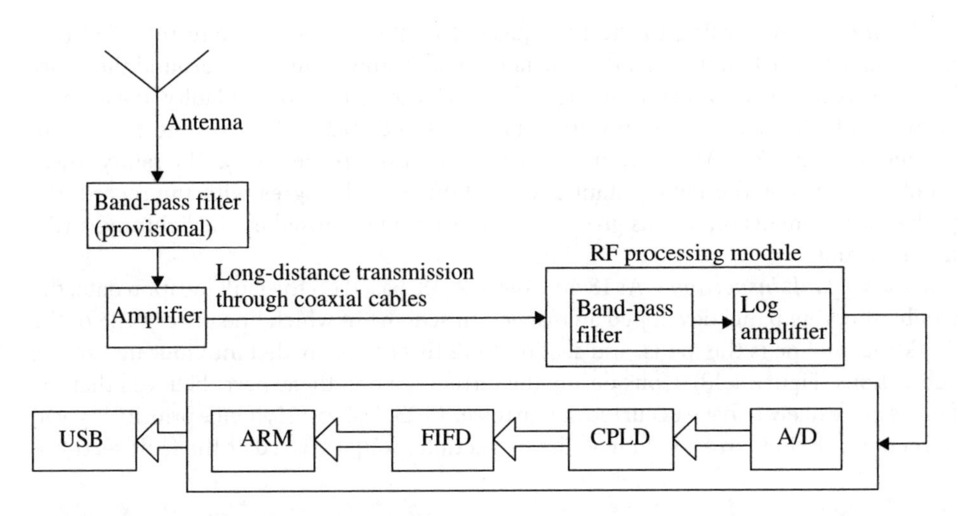

Figure 4.46 System flow chart.

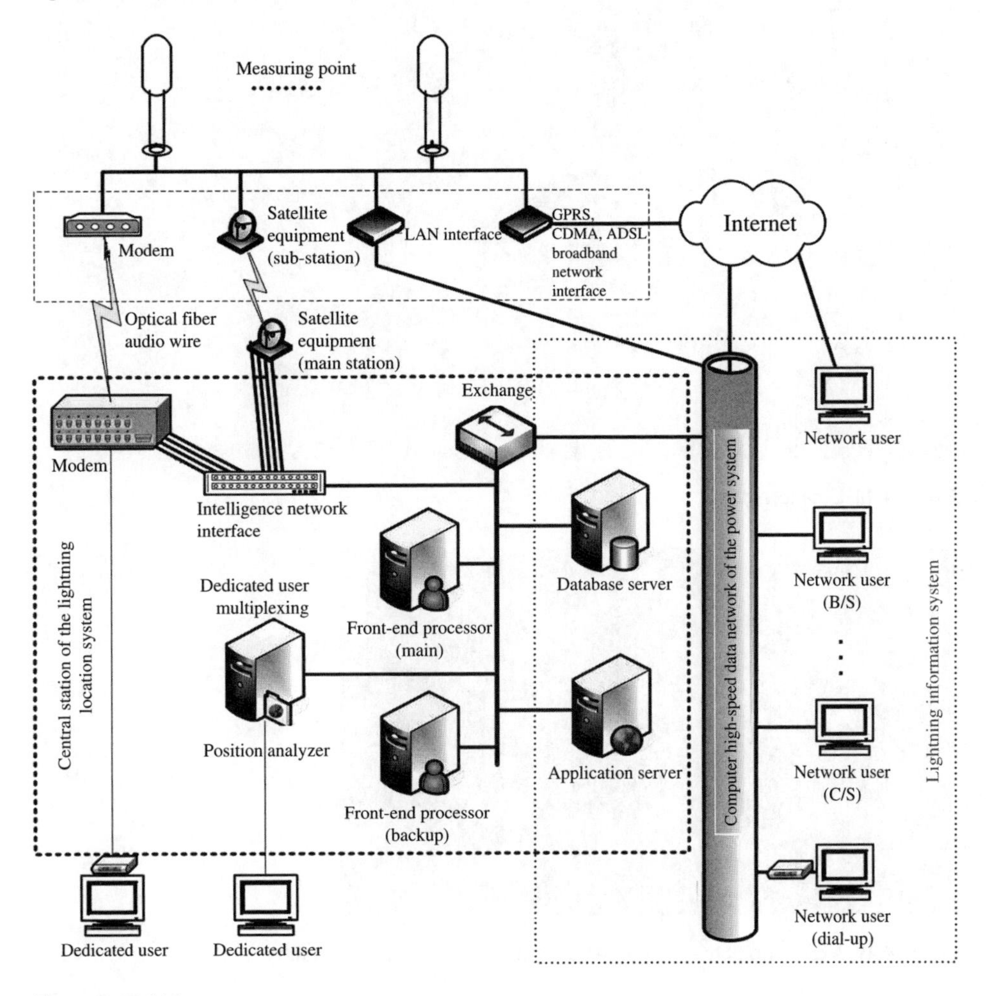

Figure 4.47 LLS system structure.

LLS query – According to the LLS query, 15 lightning events were recorded that occurred within 10 minutes around the fault instant, and within 5 km around the faulty line. One record shows that a thunderbolt struck the vicinity of the faulty line at 18 hr 25 min with the nearest line towers being #123 and #124. The two line towers are 134.583 km and 134.206 km from the main substation, respectively. The faulty tower number as well as the fault instant and location basically agree with the record; the preliminary conclusion is thus given that the tripping is caused by the lightning strike in this section.

Results of fault inspection – At 18:40, June 3, 2013, based on the fault location data, the troubleshooting panel developed an inspection scheme in which line tower #125 of the 500 kV faulty line is the focus, and section #112 to #130 is divided into four inspection subsections (Figure 4.48). Considering the terrain factors, the experts believed that the fault is more likely to have occurred at line tower #123, 124 or 125, hence priority is given to these three sites. At 8:50, June 4, the inspection groups arrived at the fault sections,

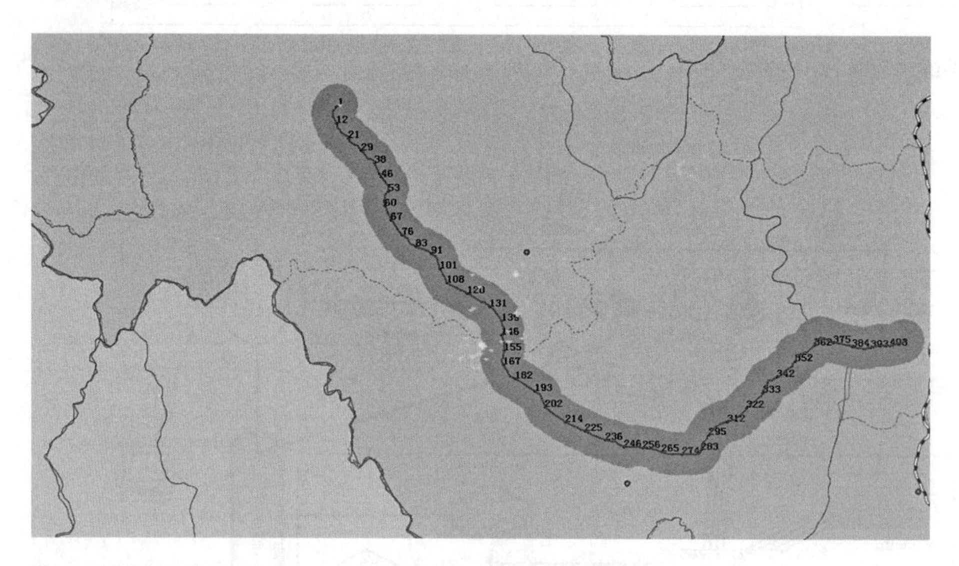

Figure 4.48 Location map.

Figure 4.49 A close-up of the discharge traces of Insulator #123 of the faulty line.

carried out ground inspection and fault screening and asked the locals about the weather condition around 18:00 on June 3rd (which was said to be a thunderstorm day). No permanent grounding fault had been observed and reported till 18:46, June 5.

According to the photo of the typical discharge traces taken during the inspection (Figure 4.49), it is indicated that the flashover discharge channel belongs to that of typical lightning discharge. By comparing the discharge channel with typical ones, by considering the lightning withstand levels of the fault section under shielding failures and back flashover, by analyzing the ground wire shielding angle, terrain, tower type and insulation configuration and by combining with the fault lightning current from the LLS query, this lightning strike was eventually identified as a shielding failure.

5

Typical Field Tests and Waveform Analysis in UHVDC Transmission Systems

5.1 Waveform Acquisition and Analysis in Typical Tests

5.1.1 Classification of Overvoltages in Converter Stations

Currently, the research on overvoltages in converter stations mainly focuses on the overvoltages from the substation ac or dc side, and from the dc lines. Each type of overvoltage may be further subdivided into temporary overvoltages, switching overvoltages and lightning overvoltages.

5.1.1.1 Overvoltages from the Substation AC Side

Temporary overvoltages include the overvoltages caused by load rejection (especially when reactive loads disappear), the saturation overvoltages due to transformer energization or fault clearing, and the pulse loss and commutation failures of converters, and it will allow ac fundamental waves to reach the dc side.

Switching overvoltages are interphase switching overvoltages caused by switching or faults at the ac side and may be transferred to the valve side through the converter transformer. They can occur in the case of circuit closing and reclosing, putting and reputing ac filters or shunt capacitors into service as well as faults to ground.

Lightning overvoltages primarily refer to the incoming surges on ac lines. Such an overvoltage has a little impact, as converter stations are equipped with quite a few incoming lines and with equipment of damping properties plus the shielding effect of converter transformers.

5.1.1.2 Overvoltages from the Substation DC Side

During converter operation, the temporary overvoltages present on the ac buses in the converter station for a variety of reasons are transferred via the converter transformer to the converter valve. The converter valve is thus subject to temporary overvoltages as well.

The switching overvoltages of the ac side are transferred to the converter through the converter transformer, and high overvoltages would be generated at the dc side if short circuits occur within the converter.

Lightning overvoltages in this case are generally not taken into account due to the shielding effects of converter transformers and smoothing reactors.

Steep wave overvoltages are generated on the converter valve when a short circuit to ground occurs between the valve and the valve-side output terminal of the converter transformer at high electric potential. Lightning or switching overvoltages may

Measurement and Analysis of Overvoltages in Power Systems, First Edition. Jianming Li.

be further caused when the dc filter discharges through the smoothing reactor with its voltage applying on the de-energized valve.

5.1.1.3 Overvoltages from DC Lines

Lightning overvoltages may be caused by direct lightning strokes and back flashover on dc lines and would propagate along the lines and reach dc switchyards or the converter station.

In the case of bipolar operation with a pole short-circuited to ground, switching overvoltages would be generated on the non-fault pole.

5.1.1.4 Switching Overvoltages in UHVDC Transmission System

AC system faults – These are one of the most common faults in the power system. High overvoltages could be generated when faults occur on the ac outgoing line side near the converter bus. Hence, the requirements for the protection system are: (a) the protection for dc systems must not be actuated in the case of ac grounding faults; (b) when the open-circuit fault in the ac system is cleared, the dc system has to recover the power quickly, while the ac system shall be kept stable.

Emergency switch-off due to the actuation of dc system protection – DC system switch-off includes normal switch-off and emergency switch-off.

The dc transmission system is energized in a gradual step-up manner to avoid overvoltages, i.e. the rectifier current is increased by gradually increasing the current setting value of the current regulator. The switch-on process is as follows: first, the inverter is initiated and β is made to be the upper limit (smaller than or equal to 90°); the rectifier is triggered with $\alpha = 90°$, while the current setting value of the regulator is made to rise exponentially. The direct current of the rectifier also rises with the effect of the current regulator. At the inverter side, the switch-on device automatically enables β to decrease when the direct current attains a certain value before reaching the rated value. This type of switch-on mode is called soft switch-on, and the starting time is generally 100–200 ms.

DC system switch-off can be achieved by adopting a method similar to that of soft switch-on: decreasing the direct current along with the current setting value through the current regulator at the rectifier side. At this time, the inverter-side current regulator would increase the trigger angle until it reaches the upper limit. When the direct current is zero, the triggering pulses sent to the converter are stopped and the switch-off process is completed.

DC system emergency switch-off (ESOF) includes emergency switch-off of rectifier stations and inverter stations. In the case of rectifier station emergency switch-off, the converter valve is locked up immediately, and the dc side is free of overvoltages at this time. If the inverter station is switched off quickly, bypass pairs in the inverter station are first put into service, then the valve of the rectifier station is locked up (the rectifier is made to operate in the inversion state by increasing its triggered phase rapidly; thereupon, the energy stored in the smoothing reactor as well as the circuit inductance and capacitance is sent back to the ac system rapidly; the rectifier-side valve is locked up when the current declines to a certain level). High overvoltages can occur on the ac side in both emergency switch-off conditions because the reactive compensation equipment on the ac bus is unable to be switched off instantly.

Energizing ac and dc filters – In dc system operation, the converters at both sides need to absorb a certain amount of reactive power. An ac filter bank is first energized in the dc system, a short period before the converters at both sides get unlocked, and another ac filter bank is put into the system immediately after the unlock, which is referred to as the minimum filter bank configuration. The number of energized ac filter banks varies in accordance with the changes in dc transmission power and dc voltage to meet the demands of the dc system for reactive power and harmonic waves.

Transient overvoltages may be generated on ac buses and ac filters when ac filters are switched on, and the overvoltage magnitude is related to such factors as the arrester protection level, input phase angle of circuit breakers and filter parameters. For instance, closing overvoltages can be reduced if adopting circuit breakers with phase-selection functions. Take a +500 kV DC project that adopts the SF6 circuit breakers from ABB as an example; the maximum phase-to-earth overvoltage amplitude on the ac bus is 1.22 p.u. when closing the ac filter.

Switching on/off dc filters can result in overvoltages on dc equipment. For example, the overvoltage on the dc pole line due to switching on/off the dc filter is 1.29 p.u. in some ±500 kV dc projects (the maximum rated operating voltage at the dc side 1 p.u. = 500 kV).

Trigger pulse losses (commutation failures) – Commutation failures may take place at the inverter station due to switching operations. Moreover, pulse losses or commutation failures can sometimes occur on the converter valves on both sides of the station due to certain causes.

During normal operation of the dc system, power frequency ac voltages will be introduced when commutation failures are caused by the trigger pulse loss of the converter valve, leading to high overvoltages on pole lines, dc filters and neutral buses. Hence overvoltages caused by loss of trigger pulses in rectifier stations and inverter stations need to be studied, and the overvoltage generated from dc blocking caused by long-time trigger pulse loss needs to be considered.

DC-side grounding short-circuit faults – In this case, overvoltages are caused by the faults between the ground and the converter station exit or the dc line midpoint, by the faults on the valve bridge and by faults on the valve bridge leads of the converter transformer.

Ground operation to metallic operation – When the dc system operates in the ground operation mode, the neutral buses on the terminals of the converter station are connected to ground poles through tens of kilometers of grounding pole lines. Therefore, the potentials of these neutral buses are voltage drops generated by the direct current flowing through the grounding pole lines, and are generally not high. To conclude, for the neutral buses of the converter stations, their voltage magnitude is low under the monopole ground operation mode.

When the system operates in the monopole metallic operation mode, the inverter station is grounded via a single site, and the potential of the neutral bus of the rectifier station is the voltage drop generated by the pole line current flowing through the dc line. The potential of the rectifier station neutral bus would rise slightly due to the great length of the dc line, and the potential rise is related to the pole line resistance and the magnitude of the current flowing through the pole line. Therefore, overvoltages would be produced during the transition from the monopole ground operation mode to the metallic operation mode.

Last breaker trip at the inverter side – The dc protection needs a period of time to be actuated during which the ac bus is connected with the ac filter and the dc system is still operating. Therefore, very high overvoltages will be generated on the ac field equipment and the dc converter, presenting major challenges to this equipment. In system commissioning, this overvoltage will be tested as a priority.

5.1.2 Overvoltage Test Methods and Principles

The mature overvoltage test methods in current use include fault wave recording at PT secondaries and fault wave recording by the voltage dividers at transformer bushing taps. However, both methods have their own problems. For the former, filters exist between the PTs and the fault recorders; the sampling rate of the fault recorder, 4 points per ms, is too low; even if bypassing the filter, lightning overvoltages and arc reignition (very fast overvoltages) will arise due to the poor frequency response of the PT iron core, the saturation property and the impact of the high-frequency stray capacitance upon the PT. The primary issue for the latter method is that filters exist between the transformers and the fault recorders. (The divider at the bushing tap under this condition fails to obtain transient overvoltages during the commissioning of some ±800 kV converter station. Theoretically, transient overvoltages may be recorded by bypassing the filters, and other dividers will be selected to monitor transient voltages if the bushing tap divider fails to be installed on the line.)

Tests of overvoltages across the arrester counter – As illustrated in Figure 5.1, since the silicon bridge connects in parallel with the arrester counter, the silicon bridge will smooth the voltage waveform when the voltage exceeds a certain value, and the dynamics of the overvoltage cannot to be truly reflected. One option is to control the voltage across the counter below the chopped voltage by connecting a resistor or capacitor in parallel with the counter. In this way, overvoltages can be reflected in real time. The zinc oxide valve plates in the counter are mainly composed of the units where

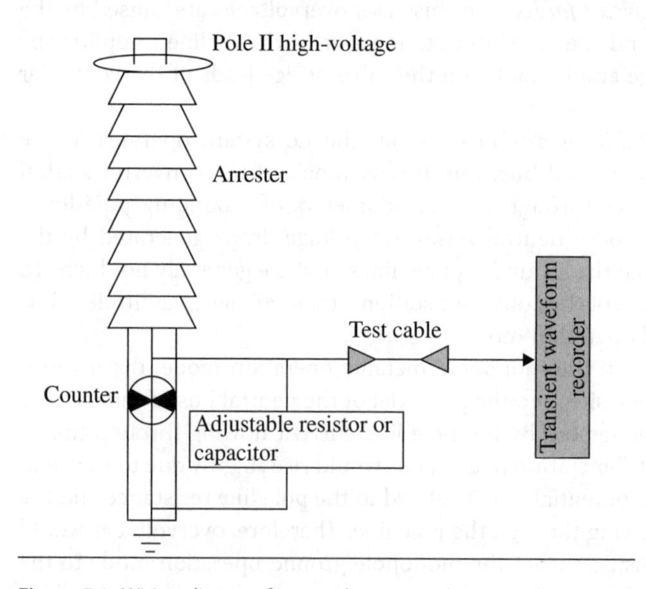

Figure 5.1 Wiring diagram for sampling overvoltages at the terminals of the arrester counter.

the variable resistor and capacitor are connected in parallel, and the impedance of the valve plate is large (in the order of megohm) under small current. So it is possible to estimate the overvoltage multiple according to the leakage current under normal conditions, and hence work out the variable resistance and capacitance. In this test, the variable resistance is 230 Ω and the capacitance is 4 μF.

Overvoltage tests at the converter transformer bushing tap – As shown in Figure 5.2, the transformer capacitance graded bushing is taken as the high voltage arm of the voltage divider, while the standard capacitance installed at the measuring tap of the bushing tap is taken as the low-voltage arm, and hence forming the capacitance-graded voltage-dividing transducer. The capacitance of the divider low-voltage arm has to be suitably chosen to ensure that the normal operating voltage of the capacitor unit in the low-voltage arm is not higher than the safety voltage of the monitoring equipment during normal operation. This measuring method is simple in operation and requires no additional ancillary equipment; and also, capacitive voltage-dividing can ensure distortion-free test overvoltages and a stable ratio of the capacitive voltage divider.

Overvoltage tests using non-contact passive optical fibers – As shown in Figure 5.3, non-contact overvoltage transducers sample the overvoltages mainly through capacitance-graded dividers. The stray capacitance between the plate and the transmission line could be used as the high-voltage arm of the divider. The voltage is then sampled from the low-voltage arm mounted between the plate and the ground. The capacitance of the low-voltage arm can be easily adjusted to change the output voltage. Meanwhile, based on the calculation formula for the capacitance, the distance and medium between capacitor plates as well as the plate area may be altered in order to change the capacitance. This method is simple in principle, accurate in testing overvoltages and has no

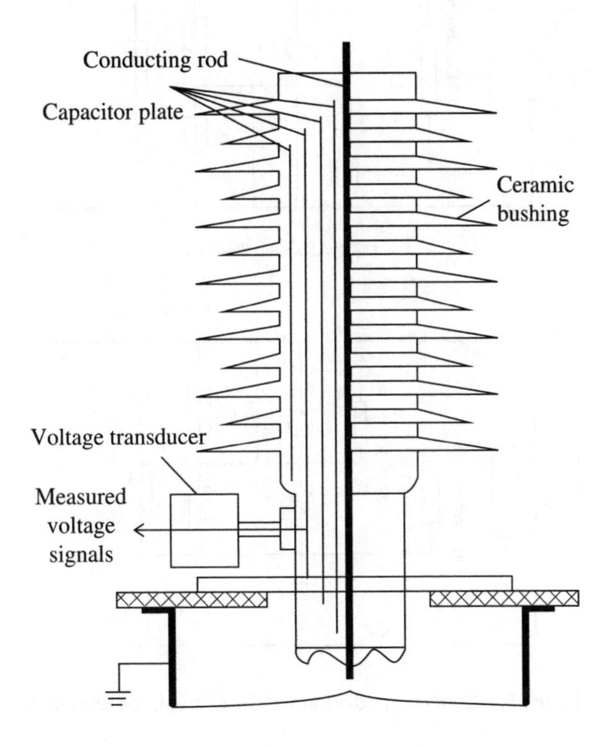

Figure 5.2 Wiring diagram of overvoltage sampling from the converter transformer bushing tap.

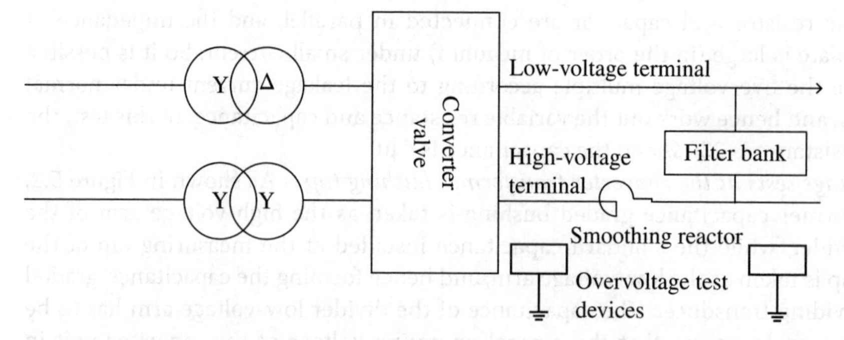

Figure 5.3 Arrangement plan for non-contact passive optical fiber overvoltage tests.

direct impact on the transmission lines. But issues including interphase coupling and interference due to spatial electromagnetic fields have to be addressed during the test.

Overvoltage tests for DC filters – As shown in Figure 5.4, the overvoltage at each point of the dc filter is tested by the dc divider. Both poles are installed with two double-tuned dc filter banks, HP12/24 and HP2/39. In the commissioning stage, HP12/24 on both poles are connected to the test equipment with their location displayed in Figure 5.4. The overvoltage test sites include: neutral bus voltage, voltage between dc filter L1 and

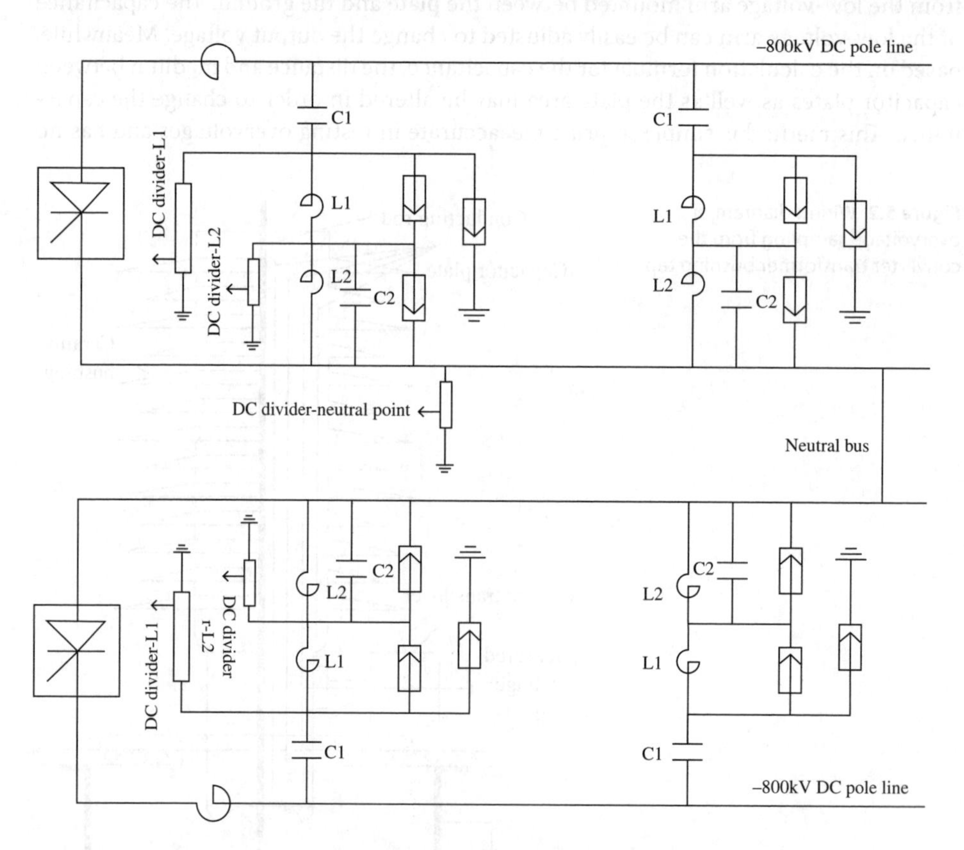

Figure 5.4 Connection diagram of the overvoltage tests for DC filters.

the ground, voltage between dc filter L2 and the ground, the voltage across dc filter L1 and the voltage across dc filter L2, of which, the overvoltages across dc filters L1 and L2 are acquired through waveform processing.

5.2 Typical Field Tests for the UHVDC Transmission System

5.2.1 Disconnecting Converter Transformers

The test results of disconnecting convertor transformers are presented in Figure 5.5. The peak value of the ac voltage across the arrester counter under steady-state operation is 1 V. Switching overvoltages appear when disconnecting convertor transformers; the overvoltages of all three phases last 10.02 ms and simultaneously reach the peaks (0.854 V, 1.296 V and 2.002 V for phases A, B and C) 5 ms after the switching operation. The three-phase voltage signals decline to zero another 5 ms later.

Figure 5.6 presents the overvoltage waveforms when closing the convertor transformer for the second time. The peak value of the ac voltage across the arrester counter under steady-state operation is 1 V. Switching overvoltages appear when disconnecting the convertor transformer; the overvoltages of all three phases last 12.02 ms and reach the peak values of 2.048 V, 1.942 V and 1.282 V, respectively 5 ms after the switching operation. The voltage signals decline to zero 7.16 ms later.

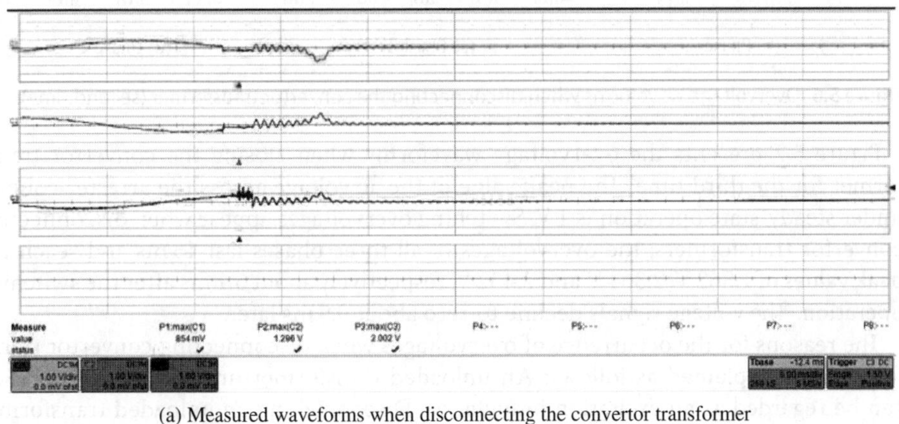

(a) Measured waveforms when disconnecting the convertor transformer

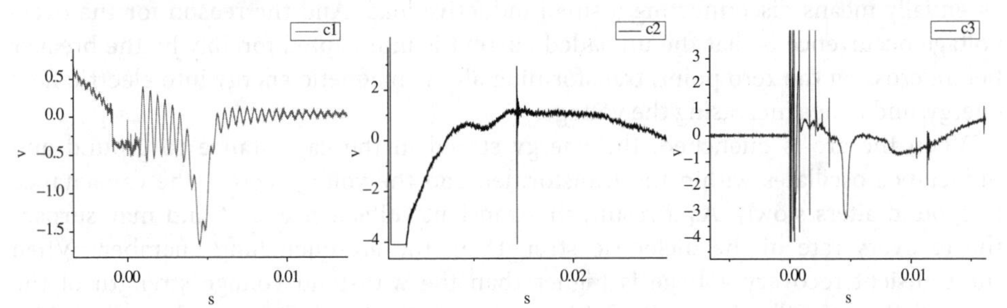

(b) Partial enlarged view of phase A (c) Partial enlarged view of phase B (d) Partial enlarged view of phase C

Figure 5.5 Overvoltage waveforms when disconnecting the convertor transformer (first time).

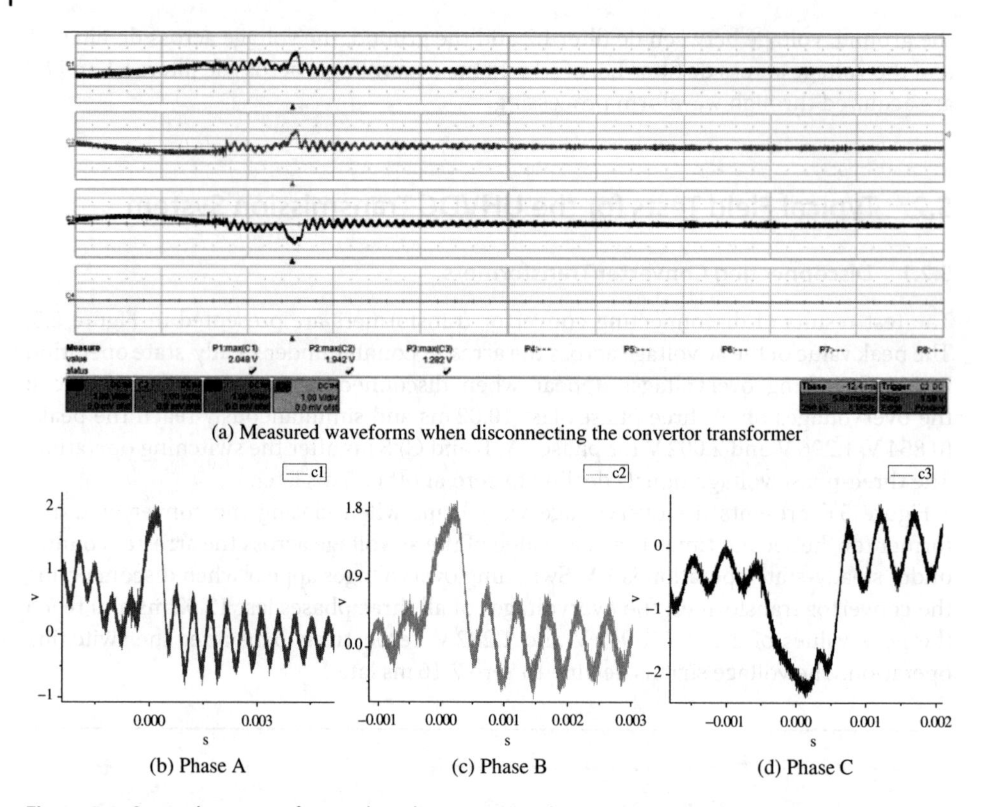

(a) Measured waveforms when disconnecting the convertor transformer

(b) Phase A (c) Phase B (d) Phase C

Figure 5.6 Overvoltage waveforms when disconnecting the convertor transformer (second time).

Figure 5.7 presents the overvoltage waveforms when closing the convertor transformer for the third time. The peak value of the ac voltage across the arrester counter under steady-state operation is 1 V. Switching overvoltages appear when disconnecting convertor transformers; the overvoltages of all three phases last 15 ms and reach the peak values of 1.362 V, 2.364 V and 2.432 V, respectively about 5.1 ms after the switching operation. The voltage signals decline to zero about 9.9 ms later.

The reasons for the occurrence of overvoltages when disconnecting convertor transformers are explained as follows: An unloaded transformer under normal operation can be regarded as an excitation inductance. Disconnecting an unloaded transformer essentially means disconnecting a small inductive load. And the reason for the overvoltage occurrence is that the unloaded current is interrupted forcibly by the breaker before crossing the zero point, transforming all the magnetic energy into electric field energy, and hence increasing the voltage.

When the arc is quenched, the energy stored in the capacitance to ground and inductance oscillates within the transformer, and the voltage across the capacitance to ground alters slowly. As a result, the transient voltage rises fast and may surpass the recovery rate of the dielectric strength in the arc-quenching chamber. When the transient recovery voltage is higher than the withstand voltage strength of the contact, the arc will reignite. Reignition connects the breaker terminals, and enables the recovery voltage to develop from zero again. As vacuum has high arc-quenching

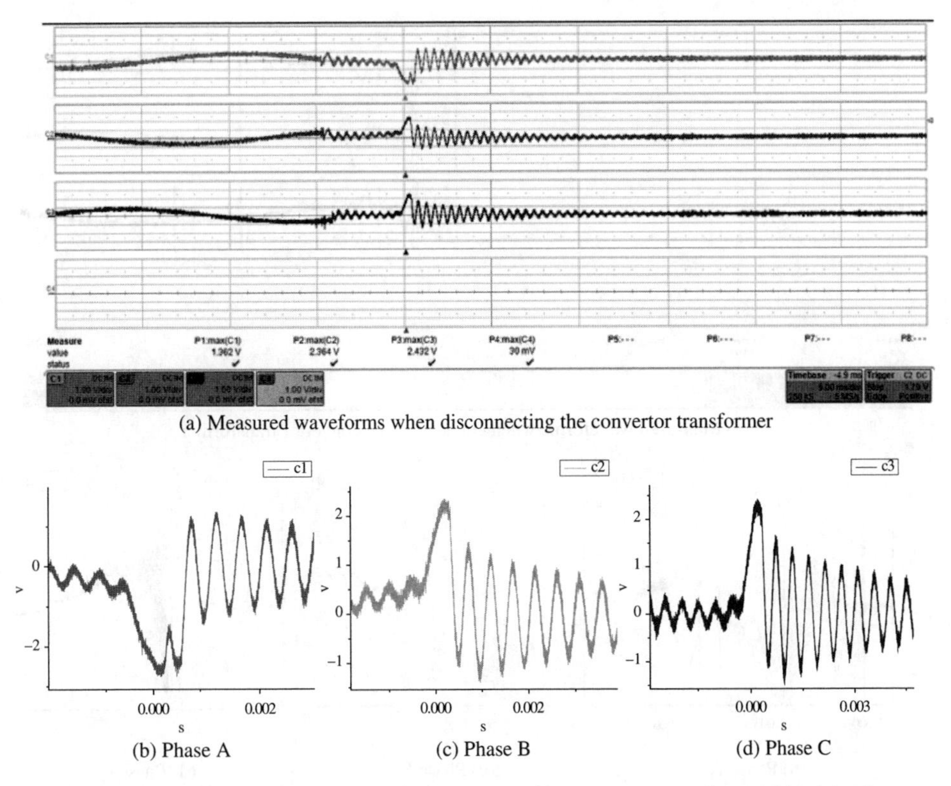

(a) Measured waveforms when disconnecting the convertor transformer

(b) Phase A (c) Phase B (d) Phase C

Figure 5.7 Overvoltage waveforms when disconnecting the convertor transformer (third time).

capabilities, the arc would be quenched the instant the reignited current is around zero. The above process would be repeated until the dielectric strength is greater than the transient recovery voltage with the increase of the contact gap. The arc will not be reignited anymore and the disconnection of the transformer is completed. The amplitudes of the overvoltages under the same operation are very likely not the same due to different initial phase positions when the disconnection begins.

5.2.2 Connecting Converter Transformers

As shown in Figure 5.8, the peak value of the ac voltage across the arrester counter under steady-state operation is 1 V, and switching overvoltages appear when connecting the convertor transformer. Oscillating voltage signals are generated within the frequency range of 0.6–1.5 kHz; the amplitude of the maximum overvoltage is approximately 4.2 times that of the normal bus voltage, and the three-phase voltages tend to be stable around 8.2 ms later.

As shown in Figure 5.9, the peak value of the ac voltage across the arrester counter under steady-state operation is 1 V, and switching overvoltages appear when connecting the convertor transformer. Oscillating voltage signals are generated; the maximum overvoltage is measured at phase C with the amplitude reaching approximately 3 times the operating voltage; the three-phase voltages tend to be stable around 13.4 ms later.

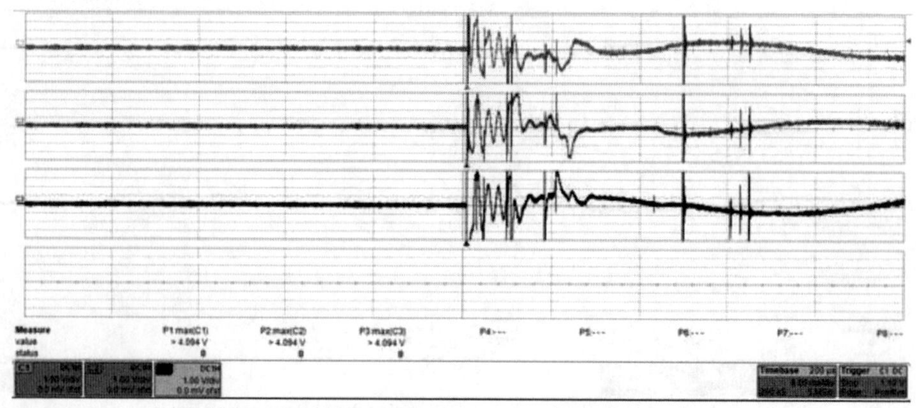

(a) Measured waveforms when connecting the convertor transformer

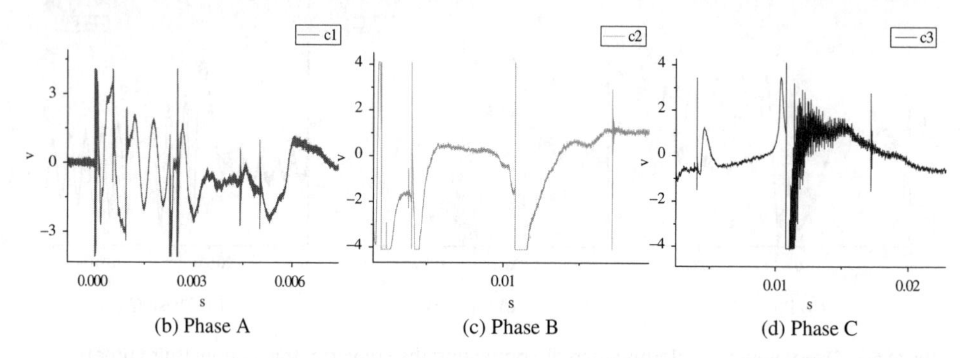

(b) Phase A (c) Phase B (d) Phase C

Figure 5.8 Overvoltage waveforms when connecting the convertor transformer (first time).

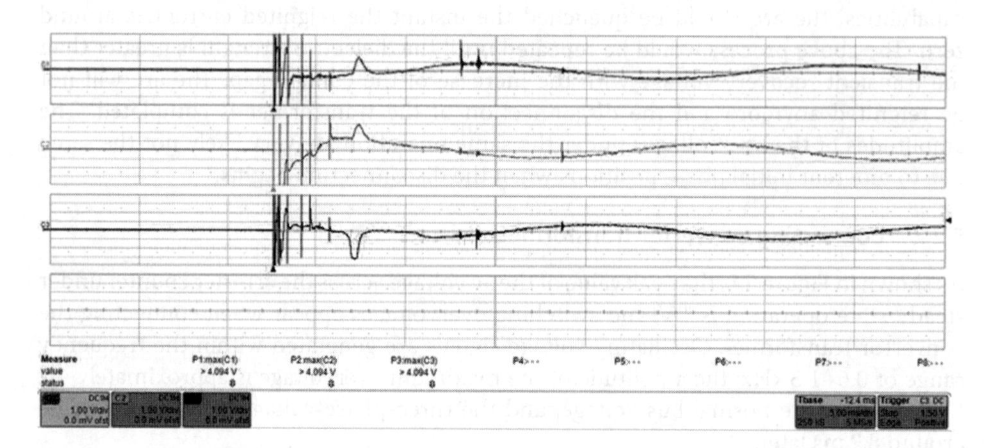

Figure 5.9 Overvoltage waveforms when connecting the convertor transformer (second time).

The reasons for the overvoltage occurrence during the connection of the transformer are as follows: Inrush current, generated by connecting unloaded transformers, has a magnitude as high as 6–8 times the rated current, and up to 100 times the unloaded current. At the worst closing instant, the maximum flux density of the iron core can reach $2\Phi_m$, meaning that the iron core is severely saturated. The more saturated, the higher the inrush current required to produce a given amount of flux. Besides, inrush current during the closing would be attenuated over time due to the presence of transformer internal resistance and line resistance. The overvoltages caused by connecting the transformer thus relate to the saturated degree and residual flux of the iron core as well as the supply phase angle at the closing instant. Ferro-resonance was not observed in this test. In the meantime, as the unloaded transformer can be seen as a capacitor, fast transient overvoltages characterized by high frequency and high magnitude would be generated during the closing due to capacitor charging/discharging and reflection/refraction of the voltage waves.

5.2.3 Converter Valve Deblocking

The results of the deblocking test for the converter at the low-voltage terminal of pole I is presented in Figure 5.10. The peak value of the voltage across the arrester counter under steady state operation is 1 V. Orders were given to deblock the converter at the low-voltage terminal of pole I. The current control mode with a fixed value of 500 A (200 Mw) is adopted. The overvoltage of each phase lasts about 4.8 ms, and continuous oscillation of high frequency is generated on all three phases simultaneously at the operation instant. The overvoltage magnitudes drop gradually to a stable state.

5.2.4 Emergency Switch-Off

The measured waveforms during emergency switch-off are given in Figure 5.11. The peak value of the voltage across the arrester counter under steady state operation is 1 V. Orders were given to shut down pole I and transfer the power to pole II. The overvoltages of all three phases last about 7.9 ms with their magnitudes reaching 3.05 V, 1.84 V and 2.24 V, respectively 1.1 ms after the operation, and subsequently dropping to 0 about 6.8 ms later.

Figure 5.12(a) shows the shape of the overvoltage across the arrester counter in the case of simulated single-phase grounding faults, and a 4 μF capacitor in parallel with the counter is used to sample the overvoltages. The fault lasts for about 2.5 cycles. Because the system is grounded via the neutral point, the non-faulty phase voltages operate in normal conditions, or the non-faulty voltages may show slight fluctuation due to neutral

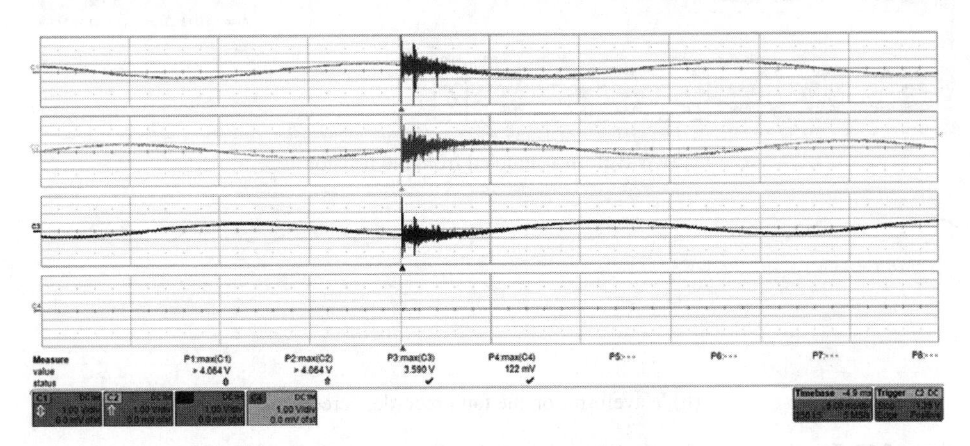

Figure 5.10 Measured waveforms during the deblocking of the inverter at the low-voltage terminal.

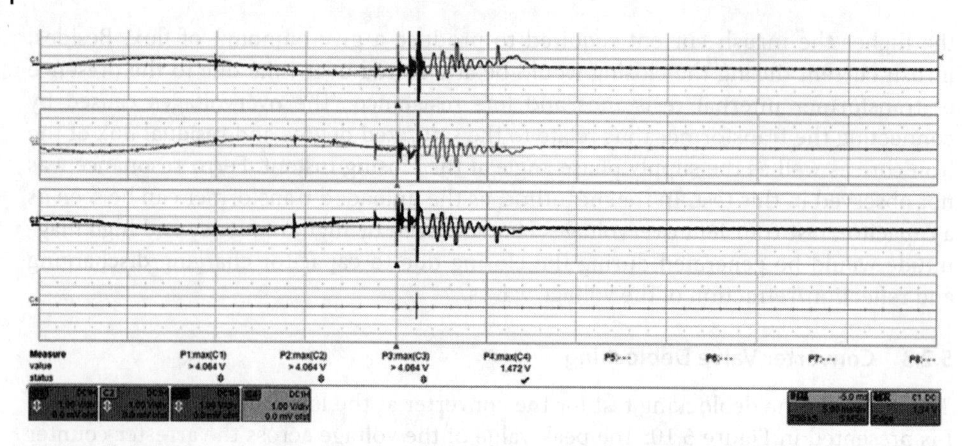

Figure 5.11 Measured waveforms during emergency switch-off of pole I.

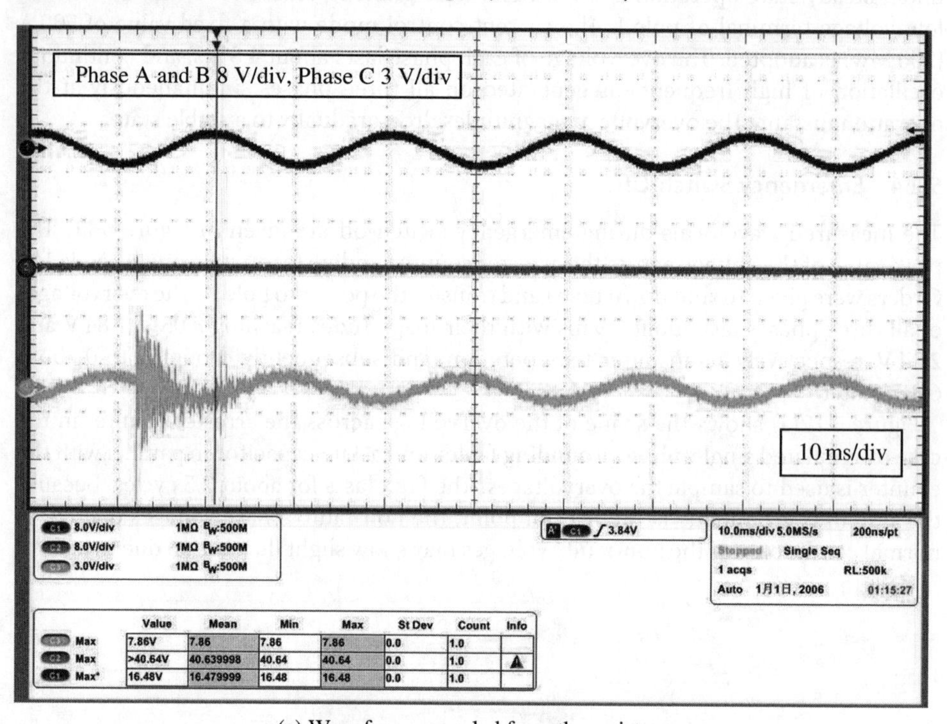

	Value	Mean	Min	Max	St Dev	Count	Info
C1 Max	7.86V	7.86	7.86	7.86	0.0	1.0	
C2 Max	>40.64V	40.639998	40.64	40.64	0.0	1.0	⚠
C3 Max*	16.48V	16.479999	16.48	16.48	0.0	1.0	

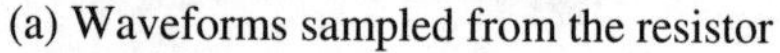
(a) Waveforms sampled from the resistor

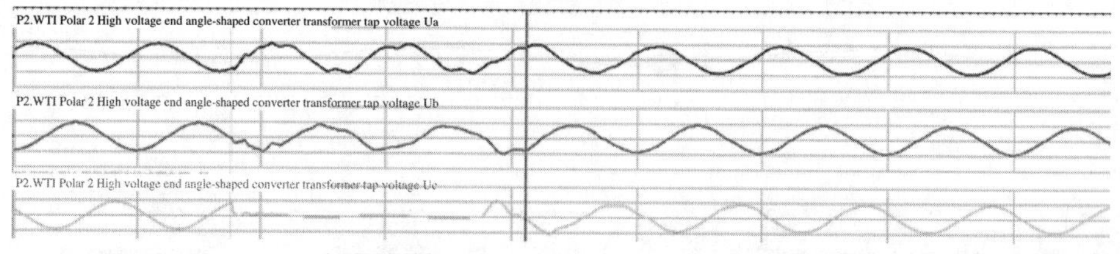

(b) Waveforms on the fault recorder screen

Figure 5.12 Transient grounding fault test of the ac line (sampled from the resistor).

displacement. Figure 5.12(b) shows the waveforms of the AC bus collected by the fault recorder screen when artificial grounding faults are simulated. It is observed that the waveforms sampled from the capacitor are virtually the same as those on the recording screen, which demonstrates good consistency.

5.2.5 Simulated single-phase grounding faults on ac lines

Figure 5.13(a) is the shape of the voltage across the arrester counter when an artificial single-phase grounding fault is simulated on the ac line, and a 4 μF capacitor in parallel with the counter is adopted to sample the overvoltages. The fault lasts for around 2.5 cycles and the peak current is 4701 A at the dc side. As the neutral point of the system is grounded, the voltage waveforms of non-fault phases basically remain in the normal operating state, or display slight fluctuation due to neutral displacement. Figure 5.13(b) shows the waveform of the ac bus voltage acquired by the wave recording screen when

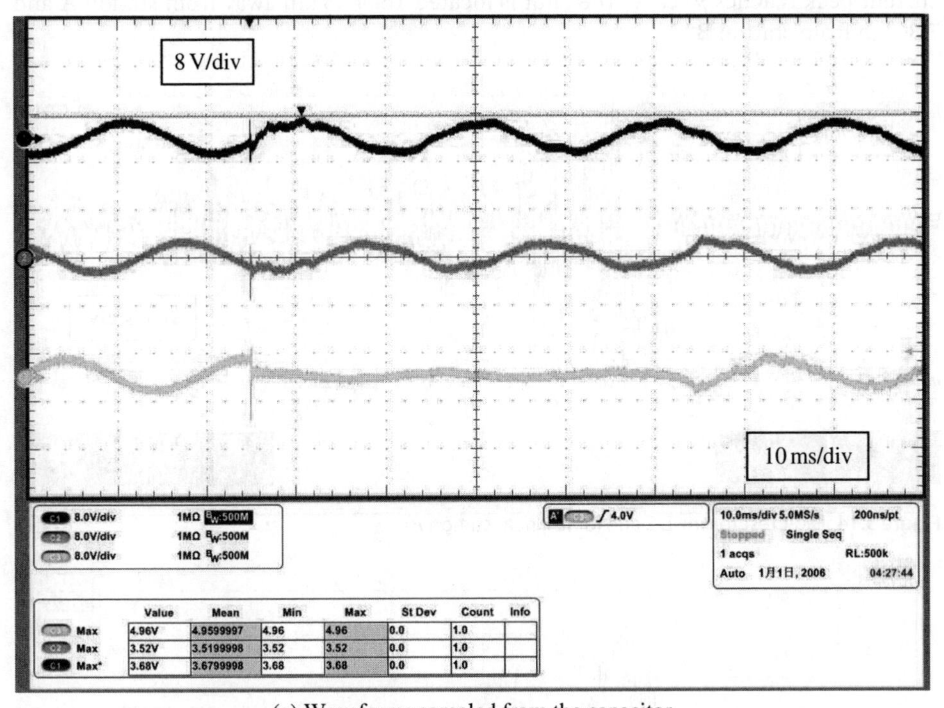

(a) Waveforms sampled from the capacitor

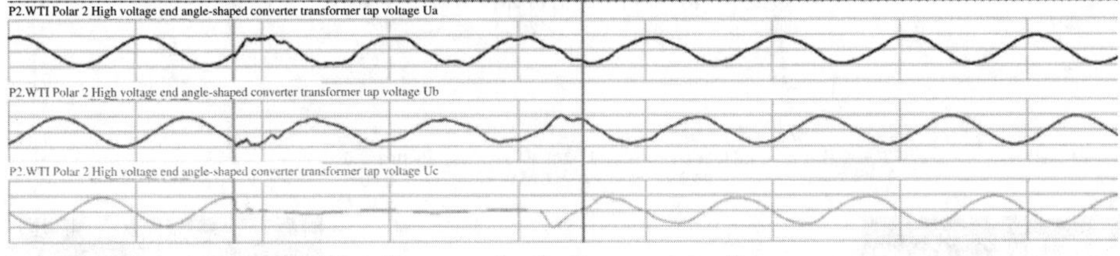

(b) Waveforms on the fault wave recording screen

Figure 5.13 Transient grounding fault test of the ac line (sampled from the capacitor).

artificial grounding faults are simulated. By comparing the two figures, it is observed that the waveforms acquired by the capacitor basically agree with those by the wave recording screen, which demonstrates good consistency.

As illustrated in Figure 5.14, forced phase shift is adopted; for protection system of pole I, travelling wave protection and voltage saltation protection for all three phases are actuated; the line is reclosed by the protection system as well. The simulated fault is placed 0.66 km away from station A and 1651.53 km away from station B with a current peak of 3984 A.

As illustrated in Figure 5.15, for the protection system for pole II, all three phases actuate the travelling wave protection and voltage saltation protection; the forced phase shift is adopted. The simulated fault is placed 0.66 km away from station A and 1652 km away from station B with a current peak of 3233 A.

As shown in Figure 5.16, for the protection system for pole II, all three phases actuate the travelling wave protection, voltage jump protection and forced phase shift. The current peak reaches 9717 A. The fault is located 1644.33 km away from station A and 7.86 km from station B.

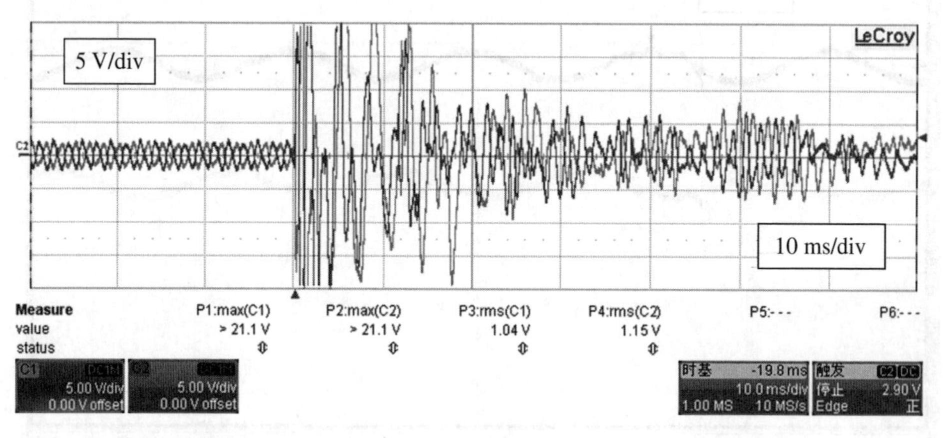

Figure 5.14 Fault test for the DC line (pole I, near station A).

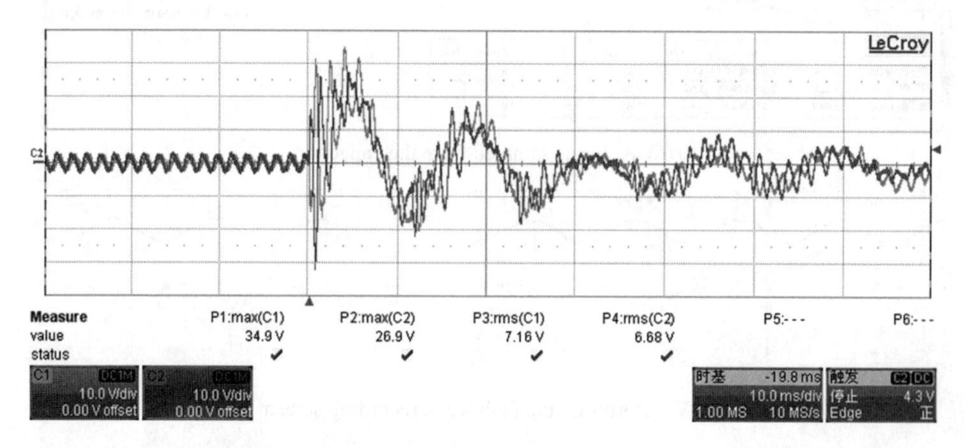

Figure 5.15 Fault test for the DC line (Pole II, near Station A).

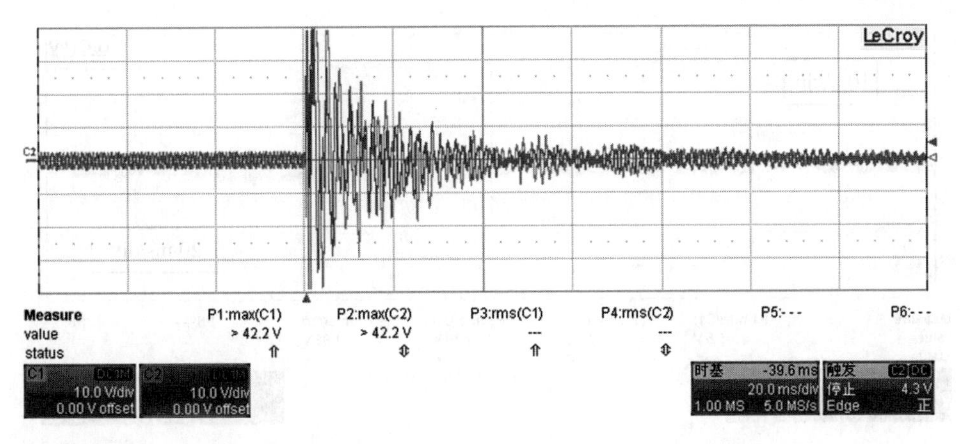

Figure 5.16 Fault test for the DC line (pole II).

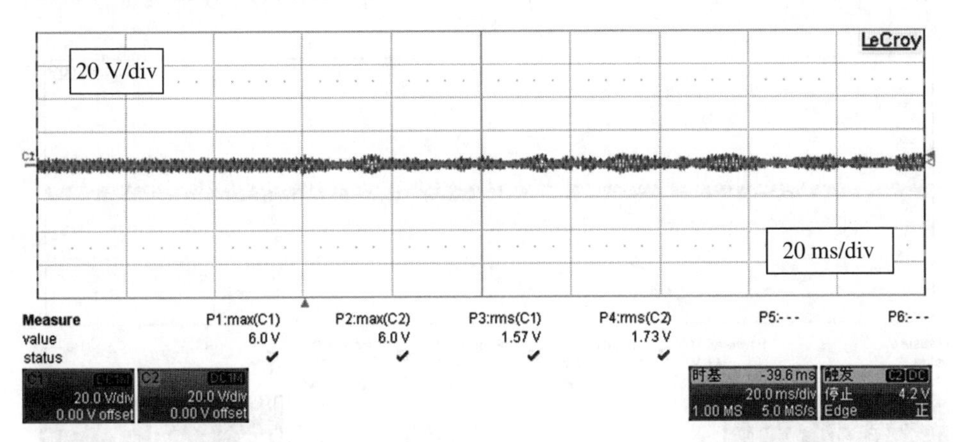

Figure 5.17 Fault test for the ac line at the converter side (sampled from the dc filter).

Figure 5.17 shows the voltage between L1 and the ground acquired by the dc filter through the dc divider. It indicates that when single-phase grounding faults are present on ac lines, no obvious overvoltage will appear at the dc side because fault waves have already been filtered by the ac switchyard.

As shown in Figure 5.18, for the protection system for pole I, the travelling wave protection, voltage saltation protection and forced phase shift for all three phases are actuated. The simulated fault is placed 1644.33 km away from station A and 7.86 km away from station B with a current peak of 9446 A.

The measured waveforms under the monopolar power mode is demonstrated in Figure 5.19.

The bipolar mode is shown in Figure 5.20 with the bipolar power of 400 MW and rising/decay rate of 50 MW/min. The control system for pole I adopts the power mode.

The shape of the wave on the metallic return of pole II is illustrated in Figure 5.21.

Figure 5.22 shows the disturbance test for the dc line. For the protection system for pole II, travelling wave protection, voltage saltation protection and forced phase shift for all three phases are actuated. The simulated fault is placed 0.52 km away from station A and 1651.67 km away from station B with a current peak of 4346 A.

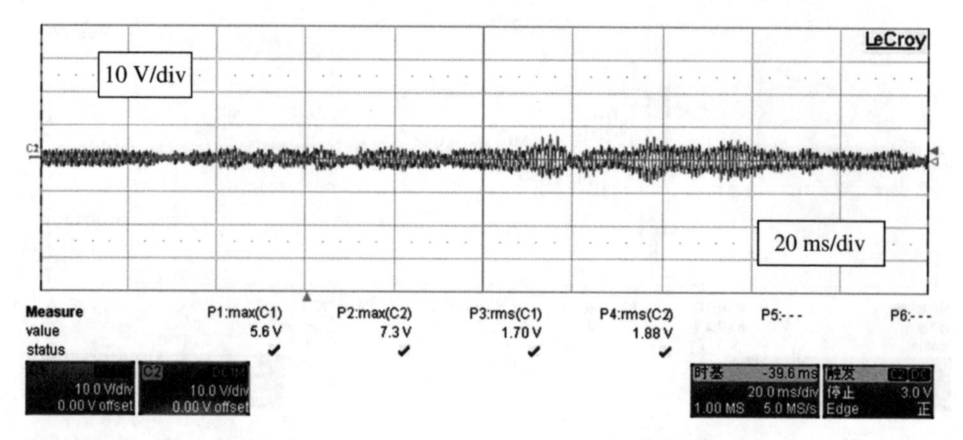

Figure 5.18 Fault test for the dc line (forward power transmission, near station B, pole I).

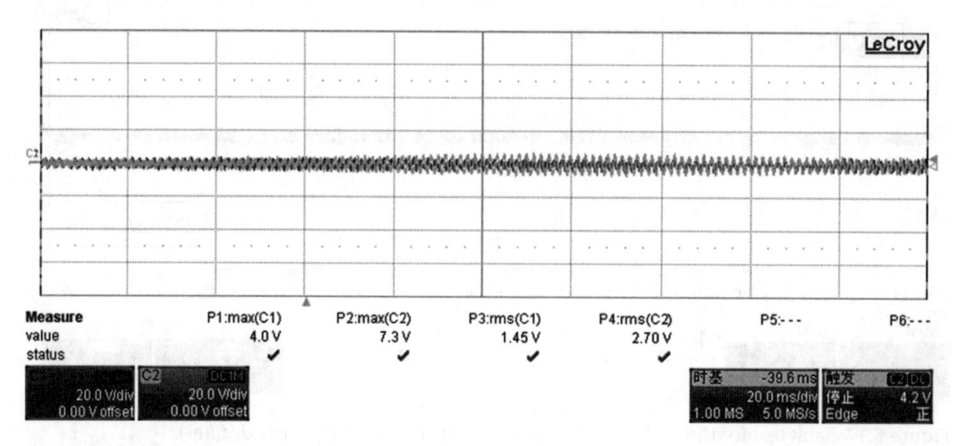

Figure 5.19 Monopolar power mode.

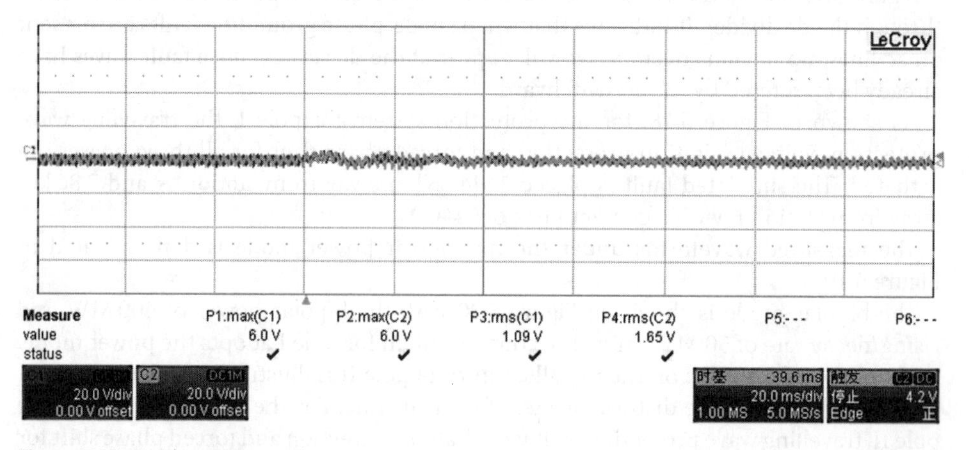

Figure 5.20 Bipolar power mode.

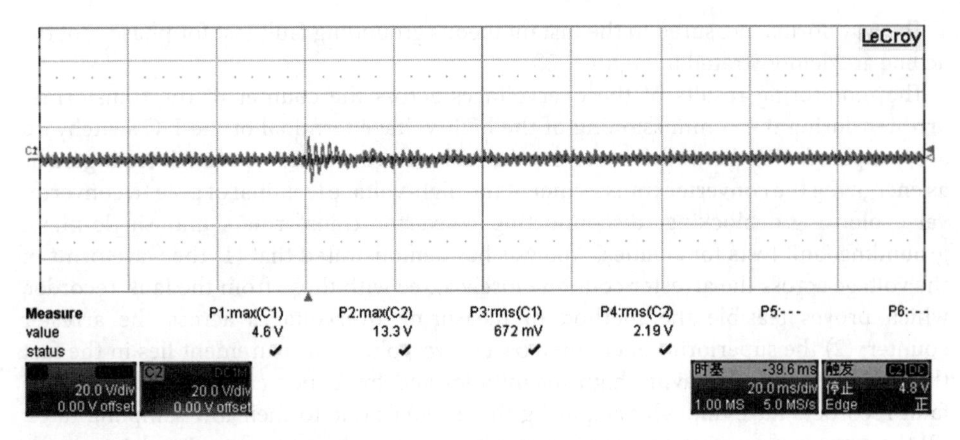

Figure 5.21 Metallic return of pole II.

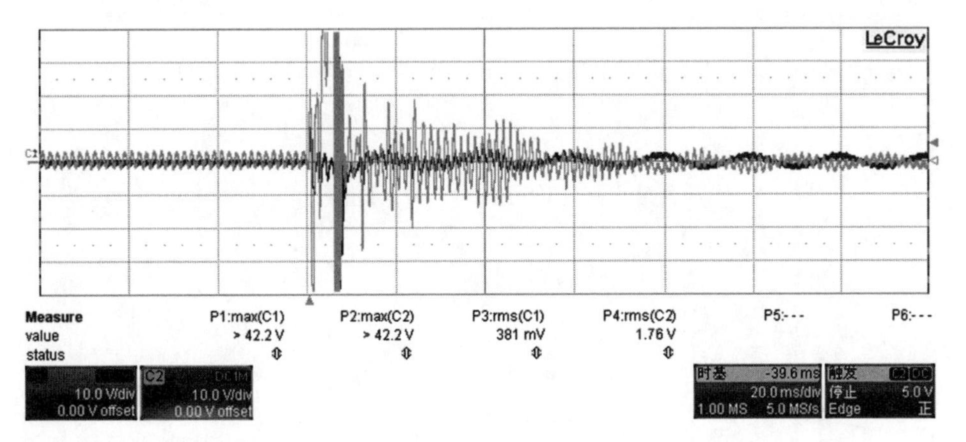

Figure 5.22 Disturbance fault test for the dc line (near station A, pole II).

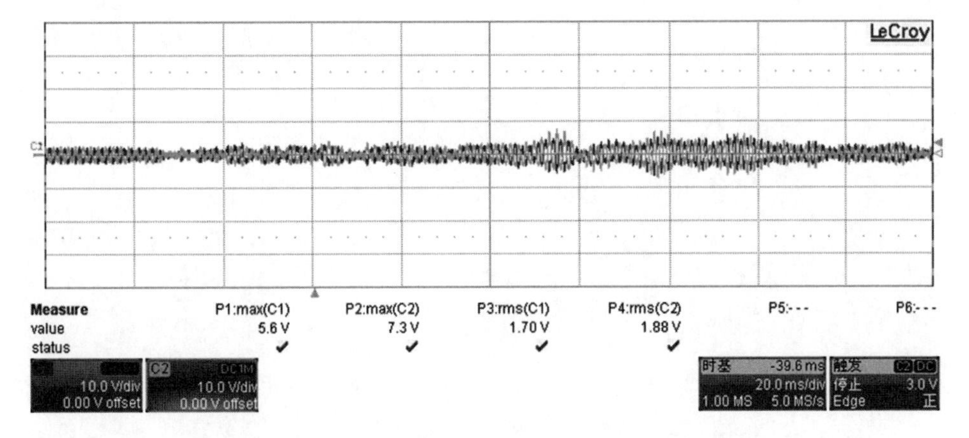

Figure 5.23 Instantaneous grounding fault test for the ac line (phase C).

The waveforms measured in the instantaneous grounding fault test for phase C of the ac line are demonstrated in Figure 5.23.

The monitoring results of the overvoltages across the counter of the transformer arrester during the commissioning of the high-voltage terminal of the DC switchyard are compared with the waveforms from the fault recorder in terms of such testing items as energizing the converter transformer at the high-voltage terminal of pole II, converter valve blocking/deblocking, disconnecting convertor transformers and single-phase grounding fault tests for ac lines. Analysis has demonstrated that (1) the waveforms of the voltage across the arrester counter closely agree with those from the fault recorder, which proves feasible the method of measuring overvoltages across the arrester counter; (2) the superiority of counter-based overvoltage measurement lies in the fact that the pulse signals having high magnitudes and frequency can be collected while fault recorders have difficulty acquiring these signals due to their low sampling rates; (3) the measured overvoltages are found not exceeding the equipment insulation level.

6
Overvoltage Digital Simulation

6.1 Overvoltage Digital Simulation Software

EMTDC (ElectroMagnetic Transient in DC System) is currently the most widely used software for power system analysis. In order to study HVDC systems, Dr Dennis Woodford developed the first version of EMTDC at Manitoba Hydro in 1976 and later established Manitoba HVDC Research Center at the University of Manitoba. This research center has been committed to improving component model bases and functions under the leadership of Dennis Woodford for years, enabling EMTDC to become a multifunctional tool not only for ac\dc power system research but also for power electronics simulation and nonlinear control. The successful development of PSCAD (power system computer aided design) in 1988 allowed users to input electrical schematic diagrams as well as component parameters and to configure relevant system parameters conveniently, making it possible for complex parts of the power system to be visualized. EMTDC has since become the simulation engine (also referred to as the computing kernel), calculating data files generated by PSCAD and presenting results to PSCAD for display. Up to now, owing to its advantages of large-scale calculation capacity, complete and accurate component model base, stable and efficient computing kernel, user-friendly interface and great openness, PSCAD/EMTDC has been adopted by research institutions, colleges and universities all over the world. PSCAD/EMTDC subdivides the parts involved in network research into different function blocks with simple and concise information transferred in between.

PSCAD/EMTDC is divided into two main parts, in which EMTDC, an offline electromagnetic transient simulation program, is the simulation kernel, and PSCAD, as the graphic user interface, completes the tasks including construction, simulation and interpretation of the network diagrams of the systems under research. Thanks to its help, system modeling is considerably simplified, and modification and error correction is also facilitated, thereby improving and ensuring research quality and efficiency. PSCAD and EMTDC are correlated, supporting one another while combining together. They have become flexible aids for digital simulation of the power system.

ATP (Alternative Transients Program) is the most widely used version of EMTP, which can be used by computers of most types. The basic function of ATP is to perform power system simulation and evaluation; for instance, one typical application is to predict time dependence of the variable of interest after the power system is subject to some disturbance (such as grounding faults and switching operations). Currently, ATP mathematical models include RLC circuits with lumped elements, multi-phase PI

Measurement and Analysis of Overvoltages in Power Systems, First Edition. Jianming Li.

equivalent circuits, multi-phase distributed transmission lines, nonlinear resistors and inductors, time-varying resistors, switchgear, voltage and current sources, etc.

The graphics input program ATP-Draw is also provided, by which simulation circuits can be drawn directly, model parameters can be set and simulation input files can be generated. In this way, users need not consider strict formats, which makes it convenient to generate and modify input files.

ATP simulation models and the evaluation results are widely accepted in academic circles following years of testing. In the high voltage area, ATP is often used for research on transient overvoltages such as ferro-resonance overvoltages and lightning overvoltages.

6.2 Evaluation of Switching Overvoltages

Switching overvoltages are classified as overvoltages due to opening unloaded lines, overvoltages due to disconnecting unloaded transformers, overvoltage due to closing unloaded lines and overvoltage due to connecting unloaded transformers. Due to the differences in overvoltage process and mechanism, different parameters and characteristics are adopted in simulation and evaluation. Parameter settings and result interpretation of various overvoltages are detailed below.

6.2.1 Evaluation of Opening Overvoltages

When evaluating overvoltages due to opening unloaded lines, the influencing factors first need to be considered, including arc-quenching performances, neutral grounding methods, bus capacitance rise due to other outgoing lines and electromagnetic PTs mounted on the line side.

Example 1 Simulate and analyze the opening overvoltage of a 500 kV overhead transmission line of 100 km long.

JMarti line model for the 500 kV overhead transmission line: [Lines/Cables]→auto compute lines/cables model [LCC]; double click LCC model and set the parameters as shown in Figure 6.1.

Figure 6.1(a) shows the model parameter setting, where Overhead line is chosen for System type; Transposed, a check item used in π-type equivalent lines, is not picked while all the other boxes such as Auto bundling and Skin effect are selected. JMarti is chosen for Model/Type; for Standard data, soil resistivity, initial frequency for fitting and line length shall be set as 50 Ω·m, 0.005 Hz and 100 km, respectively.

Figure 6.1(b) is the window displaying line model data, where Ph.no refers to the phase number of the conductor; Rin is the conductor inner diameter; Rout is the conductor outer diameter; Resis is the conductor dc resistance; Horiz is the horizontal distance from the specified reference phase; Vtower is the nominal height; Vmid is the line height at midspan; Separ is the conductor separation distance; Alpha is the shielding angle of the ground wire; NB is the number of bundles.

After filling in the geometric and electrical data of the 500 kV unloaded line, click RunATP, and click OK after entering the LCC file name. In this way, the model for the 500 kV unloaded line is built, which includes a pch file, a lib file and a dat file.

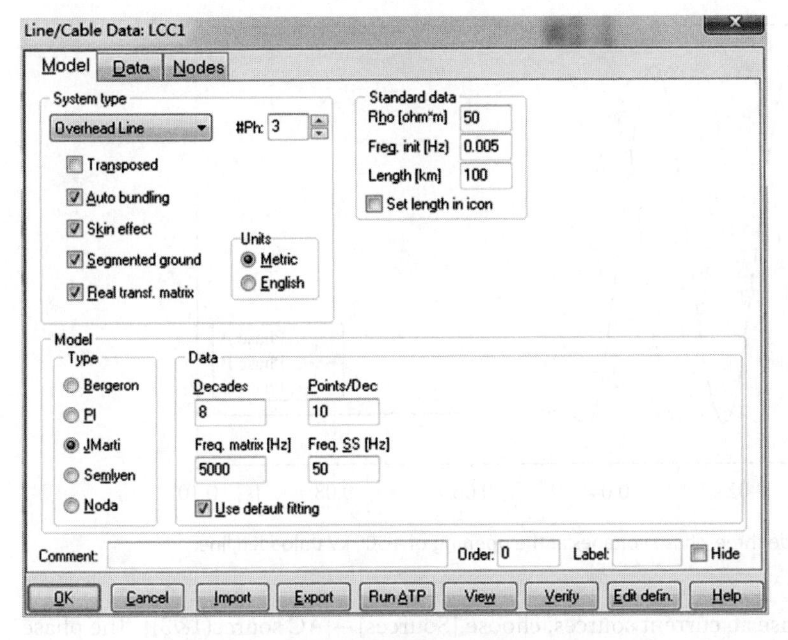

(a) Model parameter settings

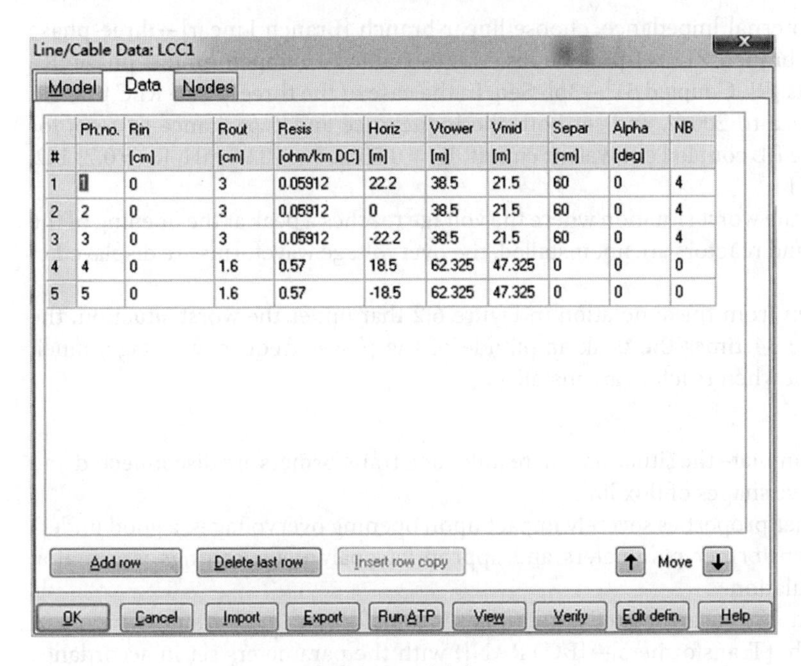

#	Ph.no.	Rin [cm]	Rout [cm]	Resis [ohm/km DC]	Horiz [m]	Vtower [m]	Vmid [m]	Separ [cm]	Alpha [deg]	NB
1	1	0	3	0.05912	22.2	38.5	21.5	60	0	4
2	2	0	3	0.05912	0	38.5	21.5	60	0	4
3	3	0	3	0.05912	-22.2	38.5	21.5	60	0	4
4	4	0	1.6	0.57	18.5	62.325	47.325	0	0	0
5	5	0	1.6	0.57	-18.5	62.325	47.325	0	0	0

(b) Line model data

Figure 6.1 Dialog box for parameters of the LCC model for 500 kV overhead transmission lines.

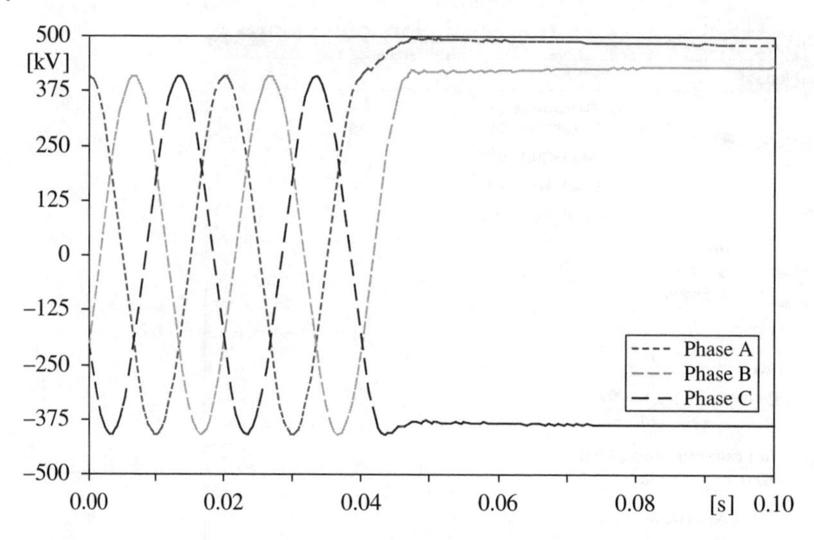

Figure 6.2 line-side three-phase voltages at the opening of a 500 kV unloaded line.

For three-phase ac current sources, choose [Sources]→[AC source(1&3)]. The phase voltage amplitude shall be $500 \times \frac{\sqrt{2}}{\sqrt{3}} \times 10^3 = 408248(\text{V})$.

For supply internal impedance, choose linear branch [Branch Linear]→three-phase RLC coupled branch[RLC 3-ph]; choose Lines/Cables→Lumped→multi-phase RL coupled circuits [RL Coupled 51]→3ph.Seq. In the case of the three-phase RLC branch, set the resistance to 200 Ω, and set both the inductance and capacitance to zero; for the three-phase RL coupled equivalent circuit, $R_0 = 0.55\ \Omega$, $L_0 = 8.98$ mH, $R_+ = 0.711\ \Omega$, $L_+ = 11.857$ mH.

Considering the worst situation where the voltage reaches a peak at the opening of the unloaded line and reactors are not installed, the overvoltage waveforms are displayed in Figure 6.2.

It can be seen from the simulation in Figure 6.2 that under the worst situation, the overvoltage is 1.34 times the peak amplitude of the power frequency voltage, much higher than that when reactors are installed.

Example 2 Simulate the situation where unloaded transformers are disconnected and observe the wave shapes of flux linkage.

As transformer properties severely impact upon opening overvoltages, a good understanding of transformer parameters and appropriate parameter settings is essential before the simulation.

Solution: Choose user-defined three-phase transformer evaluation component BCTRAN (path: [Transformers]→[BCTRAN]) with the parameters set in accordance with Table 6.1.

The dialog box of parameter setting is shown in Figure 6.3.

Choose 3 for Number of phases and Number of windings; set test frequency as 50 Hz; set HV, LV and TV line-to-line voltage as 230 kV, 121 kV and 38.5 kV; set HV, LV and TV power as 150 MVA, 150 MVA and 75 MVA; the connection symbol is YYD with a

Table 6.1 Transformer parameters.

Transformer model		SFPS9-150000/220	
Rated voltage	(230 ± 2 × 2.5%)/121/38.5	Rated current	1124.7 (low voltage)
No-load losses	125.77 kW	Unloaded current	0.51%
Impedance voltage	High to medium: 13.4% High to low: 22.79% Medium to low: 7.38%	Load loss	High to medium: 437.51 kW High to low: 127.14 kW Medium to low: 105.97 kW

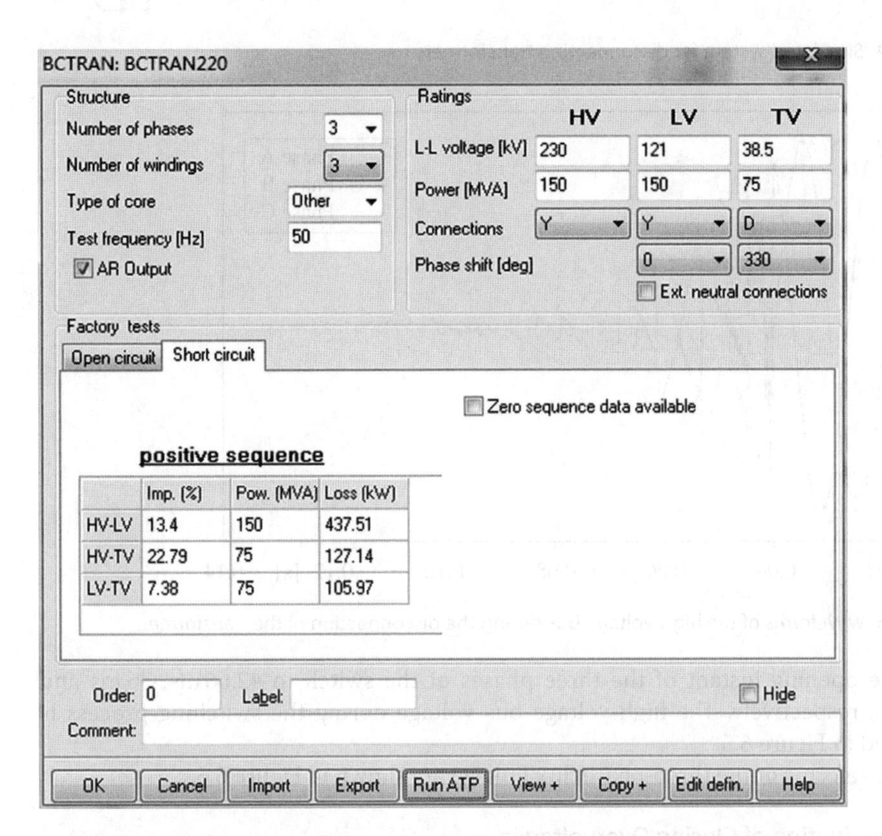

Figure 6.3 Dialog box of parameter setting for disconnecting unloaded transformers.

phase shift of 330°. Click RunATP after completing the setting, enter BCT file name and click OK after confirmation.

As the BCTRAN component does not consider the effect of hysteresis saturation, a nonlinear inductive branch has to be added to the low-voltage winding (path:[Branch Nonlinear]→[L(i)Type96]). The hysteresis loop of the transformer core can be evaluated either by using EMTP supporting subroutines Hysteresis Routine and excitation test data provided by the manufacture or by using Saturation and the *U*–*I* curves provided by the manufacturer. The simulated system for disconnecting unloaded transformers is shown in Figure 6.4.

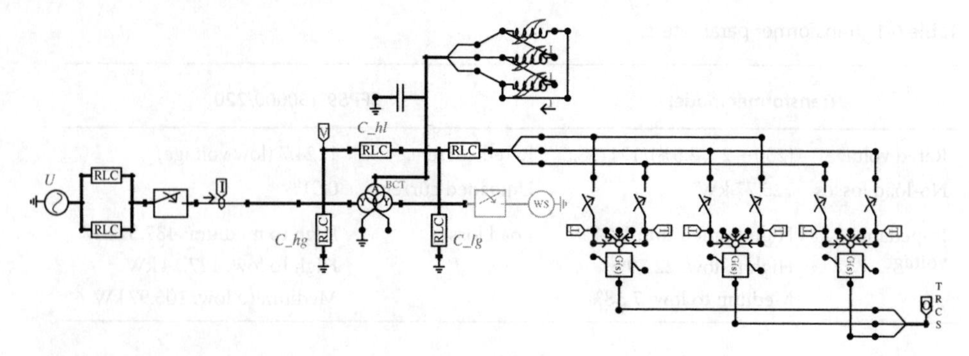

Figure 6.4 Simulated system for disconnecting transformers.

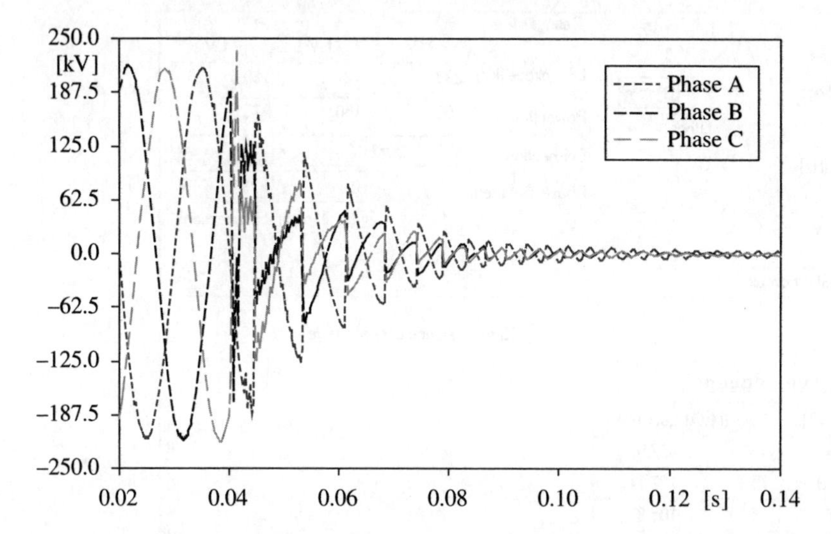

Figure 6.5 waveforms of the high-voltage bus during the disconnection of the transformer.

Set the opening instant of the three phases of the switch to 42.66 ms, 36 ms and 39.33 ms, respectively. The high-voltage bus voltage during the switching process is illustrated in Figure 6.5.

The waveforms of the three-phase flux linkage are shown in Figure 6.6.

6.2.2 Evaluation of Closing Overvoltages

Closing overvoltages are mainly divided into overvoltages due to closing unloaded lines and overvoltages due to connecting unloaded transformers, and influencing factors include the closing phase angle, line residual voltage changes and line loss.

Example 3 Simulate and analyze the closing overvoltage of a 500 kV overhead transmission line of 100 km long.

In the case of JMarti line model for the 500 kV overhead transmission line: Choose [Lines/Cables]→auto compute lines/cables model[LCC]; double click LCC model and adopt the same parameter setting as that in Example 1. The simulation scheme considers both conditions with and without closing resistors.

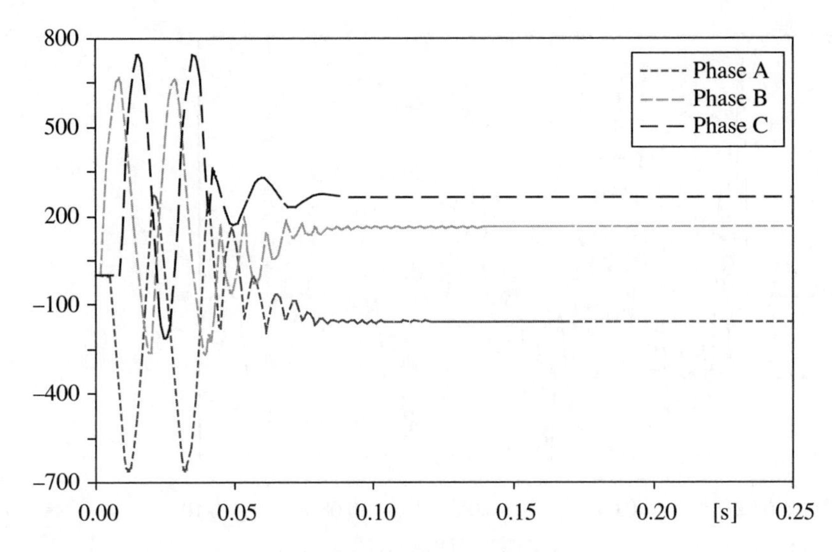

Figure 6.6 Three-phase waveforms of the flux linkage.

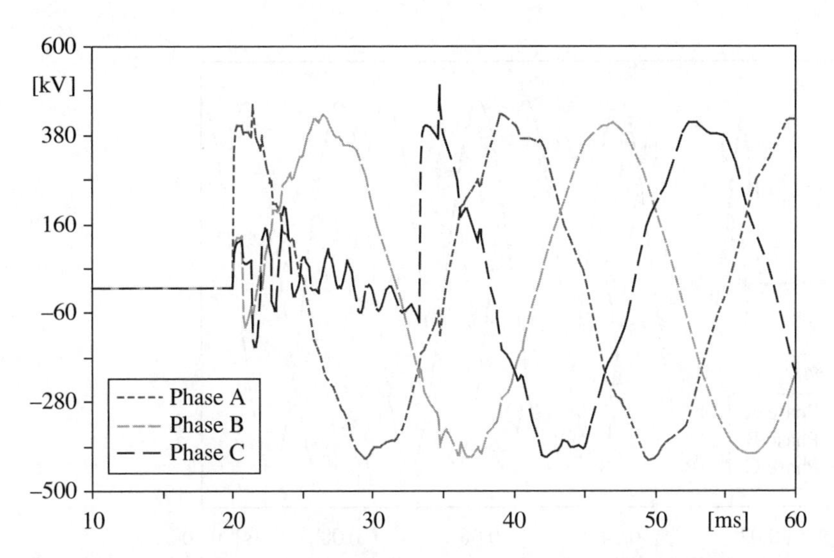

Figure 6.7 line-side three-phase voltages at the closing of a 500 kV unloaded transmission line.

Under the condition where the closing resistor is not installed and the line is closed at $\theta = 0°$, the simulation result at the closing site is as follows:

As can be seen from Figure 6.7, the maximum overvoltage magnitude of phase A is 456 kV while the overvoltages of phase B and phase C are induced, in which phase C can reach up to 502 kV.

The simulation result of phase A voltage at the line end is presented in Figure 6.8.

As shown in Figure 6.8, the maximum overvoltage amplitude of phase A at the line end reaches 920 kV, which far exceeds the power frequency voltage class.

The line is simulated again with the addition of the closing resistor to analyze its benefits. When the closing resistor is 300 Ω and the connection time is 6 ms while the

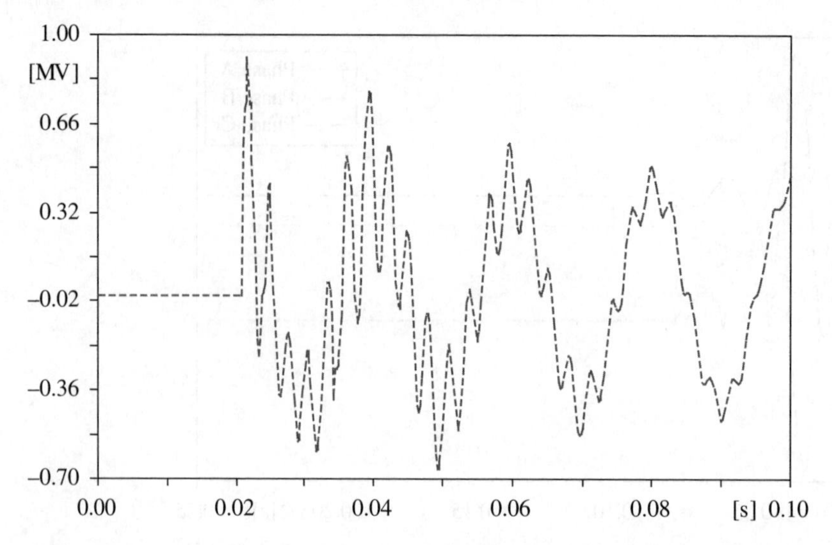

Figure 6.8 Phase A voltage at the line end when closing a 500 kV unloaded transmission line.

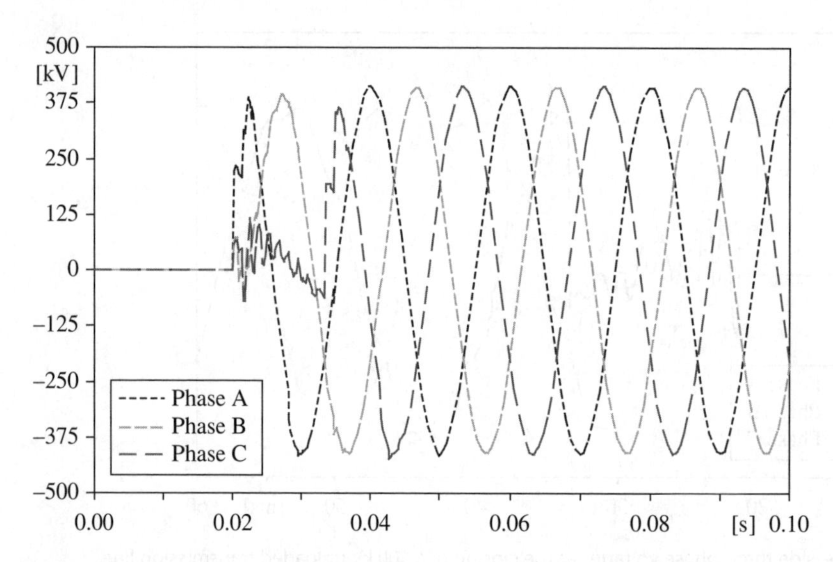

Figure 6.9 Line-side three-phase voltages after closing a 500 kV unloaded transmission line with a closing resistor installed.

other parameters remain constant, the simulation result at the closing site is illustrated in Figure 6.9.

Meanwhile, the phase A voltage at the line end is simulated and shown in the following. As can be seen from Figure 6.10, the maximum voltage amplitude of phase A at the line end is 500 kV when a 300 Ω closing resistor is installed. Overvoltage does not occur at the closing site, and the phase A voltage at the line end is much smaller than that in Figure 6.8 where the closing resistor is not employed. In this regard, the addition of closing resistors can greatly reduce overvoltage harm.

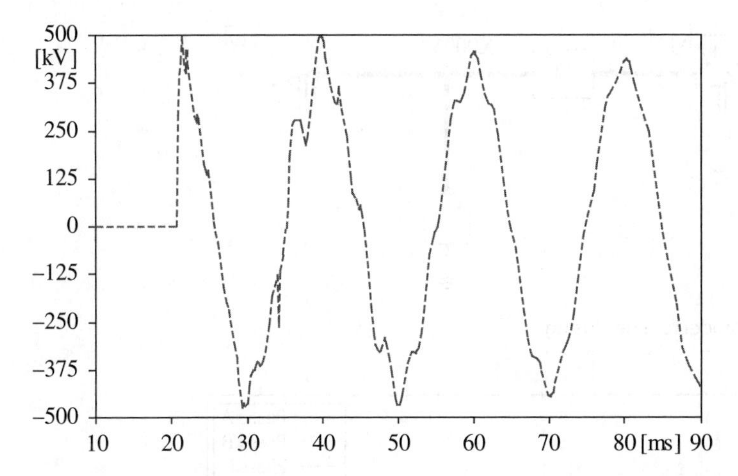

Figure 6.10 Phase A voltage at the line end when closing a 500 kV unloaded transmission line with a closing resistor installed.

Example 4 Simulate the overvoltage caused by closing a 220 kV substation cable.

The fault line is closed at a distance of 760 m from the 220 kV substation. Partial discharge occurs due to insulation defects in the cable joint, resulting in a short-circuit fault on phase B joint, which is later followed by a similar fault occurring to the phase C joint. Ultimately, joints of phases B and C are burnt and damaged.

The cable model utilized is YJLW02-Z-1 with a conductor sectional area of 2000 mm^2. The LCC cable model is established with the arrangement illustrated in Figure 6.11.

The peak value of the phase voltage is $220{,}000 \times \sqrt{2}/\sqrt{3} = 179{,}629$(U). The length on the left-hand side of the joint is 760 m while the rear-side length reaches 3325 m. The closing time is set to 0.079 s, and the simulation diagram is as shown in Figure 6.12.

The simulation result shows that the peak value of the phase B overvoltage is 269 kV, 1.49 times the amplitude of the power frequency voltage.

Figure 6.11 Cable model.

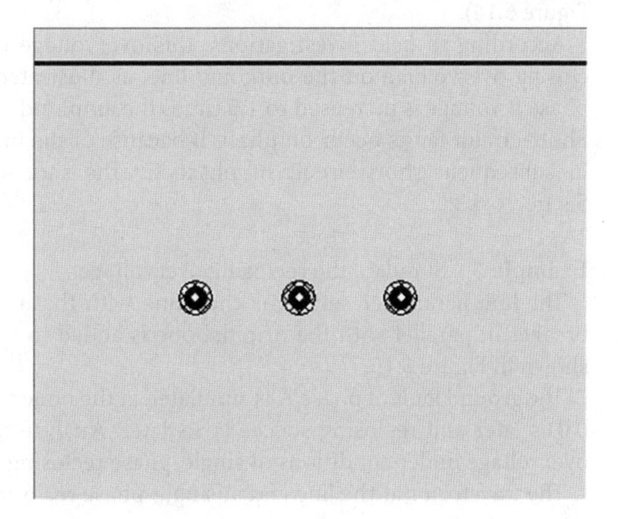

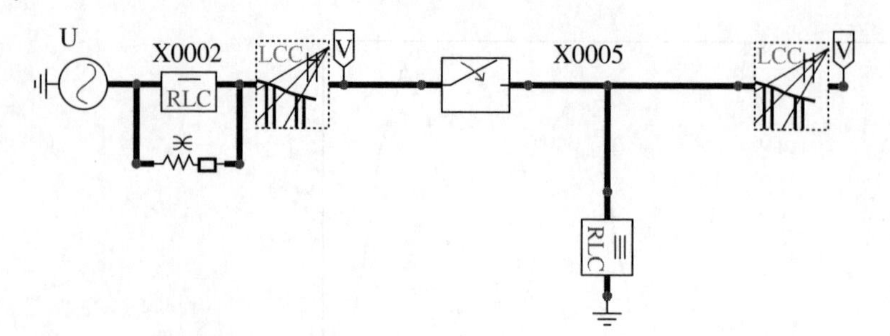

Figure 6.12 Simulation model for line closing.

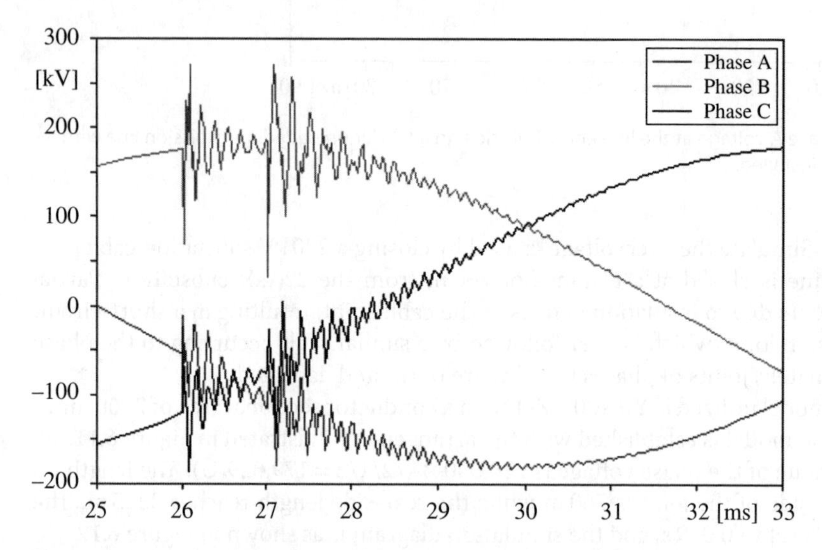

Figure 6.13 Simulated waveforms of the overvoltage due to closing lines.

The simulated waveforms basically agree with the field measured ones (shown in Figure 6.14).

According to field investigations, this overvoltage is first judged as a three-phase closing overvoltage on the outgoing line, as illustrated in Figures 6.13 and 6.14. The phase B voltage is increased to 1.5 times the amplitude of the power frequency voltage. Short-circuit faults occur on phase B because of the insulation problem, which results in subsequent short-circuit of phase C. The short-circuit faults are simulated in Section 6.4.3.

Example 5 Simulate the reclosing overvoltage.

The line parameter setting is the same with that of Example 2, and a three-phase breaker in parallel with the original one is added to simulate closing and opening, as shown in Figure 6.15.

The ground fault of phase A is simulated at the node x0007. The protection is actuated 0.01 s later and reclosing succeeds 1 s later. Analyze the waveforms of the switching overvoltage under conditions of single-phase reclosing and three-phase reclosing.

The waveforms at the line end at single-phase reclosing are shown in Figure 6.16.

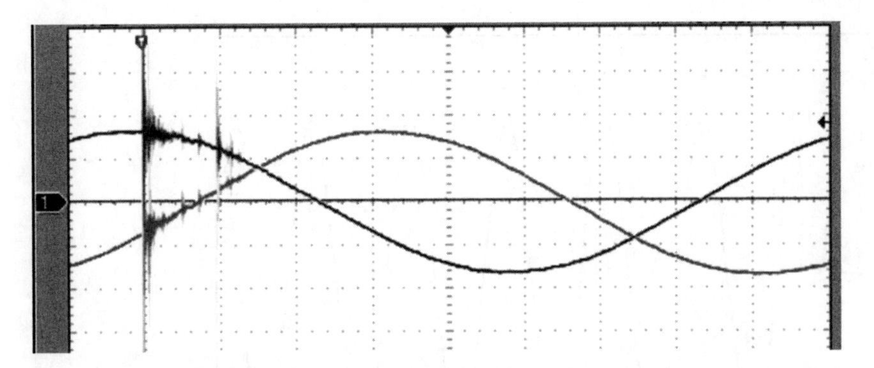

Figure 6.14 Field measured waveforms of the overvoltage caused by closing lines.

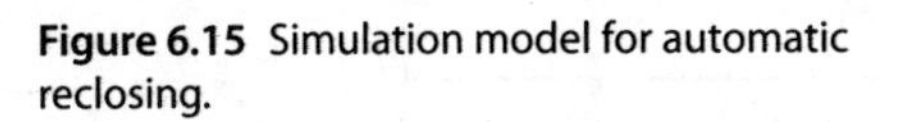

Figure 6.15 Simulation model for automatic reclosing.

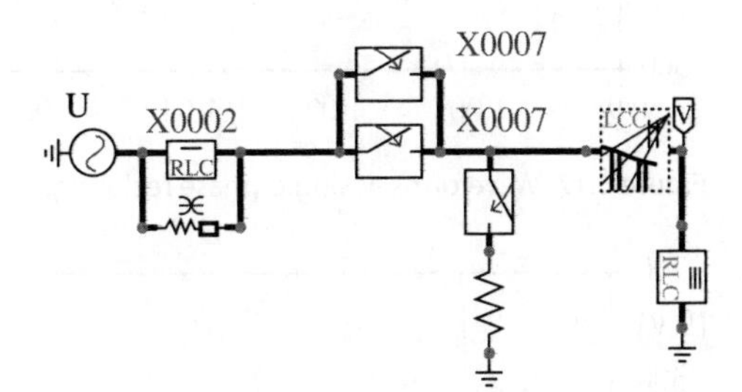

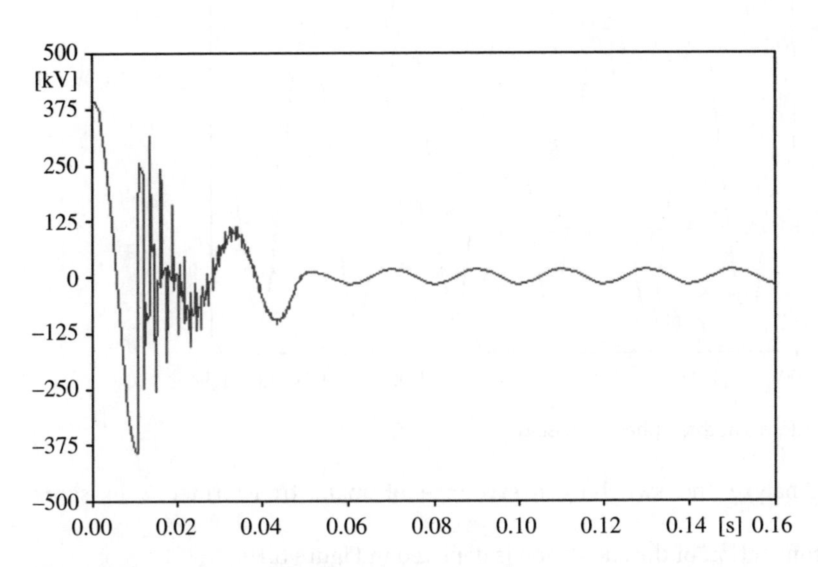

Figure 6.16 Waveforms when protection is actuated.

As can be seen from the waveforms, the overvoltage at single-phase reclosing (Figure 6.17) can reach 561 kV, 1.37 times the rated phase voltage while the overvoltage at three-phase reclosing (Figure 6.18) can reach 657 kV, 1.6 times the rated phase voltage, which indicates that the overvoltage caused by three-phase reclosing is higher than that of single-phase reclosing.

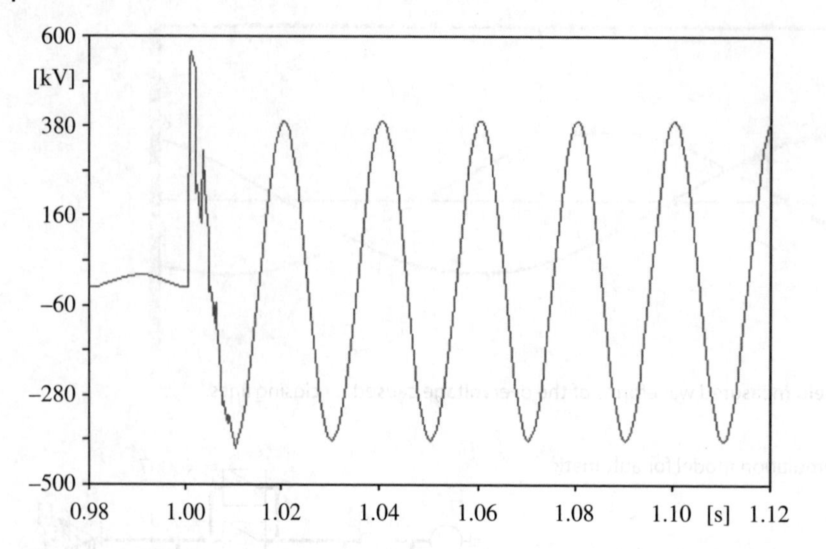

Figure 6.17 Waveforms at single-phase reclosing.

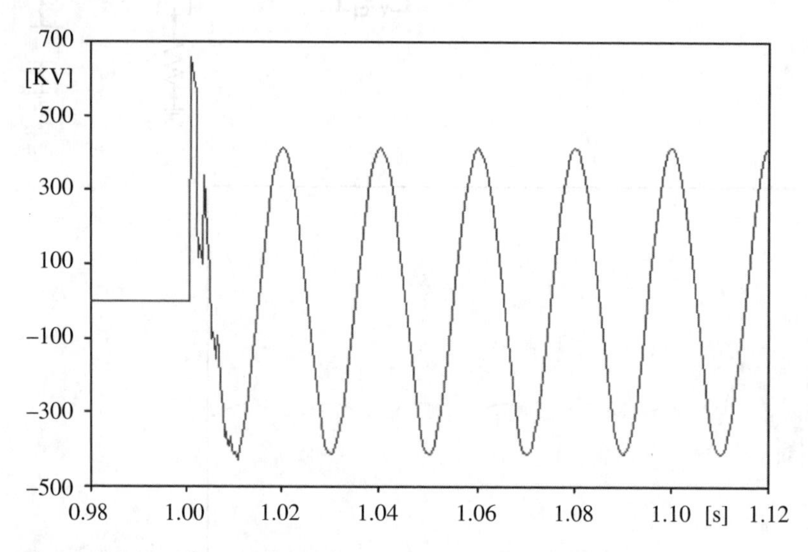

Figure 6.18 Waveforms at three-phase reclosing.

Example 6 Analyze the switching overvoltage of main transformer 2 in some substation.

The initial connection of the substation is depicted in Figure 6.19.

On April 23, 2014, main transformer 2 was disconnected from bus II and connected to bus I to allow independent power supply to line A by bus II (Figure 6.20). Due to sudden status changes at the transformer disconnection, oscillations occurred to bus II and the measured overvoltage at the arrester of bus II is shown in Figure 6.21.

The simulation of the substation bus is as shown in Figure 6.22. The voltage measuring point is located on the PT side of bus II. The waveforms of bus II when the transformers at node x0007 and x0002 are switched on and off are as shown in Figure 6.23.

As can be seen from the figure, the simulated waveforms are similar with those in the actual measurement.

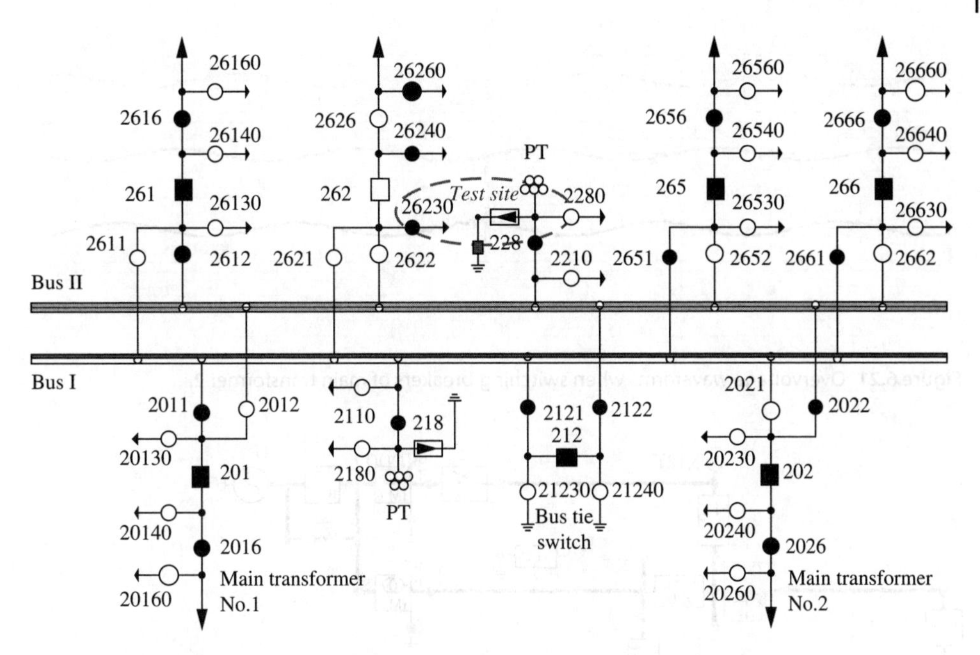

Figure 6.19 Initial connection diagram.

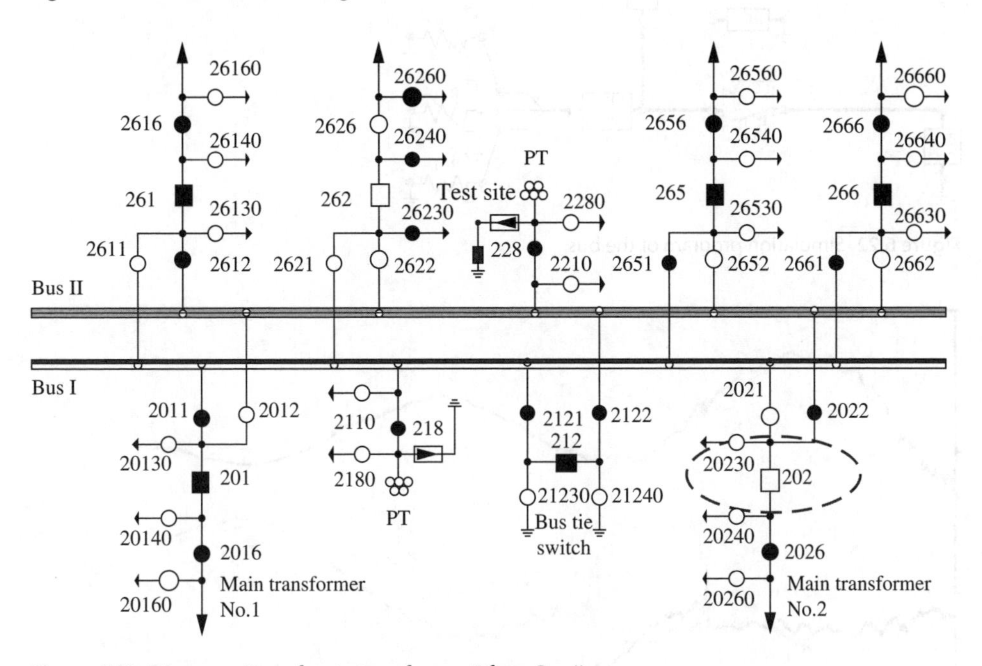

Figure 6.20 Disconnection of main transformer 2 from Bus II.

Example 7 Simulate the closing of unloaded transformers.

Inrush currents will be generated during the closing of unloaded transformers. This is because currents will flow through the primary coils and induce potentials in the secondary coils once the transformer is switched on and energized under static state. Inrush currents resisting this change will be produced due to relations between the electric and magnetic fields during the transition state. The core flux of unloaded power

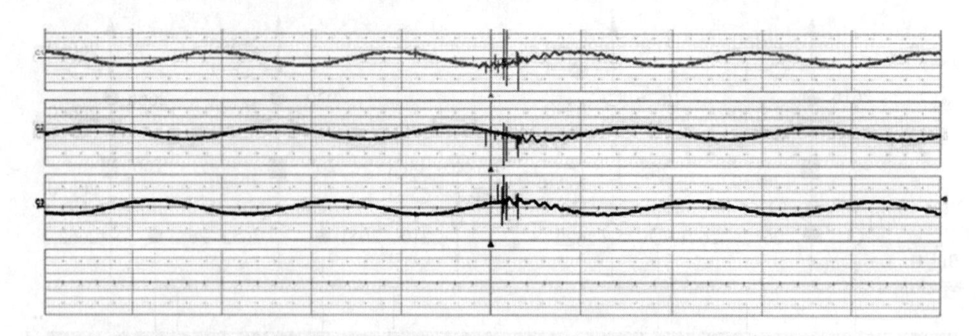

Figure 6.21 Overvoltage waveforms when switching breakers of main transformer 2.

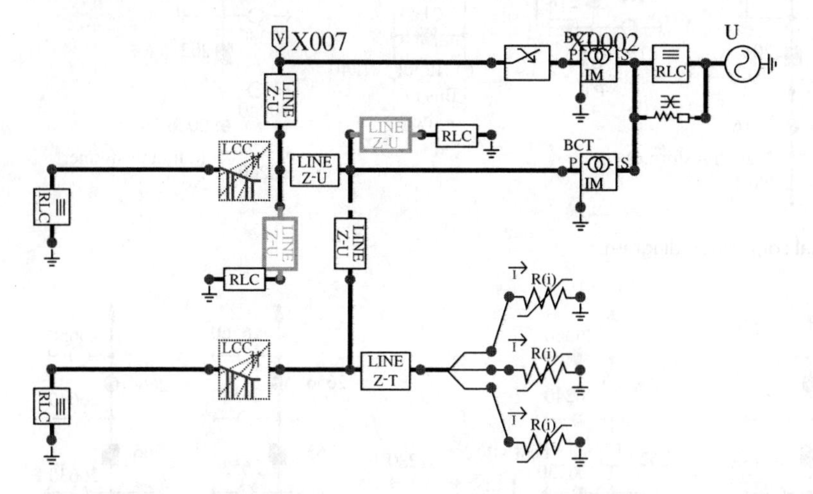

Figure 6.22 Simulation program of the bus.

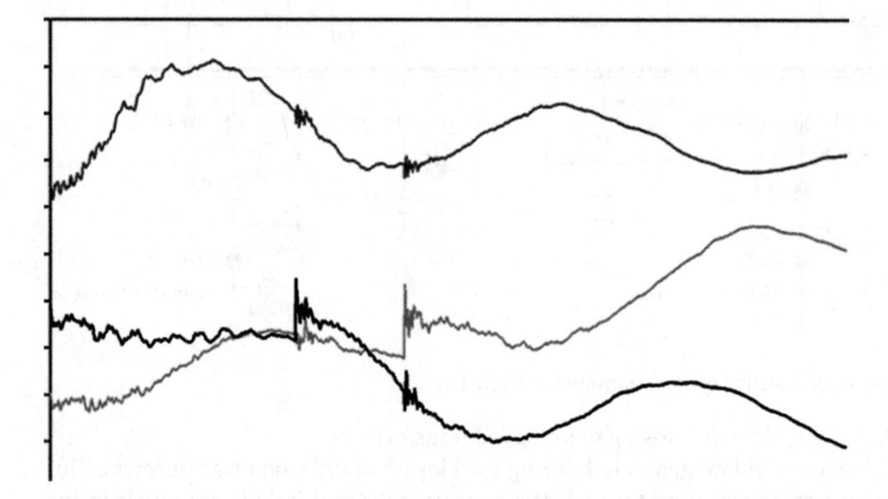

Figure 6.23 Simulated waveforms when switching the breakers of main transformer 2.

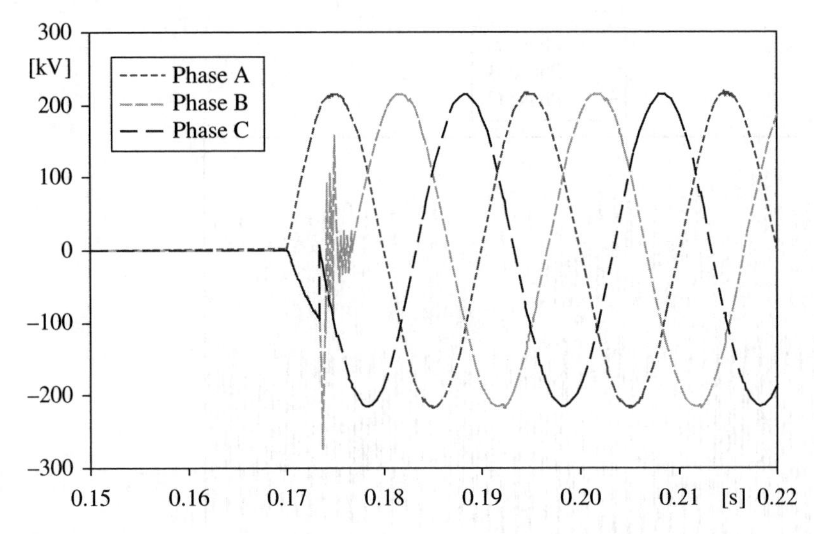

Figure 6.24 Line-side three-phase voltage after closing unloaded transformers.

transformers will undergo an instantaneous process at the closing instant, giving rise to three-phase asymmetrical flux, and resulting in fast winding coil saturation. This type of current, however, mainly consists of non-periodical dc components, which can lead to severe core saturation. Under the worst situation, the closing main flux can jump abruptly to three times its stable value and the inrush current may reach several hundreds of times the stable unloaded current.

The three-phase voltage waveforms simulated under conditions in Example 2 are shown in Figure 6.24.

The three-phase inrush current waveforms are shown in Figure 6.25.

As can be seen from the simulation results, the primary cause of the inrush current at the closing of unloaded transformers is that non-periodical flux linkage which is produced as flux linkage cannot change instantaneously will lead to transformer core saturation. Large inrush currents will occur due to nonlinear characteristics of the magnetization curve of the transformer magnetic material.

Example 8 Simulate the overvoltage caused by closing capacitors.

The substation is powered by two 110 kV incoming lines, and the internal impedance solved based on the short-circuit current is 1.9 Ω. The parameters of the two main transformers are set as follows: U_{1N} $/U_{2N}$ $/U_{3N}=110/38.5/6.3$, $P_0=37.2\,\text{kW}$, $I_0\%=0.45\%$, $P_{k1\text{-}2}=172\,\text{kW}$, $P_{k1\text{-}3}=191\,\text{kW}$; $P_{k3\text{-}2}=143\,\text{kW}$, $V_{s1\text{-}2}\%=10.5\%$; $V_{s1\text{-}3}\%=17.1\%$, $V_{s3\text{-}2}\%=6.2\%$. The ratings of the shunt capacitor is $U_n=6.6/\sqrt{3}\,\text{kV}$ and $C=400\,\mu\text{F}$. The reactor $L=8$ mH.

The simulation model is shown in Figure 6.26.

The waveforms of the 6.3 kV bus voltage when the capacitor is switched on at 0.02 s is as shown in Figure 6.27.

The highest phase voltage of the bus is 7100 V, i.e. $1.38U_N$.

6.2.3 Restrictive Measures for Switching Overvoltages

Switching overvoltages refer to system overvoltages caused by maloperations, faults or changes of operation modes, and are usually in the form of cutoff overvoltages and

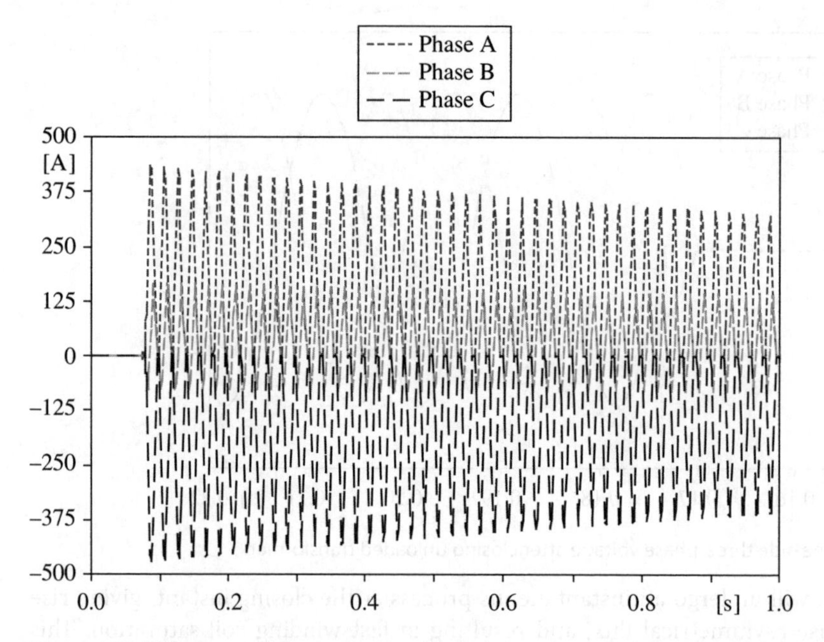

Figure 6.25 Waveforms of the three-phase line-side inrush current after closing unloaded transformers.

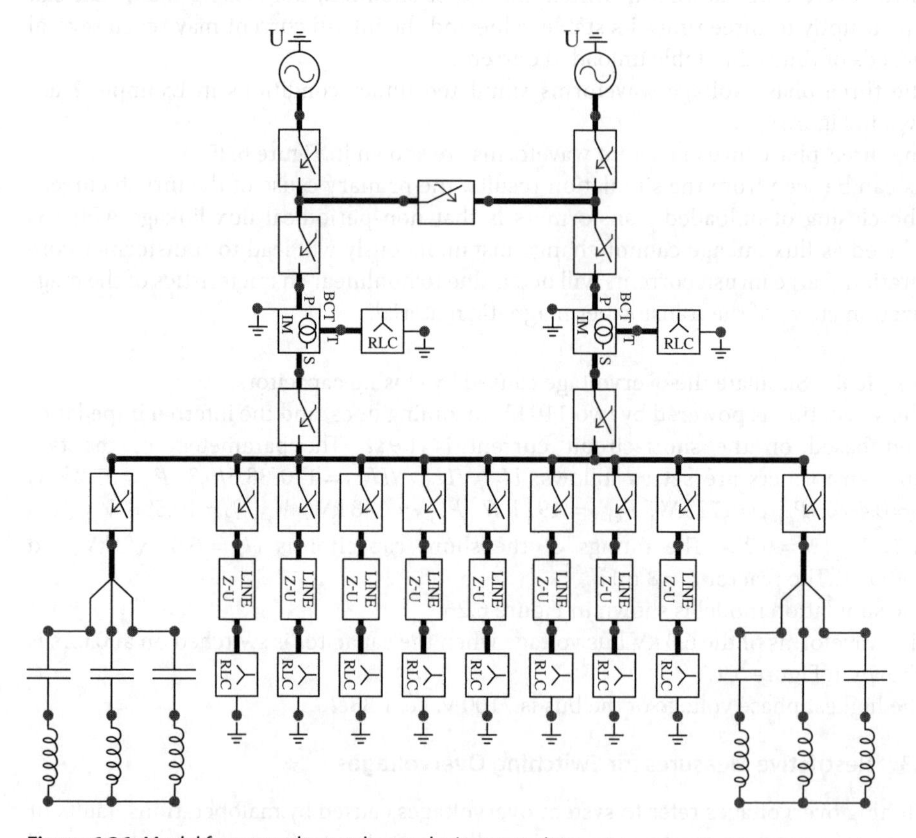

Figure 6.26 Model for overvoltages due to closing capacitors.

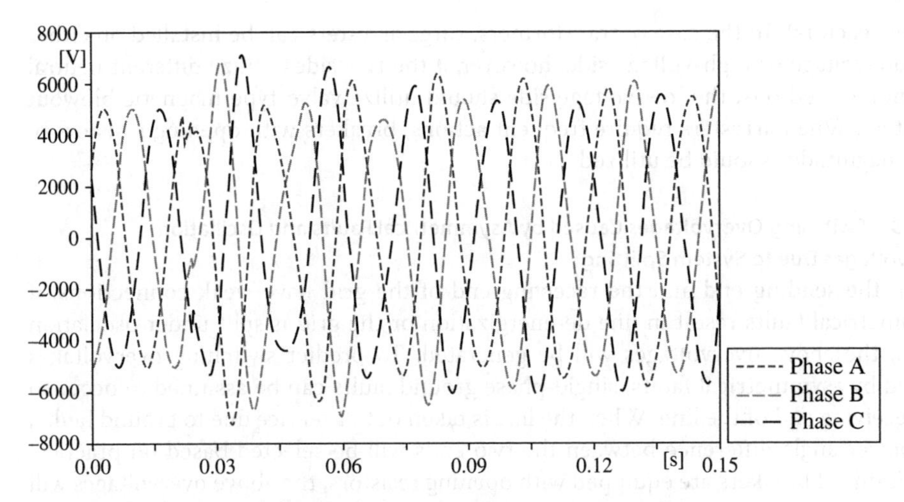

Figure 6.27 Waveforms of the 6.3 kV bus voltage when switching on the capacitor.

arc reignition overvoltages. As these overvoltages will bring considerable damage to power system generation and operation, it is essential to take measures to restrict them. Restrictive measures can be classified into the following groups.

6.2.3.1 Overvoltages Caused by Closing or Reclosing Unloaded Lines

When unloaded lines are closed, closing overvoltage will occur due to LC oscillation of the lines. When lines are reclosed, the electromagnetic oscillation is aggravated due to higher supply potentials and the presence of residue charges on the lines, which allows a further rise in overvoltages. For this reason, closing resistors have to be installed for breakers to effectively reduce closing and reclosing overvoltages.

Based on the predictive conditions, overvoltage distribution at the closing of unloaded lines, at single-phase reclosing and at successful and unsuccessful three-phase reclosing will be determined, and the statistical phase-to-ground and interphase switching overvoltages on the receiving end will also be determined. The predictive conditions for this type of switching overvoltages are: (a) in the case of closing unloaded lines, the bus voltage of the power source will be the highest grid voltage before closing the line breaker; (b) single ground faults occurred to the receiving end of the line prior to successful three-phase reclosing; single ground faults are present on the receiving end of the line at unsuccessful three-phase reclosing; (c) in the case of closing unloaded lines, single-phase reclosing and successful three-phase reclosing (performed during operation), the statistical phase-to-ground switching overvoltage on the receiving side of the line will be no greater than $2.2U_{xg}$.

6.2.3.2 Switching Overvoltages Caused by Disconnecting Unloaded Transformers and Shunt Reactors

The switching overvoltage caused by forced arc extinction when reactive currents of unloaded transformers and shunt reactors are interrupted by breakers will be determined in accordance with breaker structures, circuit parameters, connection and characteristics of the equipment (transformers or shunt reactors). The above overvoltage can generally be restricted by installing arresters between breakers and transformers (shunt reactors). In the case of transformers, surge arresters can be installed on either the low-voltage or high-voltage side; however, if the two sides utilize different neutral grounding methods, the low-voltage side should utilize valve-type magnetic blowout arresters. When arresters require frequent actions, breakers with opening resistors of high magnitudes should be utilized.

6.2.3.3 Switching Overvoltages Caused by Asymmetrical Faults and Oscillation Overvoltages Due to System Splitting

When the sending end and the receiving end of the grid have weak connections, if asymmetrical faults result in line de-energization or the grid is split under oscillation states, the above overvoltages will be generated. To predict switching overvoltages caused by asymmetrical faults, single-phase ground faults can be assumed to occur to the receiving end of the line. When the line is taken out of service due to ground faults, the power angle difference between the two ends will be selected based on practical conditions. If breakers are equipped with opening resistors, the above overvoltages will be reduced. Otherwise, surge arresters are recommended to be installed on lines for overvoltage restriction. In the case of overvoltages caused by opening unloaded lines, breakers should be utilized which will not restrike when tripping under the condition that the source-to-ground voltage is $1.3U_{xg}$.

Surge arresters will be installed in substations to prevent switching overvoltages from damaging electrical equipment. Appropriate installation positions include: (a) entrances on the line side of the outgoing line breakers, at which surge arresters installed are referred to as line arresters; (b) substation side of the outgoing line breakers, at which surge arresters installed are referred to as substation arresters. The specific installation positions and the surge arrester number will be determined in combination with the description in Section 4.4.2. Note that when shunt reactors are not installed at line entrances, surge arresters are not needed if the predicted overvoltage (consider single-phase failure of closing resistors when closing breakers) does not exceed the switching overvoltage protective level of the arrester. The rated voltage of arresters with series gaps will be no less than the power frequency overvoltage level of the installation site.

When adopting metal oxide arresters to restrict switching overvoltages, product the manual should be consulted, to meet the requirements of allowable durations of power frequency overvoltages and resonance overvoltages as well as long-term operating voltage values.

The current capacity and the allowable absorbed energy of surge arresters will meet the requirements of the grid. Additionally, the arrester voltage will be checked to see whether it exceeds the specified protection level. If so, the impact on insulation coordination will be taken into account.

6.3 Evaluation of Power Frequency Overvoltages

6.3.1 Calculation of the Ferranti Effect

The power frequency overvoltage which belongs to internal overvoltages is an electromagnetic transient phenomenon and a kind of temporary overvoltage in the power system. Generally, the amplitude of the power frequency overvoltage is not large while it may last for a long time so that it is harmless to electrical equipment with normal insulation in systems that are rated less than 220 kV and where lines are not long. However, power frequency overvoltages have a decisive effect in determining the insulation level of extra- (ultra-)high-voltage long-distance transmission system. This is because power frequency overvoltage amplitudes have direct connections with switching overvoltage amplitudes; power frequency overvoltages serve as important evidences for determining the arrester rated voltage; and power frequency overvoltage can harm the safe operation of the equipment and the system.

6.3.1.1 Capacitance Effects of Unloaded Long Lines

The unloaded lossless conductor of length l is shown in Figure 6.28, where E is the emf of the power source; U_1 and U_2 are the voltages at the head end and the end of the line, respectively; X_S is the inductive reactance of the power source; $Z_C = \sqrt{L_0/C_0}$;$\alpha = \omega\sqrt{L_0 C_0}$; a is the phase shift coefficient per kilometer, and in normal power frequency condition, $a = 0.006°/\text{km}$. The relation between the currents and the voltages at the two ends of the line is given by

$$\begin{bmatrix} U_1 \\ I_1 \end{bmatrix} = \begin{bmatrix} \cos al & jZc\sin al \\ j\dfrac{1}{Zc}\sin al & \cos al \end{bmatrix} \begin{bmatrix} U_2 \\ I_2 \end{bmatrix} \tag{6.3.1}$$

For unloaded lines ($I_2 = 0$), the voltage of each point on the line is distributed in the cosine form:

$$U_x = U_2\cos ax = \frac{U_1}{\cos al}\cos ax \tag{6.3.2}$$

The impedance of the power source will also affect the capacitance effect of unloaded long lines:

$$\begin{bmatrix} U_0 \\ I_0 \end{bmatrix} = \begin{bmatrix} 1 & jX_s \\ 0 & 1 \end{bmatrix} \begin{bmatrix} U_1 \\ I_1 \end{bmatrix} \tag{6.3.3}$$

$$\begin{bmatrix} U_0 \\ I_0 \end{bmatrix} = \begin{bmatrix} 1 & jX_s \\ 0 & 1 \end{bmatrix} \begin{bmatrix} \cos al & jZc\sin al \\ j\dfrac{1}{Zc}\sin al & \cos al \end{bmatrix} \begin{bmatrix} U_2 \\ I_2 \end{bmatrix}$$

$$= \begin{bmatrix} \cos al - \dfrac{X_S}{Zc}\sin al & j(Zc\sin al + X_S\sin al \\ j\dfrac{1}{Zc}\sin al & \cos al \end{bmatrix} \tag{6.3.4}$$

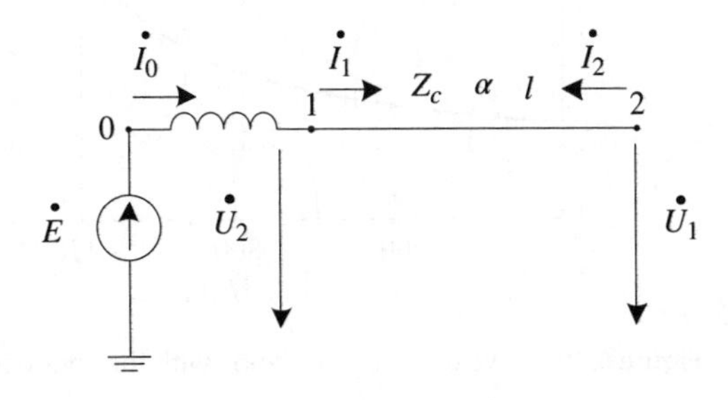

Figure 6.28 Diagram of unloaded undamaged wire.

With the boundary condition

$$\dot{U}_0 = E\,\dot{I}_2 = 0 \tag{6.3.5}$$

The power capacity is infinite (supply reactance $X_S = 0$), the ratio between the voltage at the unloaded line end and the supply potential is expressed by

$$K_{12} = \frac{\dot{U}_2}{\dot{U}_1} = \frac{\dot{U}_2}{\dot{E}} \tag{6.3.6}$$

If the supply capacity is limited (supply reactance $X_S \neq 0$), the ratio between the voltage at the unloaded line end and the supply potential is expressed by

$$K_{02} = \frac{\dot{U}_2}{E} = \frac{1}{\cos \alpha l - X_s \sin \alpha l / Z_C} \tag{6.3.7}$$

Assuming

$$\varphi = \tan^{-1}\frac{X_S}{Z_C},\ K_{02} = \frac{\cos \varphi}{\cos(\alpha l + \varphi)} \tag{6.3.8}$$

the following conclusions can be made that

1. X_s aggravates the voltage rise at the line end, and the head end voltage is higher than the supply potential.
2. The end voltage rises more as the power capacity gets smaller (i.e. the internal reactance X_s becomes larger), as shown in Figure 6.29, where curve 1 shows the situation when the supply impedance is zero, while curve 2 shows the situation where the supply impedance is not zero. Therefore, power frequency voltage rise in the worst case must be estimated according to the operation mode of the minimum power supply capacity.

Example 9 Given that a 500 kV transmission line is 500 km long in length; the supply potential is E; the positive sequence reactance of the power source is 100 Ω; the positive

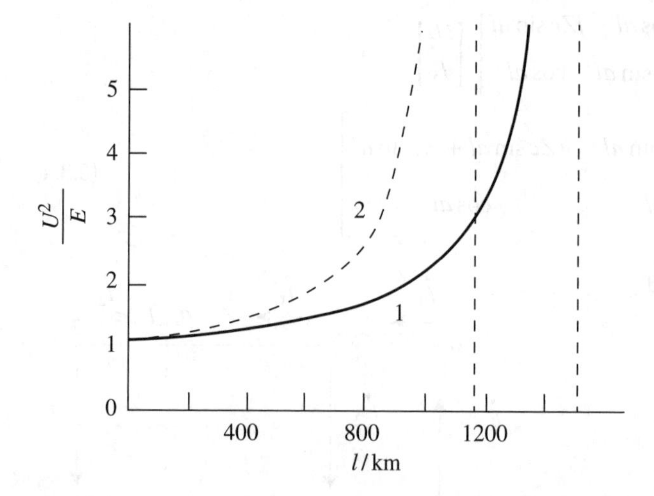

Figure 6.29 Curve of voltage variation with the conductor length.

sequence inductance L_0 and capacitance C_0 per unit length of the line are 0.9 mH/km and 0.0127 μF/km, respectively, solve the highest voltage on the line. Provided that the line end is connected to a reactor $X_P = 1034\,\Omega$, solve the highest voltage on the line again.

Solution: circuit wave impedance: $Z_C = \sqrt{L_0/C_0} = 265.7\,\Omega$

Wave velocity: $v = \sqrt{1/L_0C_0} = 2.95 \times 10^5 \text{km/s}$

The overhead transmission line can be simulated using a distributed parameter line model with lumped resistors thereof. Choose [overhead lines/cable]→distributed parameter line with lumped resistors [Distributed]→Clarke model used for transposed lines. Other components have to be chosen to build the following circuit model, as illustrated in Figure 6.30.

Figure 6.32 is obtained based on the simulation parameters in Figure 6.31, where the waveforms of the line end voltage and the supply potential without installation of reactors are displayed. The voltage at the line end is 462 kV and the supply voltage amplitude is 408 kV.

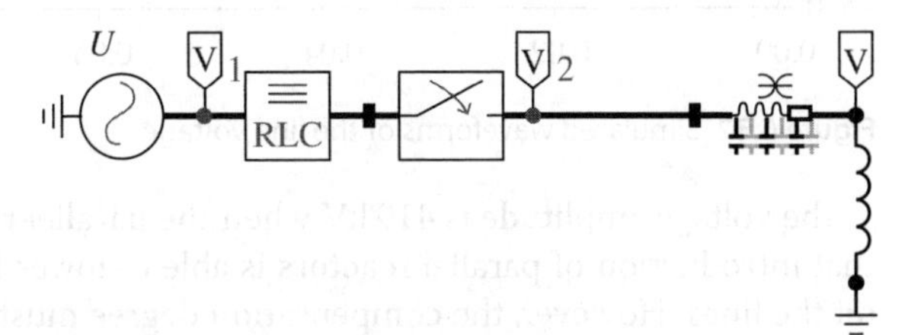

Figure 6.30 Circuit model.

Component: LINEZT_3

Attributes

DATA	UNIT	VALUE
R/l+	Ohm/m	0
R/l0	Ohm/m	0
Z+		265.7
Z0		500
v+		295200
v0		300000

NODE	PHASE	NAME
IN1	1	
OUT1	1	

Copy Paste entire data grid Reset Order: 0 Label

Comment:

Lines

Length 500 [m] ILINE: L',C' / Z, v / Z, tau Conductance: G=0 / G=R*C/L Hide $Vintage,1

Output No

Edit definitions OK Cancel Help

Figure 6.31 Parameter settings.

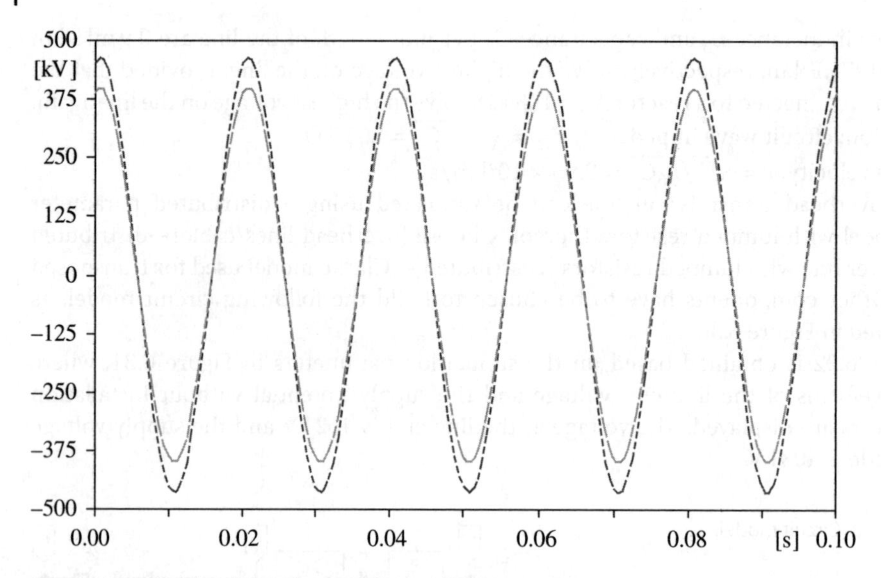

Figure 6.32 Simulated waveforms of the line voltage.

The voltage amplitude is 419 kV when the parallel reactor is installed. It is thus clear that introduction of parallel reactors is able to lower the power frequency overvoltage on the lines. However, the compensation degree must not be too high in case it affects the reactive compensation and voltage control in normal operation.

6.3.2 Evaluation of Asymmetrical Short-circuit Faults

The asymmetrical short circuit is a common fault encountered in transmission lines. The zero-sequence component of the short-circuit current can cause a power frequency voltage rise to the sound phases, often referred to as the asymmetrical effect. In the case of asymmetrical short-circuit faults, single-phase grounding faults are the most common.

Assuming that a single-phase grounding fault occurs on phase A, the voltages of the sound phases, phases B and C, can be given by the following equations by applying the symmetrical component method:

$$\dot{U}_B = \frac{(a^2 - 1)Z_0 + (a^2 - a)Z_2}{Z_1 + Z_2 + Z_0}\dot{E}_A \tag{6.3.9}$$

$$\dot{U}_C = \frac{(a - 1)Z_0 + (a^2 - a)Z_2}{Z_1 + Z_2 + Z_0}\dot{E}_A \tag{6.3.10}$$

where $\dot{E}_A$ is the electromotive of phase A at the fault site in normal operation; Z_1, Z_2, and Z_0 are the positive sequence, negative sequence and zero-sequence impedance seen from the fault site, respectively; the operational factor $a = -\frac{1}{2} + j\frac{\sqrt{3}}{2}$.

If K stands for the voltage rise of the sound phase when a single-phase grounding fault happens, the equation $\dot{U}_B = K\dot{E}_A$ is given. For systems with large supply capacity,

$$K = -\frac{1.5\frac{X_0}{X_1}}{2 + \frac{X_0}{X_1}} \pm j\frac{\sqrt{3}}{2} \tag{6.3.11}$$

The module value of K is given by

$$|K| = \sqrt{3} * \frac{\sqrt{\left(\frac{X_0}{X_1}\right)^2 + \frac{X_0}{X_1} + 1}}{2 + \frac{X_0}{X_1}} \tag{6.3.12}$$

Example 10 Given that the length of a 500 kV transmission line is 400 km; the supply potential is E; the positive sequence reactance of the power supply $X_{S1} = 100\,\Omega$; the zero-sequence reactance of the power source $X_{S0} = 50\,\Omega$; the positive sequence characteristic impedance of the line $Z_{C1} = 260\,\Omega$; the zero-sequence characteristic impedance $Z_{C0} = 500\,\Omega$; the positive sequence wave velocity $v = 3 \times 10^5$ km/s; the zero-sequence wave velocity $v_0 = 2 \times 10^5$ km/s, work out the voltage rise multiple of the sound phases at the line terminal when the terminal of phase A is grounded.

Solution:

$$\Phi = \tan^{-1}\frac{X_{S1}}{Z_{C1}} = 21°$$

$$\Phi_0 = \tan^{-1}\frac{X_{S0}}{Z_0} = 5.71°$$

$$\beta l = \frac{0.06°}{\text{km}} \times 400\,\text{km} = 24°$$

$$\beta_0 l = \beta l \frac{v}{v_0} = 36°$$

According to the above equations, the equivalent positive and zero-sequence entrance impedances seen from the line terminal to the power source are respectively given by

$$Z_{R1} = jZ_{C1}\tan(\beta l + \phi) = 260j(\Omega)$$

$$Z_{R0} = jZ_{C0}\tan(\beta_0 l + \varphi_0) = 445.6j(\Omega)$$

$$\frac{X_0}{X_1} = 1.714$$

The voltage rise multiple of the sound phases when a single-phase grounding fault occurs is expressed as

$$|K| = \sqrt{3} * \frac{\sqrt{\left(\frac{X_0}{X_1}\right)^2 + \frac{X_0}{X_1} + 1}}{2 + \frac{X_0}{X_1}} = 1.109$$

Prior to the fault, the voltage rise coefficient at the terminal of the unloaded line (phase A) is given by

$$K_{02} = \frac{\cos\varphi}{\cos(\beta l + \varphi)} = 1.32$$

When phase A encounters the grounding fault, the voltage rise coefficient of the sound phases is given by

$$\frac{U_B}{E} = \frac{U_C}{E} = K_{02}K = 1.464$$

6.3.3 Simulation of Asymmetric Grounding Faults

When systems with ungrounded neutrals encounter asymmetrical short-circuit faults, the resulting zero-sequence currents will lead to power frequency overvoltage rise on the outgoing lines of non-fault phases. High overvoltages may appear on non-fault phases in the case of single-phase grounding faults. If the arresters of sound phases are actuated, the arresters are required to extinguish the follow current with high power frequency voltage.

Example 11 Simulate single-phase grounding faults.

When an asymmetric short-circuit fault happens in the system with an ungrounded neutral, the voltages of non-fault phases will have a corresponding rise. Next, we will work on overvoltages on the non-fault phases when an asymmetric short-circuit fault occurs on a 35 kV distribution line. The LCC model without ground wires is chosen for the 35 kV to simulate the power frequency overvoltages of the non-fault phases when phase C encounters a short-circuit grounding fault. The built model is shown in Figure 6.33.

The phase-to-ground voltages of the non-fault phases by simulation is displayed in Figure 6.34.

As shown in Figure 6.34, when a grounding short-circuit fault is present on phase C, overvoltages are induced on phases A and B with magnitudes of 1.72 U_N, approaching $\sqrt{3}\,U_N$.

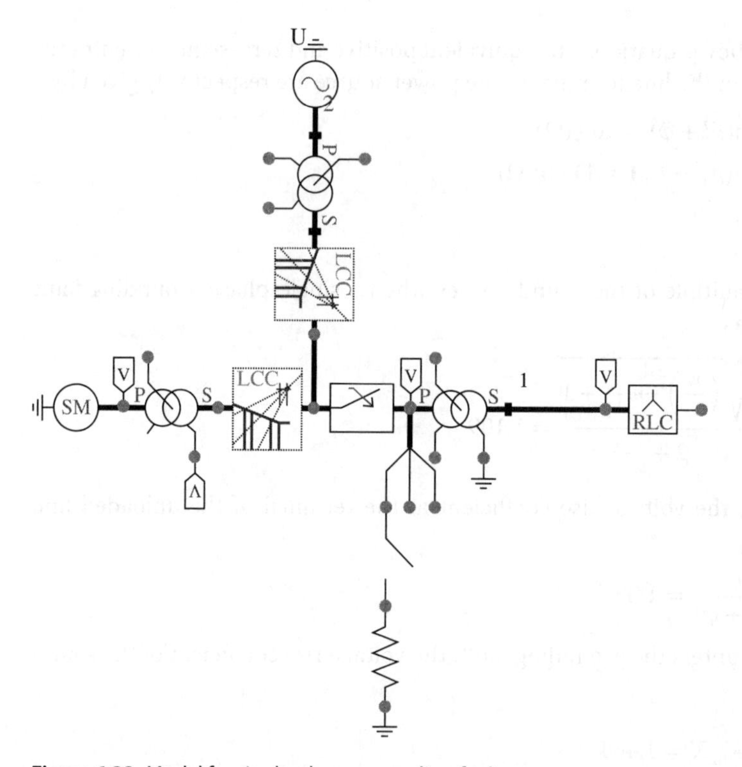

Figure 6.33 Model for single-phase grounding faults.

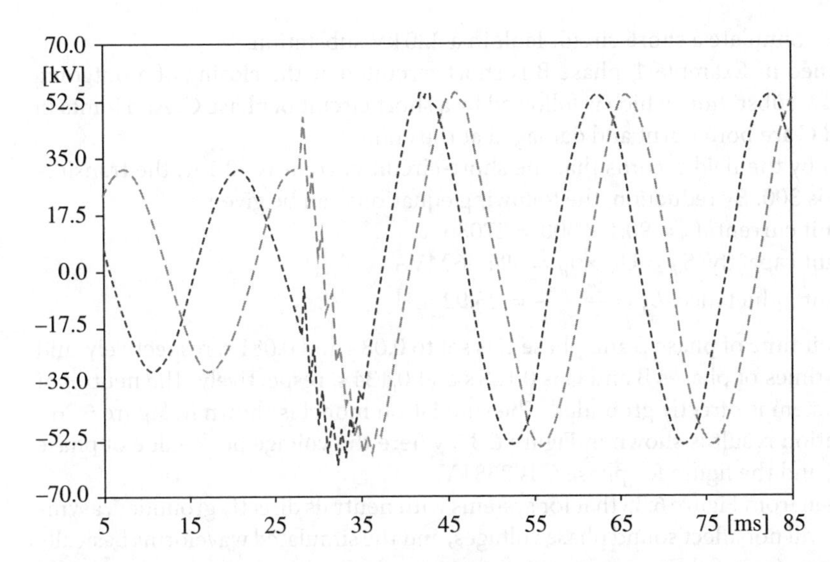

Figure 6.34 Waveforms of phase-to-ground voltages of non-fault phases in the case of single-phase grounding.

Example 12 Simulate two-phase grounding faults.

The line parameters are the same as those in Example 11. The voltage rise in phase A is simulated in a 35 kV distribution grid with an ungrounded neutral when phases B and C have grounding short-circuit faults.

The phase-to-ground voltage of the non-fault phase by simulation is shown in Figure 6.35.

As shown in Figure 6.35, when a grounding short circuit happens in phases B and C, the induced overvoltage is generated on phase A with a peak value of 42 kV, approximately 1.5 U_N.

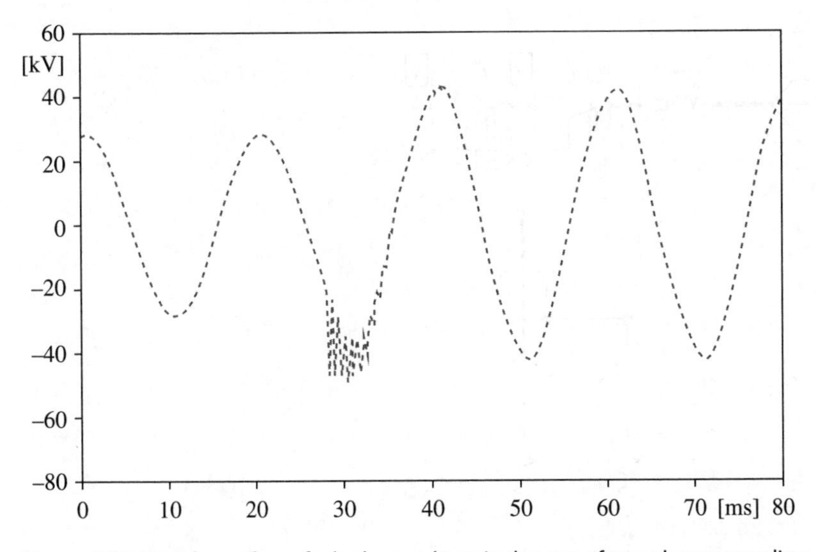

Figure 6.35 Waveform of non-fault phase voltage in the case of two-phase grounding.

Example 13 Simulate a short-circuit fault in a 220 kV substation.

As mentioned in Example 4, phase B is short-circuited at the closing of a outgoing line of a 220 kV substation, which is followed by a short circuit of phase C. And joints in phases B and C are both burnt and damaged at the end.

It is shown by the field records that the short-circuit current is 90.1 A, the transformation ratio is 300. By reduction, the following equations can be given.

Short-circuit current: $I_d = 90.1 \times 300 = 27030\,A$

Short-circuit capacity: $S_d = U_N \times I_d = 5946.6\,MVA$

Short-circuit inductance: $L_d = \frac{U_N^2}{S_d} \cdot \frac{1000}{2\pi f} = 25.92\,mH$

Short-circuit time of phase B and phase C is set to 0.08 s and 0.081 s, respectively, and the recovery times of phases B and C is 0.128 s and 0.134 s, respectively. The neutral of the 220 kV system is directly grounded. The simulation model is shown in Figure 6.36.

The simulation result is shown in Figure 6.37, where the voltage peak value of phase B is −272 kV, and the figure for phase C is 268 kV.

It can be seen from Figure 6.38 that for systems with neutrals directly grounded, asymmetric faults will not affect sound phase voltages, and the simulated waveforms basically agree with those from field measurement.

6.3.4 Restrictive Measures for Power Frequency Overvoltages

For systems rated 330, 500 and 750 kV, it is specified that the transient power frequency overvoltage rise on the bus and on the line shall be no larger than 1.3 times and 1.4 times the maximum phase voltage, respectively.

For high-voltage long-distance unloaded lines, parallel reactors are usually utilized to compensate for the capacitance effect. When the capacitive current flows through the

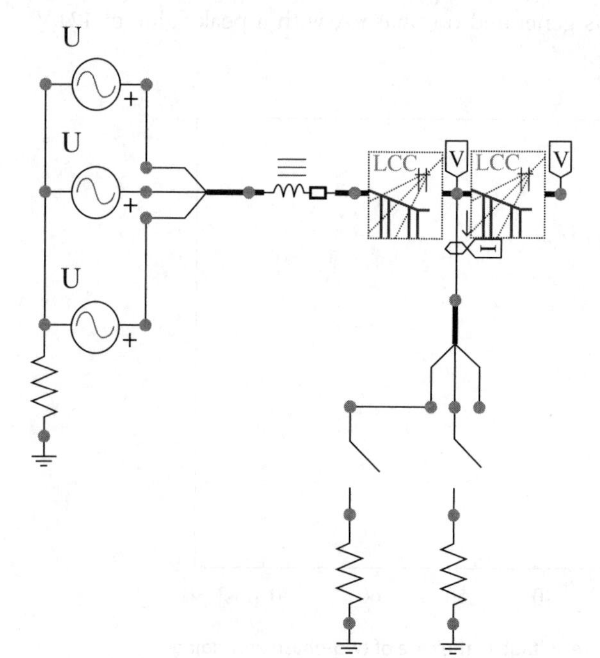

Figure 6.36 Simulation model for short-circuit faults in a substation.

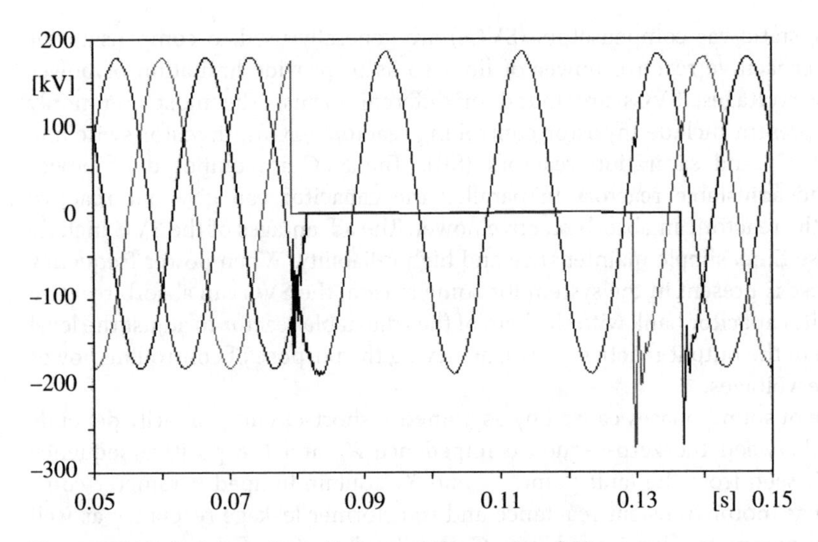

Figure 6.37 Voltage waveforms in the case of grounding faults.

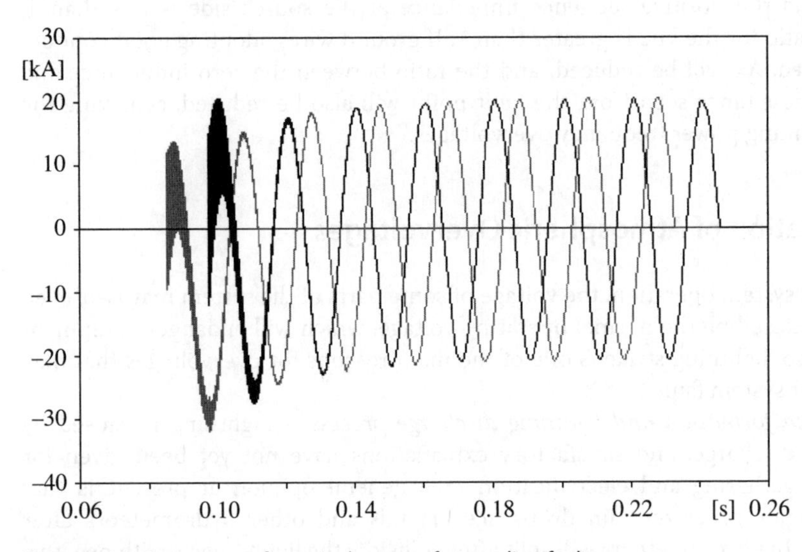

Figure 6.38 Current waveforms in the case of grounding faults.

source inductive reactance, the capacitance effect will be aggravated as well, as if increasing the line length of line. Apparently, the smaller the supply capacity, the more severe the capacitance effect. The introduction of reactors at the line terminal is equivalent to decreasing the line length and lowering the voltage transfer coefficient, which helps to reduce the terminal voltage.

The natural power and the charging power of a high-voltage transmission line are proportional to the square of the line operating voltage. When the transmission capacity is large (heavily-loaded), high voltage reactors must operate at low compensation degrees to ensure the balance between inductive and capacitive reactive powers. When the UHV or EHV line is unloaded or lightly loaded, reactors must have higher compensation degrees to limit power frequency voltage rise.

These days, static var compensators (SVCs) are generally used to compensate for inductive or capacitive reactive power of lines so as to provide limitation of power frequency overvoltages. SVCs are found in different forms. The most commonly used ones at present include thyristor controlling reactors (TCR), thyristor switching capacitors (TSC) and saturation reactors (SR). The SVC is composed of power capacitors and adjustable reactors in parallel; the capacitor can give out reactive power while the reactor can absorb reactive power. The advantages of the SVC include quick response time, simple maintenance and high reliability. When power frequency overvoltage rise is present in the system for some reason, the SVC can absorb reactive power from the capacitor bank with the help of the adjustable reactor to adjust the level and direction of the output reactive power, achieving the purpose of controlling power frequency overvoltages.

Voltage rise of sound phases caused by asymmetric short-circuit primarily depends on the ratio between the zero-sequence impedance X_0 and the positive sequence impedance X_1 seen from the fault point. X_0 and X_1 contain lumped parameter components such as motor transient reactance and transformer leakage reactance as well as distributed parameter line impedance. Generally, the ratio of the zero-sequence impedance and the positive sequence impedance at the source side is less than 1, whereas the ratio for the line is greater than 1. If ground wires adopting good conductors are utilized, X_0 will be reduced, and the ratio between the zero inductance and the positive inductance seen from the fault point will also be reduced, achieving the purpose of limiting power frequency overvoltages.

6.4 Evaluation of Atmospheric Overvoltages

During power system operation, the voltage of some parts of the system may rise, even to a level far exceeding the normal operating voltage, which will endanger equipment insulation. Also, lightning strike is one of the main reasons for overvoltages that may result in power system faults.

Thundercloud formation and lightning discharge process – Lightning is caused by thundercloud discharge, and satisfactory explanations have not yet been given for thundercloud gathering and electrification. The general opinion at present is that the rising hot air generates rain drops, ice crystals and other hydrometeors after condensation. The hydrometeors will split after collision; the lighter parts with negative charge are blown away by the wind and form large thunderclouds, while the heavier parts with positive charge may condense into water drops, falling down or being suspended in the air to form locally positive cloud areas. Practical measurements have shown that clouds at heights of 5–10 km mainly carry positive charges, while clouds at heights of 1–5 km mainly carry negative charges. The entire cloud can have several intensive charge centers that are located at the thundercloud bottom and can induce a lot of positive charge on the ground. As a result, strong electric fields are formed between thunderclouds carrying a great many charges of opposite polarities or different quantity, or between the thundercloud and the ground.

With the development and movement of the thundercloud, once the spatial electric field intensity exceeds the critical electric field for atmospheric free discharge (about 30 kV/cm in the atmosphere, and 10 kV/cm when water-drops are present), intracloud or cloud-to-ground spark discharge will take place. In lightning protection engineering, the main concern is the discharge between the thundercloud and the ground.

The cloud-to-ground discharge is usually divided into two stages: the leader discharge and the main discharge. The cloud-to-ground linear lightning tends to initiate from the thundercloud edge to the ground, progressing downward in a step-by-step manner. The step length is 10–200 m; the propagation rate is about 10^7 m/s, and the time interval between steps is 10–100 μs, so the average development rate is only $1–8 \times 10^5$ m/s. Such discharge is known as the leader discharge. When leaders approach the ground, the electric field at some grounded structure tops increases to such a level that air ionization is caused, which produces upward-moving connecting leaders. When the upward-moving connecting leader meets the downward stepped leader, strong charge neutralization will happen, generating large currents, accompanied by thunder and lightning flashes. Such a process is called the main phase of lightning discharge which is of very short duration: 50–100 μs. The main discharge is developed inversely in a bottom-up manner with speeds of 20,000–150,000 km/s. The downward moving lightning flashes resulting from thunderclouds, carrying positive charges, account for a small percentage, and the development process is similar to the above description.

Observations show that most cloud-to-ground lightning strokes are repeated, i.e. there are multiple discharges in the channel formed by the first stroke. The time interval between discharges is about 0.5–500 ms. The main reason is that the first leader-main discharge primarily releases charges of the first charge center. Because the main discharge channel still has a conductivity higher than that of the ambient air, discharges can occur between other charge centers and the ground, through the existing channel. Multiple lightning strokes are consequently generated. In the subsequent discharges, leaders develop continuously from up to down without any pause, and the average discharge number is 2–3 times. Generally, the first discharge current is the largest, and any subsequent current is smaller in amplitude.

6.4.1 Lightning Parameters

Lightning parameters are a series of characteristic quantities regarding thundercloud discharge, including wave impedance of the main discharge channel, lightning current waveforms, probability distribution of lightning current amplitudes, lightning polarity, repeated discharge times and transported charge quantity to ground.

Wave impedance of the main discharge channel – From the perspective of practical engineering and the real effect on the ground, the leader channel approximates to a uniformly distributed parameter conducting channel composed of inductors and capacitors. Its wave impedance is given by

$$Z = \sqrt{\frac{L_0}{C_0}}\,\Omega$$

where L_0 is per-unit-length inductance; C_0 is per-unit-length capacitance.

The wave impedance of the main discharge channel is related to the lightning current of the main discharge channel; the larger the lightning current, the smaller the impedance. Generally, Z is in the range 300–3000 Ω.

Lightning current waveforms – Measurements from all over the world have shown that waveforms of lightning current in cloud-to-ground discharge are basically the same, of

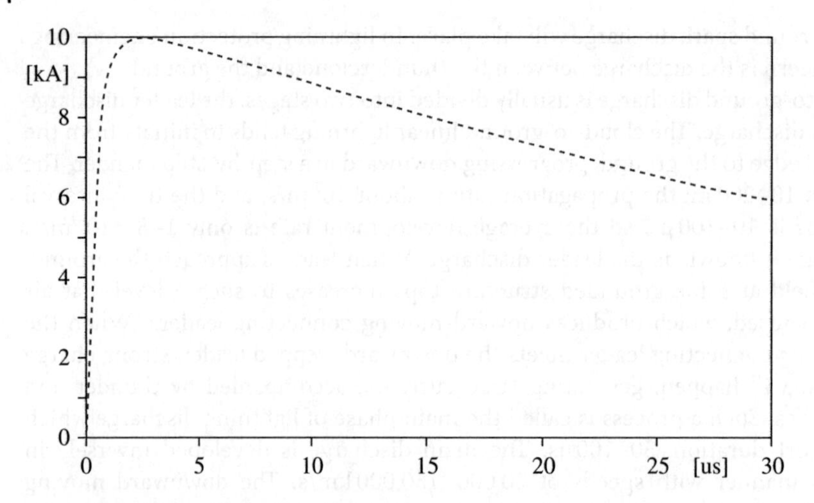

Figure 6.39 A typical lightning current wave.

which the majority are unipolar repetitive pulses, and only a few are negative overshoots. A discharge process usually contains multiple leader-to-main discharge processes and the subsequent current process. A typical lightning current wave is shown in Figure 6.39.

The typical lightning current wave is usually described using double exponential functions.

$$i = I_0(e^{\alpha t} - e^{\beta t})$$

where I_0 is lightning current amplitude; α, β are constants.

Combining observations from around the world, about 85% of the lightning current wavefront time is 1–5 μs with an average of 2.5 μs, the recommended value in engineering is 2.6 μs. The lightning current duration is 20–100 μs with an average value of approximately 50 μs.

Thunderstorm day and thunderstorm hours – Indicators of lightning activity frequency in different areas.

Probability distribution of lightning current amplitude – The amplitude of a lightning current is random, and the probability distribution curve for lightning current amplitudes can be obtained by statistical analysis, based on substantial field measurements. The probability distribution of lightning current amplitudes varies with regions and is mainly related to latitude, terrain, weather and thunderstorm intensity.

It is recommended by DL/T620-1997 *Overvoltage Protection and Insulation Coordination for AC Electrical Installation* that the probability distribution of lightning current amplitudes in the area having more than 20 thunderstorm days is

$$\lg P = -\frac{I}{88}$$

For an area with less than 20 thunderstorm days, the probability distribution decreases to

$$\lg P = -\frac{I}{44}$$

where P is probability that the lightning current amplitude exceeds I; and I is lightning current amplitude in kA

Lightning current polarity – When thunderclouds carry positive charges, thundercloud discharge is of positive polarity, the lightning current thus is also of positive polarity; otherwise, the lightning current is of negative polarity. Field measurement data has indicated that the percentages that currents of positive and negative polarities account for vary with terrain, and currents of negative polarity account for 75–90% of the total.

Repeated discharge times and transported charge quantity to ground – There are multiple intensive charge centers in a single thundercloud. As a result, one thundercloud discharge usually contains multiple discharge pulses, known just as multiple discharge. According to field measurements of 6000 times, the average number of repeated discharge times is 2 to 3 and the maximum is 42. The interval time between discharges is usually 30–50 ms with the shortest being 15 ms and the longest 700 ms, and the interval time is increased with increases in discharge times. The transported charge quantity to ground in each discharge process is referred to as the discharge charge, while the transported charge quantity to ground in each stroke is referred to as the flash charge.

6.4.2 Equivalent Circuit of Lightning Discharge

When a lightning main discharge occurs, a lot of positive charges move reversely along the leader channel and neutralize negative charges in the thundercloud. Since currents are generated due to charge movement, the strike point potential will change suddenly. The current magnitude depends upon the charge density in the leader channel and the development speed of the main discharge, and it is also affected by the impedance Z.

One of the main concerns in lightning protection research is the potential rise of the strike point, while main discharge speed, leader charge density and concrete lightning mechanisms may not be considered. Hence, the process of lightning discharge can be simplified to a mathematical model with its equivalent circuits shown in Figure 6.40, where Z_0 represents the wave impedance of the lightning channel, and Z represents the impedance between the struck object and the ground or the wave impedance of the struck object.

When lightning strikes an object, it can be considered as a process in which an incoming current wave i_0 propagates along the lightning channel to the struck object. If $Z = 0$, $i = 2i_0$; if $Z \ll Z_0$, $i \approx 2i_0$. So, it is an international norm to refer to the current that flows through the struck object with a zero characteristic impedance as the lightning

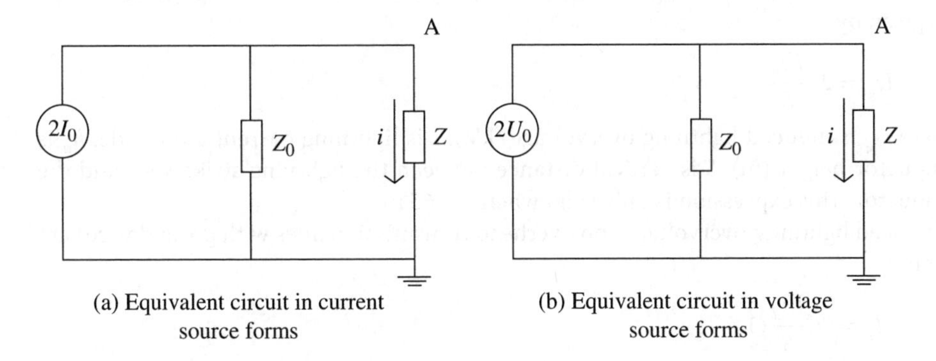

Figure 6.40 Equivalent circuit of lightning discharge.

current. The lightning current amplitude, by definition, is two times as large as the current wave i_0 along the channel Z_0.

6.4.3 Lightning Overvoltages

6.4.3.1 Direct Lightning Overvoltages

The overvoltage resulteding from lightning that directly hits the grid is called the direct lightning overvoltage. When lightning directly strikes the line, it can be seen as an incoming current wave with the amplitude of I along the main discharge channel. The current wave flows towards both sides of the conductor, and the strike point voltage $U = IZ/2$, with Z being the conductor characteristic impedance. If $Z = 300\ \Omega$ and $I = 50$ kA, the overvoltage can reach up to 7500 kV. It is thus clear that even for UHV transmission lines with high insulation strength, lightning directly striking the conductor can easily lead to insulator string flashover.

Therefore, the measures to suspend ground wires is adopted by almost all overhead transmission lines rated above 110 kV. When lightning directly strikes electrical equipment, if no arresters are employed, extremely high overvoltages will occur, causing equipment insulation damage. Electrical equipment thus must be connected in parallel with arresters to limit lightning overvoltages. The lightning overvoltage level of equipment employed with arresters shall be determined by arrester protection properties.

6.4.3.2 Induced Lightning Overvoltages

In the leader stage of thundercloud discharge, the leader channel is filled with charges by which the conductor is induced. As for the negative leader, positive bound charges are accumulated on the nearby conductor and negative charges on the conductor are rejected to the distant end. Since the developing speed of the leader is low, the conductor current in the above process is sufficiently small that it can be ignored, and the conductor will keep a zero potential through the system neutral or the leakage resistance. When the leader arrives at the nearby ground, the main discharge is initiated, charges in the leader channel are neutralized and the corresponding bound charges of opposite signs on the conductor are released, flowing towards the conductor ends in the form of waves.

6.4.4 Evaluation of Induced Lightning Overvoltages

Induced lightning overvoltages on overhead transmission lines without ground wires are given by

$$U_g = 25\frac{Ih_d}{S}$$

where U_g *is* induced lightning overvoltage (kV); I is lightning current amplitude; h_d is conductor height (m); S is vertical distance between the lightning strike spot and the conductor. This expression is only valid when $S > 65$ m.

Induced lightning overvoltages on overhead transmission lines with ground wires are given by

$$U_g = 25\frac{Ih_d}{S}(1 - K_0)$$

K_0 is coupling coefficient between the conductor and the ground wire.

When $S < 65$, lightning will directly strike the top of the overhead transmission line tower because of connecting leaders. Affected by the connecting leader initiated from the tower top, U_g cannot be calculated by the two equations above.

6.4.5 Evaluation of Direct Lighting Overvoltages

Example 14 This example demonstrates the ATP-EMTP simulation for the lightning transient process in a substation. Figure 6.41 is the single-line diagram for a 400 kV substation. The figure labelled beside the bus shows the length of each segment (m), and the empty box means that the circuit breaker is disconnected. Therefore, only two lines are connected to the transformer protected by an arrester under this operation mode. The simulated event is single-phase flashover caused by the lightning accident 0.9 km away from the substation.

The first step is to build a model. The fault is assumed to be caused by lightning directly striking the ground wire. The parameters of the lightning current include amplitude $I_m = 120{,}000$ A, front time is 4×10^{-6} s, and the half-amplitude decay time is 5×10^{-5} s. The simulated lightning waveform is as shown in Figure 6.42. The characteristic impedance of the lightning channel is 400 Ω.

Next, the selection of the lightning strike spot will be introduced in detail. In the calculation, the substation and the incoming line are combined as a unified network. Tower #1–#6 on the incoming line are chosen as the lightning strike spots, of which lightning striking tower #6 is considered as far zone lightning, whereas the rest is seen as nearby lightning. It is specified in China's regulations that only far zone lightning occurred 2 km outside the substation has to be calculated, while nearby lightning within 2 km of the substation need not be considered. This is a convention in MV and

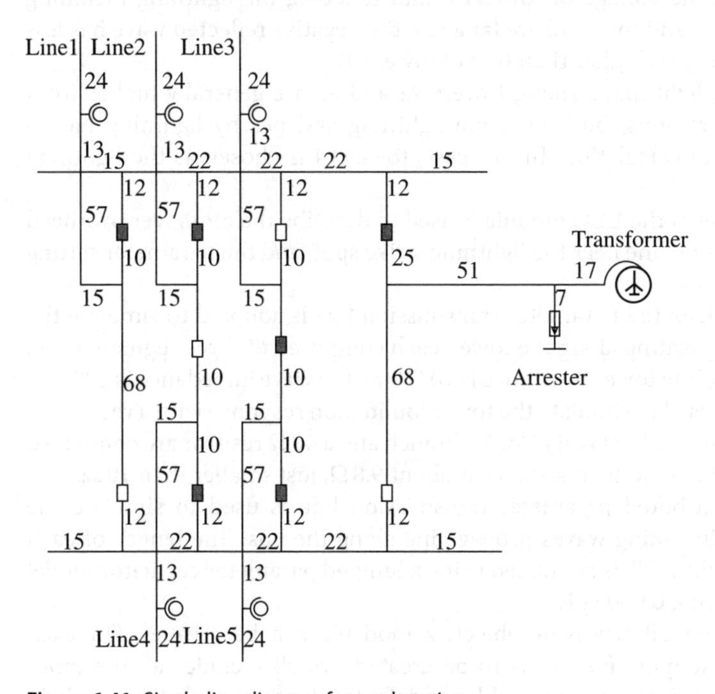

Figure 6.41 Single-line diagram for a substation.

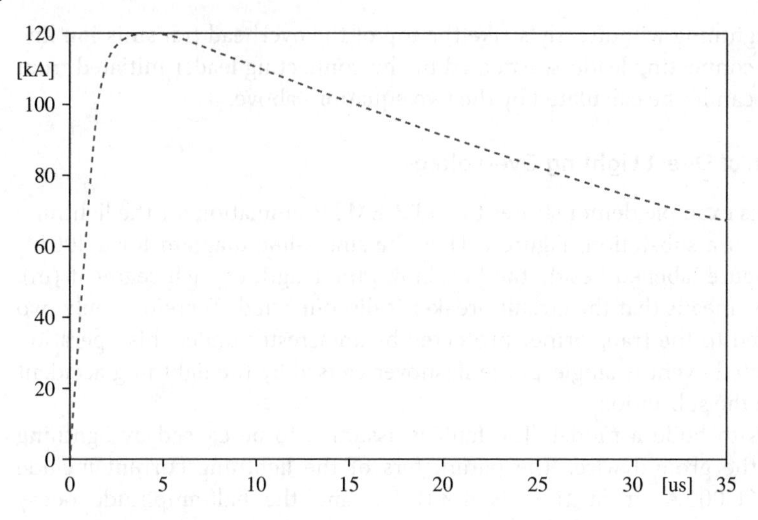

Figure 6.42 Simulated lightning waveform.

HV systems, where incoming waves caused by lightning outside the incoming lines are made the research focus. In practice, the main threat to the equipment in the substation is nearby lightning. Tower #1 and the gate-type terminal structure of the substation is generally close in distance, and the impulse grounding resistance of the gate-type structure is small. For this reason, when lightning hits tower #1, the negative reflected waves are returned from the gate-type structure to the tower #1 through the ground wire quickly, reducing the voltage of tower #1, and reducing the lightning incoming overvoltage. As tower #2 and tower #3 are far away, the negative reflected wave has less effect, and the overvoltage is higher than that of tower #1.

Overvoltages due to lightning striking towers #2 and #3 are generally higher from experience. For these reasons, both far zone lightning and nearby lightning has to be taken into account in calculation. In this case, tower #4 is chosen as the lightning strike spot.

The JMarti line model in the LCC module is used to describe the crossover overhead single-circuit transmission line near the lightning strike spot, and the parameter setting is as shown in Figure 6.43.

The single-phase distributed parameter transmission line is adopted to simulate the response of surges propagating along the tower, each composed of three segments 8 m, 7 m, and 18 m in length. The tower resistance is 10 Ω/km, the wave impedance is 200 Ω; a lumped inductor RL is used to simulate the tower foundation response with a resistance of 13 Ω and an inductance of 0.005 mH. An RL branch and a 40 Ω resistor are connected in parallel; the impulse grounding resistance is about 9.8 Ω, just smaller than 10 Ω.

The three-phase distributed parameter transmission line is used to simulate the response of lightning incoming waves propagating along the bus. The length of each bus is shown in Figure 6.43. PT is simulated using a lumped parameter capacitor model with the capacitance being 0.0005 μF.

In ATP-EMTP's menu Library-New object, a mod file can be created. For each component, a unique support file needs to be created, which includes all the input data information, node objects, input variables of default values, icons and related help

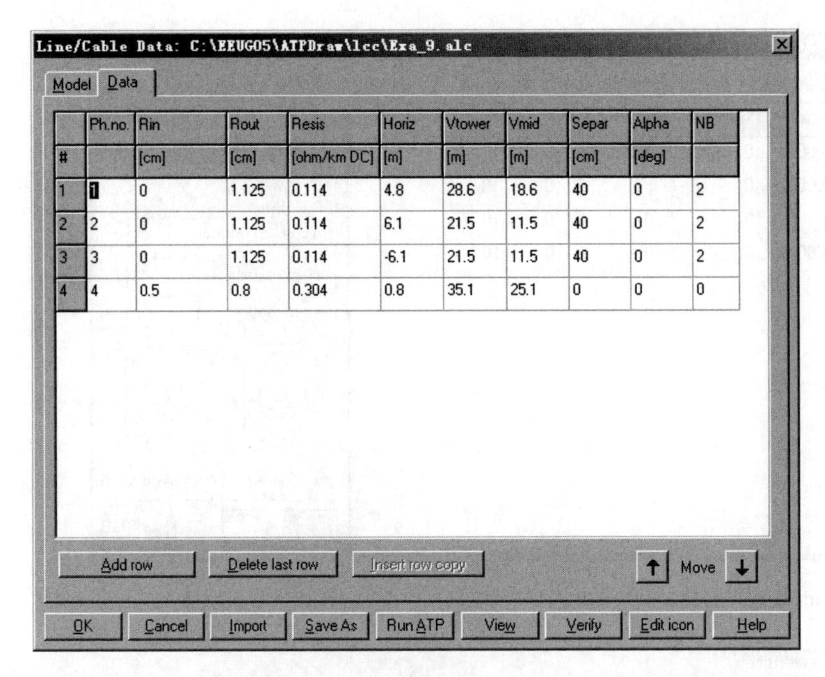

#	Ph.no.	Rin [cm]	Rout [cm]	Resis [ohm/km DC]	Horiz [m]	Vtower [m]	Vmid [m]	Separ [cm]	Alpha [deg]	NB
1	1	0	1.125	0.114	4.8	28.6	18.6	40	0	2
2	2	0	1.125	0.114	6.1	21.5	11.5	40	0	2
3	3	0	1.125	0.114	-6.1	21.5	11.5	40	0	2
4	4	0.5	0.8	0.304	0.8	35.1	25.1	0	0	0

Figure 6.43 Parameter setting for overhead lines.

files. The Mod component can be freely controlled without independent support files, because a default support file can be automatically created and named after the mod text title. The Mod support files can be edited in the menu Library-Edit object.

If users need a different icon or other node position icons, the default support files can be freely modified, or new support files can be created. The edit dialog box is shown in Figure 6.44.

A simple voltage threshold model is chosen to model flashover. Flash.sup built in ATPDraw is called. UINF = 1.4 MV, U0 = 3 MV, the decay time constant TAU = 8e-7 s,UINIT = 350 kV. Flashover occurs and the switch is closed.

The metal-oxide surge arrester is selected for arresters, with its parameter settings shown in Figure 6.45.

In Figure 6.45, the units of current and voltage is ampere (A) and volts (V), respectively. Figure 6.46 shows the characteristic curve of the metal-oxide surge arrester, with its unit being kA and MV.

This curve is not unique and fixed; it varies with voltage endurance and the characteristics of different arresters. However, the overall trend is similar to that in the figure.

Figure 6.47 is the complete ATPDraw simulation circuit.

Assuming that the power frequency voltage reaches the maximum at the reversal polarity when lightning strikes the fault phase, insulator string flashover will take place if the voltage exceeds the flashover overvoltage of the insulator.

Figure 6.48 shows the overvoltage waveforms on three-phase conductors of tower #4. The incoming wave amplitude reaches 2.25 MV.

Figure 6.49 shows the voltage waveforms of the three-phase conductors at the substation entrance of line 1; the incoming wave peak reaches 1.29 MV.

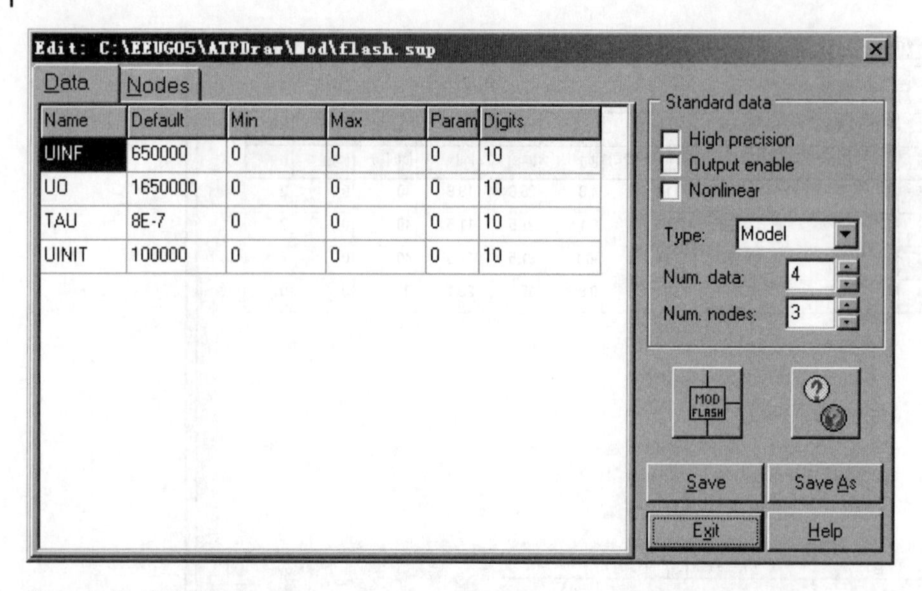

Figure 6.44 Mod file edit page.

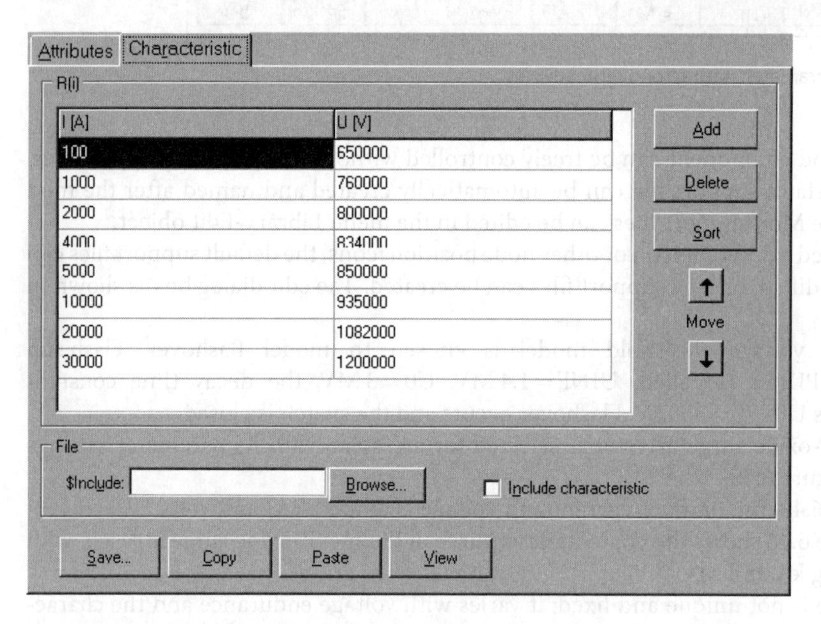

Figure 6.45 Arrester characteristic data.

Figure 6.50 shows the overvoltage waveforms of the three-phase conductor at the site where the instrument transformer in the substation is installed with an amplitude as high as 1.38 MV. It is thus clear that the overvoltage amplitude is increased compared with that at the substation entrance.

Figure 6.51 displays the voltage waveforms of the three-phase leads at the transformer with its peak amplitude reaching 1.1 MV. From the figure, it is apparent that the lightning incoming surge steepness shows an obvious decrease.

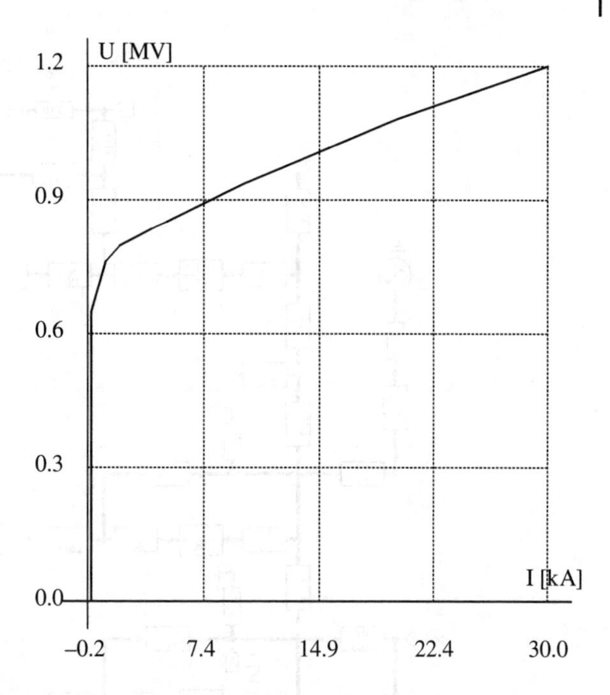

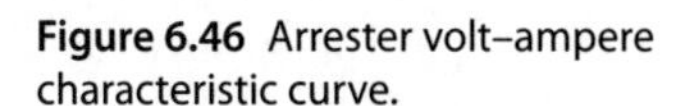

Figure 6.46 Arrester volt–ampere characteristic curve.

6.5 Evaluation of Ferro-Resonance Overvoltages

In the system where the neutral point is not directly grounded, ferro-resonance overvoltages may be generated in the zero-sequence circuit. However, such systems are mainly distribution networks, of which the incoming and outgoing lines are generally free of commutation, hence the three-phase phase-to-ground parameters are not symmetric, and the components of each sequence in the symmetric component method is not independent. As a result, parameters such as outgoing line loads and interphase capacitance may affect triggering of the resonance to a certain extent, which cannot be ignored in the stimulation model. Moreover, in the system with indirectly grounded neutrals, ferro-resonance occurs simultaneously in the three-phase circuits; so a simple single-phase equivalent simulation model is not suitable for analysis. The model in this book is a three-phase simulation model for a substation having three voltage classes established based on field and test data. Aiming at the asymmetric characteristics of phase-to-ground parameters in distribution networks, the zero and positive sequence parameters of buses as well as the interphase parameters of incoming and outgoing lines and their distribution characteristics are considered in the model. The overhead line uses the JMarti model in module LCC in ATP, and the bus uses module LINEZT 3, which increases evaluation accuracy. The basic connections in the substation are shown in Figure 6.52.

6.5.1 Electromagnetic PT Module

Electromagnetic PT is one of the important components that can generate ferro-resonance, and the nonlinear excitation characteristics of the PT iron core

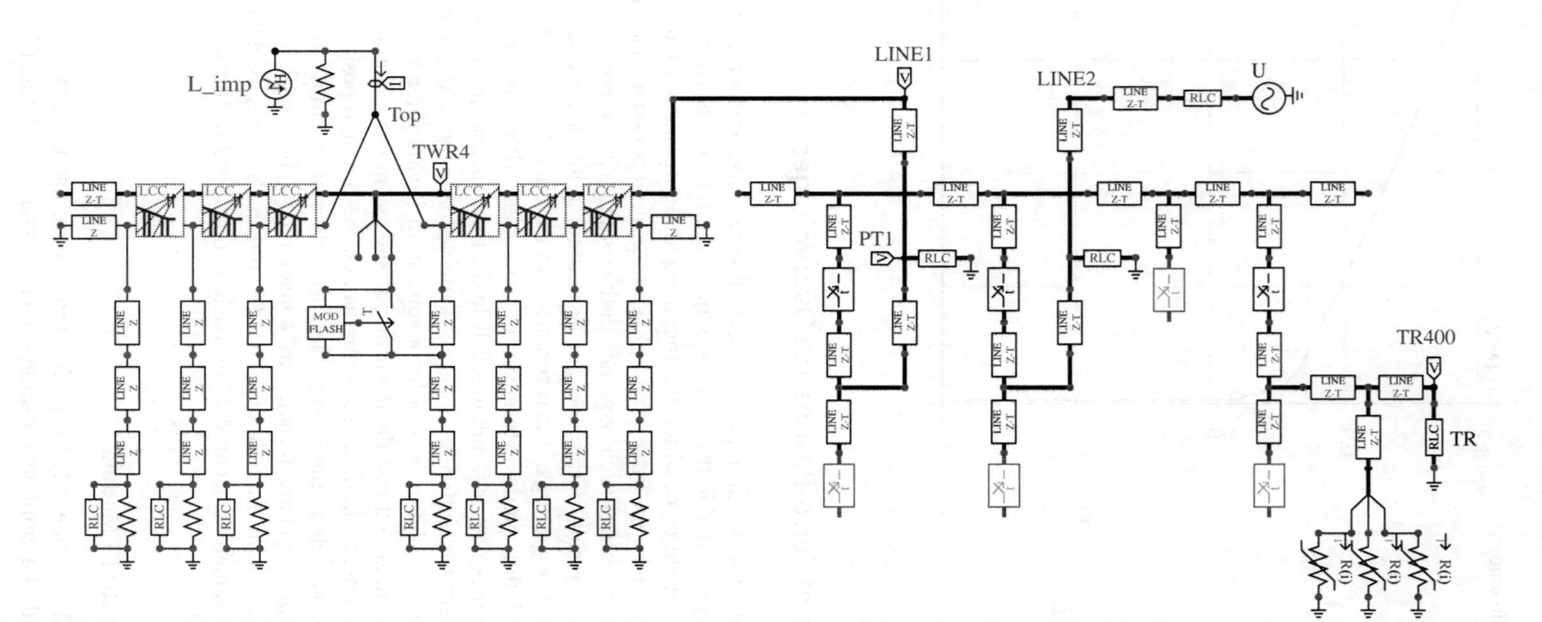

Figure 6.47 Circuit model.

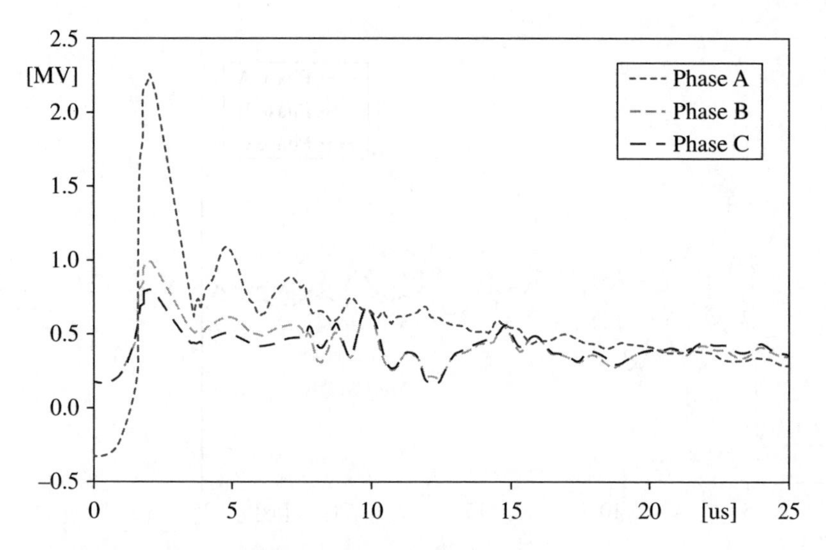

Figure 6.48 Voltage waveforms of the three-phase conductors of tower #4.

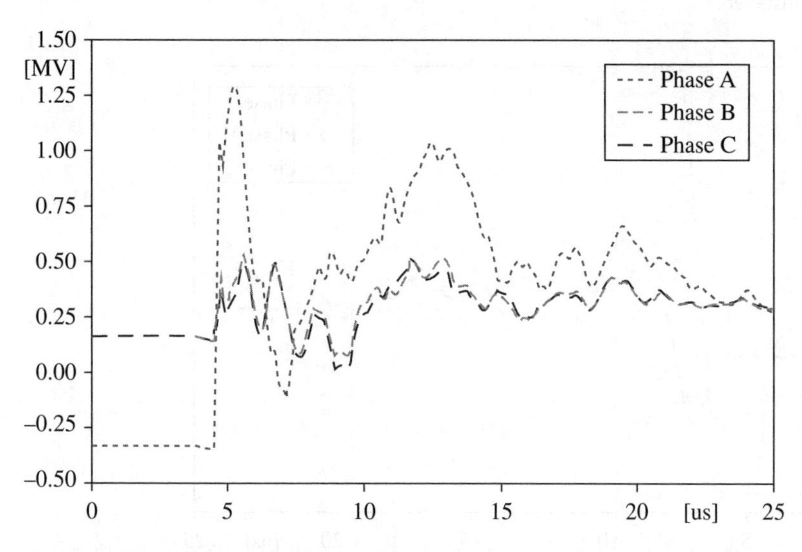

Figure 6.49 Overvoltage waveforms of three-phase conductors of line 1.

are the root cause for ferro-resonance. It is of great significance to obtain accurate and reliable PT excitation parameters in numerical calculation.

6.5.1.1 Acquisition of PT Excitation Characteristic Curve

Ferro-resonance in the substation is mainly caused by saturation of electromagnetic PT with nonlinear inductors. The characteristics of this nonlinear inductor is often described by magnetization characteristics, and in practical tests, by volt–ampere curves of the iron core. The PT excitation parameters in the simulation model are calculated via a point-by-point conversion method with high precision. The specific conversion principles are as follows: for inductor coils having iron cores, the nonlinear

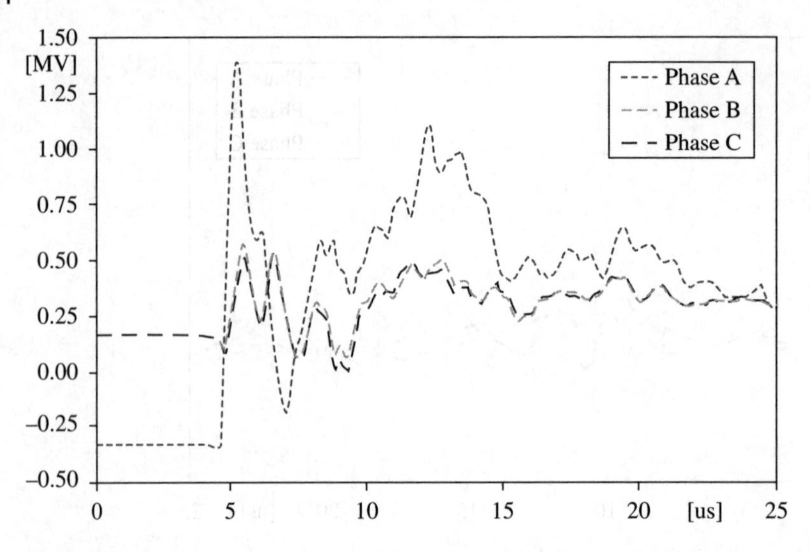

Figure 6.50 Overvoltage waveforms of three-phase conductors at the place where instrument transformers are installed.

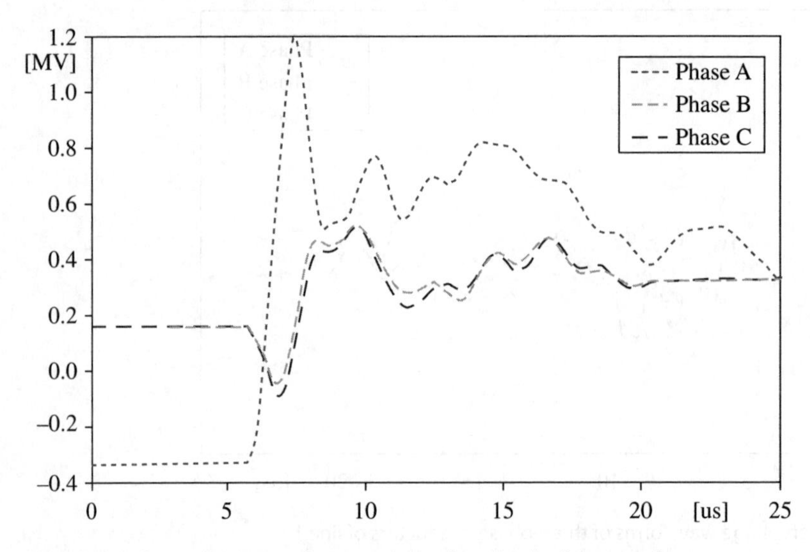

Figure 6.51 Overvoltage waveforms at three-phase leads of the transformer.

excitation characteristic curve $\varphi(i)$ is caused by iron core saturation. As the power source voltage is a sine wave, and based on the relation between the iron core magnetic flux $\varphi(t)$ and the applied voltage $u(t)$ $\frac{d\varphi}{dt} = u(t) = \sqrt{2}U\cos(\omega t)$:

$$\varphi(t) = \frac{\sqrt{2}U}{\omega}\sin(\omega t) \tag{6.5.1}$$

The steady-state amplitude of $\varphi(t)$ is $\varphi_m = \frac{\sqrt{2}U}{\omega}$. It is thus clear that φ_m has a direct connection with the voltage effective value U, but i in curve $\varphi(i)$ does not show a direct

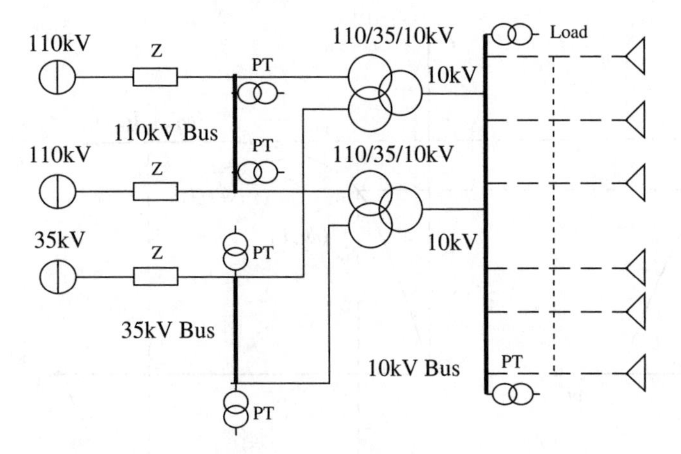

Figure 6.52 Substation structure diagram.

connection with the current effective value I. Figure 6.53 shows the relations between $\varphi(t)$, $i(t)$ and $\varphi(i)$ under the applied sine power source $u(t)$.

Provided that the start point of U–I (0,0) corresponds to the point (0,0) of $\varphi(i)$, solve (φ_1,i_1) according to (U_1,I_1), $\varphi_1 = \frac{\sqrt{2}U_1}{\omega}$, the linear equation from point (0,0) on curve $\varphi(i)$ to (φ_1,i_1) is given by

$$i(t) = \frac{i_1}{\varphi_1}\varphi(t) \tag{6.5.2}$$

$$I_1 = \frac{2}{\pi}\int_0^{\pi/2} i^2(\omega t)d(\omega t) = \frac{2}{\pi}\int_0^{\pi/2} \frac{i_1^2}{\varphi_1^2}\left[\frac{\sqrt{2}U_1}{\omega}\right]\sin^2(\omega t)d(\omega t)$$
$$= \frac{2}{\pi}\int_0^{\pi/2} i_1^2\sin^2(\omega t)d(\omega t) \tag{6.5.3}$$

Calculated from Equation (6.5.2) and (6.5.3),

$$i_1 = \sqrt{2}I_1 \tag{6.5.4}$$

While solving (φ_2,i_2) from (U_2,I_2) and (φ_1,i_1) based on Equation (6.5.1), $u(t) = \sqrt{2}U_2\cos(\omega t)$, $\varphi_2 = \frac{\sqrt{2}U_2}{\omega}$. Linearizing (0,0) and (φ_1,i_1) to (φ_2,i_2) gives the equation:

$$i_{21} = \frac{i_1}{\varphi_1}\varphi = \frac{i_1}{\varphi_1}\frac{\sqrt{2}U_2}{\omega}\sin(\omega t) \tag{6.5.5}$$

$$i_{22} = i_1 + \frac{i_2 - i_1}{\varphi_2 - \varphi_1}(\varphi - \varphi_1) = i_1 - \frac{i_2 - i_1}{\varphi_2 - \varphi_1}\varphi_1 + \frac{i_2 - i_1}{\varphi_2 - \varphi_1}\frac{\sqrt{2}U_2}{\omega}\sin(\omega t) \tag{6.5.6}$$

$$\frac{\pi}{2}I_2^2 = \int_0^{\omega t_1} i_{21}^2(\omega t)d(\omega t) + \int_{\omega t_1}^{\pi/2} i_{22}^2(\omega t)d(\omega t) \tag{6.5.7}$$

According to $\varphi_1 = \frac{\sqrt{2}U_2}{\omega}\sin(\omega t_1)$; $\frac{\sqrt{2}U_1}{\omega} = \frac{\sqrt{2}U_2}{\omega}\sin(\omega t_1)$:

$$\omega t_1 = \arcsin\left[\frac{U_1}{U_2}\right] \tag{6.5.8}$$

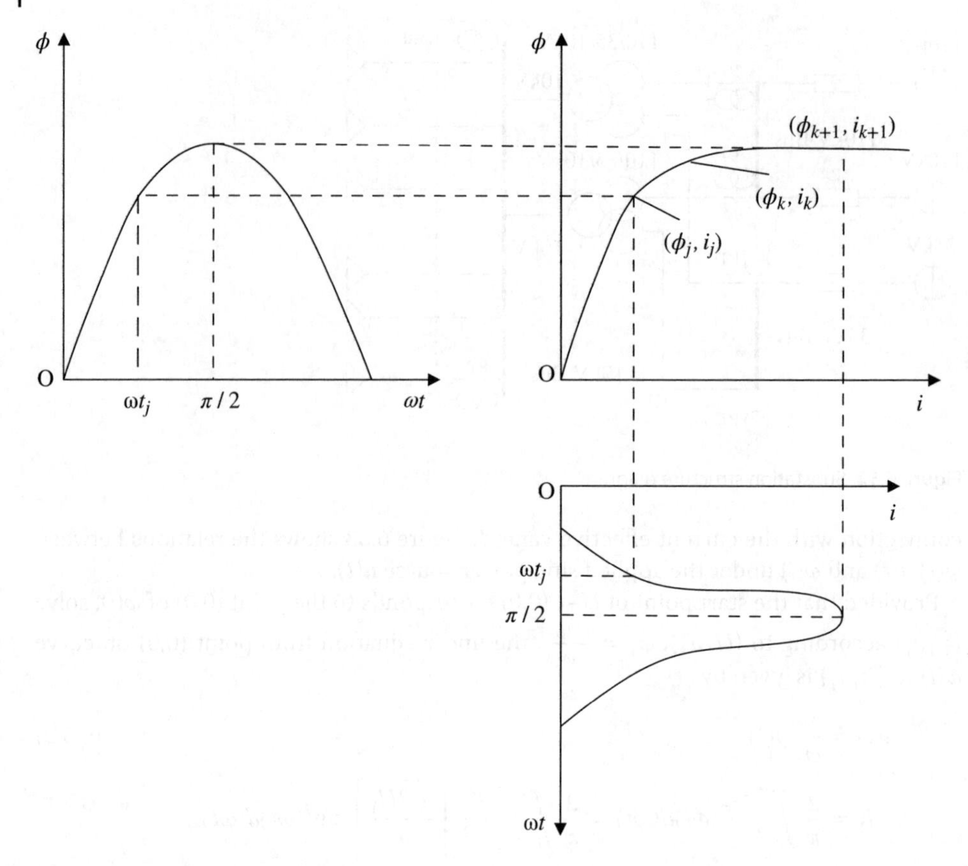

Figure 6.53 Relations between $\varphi(t)$ and $i(t)$.

Therefore, the linear equation from (φ_k, i_k) to (φ_{k+1}, i_{k+1}) is given by

$$i = i_k - \frac{i_{k+1} - i_k}{\varphi_{k+1} - \varphi_k}\varphi_k + \frac{i_{k+1} - i_k}{\varphi_{k+1} - \varphi_k}\varphi;\ \varphi_k = \frac{\sqrt{2}U_{k+1}}{\omega}\sin(\omega t_k) \tag{6.5.9}$$

A relevant conversion program is written via Matlab, based on the above principle, and the volt–ampere characteristic of the excitation curve from the test is converted into current-magnetic flux relation, as shown in Tables 6.2 and 6.3. (The original test data is the data at the PT secondary side, and the data acquired from calculation is converted into the primary side.)

The PT excitation curve corresponding to Table 6.2 after conversion (Phase C) is shown in Figure 6.54.

6.5.2 Establishment of Open Delta PT Model

The principle of PTs is similar to that of transformers, the only difference being the transformation ratio and the excitation performance. The tertiary winding of the electromagnetic PT is an open delta that cannot be found in the transformer model in

Table 6.2 Conversion data A for the PT excitation curve of an 110 kV bus.

PT Model		Voltage (V)	40	50	60	70	80	90
JDCF-110	Phase A after conversion	Current (A)	1.69	2.01	1.96	4.19	9.24	18.8
		Magnetic flux (Wb)	198.07	247.59	297.1	346.62	396.1	445.7
		Current (mA)	2.1727	2.4587	2.768	7.7919	17.08	35.24
	Phase B after conversion	Current (A)	1.71	1.99	2.62	7.46	17	
		Magnetic flux (Wb)	198.07	247.59	297.1	346.62	396.1	
		Current (mA)	2.1727	2.4587	2.768	7.79	17.08	
	Phase C after conversion	Current (A)	1.74	2.09	2.01	4.34	8.72	17.2
		Magnetic flux (Wb)	198.07	247.59	297.1	346.62	396.1	445.7
		Current (mA)	2.237	2.5587	2.739	8.11	15.67	32.03

Table 6.3 Conversion data B for the PT excitation curve of an 110 kV bus.

PT Model		Voltage (V)	50	60	70	80	90
JDCF-110	Phase A after conversion	Current (A)	1.41	2.20	8.90	21.00	
		Magnetic flux (Wb)	247.59	297.10	346.62	396.14	
		Current (mA)	1.80	3.46	17.72	39.329	
	Phase B after conversion	Current (A)	1.40	1.52	4.30	14.80	20.00
		Magnetic flux (Wb)	247.59	297.10	346.62	396.14	445.66
		Current (mA)	1.80	1.626	8.37	29.66	30.18
	Phase A after conversion	Current (A)	1.41	3.31	14.80	74.50	
		Magnetic flux (Wb)	247.59	297.12	346.62	396.13	
		Current (mA)	1.80	5.92	29.55	154.7	

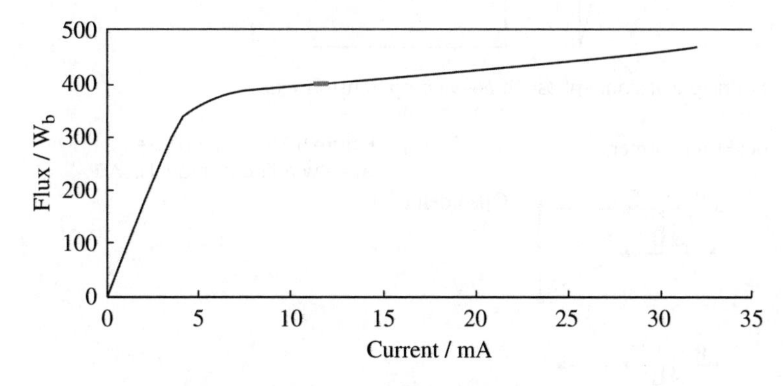

Figure 6.54 Excitation curve of an 110 kV PT (Phase C).

ATP. For this reason, it is necessary to establish a PT module with open delta on the ATP platform.

6.5.2.1 Establishment of Single-Phase Three-Winding (Y-Y_0) PT Model

The single-phase PT is designed based on the T-type equivalent circuit of the transformer. Figure 6.55 illustrates the equivalent principle of the transformer.

Figure 6.56 shows the corresponding ATP connection.

Model descriptions:

1. As the PT secondary side and the open delta side are in the open circuit status most of the time, the influence of leakage reactance of the secondary and open delta sides is not considered in the model.
2. The excitation branch in the figure is equivalent to an ATP Type 98 nonlinear inductor connecting a resistor in parallel, and the nonlinear excitation characteristics of the excitation inductor are obtained through the calculation method mentioned in Equation (6.5.1).
3. As the ideal transformer module in ATP does not allow open circuit operation of the transformer without reference potential at the terminal (without ground potential), a capacitor of 1E-10 μF is connected at the output terminal of the ideal transformer, which satisfies the software requirements, and it has been verified by stimulation that the added capacitor does not affect the model accuracy.

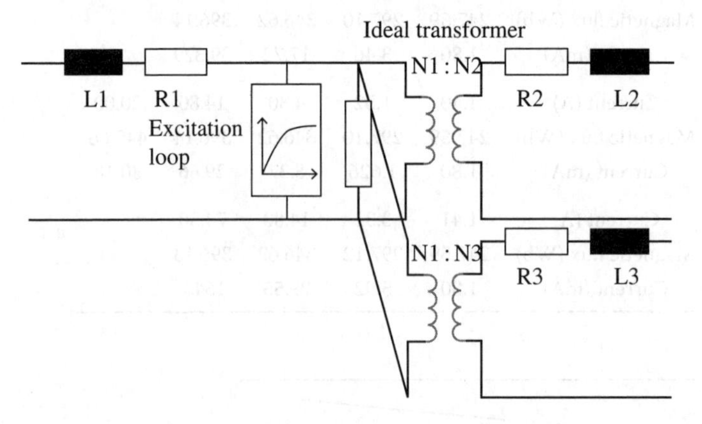

Figure 6.55 Equivalent principle of single-phase three-winding transformers.

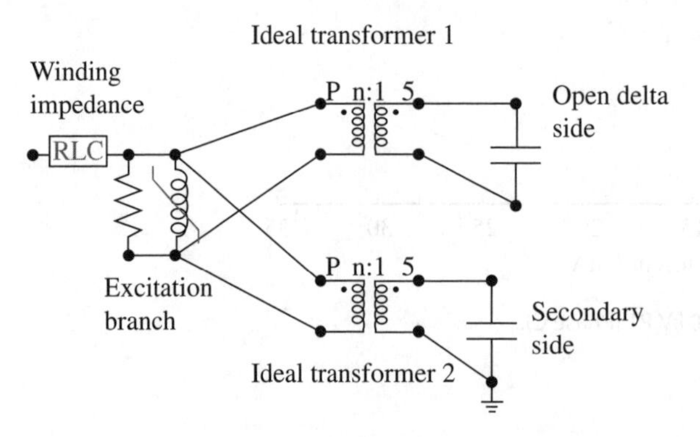

Figure 6.56 Single-phase three-winding PT model in ATP.

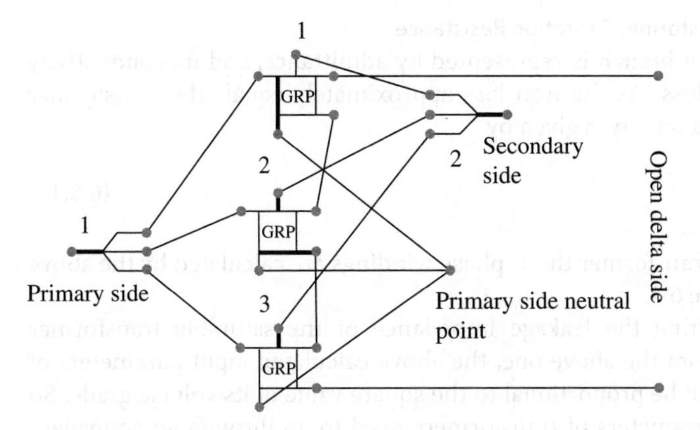

Figure 6.57 Three-phase PT connection diagram.

6.5.2.2 Establishment of Three-Phase Three-Winding PT Models

The model in Figure 6.57 is packaged via the packaging function of ATP after the single-phase three-winding PT model is established. In the packaging process, the secondary side provides only one voltage measuring terminal.

A three-phase PT with an open delta can be set up based on the actual physical connection of the PT, after the single-phase three-winding PT model has been packaged.

6.5.3 Transformer Module

6.5.3.1 Calculation Principles of Winding Parameters

For three-phase winding transformers, the relations between short-circuit loss of each winding and the tested short-circuit losses $P_{k(1-3)}$, $P_{k(1-2)}$, $P_{k(2-3)}$ are given below (parameters are all calculated to the high-voltage side):

$$\begin{aligned} P_{k1} &= 1/2(P_{K(1-3)} + P_{k(1-2)} - P_{k(2-3)}) \\ P_{k2} &= 1/2(P_{K(1-2)} + P_{k(2-3)} - P_{k(1-3)}) \\ P_{k3} &= 1/2(P_{K(1-3)} + P_{k(2-3)} - P_{k(1-2)}) \end{aligned} \tag{6.5.10}$$

Substituting the short-circuit loss of each winding obtained from Equation (6.5.10) into each winding resistance

$$R_{T1} = \frac{P_{k1}U_N^2}{1000S_N^2}, R_{T2} = \frac{P_{k2}U_N^2}{1000S_N^2}, R_{T3} = \frac{P_{k3}U_N^2}{1000S_N^2} \tag{6.5.11}$$

the short-circuit voltage drops and short-circuit losses obtained from the winding short-circuit test demonstrate similar relations:

$$\begin{aligned} U_{k1}\% &= 1/2(U_{k(1-3)}\% + U_{k(1-2)}\% - U_{k(2-3)}\%) \\ U_{k2}\% &= 1/2(U_{k(2-3)}\% + U_{k(1-2)}\% - U_{k(1-3)}\%) \\ U_{k3}\% &= 1/2(U_{k(1-3)}\% + U_{k(2-3)}\% - U_{k(1-2)}\%) \end{aligned} \tag{6.5.12}$$

Calculating the leakage reactance of each winding based on Equation (6.5.12) gives

$$X_{T1} = \frac{U_{k1}\%U_N^2}{100S_N}, X_{T2} = \frac{U_{k2}\%U_N^2}{100S_N}, X_{T3} = \frac{U_{k3}\%U_N^2}{100S_N} \tag{6.5.13}$$

6.5.3.2 Calculation of Transformer Excitation Resistance

The transformer excitation branch is represented by admittance, and its conductivity corresponds to the iron loss. As the iron loss approximately equals the transformer unloaded loss P_0, the conductivity is given by

$$G_T = \frac{P_0}{1000U_N^2} \tag{6.5.14}$$

The parameters of the transformer three-phase windings are calculated by the above method, as shown in Table 6.4.

As the method evaluating the leakage impedance of the saturable transformer component is different from the above one, the above calculated input parameters of the leakage impedance will be proportional to the square value of its voltage grade. So the calculated physical parameters of transformers need to go through an equivalent treatment to satisfy ATP requirements.

The conversion step is as follows: First, calculate the total impedance value:

$$R'_T = R_{T1} + R_{T2} + R_{T3}, X'_T = X_{T1} + X_{T2} + X_{T3} \tag{6.5.15}$$

Then keep the total impedance value constant, and distribute the leakage impedance to each voltage side according to the proportional relations between the leakage impedances and the square values of their voltage grades.

$$\begin{aligned} R'_{T1} &= \frac{R'_T}{n}; R'_{T2} = \frac{R'_T \cdot U_2^2}{nU_1^2}; R'_{T3} = \frac{R'_T \cdot U_3^2}{nU_1^2}; \\ X'_{T1} &= \frac{X'_T}{n}; X'_{T2} = \frac{X'_T \cdot U_2^2}{nU_1^2}; X'_{T3} = \frac{X'_T \cdot U_3^2}{nU_1^2} \end{aligned} \tag{6.5.16}$$

where n is the winding number, and equals 3 here (Table 6.5).

The method of calculating parameters regarding transformer excitation inductance is the same as that of PT excitation inductance, and thus will not be covered again. After obtaining the above transformer parameters, the saturable transformer module in the ATP-EMTP module base is adopted, as shown in Figure 6.58. In the transformer main body, P, S and T are HV, MV and LV connection terminals, respectively, and the left and right of the two bottom terminals are the HV and MV side neutral points.

Table 6.4 Parameters of the transformer three-phase windings.

Transformer model	R_{T1} (Ω)	R_{T2} (Ω)	R_{T3} (Ω)	X_{T1} (Ω)	X_{T2} (Ω)	X_{T3} (Ω)	R_0 (Ω)
SFSZ7-31500/110	0.88	1.29	0.74	27.50	40.88	≈0	240000
SFSZ8-31500/110	1.21	0.81	0.95	40.36	≈0	0.86	150600

Table 6.5 ATP transformer model parameters.

Transformer model	R_{T1} (Ω)	R_{T2} (Ω)	R_{T3} (Ω)	X_{T1} (Ω)	X_{T2} (Ω)	X_{T3} (Ω)	R_0 (Ω)
SFSZ7-31500/110	0.97	0.0956	0.008	22.80	2.28	0.19	260000
SFSZ8-31500/110	0.99	0.1	0.0082	13.75	1.376	0.1135	150600

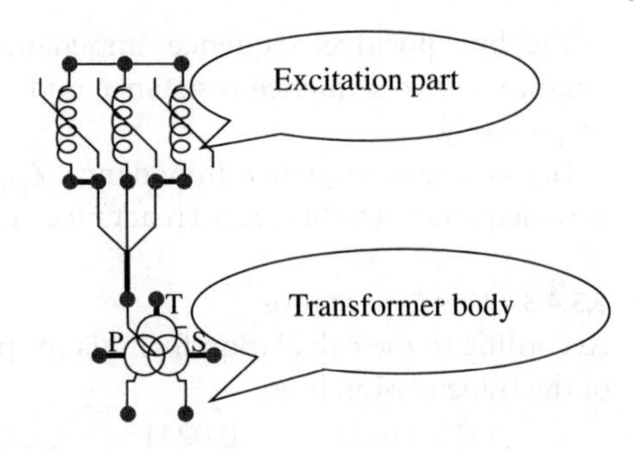

Figure 6.58 Saturable transformer model.

6.5.4 Bus Module

The buses include 110, 35 and 10 kV buses, and the 110 kV bus is taken as an example to illustrate how to calculate bus parameters. Table 6.6 shows the practical measured data of an 110 kV bus.

6.5.4.1 Line Resistance

DC resistance per unit line length: $R_a = \frac{\rho}{S} = 0.138\ \Omega/\text{km}$ where ρ is the resistivity of the current-carrying conductor of the bus ($\Omega{\cdot}\text{mm}^2/\text{km}$), and S is the section area of the bus current-carrying part (mm^2).

6.5.4.2 Line Inductance

The average geometric spacing between the three phases of the bus:

$$D_m = \sqrt[3]{r_{ab}\, r_{bc}\, r_{ca}} = 2.772(m)$$

The average geometric spacing between the three phases of the bus and their mirror images:

$$H_m = \sqrt[9]{h'_{aa}\, h'_{bb}\, h'_{cc}\, h'^2_{ab}\, h'^2_{bc}\, h'^2_{cb}} = 15.21(m)$$

Refer to GB1179-83 for main technical parameters of steel core aluminum reinforced conductor with $r = 0.8(\text{cm})$.

Equivalent radius: $r' = 0.779,\ r = 0.779 \times 0.8 = 0.64(\text{cm})$

The average geometric radius of the three-phase buses: $r_{eq} = \sqrt[3]{r' \times D_m^2} = 0.36\ \text{m}$

The conductor equivalent depth $D_g = 1000\ \text{m}$ according to the derived Carlson formula;

The mutual impedance between the three bus phases:

$$Z_m = 0.05 + j0.1445 \lg\left(\frac{D_g}{D_m}\right) = 0.05 + j0.1445 \lg\left(\frac{1000}{2.772}\right) = 0.05 + j0.37(\Omega/\text{km});$$

Self-impedance: $Z_s = R_a + 0.05 + j0.1445 \lg\left(\frac{D_g}{r}\right) = 0.188 + j0.748(\Omega/\text{km});$

Table 6.6 Measured data of an 110 kV bus.

Model	LGJ-240	Sectional area	Aluminum: 228mm^2	Steel: 43.1mm^2
Material	ACSR	Length	60–70 m	
Bus-to-ground and interphase positions		Bus-to-ground vertical position: 7.5 m Bus interphase spacing: 2.2 m		

The bus positive sequence impedance $Z_{(1)} = 0.138 + j0.378\,\Omega/\text{km}$ with the unit length positive sequence resistance and reactance being 0.138 Ω/km and 0.378 Ω/km, respectively.

The bus zero-sequence impedance $Z_{(0)} = 0.288 + j0.378\,\Omega/\text{km}$ with the unit length zero-sequence resistance and reactance being 0.288 Ω/km and 1.452 Ω/km, respectively.

6.5.4.3 Line Capacitance

According to the calculation formula for positive sequence capacitance per unit length of the transmission line:

$$C_1 = \frac{0.0241}{\lg\left(\frac{D_m}{r}\right)} = \frac{0.0241}{\lg\left(\frac{2.772}{0.00664}\right)} = 0.0092\,\mu\text{F/km} \tag{6.5.17}$$

According to the calculation formula for zero-sequence capacitance per unit length of the transmission line:

$$C_0 = \frac{0.0241}{3\lg\left(\frac{H_{eq}}{r_{eq}}\right)} = \frac{0.0241}{3\lg\left(\frac{15.21}{0.36}\right)} = 0.0045\,\mu\text{F} \tag{6.5.18}$$

Similarly, parameters of 35 kV and 10 kV buses can be calculated, as shown in Table 6.7.

6.5.5 Ferro-Resonance Overvoltage Simulation Results

Single-phase grounding occurs at $t = 0.04$ s, and the grounding fault disappears at $t = 0.08$ s, giving rise to resonance overvoltages. However, the resonance overvoltage will not appear if single-phase grounding fault disappears at $t = 0.095$ s. Figure 6.59 shows the resonance conditions when fault disappears at $t = 0.08$ s. When single-phase grounding fault occurs, except for the grounded phase, the other two phases starts to oscillate dramatically, and the maximum voltage amplitude (phase C) attains 2.45 p.u. during the oscillation, and at time $t = 0.042$ s after the oscillation, the voltages of phases B and C remain stable at the line voltage. At time $t = 0.08$ s, the grounding fault disappears suddenly and triggers the resonance, and the three-phase voltages of the 35 kV bus increases simultaneously during the resonance; the resonance voltage amplitude is slightly higher than the line voltage amplitude and phase C has the most serious overvoltage that reaches 2.5 p.u. Meanwhile, overcurrents occur on all three phases of the PT, which is typical in harmonic wave overvoltage conditions. As the excitation performances of the three-phase PT branches are not always the same, the three-phase current amplitudes are different during the resonance process. Also, as the

Table 6.7 Substation bus parameters.

Bus	Actual radius (cm)	Equivalent radius (cm)	Resistivity (Ω/km) positive/zero-sequence	Reactance (Ω/km) positive/zero-sequence	Capacitance(μF/km) positive/zero-sequence
110 kV	0.852	0.640	0.138/0.288	0.378/1.452	0.0092/0.0049
35 kV	1.070	0.834	0.0875/0.238	0.274/1.650	0.0135/0.0035
10 kV	1.784	1.390	0.029/0.179	0.217/1.670	0.0161/0.0046

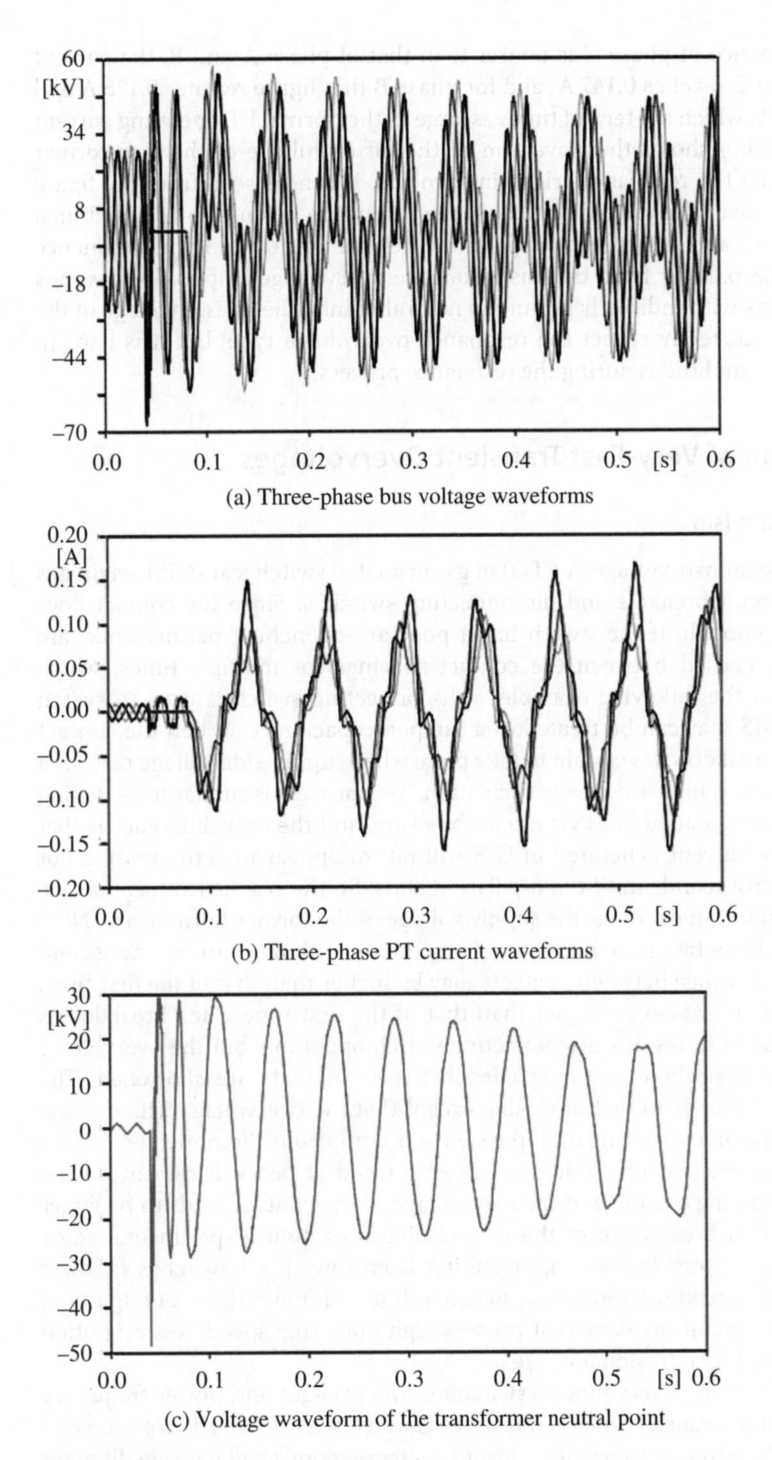

Figure 6.59 PT resonance at the 35 kV side.

excitation performance of phase C is poorer than that of phase A and B, the current amplitude of phase C reaches 0.147 A, and for phase B that figure reaches 0.125 A and for phase A 0.123 A, which are tens of times as large as the normal PT operating current 8.5 mA. Figure 6.59(c) shows the waveform of the offset voltage of the transformer neutral point during the resonance triggering process. It can be seen from the figure that the resonance overvoltage cycle is 0.06 s, the resonance belongs to 1/3 fractional frequency resonance and the offset voltage of the neutral point during the resonance is around 22 kV, 0.8 p.u., far from the bus resonance overvoltage amplitude. It is thus clear that in systems with indirectly grounded neutral points, the offeset voltage of the neutral point may correctly reflect the resonance overvoltage type, but it is hard to reflect overvoltage amplitudes during the resonance process.

6.6 Evaluation of Very-Fast Transient Overvoltages

6.6.1 VFTO Mechanism

The very-fast transient overvoltage (VFTO) in gas insulated switchgear (GIS) originates from switching circuit breakers and disconnecting switches. Since the contact does not move quickly enough, if the switch has a poor arc-quenching performance, arc reignition may be caused between the contact spacings for multiple times, which leads to VFTOs. In the following example, a disconnecting switch is used to switch on de-energized GIS that can be treated as a lumped capacitance. When the contact spacing shortens, breakdown is certain to take place when supply-side voltage reaches a peak, which also causes high-frequency oscillation. This process is similar to switching on/off capacitors or unloaded lines via circuit breakers, and the only difference is that the high frequency current generated in GIS will not disappear in zero-crossing but last for tens of microseconds until the oscillation ends. So the residual voltage on the capacitor when reignition occurs is the supply voltage of the former burning arc. Nevertheless, the switch contact is asymmetric. When the reignition occurs for the second time, the voltage difference between contacts may be higher than that of the first time, and the overvoltage may also be higher than that of the first time. Such breakdowns can occur hundreds of times in a disconnecting switch operation, but the overvoltage becomes lower on the whole as the spacing between contacts are shortened. The opening situation is similar to that of closing except that the overvoltage will increase with the increasing contact spacing until the switch is completely disconnected.

After the GIS is switched off, residual charges formed at last will remain on the capacitor for quite a long period, and the overvoltage of the contact tends to be larger than 1 p.u. in the first breakdown of the next closing operation. Experimental study has suggested that the residual voltage resulting from low-speed switches is lower compared with high-speed disconnecting switches, but with more times of reignition. This also works for circuit breakers that possess high operating speed, less reignition occurrence and thus higher residual voltage.

Different from lightning overvoltages, switching overvoltages and power frequency overvoltages, very fast transient overvoltages are characterized by short rise time and high amplitude. Wavefronts of several to dozens of nanoseconds will damage different parts of a GIS considerably, influence switchgear stability and even endanger equipment insulation properties. The generated voltage waves will be refracted and reflected constantly within the GIS, leading to a surge in transient overvoltage frequency.

Table 6.8 Equivalent models and parameters for GIS components.

Components		Descriptions	550 kV GIS	800 kV GIS	1100 kV GIS
Transformer		Entrance capacitance (pF)	5000	9000	10000
Circuit breaker	Open	Equivalent series capacitance at the fracture (pF)	350	520	540
	Closed	Equivalent to a bus portion			
Disconnecting switch	Open	Capacitance to ground (pF)	240	276	296
	Closed	Capacitance to ground (pF)	125	140	173
	Arc burning	Equivalent to the arc-burning resistance and the fracture-to-ground capacitance	$R(t) = R_0 e^{-(1/t)} + R_a$ $R_0 = 10^{12}\,\Omega$ $R_a = 0.5\,\Omega$ $T = 1\,\text{ns}$		
Capacitance to ground of the grounding switch (PF)			240	240	300
GIS pipeline wave impedance (Ω)			63	84	70
Bushing-to-ground capacitance (pF)			320	350	450
Arrester-to-ground capacitance (pF)			19	19	
PT-to-ground capacitance (pF)			400	500	1000
Wave velocity (m/μs)			296	277	270

6.6.2 VFTO Models and Parameters

According to calculation and practical measurements, VFTO amplitudes are related to many factors but primarily depend on GIS structure. The amplitude will decrease with increases in network branch numbers. In the simulation, single electric machine, single transformer and single-circuit power supply mode are generally adopted to study VFTO.

In order to ensure calculation accuracy, transient spatial electromagnetic field method will be adopted to perform analogue simulation for each component within VFTO occurrence range based on component structure, layout and connections. Table 6.8 provides the models and their parameters for key components under each voltage class.

6.6.3 VFTO Simulation and Field Tests

Example 15 Simulate VFTO at the ac side of a 500 kV converter station.

GIS is adopted at the ac side of a 500 kV converter station. Figure 6.60 shows the wiring diagram at the converter station ac side before the test.

EMTP is used to simulate the operating mode of the converter station ac side. The calculation model and relevant parameter settings are then determined. The operation mode and switch actions are given in Figure 6.61. This is a switching test for a station service transformer after the station bus is charged. Breaker 301 is turned into the hot standby state, i.e. close the disconnecting switch 3101 and close GIS 310 to charge the station service transformer 301B.

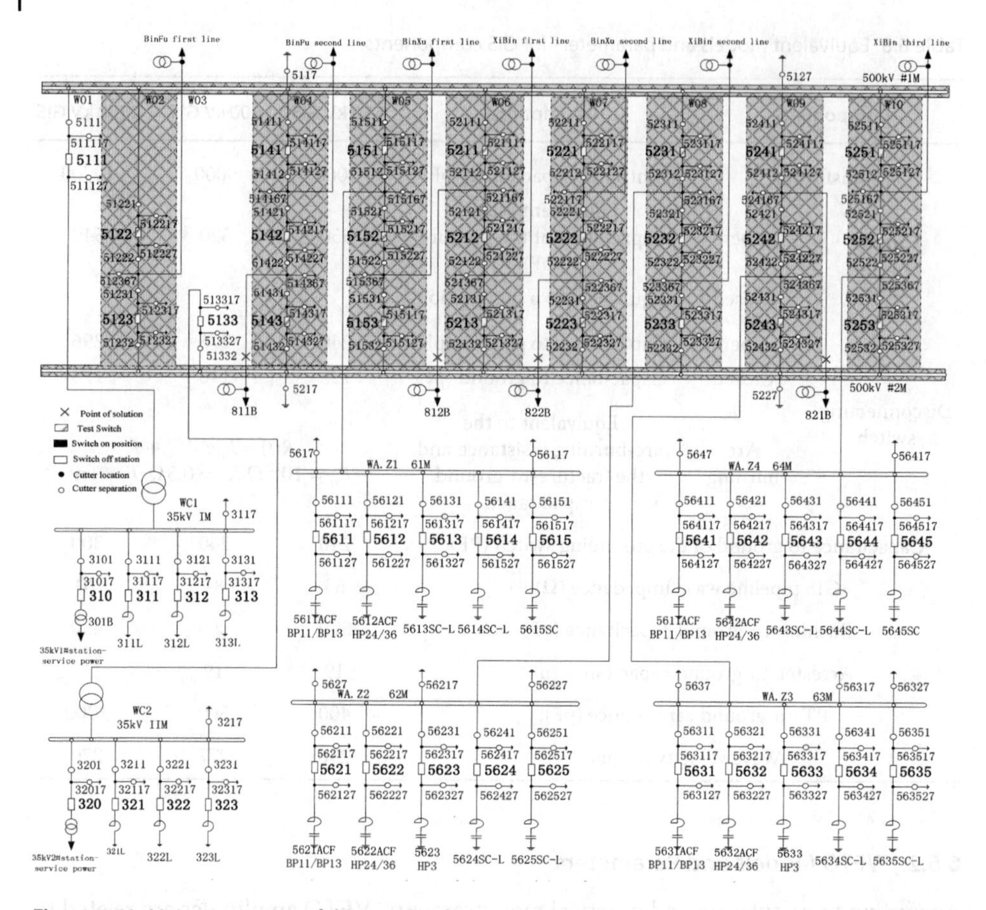

Figure 6.60 Main connection of the ac side of a converter station.

The parameters of the station and components in the line are: bushing-to-ground capacitance is 320 pF, the reactor is 6.1 H, capacitance to earth is 5000 pF, disconnecting switch capacitance to ground is 240 pF, breaker-to-ground capacitance is 320 pF. Distributed parameter simulation is adopted by the GIS with an inductance per unit length $L_0 = 3.35 \times 10^{-4}$ mH/m, a capacitance to earth $C_0 = 3.9 \times 10^{-5}$ μF/m, a wave velocity $v = 277 \times 10^6$ m/s, and a wave impedance $Z = 92.68\,\Omega$. The maximum bus operating voltage is 530 kV, and 1 p.u. $= 530 \times \frac{\sqrt{2}}{\sqrt{3}}$ kV.

The simulation model and results are presented in Figures 6.62 and 6.63. Breaker 310 has been marked in the simulation model.

Figure 6.63 shows the VFTO waveform of the high-voltage bus during the closing of switch 310. The VFTO amplitude is not high, approximately 1.5 p.u., but the wave process time is particularly short. The high-frequency component of 1 MHz can be found, generated due to constant reflection and refraction of traveling waves.

Figures 6.64 and Figure 6.65 are VFTOs measured by oscilloscopes in a station when GIS is actuated and when the bus is being charged.

Figures 6.64 and 6.65 are waveforms measured by the oscilloscope at the PT secondary side. The power frequency voltage amplitude is about 81 V, and the VFTO amplitude is 120 V, about 1.5 p.u., which are almost the same as the simulation results.

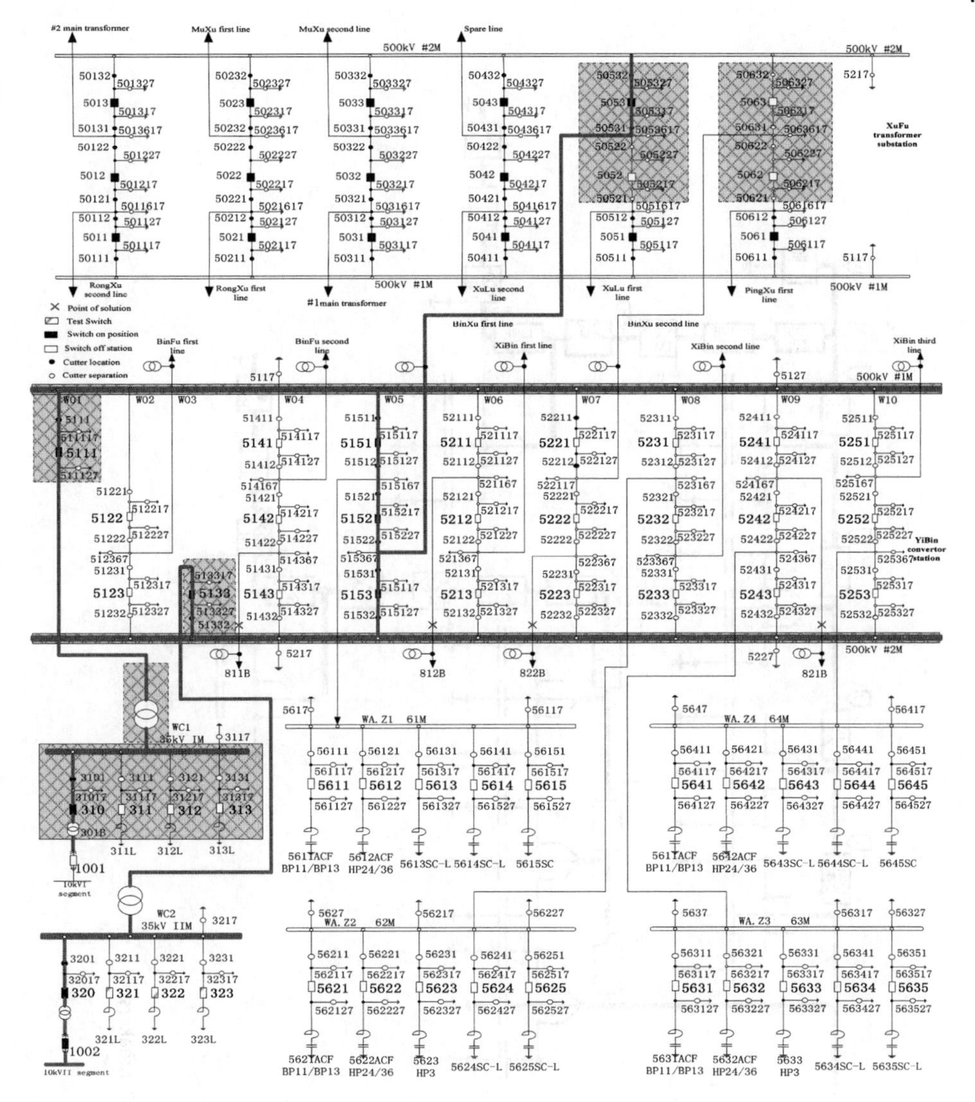

Figure 6.61 Equipment state diagram after energizing a line.

6.7 Transient Calculation for UHVDC Transmission Systems

6.7.1 PSCAD Models for UHVDC Transmission Components

The dc transmission system consists of three major parts: ac system, rectification system and dc transmission system, as shown in Figure 6.66. The main equipment includes the ac power supply, ac filters, converter transformers, converter valves, smoothing reactors and dc filters. The function of each component is given as follows.

Converter: equipment that converts either ac power into dc power, or dc power into ac power.

Rectifier: a converter that converts ac power into dc power.

Inverter: a converter that converts dc power into ac power.

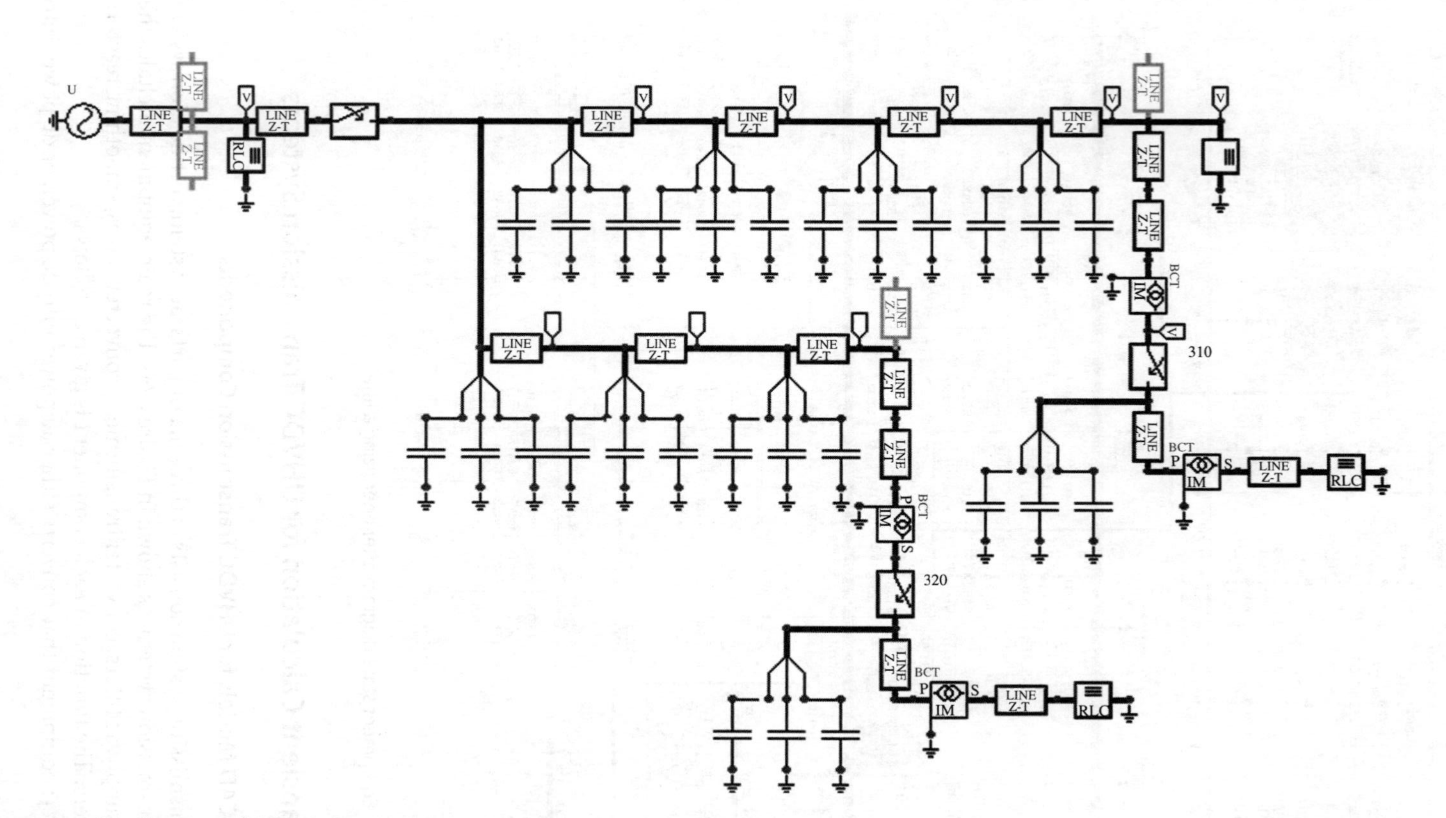

Figure 6.62 Simulation model for closing GIS breakers.

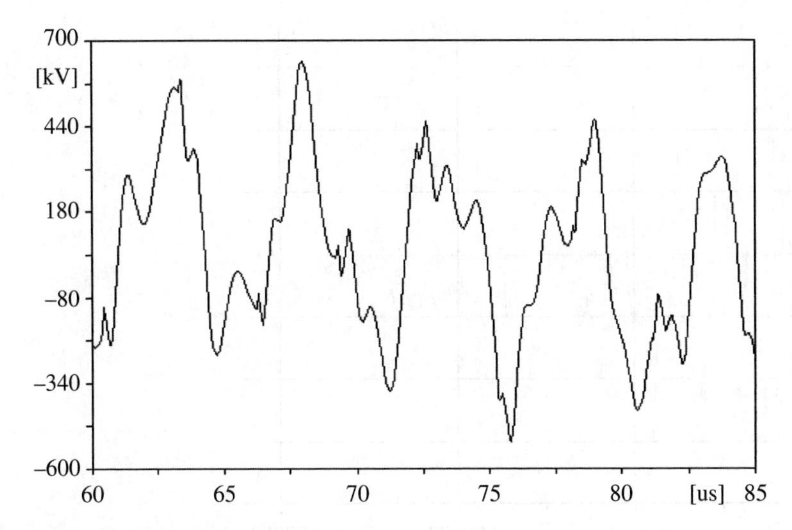

Figure 6.63 VFTO on the bus (phase A).

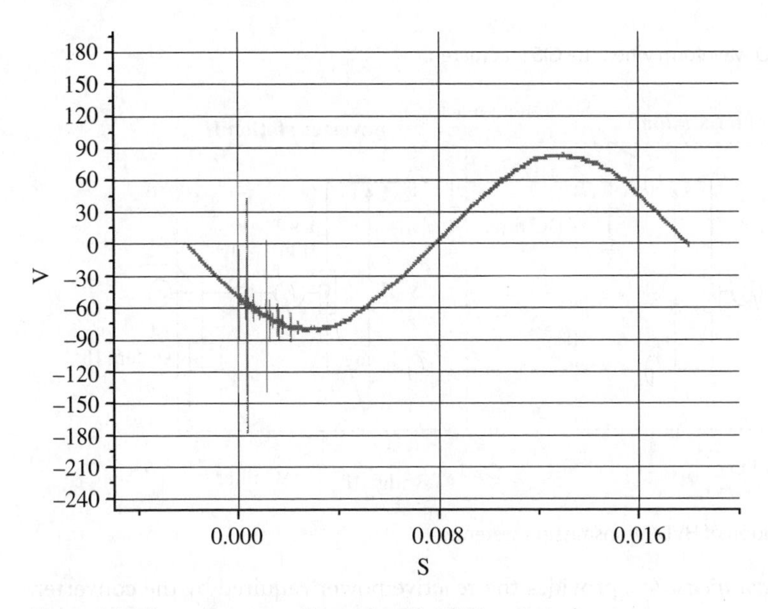

Figure 6.64 Bus voltage waveform when the GIS is actuated.

Converter transformer: equipment that provides an ungrounded three-phase voltage source with suitable voltage grades to the converter.

Smoothing reactor: reduces the harmonics injected into the dc system, reduces the probability of commutation failures, limits the dc short-circuit current peak and prevents the dc current from being interrupted under lightly loaded conditions.

Filters: equipment that can reduce harmonics injected into the ac and dc systems, and can be divided into ac and dc filters according to the installation site, into active and passive filters according to filter categories, of which passive filters can be further divided into single tuned filters, double tuned filters and high pass filters.

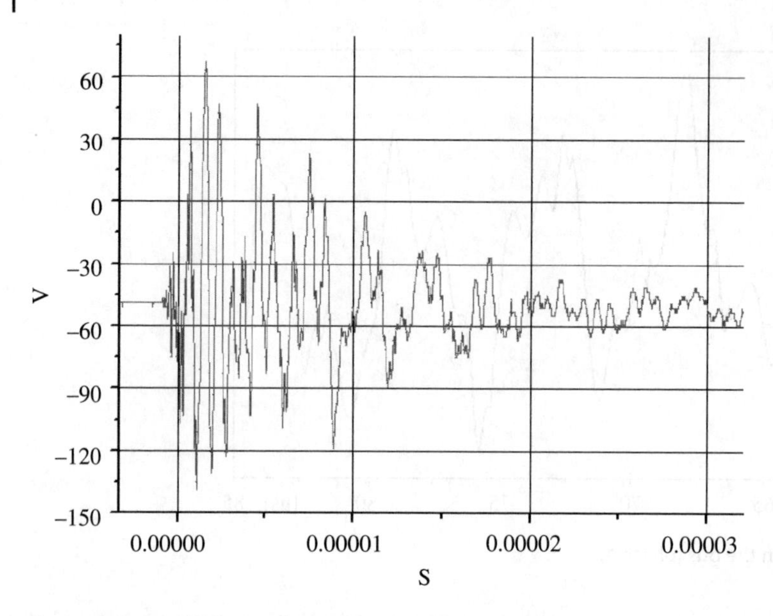

Figure 6.65 Bus VFTO waveform when the GIS is actuated.

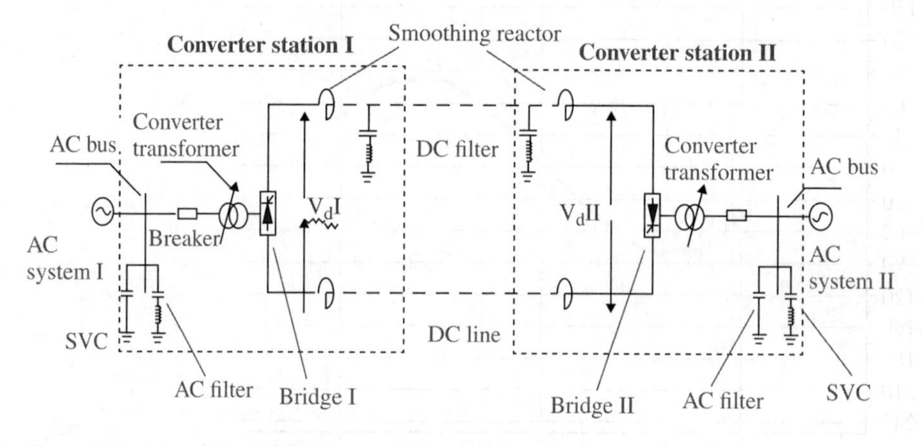

Figure 6.66 composition of HVDC transmission systems.

Reactive power compensator: provides the reactive power required by the converter, and reduces the reactive power exchange between the converter and the system.

6.7.1.1 Converter Valves

The valve model is a series-parallel connection of thyristor models matched with corresponding voltage sharing and current sharing circuits (Figure 6.67). In the valve model, both static and dynamic voltage sharing are required. Static voltage sharing refers to voltage sharing between the components of the thyristor that is subject to power frequency voltage or dc voltage under the blocking state. In this case, the front time of the voltage wave is long, and resistor voltage sharing is adopted. Dynamic voltage sharing refers to the voltage sharing between thyristors in the same bridge arm in the on/off process, namely the voltage sharing during the transition.

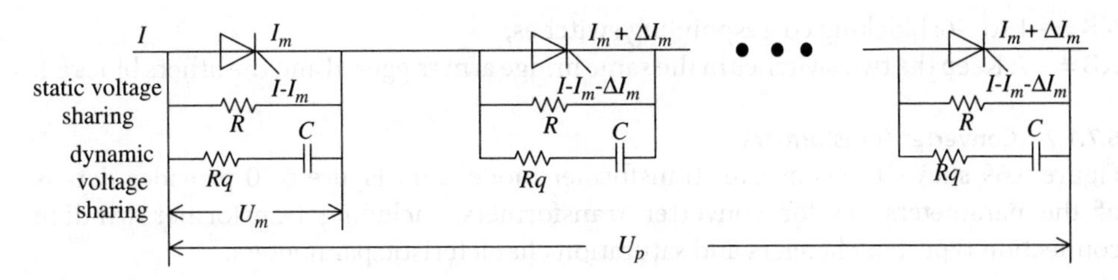

Figure 6.67 Converter valve model.

The static voltage sharing resistance is calculated by

$$R = \frac{U_P\left(\frac{1}{1-K}\right)}{(n-1)\times \Delta I_m}$$

The dynamic voltage sharing resistance is calculated by

$$C = \frac{(n-1)\times \Delta Q_r}{U_P\left(\frac{1}{1-K}\right)}$$

$$Q_r = \frac{1}{2} I_{TR} \frac{T_{r2}}{0.64}$$

As illustrated in Figure 6.68, the pins of the PSCAD converter valve have the functions as follows:

AM, GM – the trigger pulse angle and arc-quenching angle for the measurement;
AO – trigger pulse signal; KB- lock and unlock control;
KB = 0: blocking all pulses;
KB = 1: deblocking;

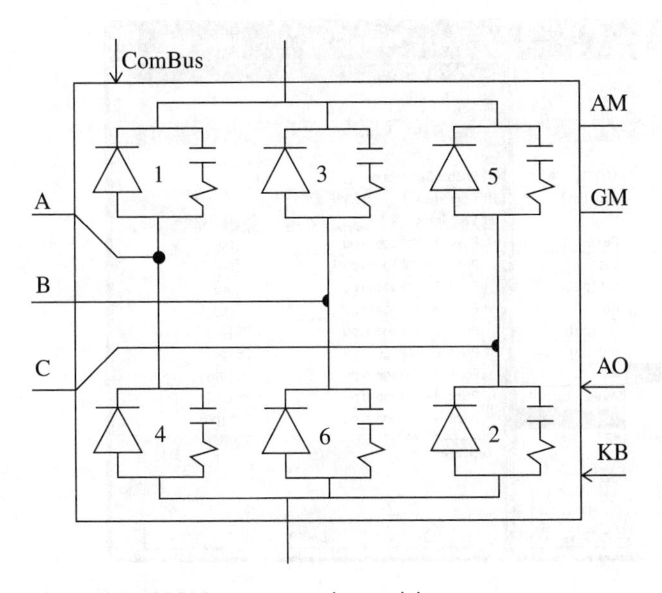

Figure 6.68 PSCAD converter valve model.

KB = −1 to −6: blocking corresponding switches;
KB = −7: Keep the two switches in the same bridge arm triggered and the others blocked.

6.7.1.2 Converter Transformers

Figure 6.69 shows the converter transformer model, and Figure 6.70 provides details of the parameters set for converter transformers, including transformer winding connection type, tap changers and saturation characteristic parameters.

6.7.1.3 Line Models

Overhead line modeling steps (Figure 6.71):

Step 1: Create the configuration elements of the transmission line (Figure 6.72).
There are two modes in PSCAD to build an overhead line: the Remote Ends mode and the Direct Connection mode. In the Remote Ends mode, the line terminal is not directly connected to other components physically, and interface components for the overhead line are needed; in the Direct Connection mode, the line terminal can be directly connected to other components, but can only be used in the single-phase, three-phase or six-phase single display system.

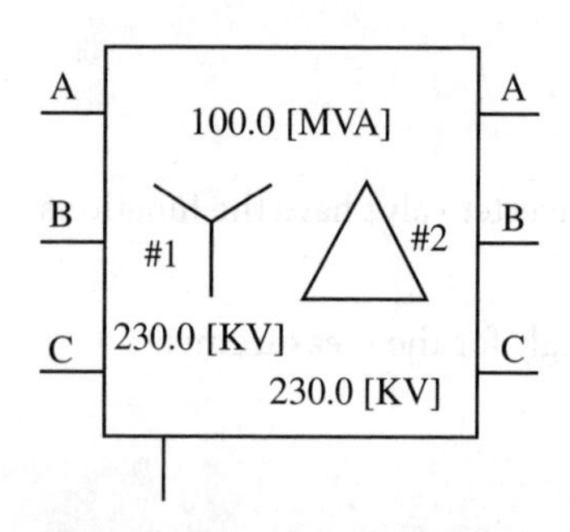

Figure 6.69 Converter transformer model in PSCAD.

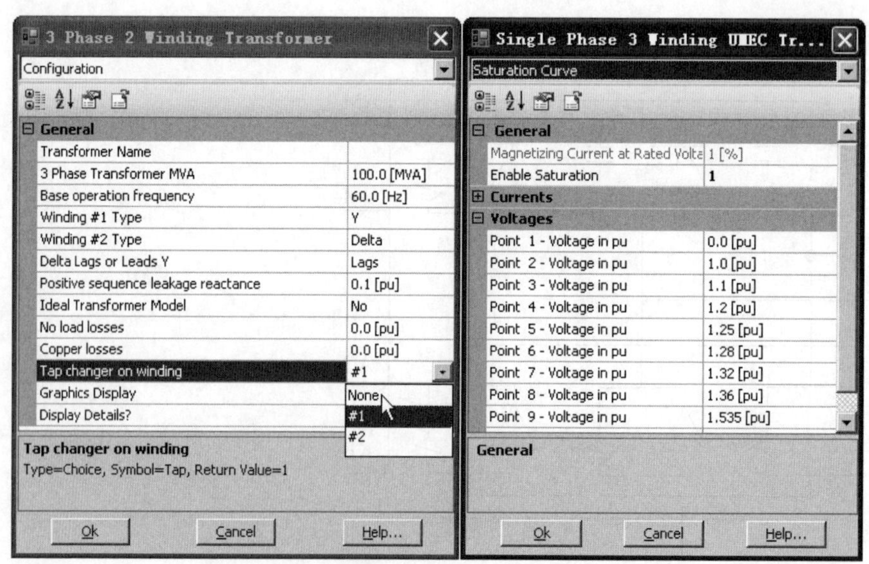

Figure 6.70 Parameter setting details for converter transformers.

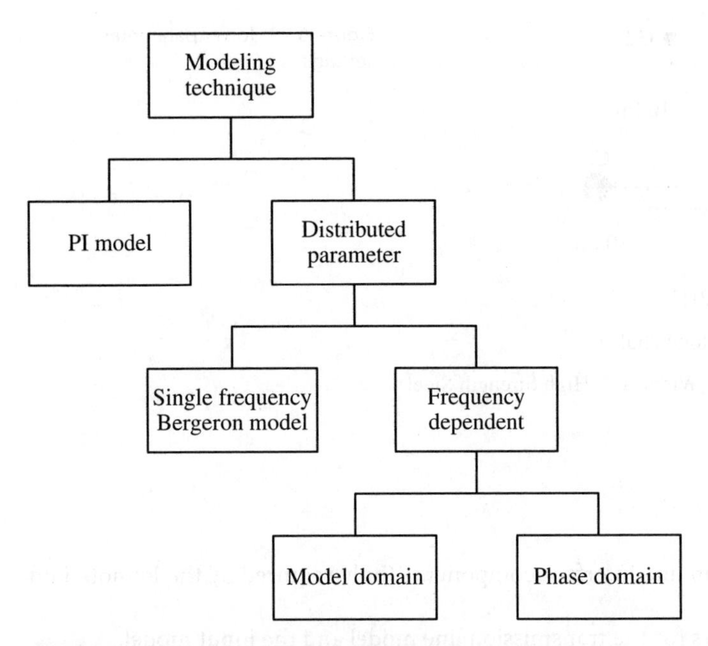

Figure 6.71 Steps to model lines in PSCAD.

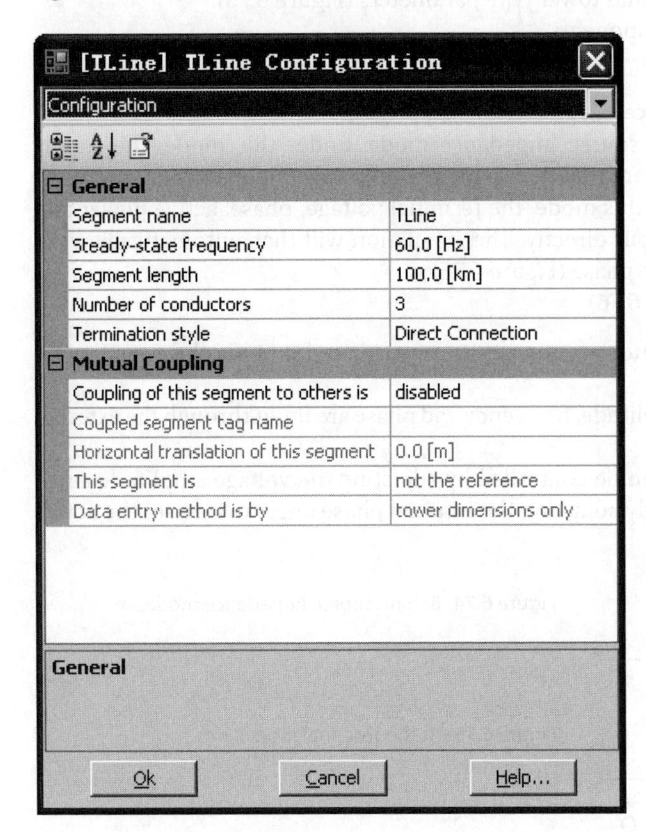

Figure 6.72 Modeling step 1.

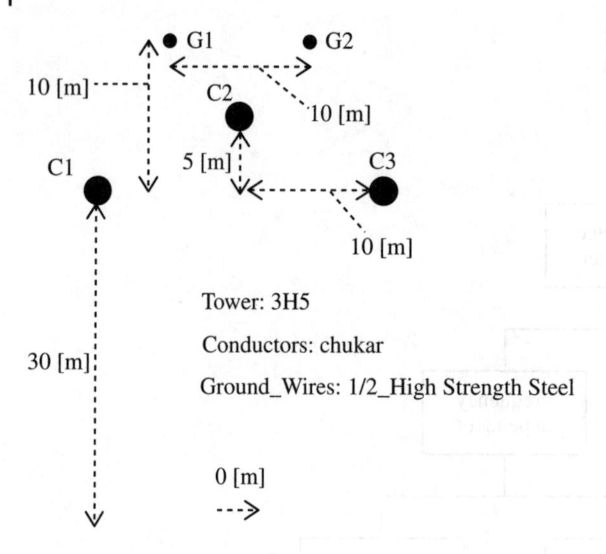

Figure 6.73 Tower parameter settings.

Step 2: Add transmission line interface components (only required by the Remote End mode).
Step 3: Select parameters for the transmission line model and the input model.
Step 4: Input line parameters and tower type parameters (Figure 6.73).
Step 5: Add ground plane components.

6.7.1.4 Equivalent Power Sources

Power source type – Behind source impedance mode: under this mode, the supply voltage, phase and frequency are required to be input directly (Figure 6.74)

At the Terminal mode – In this mode, the terminal voltage, phase, active power and reactive power need to be input directly. The simulation will then automatically work out the supply voltage and the phase (Figure 6.75).

Power control mode (Figure 6.76)

- Fixed: power supply amplitude, frequency and phase are input via the page Source Values for Fixed Control.
- External: power supply amplitude, frequency and phase are input through the external connection terminal.
- Auto: voltage on the bus can be controlled by adjusting the voltage amplitude automatically or by automatically adjusting the internal phase angle to control the active power output.

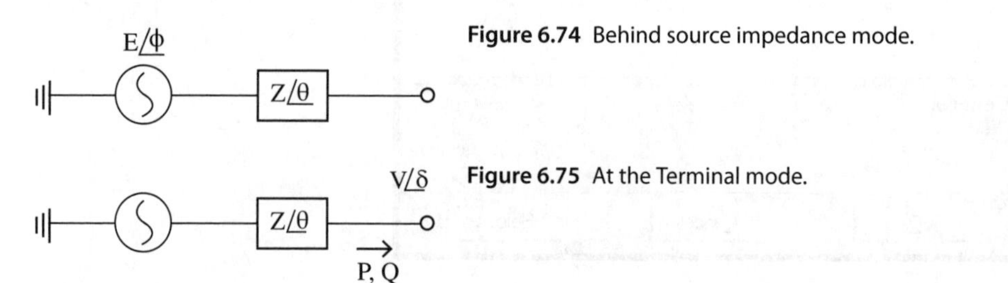

Figure 6.74 Behind source impedance mode.

Figure 6.75 At the Terminal mode.

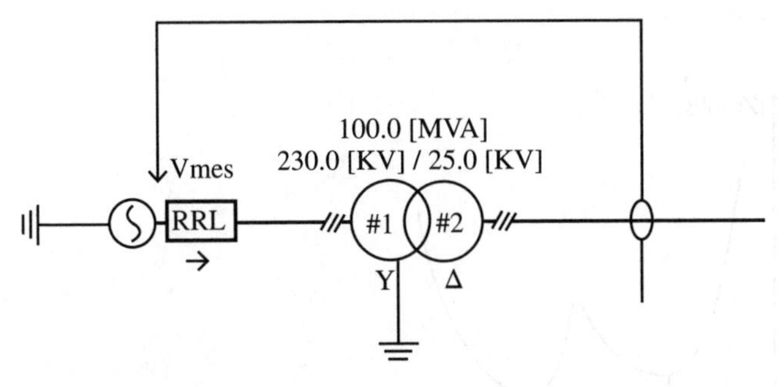

Figure 6.76 Power control mode.

Impedance data input format:

- RRL values: input R and L values directly.
- Impedance: input impedance parameters in the polar form, and impedance amplitude and phase angle are needed.

6.7.1.5 Other Component Models

Smoothing reactors – Smoothing reactors can maintain current continuity in the case of small current of the dc line. When the trigger delayed angle $10.1° < a < 169.9°$, the inductance of the reactor is given by

$$L_d = \frac{U_{d0} \times k_0 \times \sin\alpha}{\omega \times I_{dLi}} \tag{6.7.1}$$

When faults occur to the dc power supply circuit, the smoothing reactor can suppress the current rise rate so as to prevent secondary commutation failures. The reactor inductance at this time is given by

$$L_d \geq (\Delta U_d \times \Delta t)/\Delta I_d \tag{6.7.2}$$

Generally speaking, a larger reactor inductance L_d is better. However, if L_d is too large, the overvoltage generated on the smoothing reactor due to rapid changes in the current L_d(di/dt) will also be greater. Moreover, as a time delay process, a too large L_d is not good for automatic adjustment of the dc current. Therefore, the smoothing reactor inductance shall be as small as possible on the premise of meeting the above requirements.

DC-side filters – The dc-side passive filters (Figure 6.77) do not compensate reactive power, and is only used for filtering. Their parameters are determined by the line voltage, filtering requirements and economic factors. Passive filters, including single-tuned filters, double-tuned filters, C-type filters and triple tuned filters, are usually connected to the rear ends of smoothing reactors. For economical and space considerations, double-tuned filters, equivalent to two single-tuned filters connected in parallel, are more often adopted by HVDC systems. The increase in the inductance will increase the filter capacity of the smoothing reactor but lower the requirements for filter capacity of the passive reactor, and vice versa.

AC-side filters – These are as shown in Figure 6.78.

Harmonic filters are mainly used for filtering, and sometimes for compensating reactive power as well, while C-type filters are mainly used to compensate reactive power for fundamental waves.

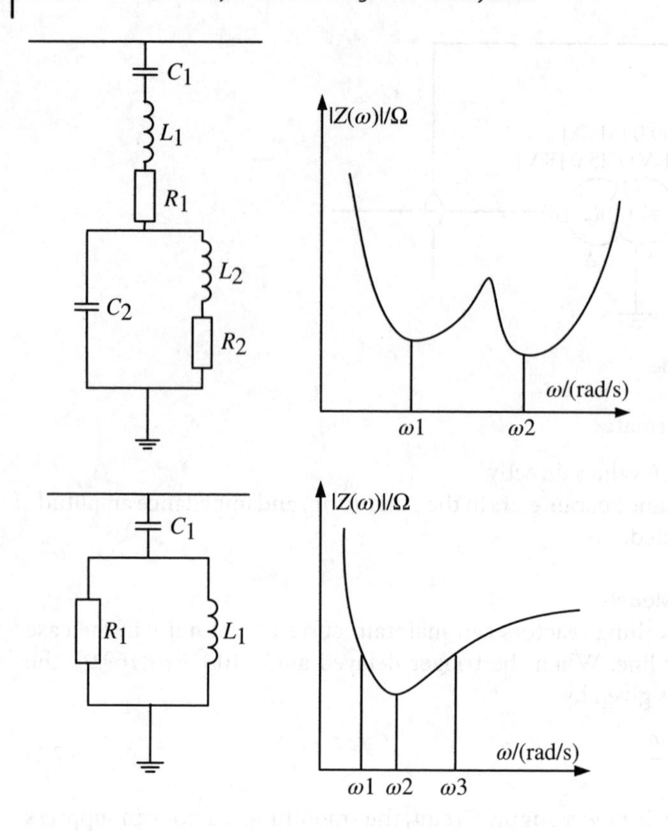

Figure 6.77 DC filter types.

6.7.2 PSCAD Simulation for UHVDC Systems

6.7.2.1 Construction of UHVDC Control Systems

The EMTDC-based controller circuit used in CIGRE dc transmission standard test system is adopted in this section. The pole control layer, as one of the hierarchical structures of the dc transmission control system, is established. As for the basic control mode of the test system, the rectifier-side control (Figure 6.79) includes the constant current control and the Mina limit, while the inverter-side control (Figure 6.80) is equipped with constant current control and constant turn-off angle control, but without constant voltage control. Additionally, both the rectifier side and the inverter side are equipped with voltage dependent current order limiters (VDCOL), and the inverter side has additional current error controllers (CEC). Under normal operating conditions, the triggering angle is 15°, and the extinction angle is 17°.

6.7.2.2 Simulated system parameter settings

AC systems – Setting are as shown in Table 6.9.

DC systems – The rectifier station and the inverter station both utilize bipolar double 12-pulse converters connected in series, and the monopole application adopts two 400 kV converters connected in series.

DC-side parameters:

DC power: 2 × 3200 MW; DC voltage: ±800 kV; DC current: 4000 A;

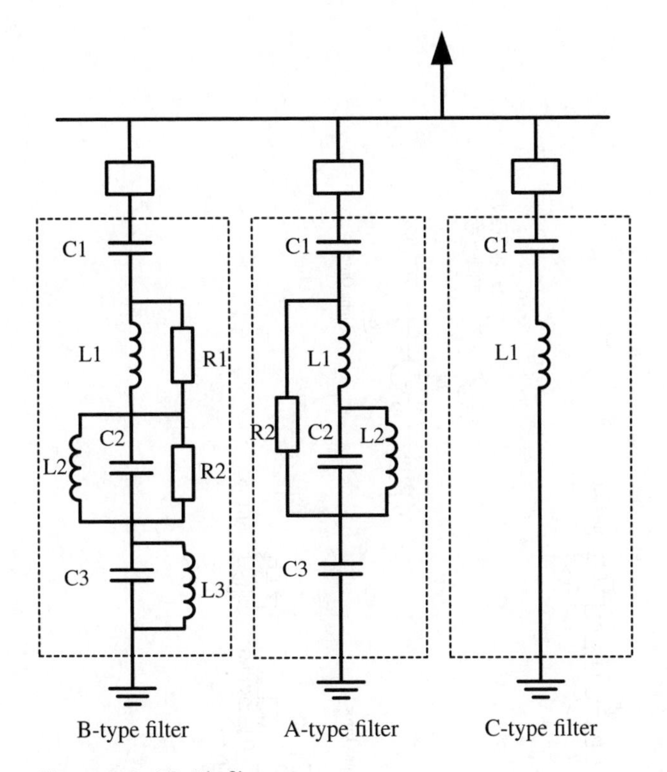

Figure 6.78 AC-side filter types.

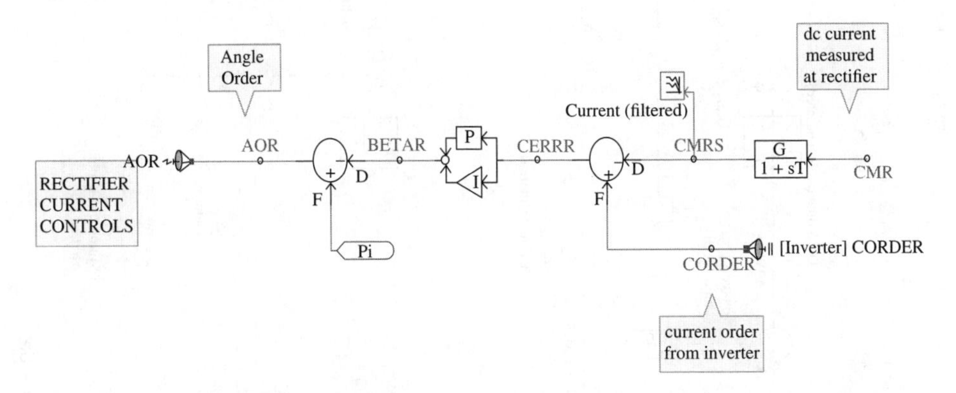

Figure 6.79 Rectifier-side control mode.

Smoothing reactor: 2 × 200 mH, installed on the dc pole line and the neutral bus;

Rectifier-side converter transformer parameters:

Rated capacity: 952.5 MW; short-circuit impedance: 18%; transformation ratio: 55/175

Inverter-side converter transformer parameters:

Rated capacity: 888MW; short-circuit impedance: 18%; transformation ratio: 525/163

Trigger angle: rated value: 15°; minimum limit angle: 5°; maximum limit angle: 20°

Extinction angle: rated value: 17°

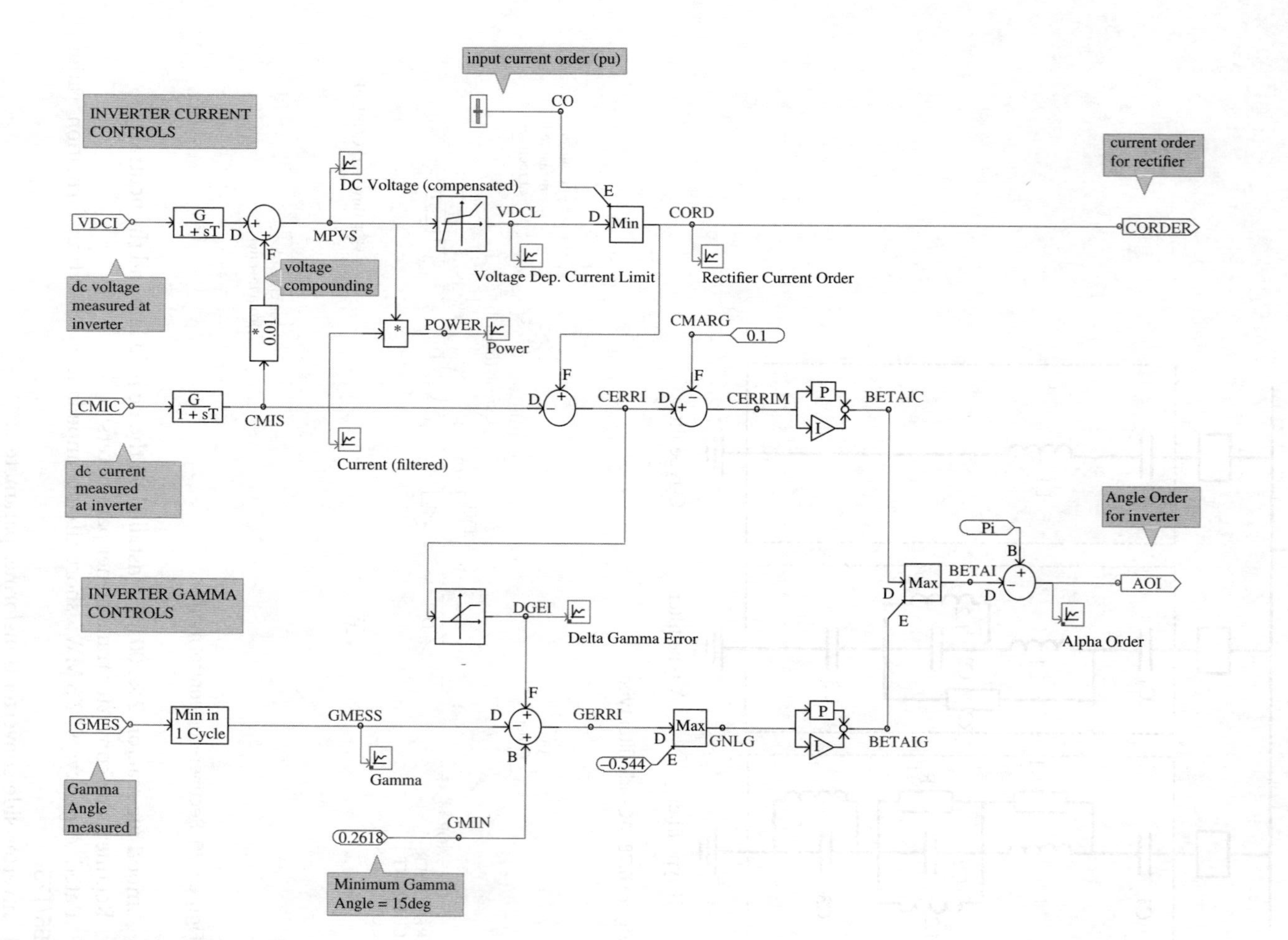

Figure 6.80 Inverter side control mode.

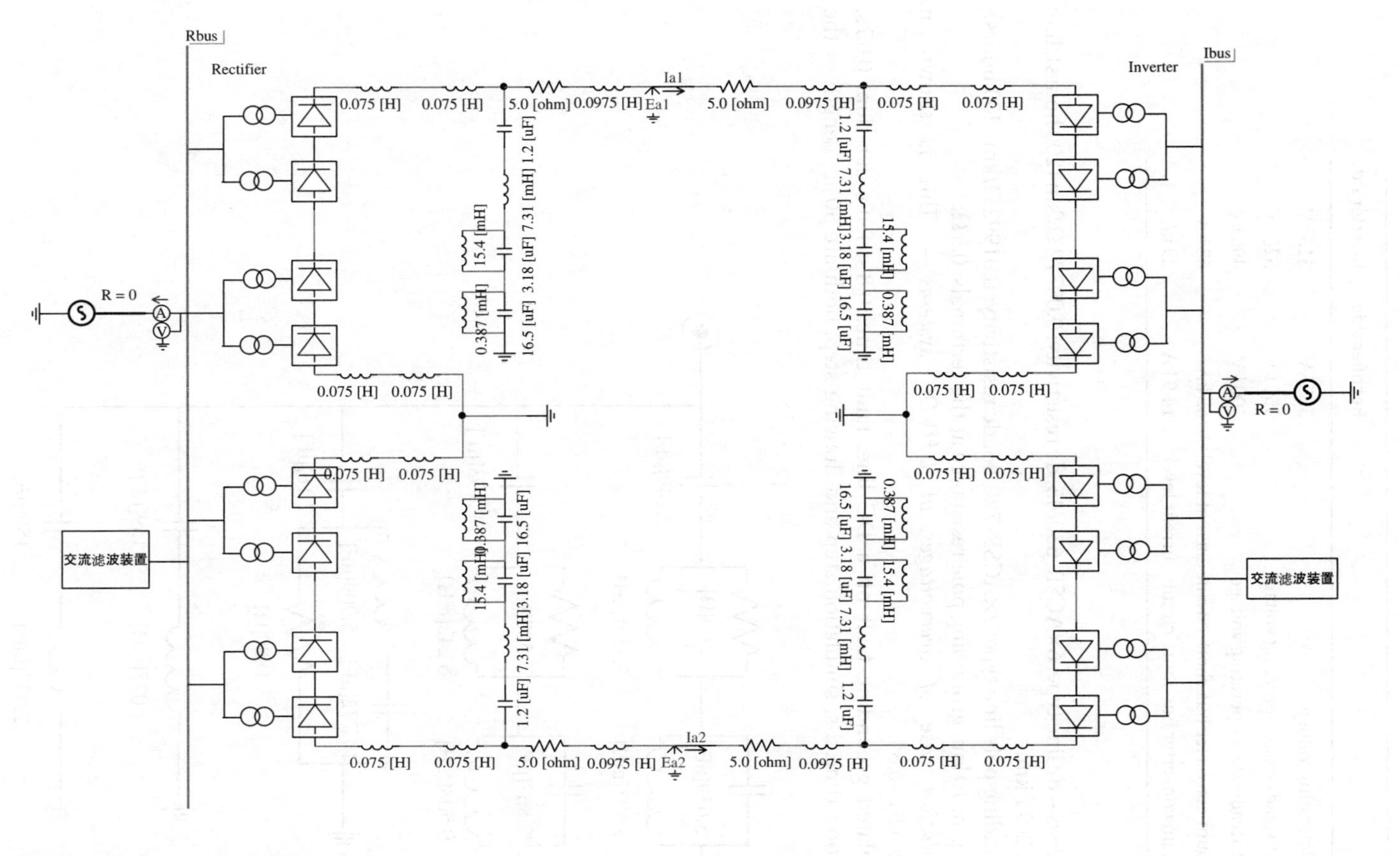

Figure 6.81 PSCAD model for UHVDC systems.

Table 6.9 AC system data in UHVDC systems.

	Rectifier side	Inverter side
Rated operating voltage	525 kV	515 kV
Highest steady-state operating voltage	550 kV	525 kV
Lowest steady-state operating voltage	500 kV	490 kV
Maximum converter bus short-circuit current level	63 kA	63 kA
Minimum converter bus short-circuit current level	14.9 kA	29 kA

DC lines – dc line type: 6×ACSR-630/45; dc resistivity: 0.00772 Ω/km (200 hours), line length: 2071 km.

Grounding pole line type: 2×ACSR-720/45; dc resistivity: 0.01992 Ω/km (200 hours); line length: 10 km; grounding pole resistance at the terminals: 0.5 Ω.

Simulation case of overvoltages in UHVDC systems – This is shown in Figures 6.81–6.84.

Simulated grounding fault on a DC line: fault time: 0.8 s; fault duration: 0.05 s. simulation time: 1 s, simulation step and drawing step: both are 50 μs. Length of the

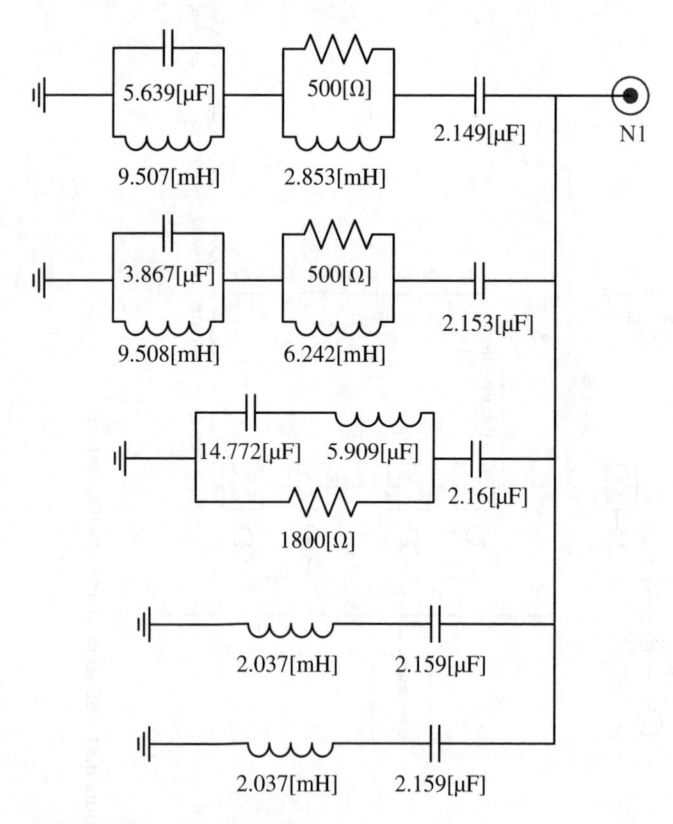

Figure 6.82 AC-side filtering equipment.

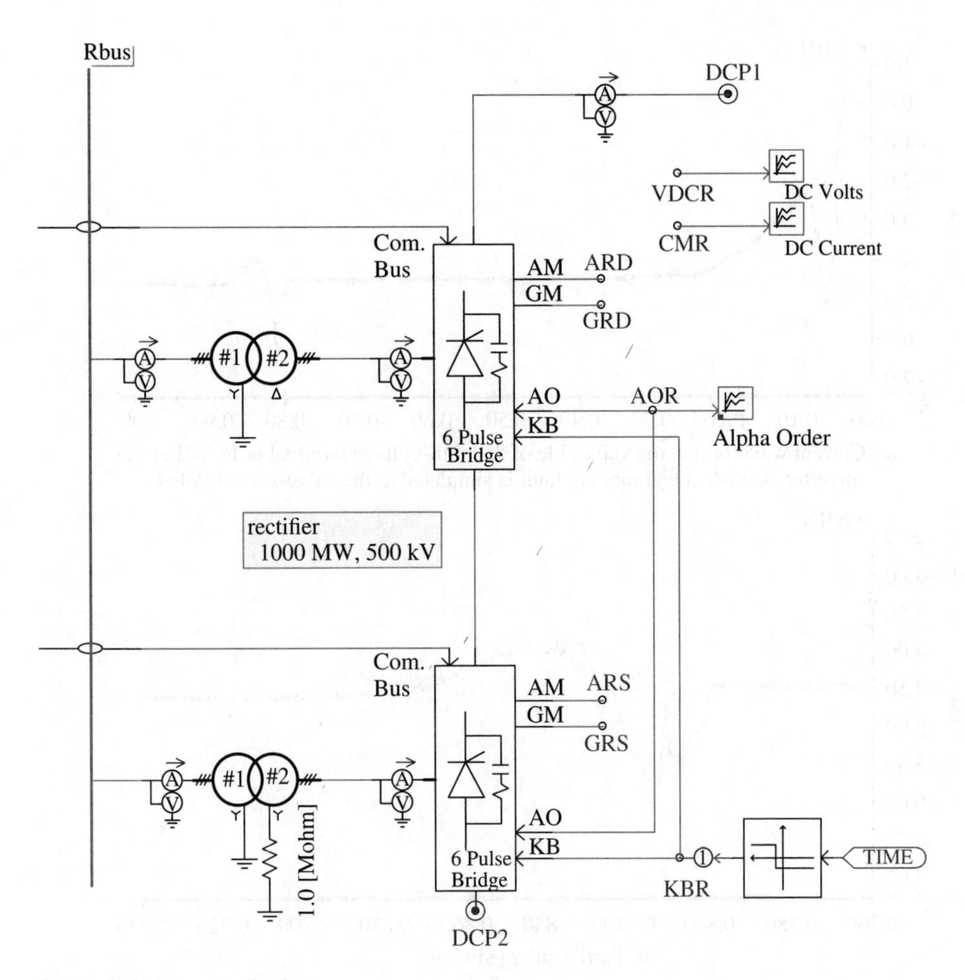

Figure 6.83 Converter bridge.

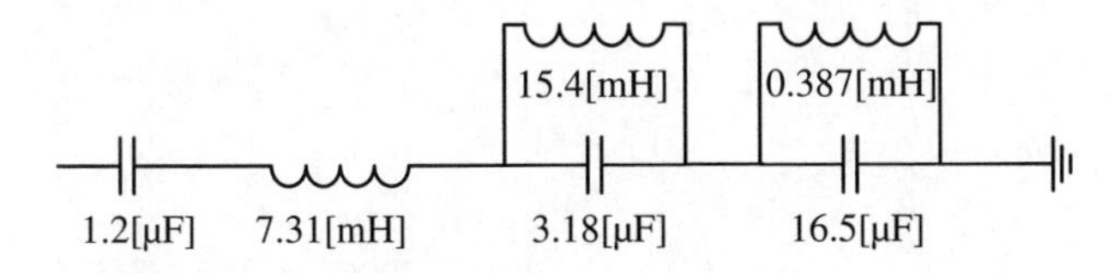

Figure 6.84 DC-side filtering equipment.

line: 2100 km. Figures 6.85 and 6.86 show the current and voltage wave shapes when simulated faults occur at different locations of the DC line.

As can be seen from Figures 6.85 and 6.86, when a grounding fault occurs at the valve side of pole I, an overcurrent of 6 kA and an overvoltage of 1100 kV appear at the high-voltage terminal of pole II at the inverter-side due to induction and coupling effects, and are followed by a continuous oscillation.

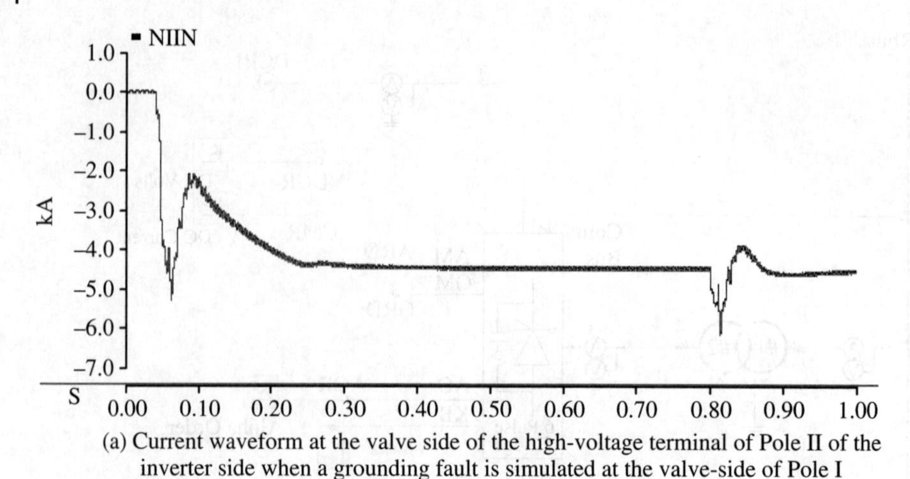

(a) Current waveform at the valve side of the high-voltage terminal of Pole II of the inverter side when a grounding fault is simulated at the valve-side of Pole I

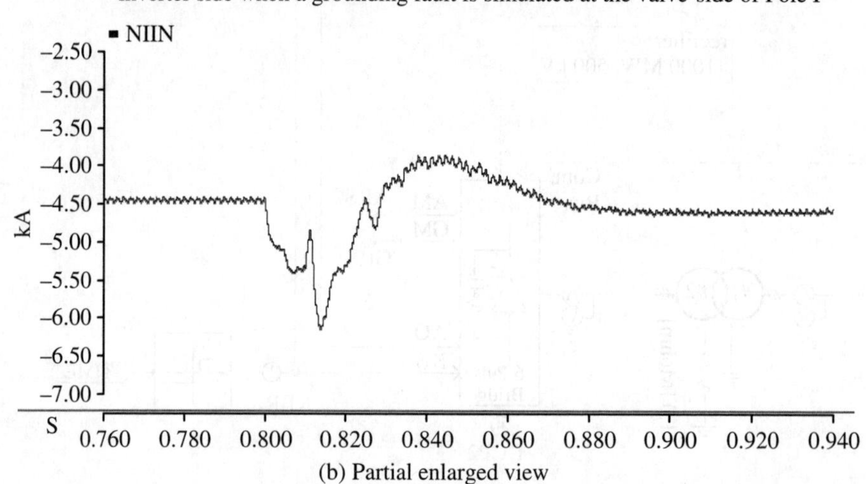

(b) Partial enlarged view

Figure 6.85 Current waveform at the high-voltage terminal of pole II at the inverter side.

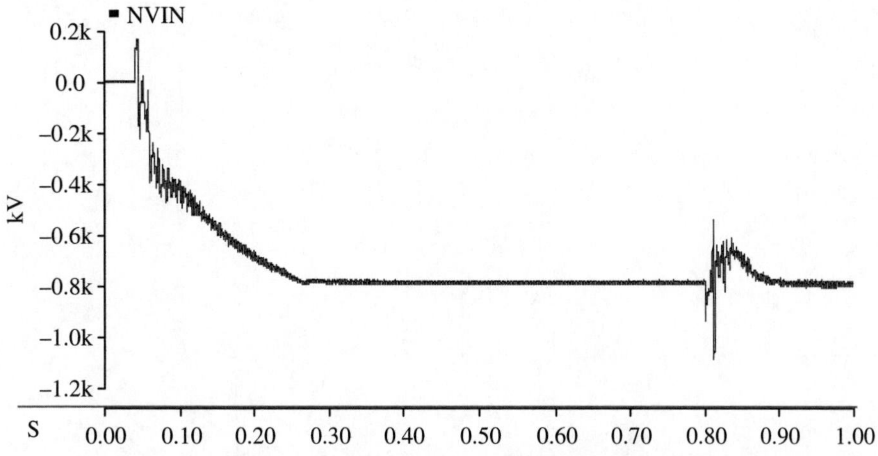

(a) Voltage waveform at the valve side of the high-voltage terminal of Pole II of the inverter side when a grounding fault is simulated at the valve-side of Pole I

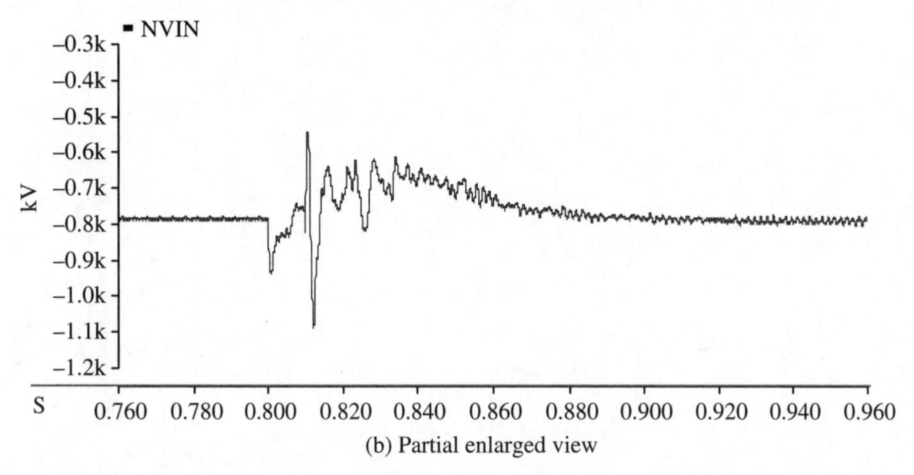

(b) Partial enlarged view

Figure 6.86 Voltage waveform at the high-voltage terminal of pole II at the inverter side.

7

Entity Dynamic Simulation of Overvoltages on Transmission Lines

7.1 Overview

The travelling wave method and EMTP are adopted internationally by most cases of back flashover calculation. A brief introduction is presented here.

Travelling wave method: each tower segment is regarded as a line segment with distributed parameters which is then turned into a lumped parameter model. Based on the nodal analytical method for circuits with lumped parameters, the voltage of each tower node can be solved, and the variation of the insulator string potential difference with time can be obtained and compared with its volt-second characteristics to judge if insulator string flashover will take place. The evaluation process reflects the process of lightning wave transmission along the tower and the influence of reflected waves on tower node potentials. This method is also known as the Bergeron method because it is based on the Bergeron mathematical model for transmission lines.

EMTP method: EMTP (electromagnetic transient program), developed by the Bonneville Power Administration (BPA), has become the most widely adopted program for studying the power system transient process.

Its technical difficulties include several issues such as:

1. tower model selection and parameter determination;
2. volt-second characteristics and flashover criteria for insulators;
3. tower impulse grounding;
4. impulse corona;
5. induced overvoltages;
6. shielding failures.

7.2 Modeling Methods for Transmission Line Lightning Channels

As shown in Figure 7.1, a 200 km 220 kV overhead transmission line model is taken as an example to study the information received from the voltage and current monitoring sites of each tower after the towers on different line locations are struck by lightning. Strike spot, lightning magnitude and rough lighting wave shape are analyzed as well.

Three adjustable positions are placed on the model which are the most convenient positions on the overhead lines for adjusting:

Measurement and Analysis of Overvoltages in Power Systems, First Edition. Jianming Li.

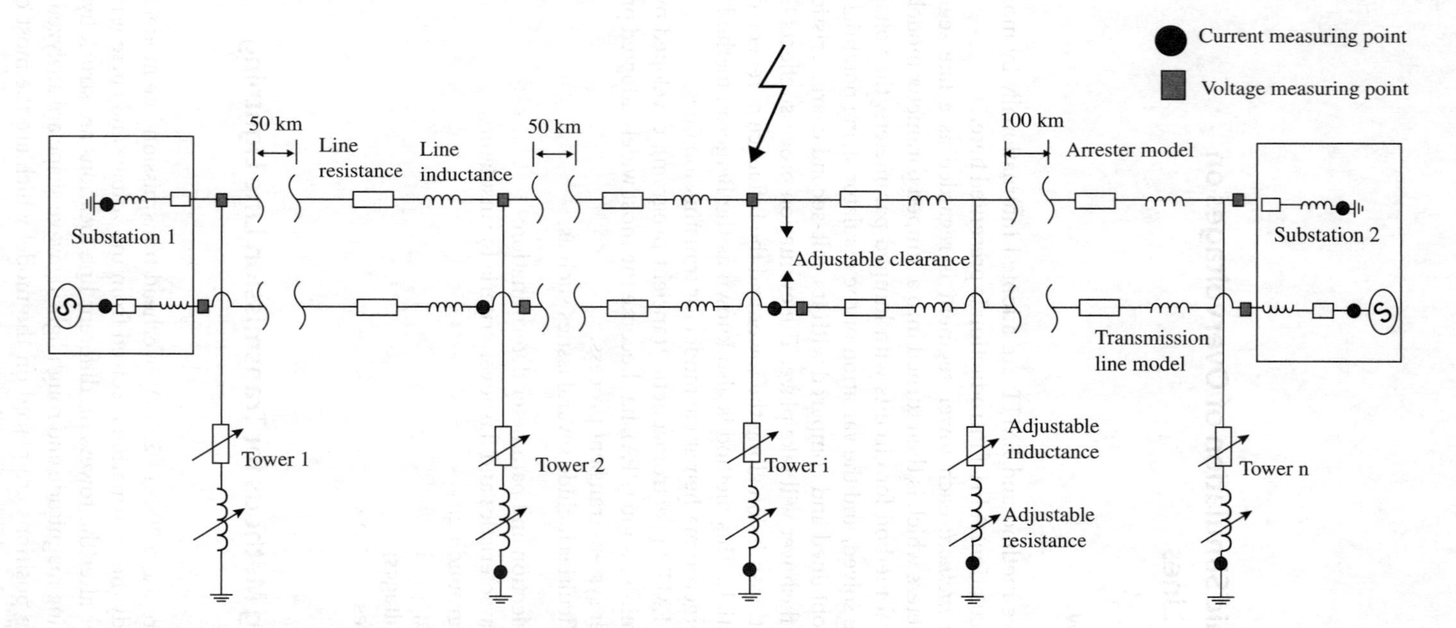

Figure 7.1 Model for lightning direct striking channel of a 200 km overhead transmission line.

1. the adjustable clearance as shown in Figure 7.1, which represents the insulator of the overhead line;
2. the adjustable resistance and inductance of the tower, which correspond to the grounding impedance of the tower ground electrode;
3. the randomness of lightning stroke current (amplitude randomness and strike position uncertainty) which is simulated by the lightning strokes with different current values and on different towers.

The waveforms obtained from 16 current and voltage monitoring sites are analyzed by changing the three variable factors above, and by combining transmission line parameters. Remote monitoring could be used to locate lightning strike spots and recognize lightning current waveforms based on the fact that lightning waves traveling towards the substation ends undergo different attenuation degrees.

7.2.1 Structure of Dynamic Simulation Testbed

Models for transmission lines and arresters – Unlike the traditional transmission line model, the physical model for ground wires is integrated, and electromagnetic coupling between the ground wire and the transmission line is taken into account. Multiple equivalent π-type circuits are used to simulate the self-impedance, mutual impedance (Figure 7.2), self-admittance and mutual admittance (Figure 7.3) regarding the ground wire and the transmission line; an instrument transformer is used to simulate the mutual impedance between the lines, and current and voltage monitoring devices are installed on the ground wire of each tower. The idea of simultaneously acquiring lightning wave data from the ground wire and the transmission line is proposed for the first time. Compared with only acquiring lightning wave data from transmission lines, comprehensive two-channel analysis is able to eliminate interference and visually recognize lightning failures (back flashover and shielding failures).

Z_{kk} is the self-impedance of each line, while the rest refer to the mutual impedances between the lines. Y_{CC} is the self-admittance at each line terminal while the rest are the mutual admittances between the lines.

Figure 7.4 shows the structure diagram of the circuit model for the self-impedances of the transmission line and the ground wire and the mutual impedances between ground wire #1 and the other lines.

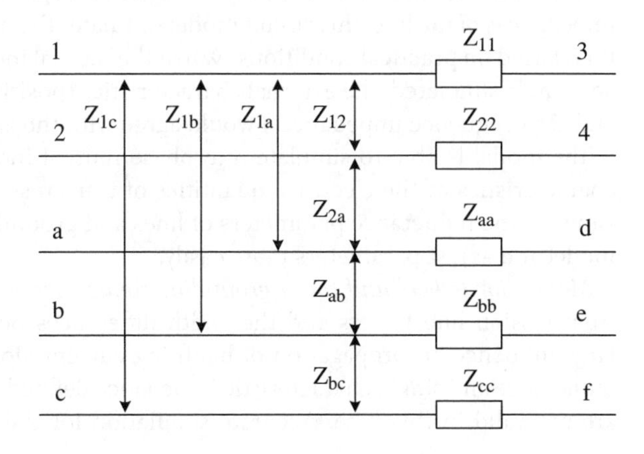

Figure 7.2 Self-impedance of each line and the mutual impedance between each line and ground wire #1.

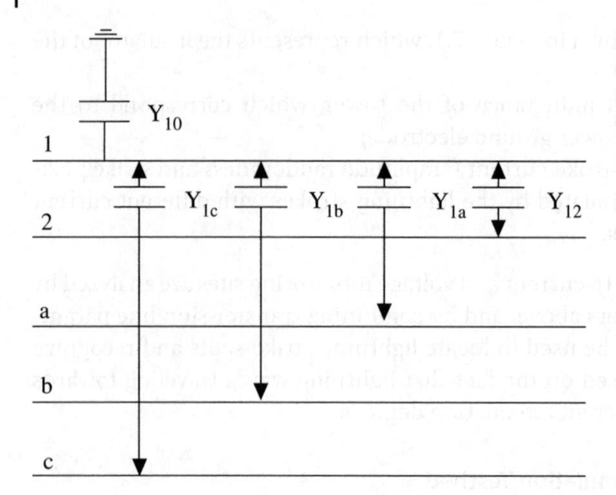

Figure 7.3 Wire-to-ground admittance of ground wire #1 and the mutual admittance between each line and ground wire #1.

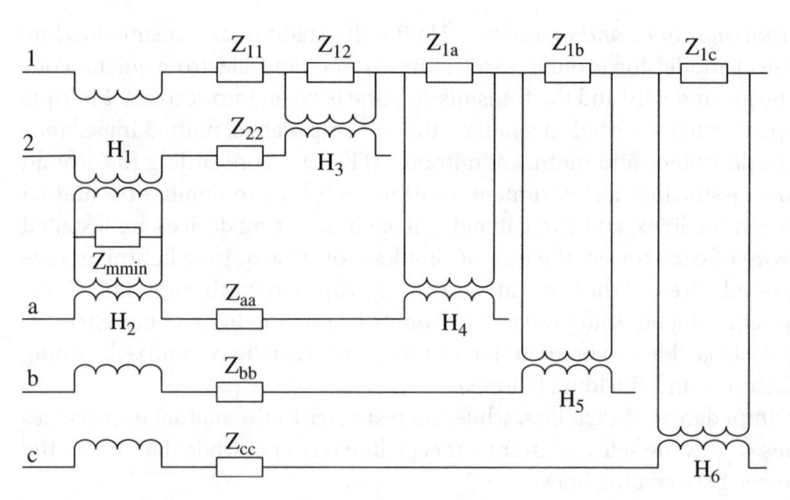

Figure 7.4 Circuit model for self-impedances and mutual impedances.

Instead of starting with the positive-sequence, negative-sequence and zero-sequence impedances of the line, this circuit model simulates the mutual inductances between the lines based on practical conditions. When the mutual inductances between the lines are accurately simulated, the external characteristics (positive sequence, negative sequence and zero-sequence impedances) would agree with the actual line.

The model is able to simulate interphase mutual inductances, and fully reflect the characteristics of the electrical quantities of transmission lines. Impedors are used to simulate the inductance parameters of lines and ground wires in order to establish the model and adjust parameters more easily.

Models for towers and tower grounding connectors – The large height of EHV/UHV transmission line towers and the width differences between tower segments have a large influence on propagation of lightning current along the tower, thus the lumped inductance and single characteristic impedance defined in the conventional regulations are not valid in this case. Accurate simulation for the lightning current propagation process depends on the simulation precision of tower characteristic impedances, and the multi-wave impedance model in parallel and non-parallel multi-conductor systems (Figures 7.5–7.7) are used.

$$Z = \frac{1}{2\pi}\sqrt{\frac{\mu}{\varepsilon}}\ln\left(\frac{r}{\sqrt{r^2+h^2-h}}\right) \tag{7.2.1}$$

where μ is the permeability of the tower material; ε is the permittivity of air; r is the radius of the cylinder; h is the height of the cylinder.

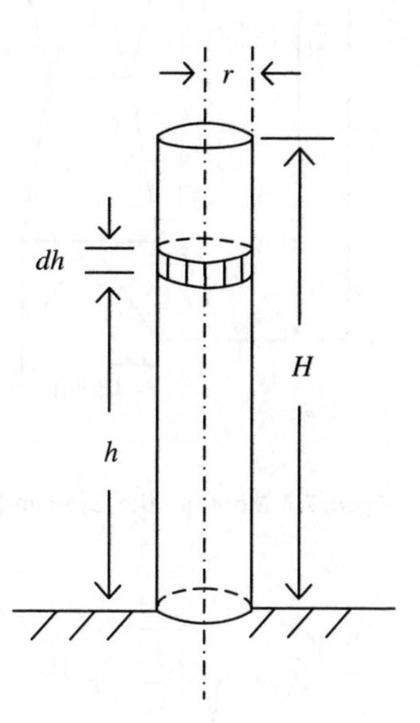

Figure 7.5 Vertical cylinder model for the tower.

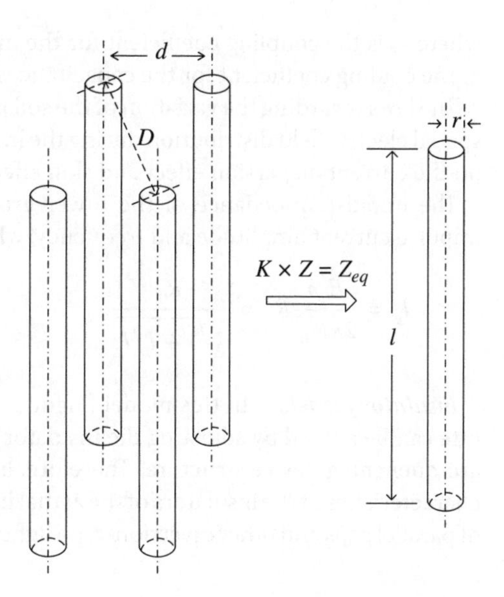

Figure 7.6 Parallel multi-conductor model for the tower.

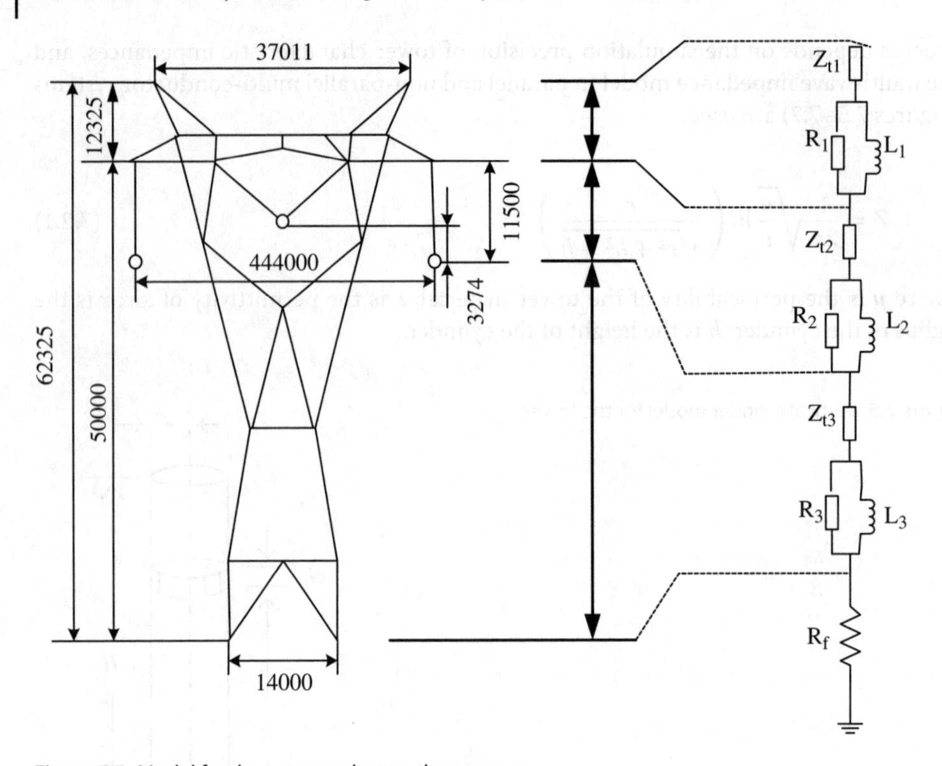

Figure 7.7 Model for the tower and ground connector.

$$Z_{eq} = \sqrt{\frac{k_l L}{k_c C}} = KZ \tag{7.2.2}$$

where k_l is the coupling coefficient for the inductances of the parallel conductors, and k_c the couling coefficient for the capacitances of the parallel conductors.

The laws regarding the variation of the soil parameter time-varying characteristic with spatial electric field distribution during the impulse current dispersion is analyzed based on the current dispersion effect and skin effect of the tower grounding connector.

The impulse impedance of the tower grounding connector may be influenced by impulse current amplitude and frequency, which is obviously nonlinear.

$$I_g = \frac{E_0 \rho}{2\pi R_0^{\ 2}} R_T = \frac{R_0}{\sqrt{1 + I/I_g}} \tag{7.2.3}$$

Insulator models – In this model (Figure 7.8), the flashover voltage and the arc over rate can be altered by adjusting the insulator string length, the parallel gap size and the arc-quenching device structure. Therefore, lightning trip-out rate can be analyzed, the characteristics of the insulators of the actual lines can be simulated and the configuration of parallel gaps and other diversion-type lightning protection devices can be worked out.

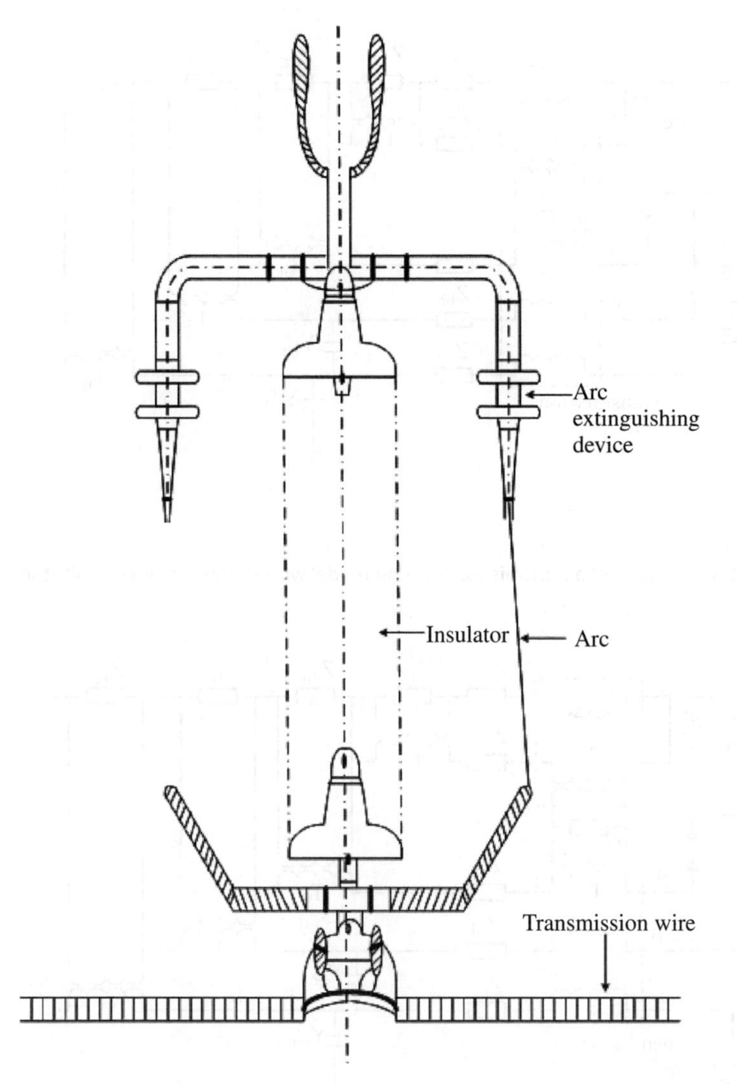

Figure 7.8 Insulator model.

7.2.2 Structural Diagram

The circuit diagram of a transmission line unit model is illustrated in Figure 7.9 when lightning hits the tower top directly.

The circuit structural diagram of a transmission line unit model is shown in Figure 7.10 when a shielding failure occurs on a single-phase conductor.

In Figure 7.10, T_1 – T_6 are current transformers with a transformation ratio of 1:1, and T_1 has three windings on the iron core, T_2 has four windings on the iron core. The iron cores are made of Mn-Zn ferrite with a maximum operating frequency of 3 MHz; R_f is the impulse resistance of the tower grounding connector.

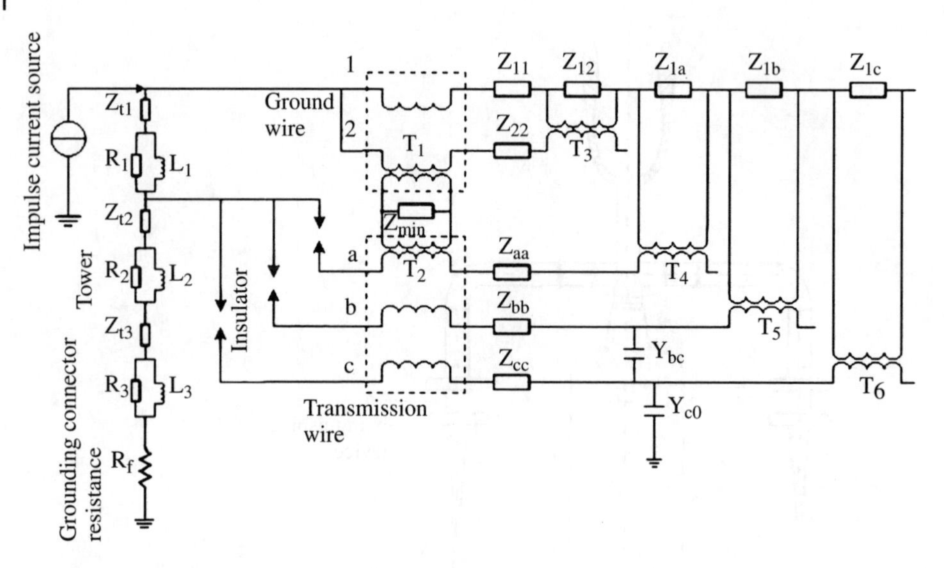

Figure 7.9 Circuit structural diagram of a transmission line unit model when lightning directly hits the tower top.

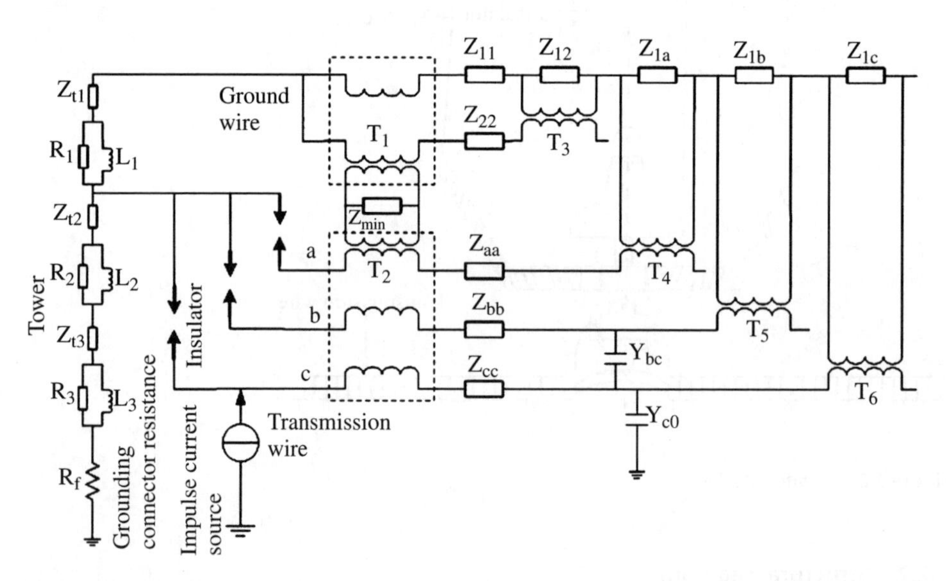

Figure 7.10 Circuit structural diagram of a transmission line unit model when shielding failure occurs.

7.2.3 Definitions of Parameters

Transmission line parameters:

$$Z_{ii} = (R_{ii} + \Delta R_{ii}) + j\left(2\omega \times 10^{-4} \ln \frac{2h_i}{GMR_i} + \Delta x_{ii}\right) \tag{7.2.4}$$

$$Z_{m\,\min} = \min(Z_{ik}) = \min(Z_{ki}) = \min\left[\Delta R_{ik} + j\left(2\omega \times 10^{-4} \ln \frac{D_{ik}}{d_{ik}} + \Delta X_{ik}\right)\right] \tag{7.2.5}$$

$$Z_{ik} = Z_{ki} = \Delta R_{ik} + j\left(2\omega \times 10^{-4} \ln \frac{D_{ik}}{d_{ik}} + \Delta X_{ik}\right) - Z_{m\min} \tag{7.2.6}$$

$$Y_{i0} = j\omega = \frac{1}{18 \times 10^6 \ln \frac{2h_i}{r_i}} \tag{7.2.7}$$

$$Y_{ik} = j\omega = \frac{1}{18 \times 10^6} \ln \frac{D_{ik}}{d_{ik}} \tag{7.2.8}$$

where j – symbol for the imaginary part of the complex;
r_i – radius of line i; (i is a, b, c, 1, 2);
R_{ii} – AC resistance of line i; (i is a, b, c, 1, 2);
h_i – average suspension height above ground of line i; (i is a, b, c, 1, 2);
D_{ik} – distance between line i and the mirror image of line k; (i, k are a, b, c, 1, 2);
d_{ik} – distance between line i and line k; (i, k are a, b, c, 1, 2);
GMR_i – average geometric distance of line i; (i is a, b, c, 1, 2);
$\omega = 2\pi f$ – angular frequency at frequency f, with the unit of rad/s;
ΔR_{ii}, ΔR_{ik}, ΔX_{ii}, ΔX_{ik} – correction items considering the effect of the ground. (i, k are a, b, c, 1, 2);
Z_{ii} – the self-impedance of line i; (i is a, b, c, 1, 2);
$Z_{m\min}$ – the minimum value of the mutual impedances between lines;
Z_{ik}, Z_{ki} – the mutual impedance between line i and line k minus $Z_{m\min}$; (i, k are a, b, c, 1, 2);
Y_{i0} – the admittance to ground of line i; (i is a, b, c, 1, 2);
Y_{ik} – the mutual admittance between line i and line k. (i, k are a, b, c, 1, 2).

Tower parameters:

$$Z_{ti} = 60\left[\ln \frac{2\sqrt{2}H_i}{2^{1/8}(r_{ti}{}^{1/3} r_B{}^{2/3})^{1/4}(R_{ti}{}^{1/3} R_B{}^{2/3})^{3/4}} - 2\right] r_{ti} R_{ti} \tag{7.2.9}$$

$$R_i = -2Z_{ti}[H_i/(H_1 + H_2 + H_3)] \ln \sqrt{\gamma} \tag{7.2.10}$$

$$L_i = \alpha R_i 2H_i / v_t \tag{7.2.11}$$

H_i – height of each tower segment (i is 1, 2, 3);
R_{ti} – radius of the tower main support (i is 1, 2, 3);
r_{ti} – radius of the tower support (i is 1, 2, 3);
Z_{ti} – characteristic impedance of each tower segment (i is 1, 2, 3);
r_B, R_B – radii of the upper and lower tower bases;
R_i – damping resistance of each tower segment (i is 1, 2, 3);
L_i – damping inductance of each tower segment (i is 1, 2, 3);
α – damping coefficient;
v_t – light velocity;
γ – attenuation coefficient.

The following purposes can be eventually achieved:

1. It is practicable to strengthen the insulation of an insulator and reduce back flashover probability by monitoring the lightning strike probability at each node on the line and by optimizing grounding at spots with higher lightning strike rate.
2. It is possible to study the original lightning waveforms at the lightning strike spot by using the monitored lightning waveforms through parameter inversion.
3. Lightning overvoltages at the substation entrance can be acquired, and arrester arrangement and equipment insulation coordination can be implemented in a reasonable manner.

7.3 Verification of Simulated Transmission Line Lightning Channels

Lightning (of 130 kA) directly striking the tower: In the case of a 220 kV transmission line, the tower 60 km away from substation 1 (Figure 7.11) is directly hit by the lightning stroke with a current of 130 kA. The current and voltage waveforms of the transmission line and the ground wire at the lightning strike spot and the substation entrances are presented in Figures 7.12–7.15.

Back-flashover from the tower to the transmission line takes place with the waveforms shown in Figure 7.12. The peak voltage at the back-flashover spot on the line is 2.6MV. When the wave reaches the terminals of substation along the line, obvious attenuation is shown. Since substation terminal 2 is farther from the strike spot, the attenuation is more significant. As shown in Figure 7.13, lightning voltage signals propagating along the ground wire attenuate quickly because of the shunt effect of the tower against the ground.

Figure 7.16 shows the voltage waves at each line point when applying lightning currents of 120kA, 130kA and 140kA on the top of the tower 60 km away from substation terminal 1.

Figure 7.16 shows that for transmission lines, back flashover does not occur when the lightning current amplitude is 120 kA, but occurs when the lightning current amplitude exceeds 130 kA.

The lightning strike spots are placed 60 km and 100 km away from substation terminal 1. The waveforms of lightning current at substation terminal 1 are shown in Figure 7.17.

When the grounding resistance of the struck tower (60 km away from substation terminal 1) is changed from 10 Ω to 5 Ω, the waveforms before and after the change are shown in Figure 7.18.

Figure 7.18 shows that the potential at the tower top is not sufficiently high to cause back flashover when the grounding resistance of the grounding connector is 5 Ω. Therefore, the amplitude of the lightning wave is quite small when it reaches substation terminal 1.

Similarly, the lightning waveforms at the terminals of the substation can be measured under different situations by adjusting the adjustable gap in the model (corresponding to the breakdown voltage of the insulator model).

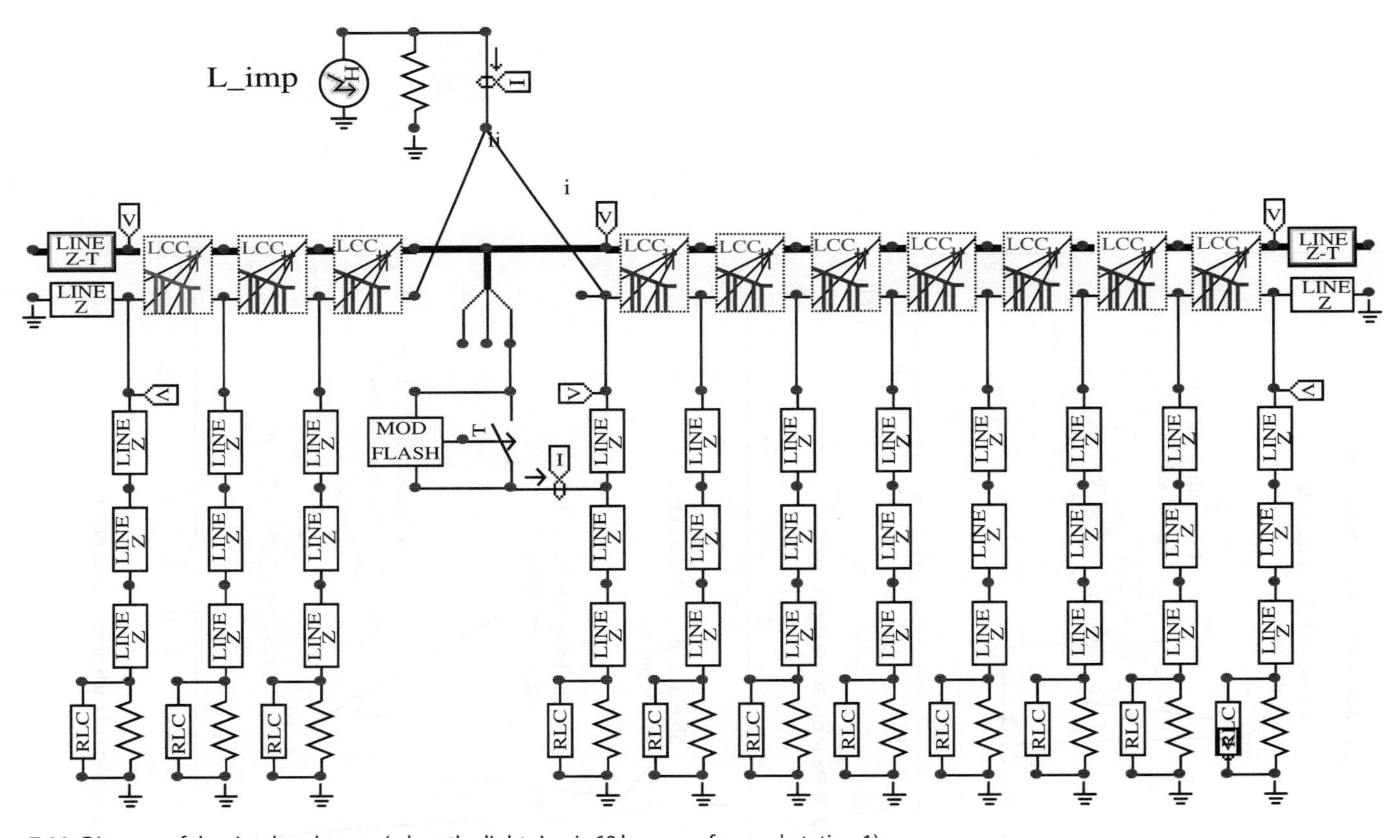

Figure 7.11 Diagram of the simulated tower (when the lightning is 60 km away from substation 1).

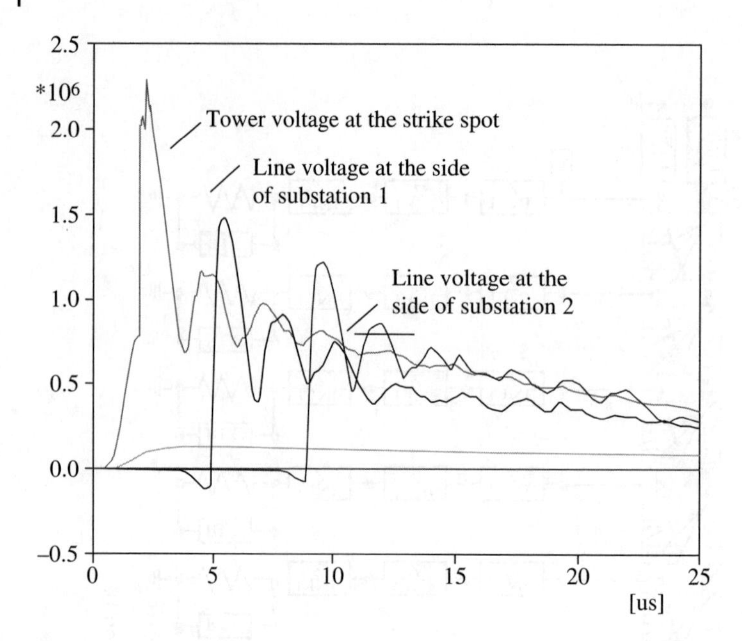

Figure 7.12 Voltage of each point on the line.

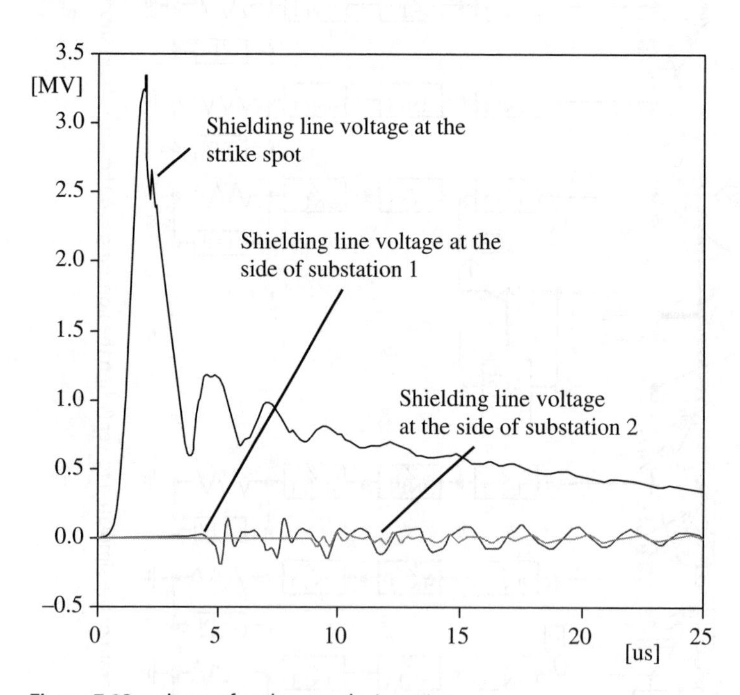

Figure 7.13 voltage of each ground wire point.

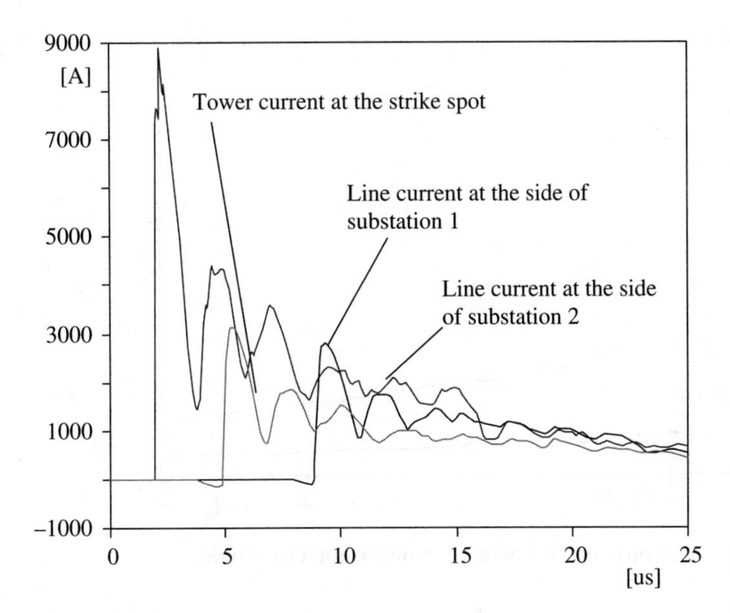

Figure 7.14 Currents of transmission line branches.

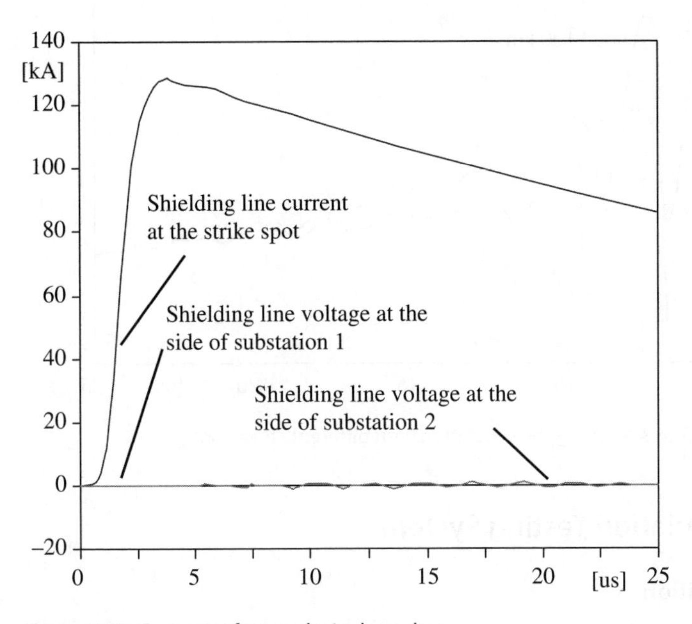

Figure 7.15 Currents of ground wire branches.

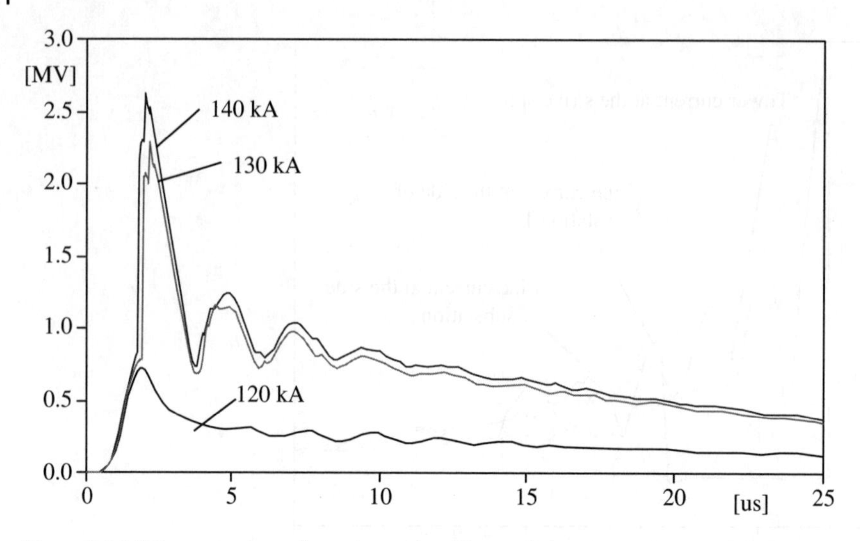

Figure 7.16 Voltages at the strike point under different lightning current amplitudes.

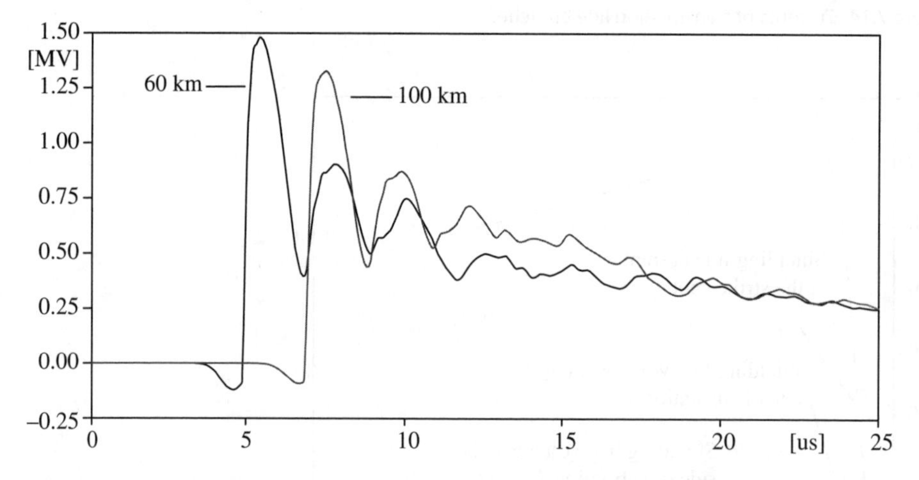

Figure 7.17 Lightning waveforms reaching the substation with different strike spots.

7.4 Dynamic Simulation Testing System

7.4.1 System Composition

The dynamic simulation system for lightning electromagnetic transients on transmission lines consists of three parts: (a) SIN, the dynamic simulation testbed for lightning electromagnetic transients on transmission lines and for signal sampling and conditioning; (b) EPC, the embedded measurement system and the signal acquisition and processing section that include a 19-inch LCD screen; (c) PCU, the simulation center that establishes the data management system and can enable remote data exchange. The schematic of the structure is shown in Figure 7.19.

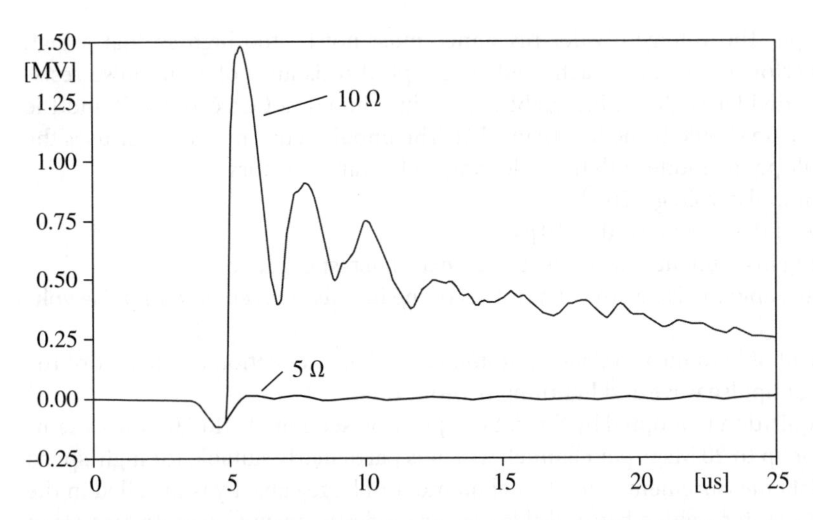

Figure 7.18 Lightning waveform reaching substations with smaller grounding resistance.

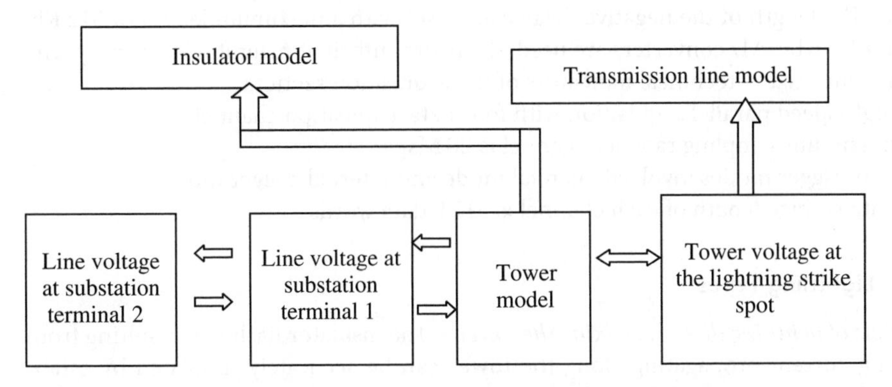

Figure 7.19 Dynamic simulation system for lightning electromagnetic transients on transmission lines.

7.4.2 Main Technical Indicators

1. This system adopts X86 as the embedded hardware platform and Windows as the software platform.
2. EPC adopts an embedded platform featuring a 1 GHz CPU and a 1 Gb memory; a 19-inch display screen is also used.
3. PCU adopts a mainstream industrial control board featuring a 2 GHz CPU and a 2 Gb memory; a 19-inch display screen is also used.
4. Modular design is adopted, which is easy to replace and upgrade.
5. The performances of current dividers and voltage dividers are the key to ensuring the reliability of the data acquisition system, thus careful design and testing is a must. The stray inductances of the current divider and voltage divider are required to be sufficiently small, and non-inductive resistors should preferably be used. Otherwise, wavefront glitch or overshoot may be caused. The current dividers need to use low value resistors with very low inductance, while the voltage dividers should be the

resistive type. The voltage divider resistance must not be too high so that a high response performance can be achieved. The typical resistance of an impulse resistive voltage divider is 10 kΩ. Preferably, the value should not exceed 20 kΩ and the minimum value should be no less than 2 kΩ. The impulse current generator uses the voltage multiplying mode with the following technical indicators:

- highest impulse voltage: 2000 V
- rise time of the discharge: 0.5–10 μs
- measuring instrument can connect to an external test capacitor
- test results able to display the waveforms of the impulse current and impulse voltage
- instrument able to measure such parameters as the attenuation coefficient of the lightning impulse wave and lightning trip-out rate.

6. Parallel acquisition is adopted by the data acquisition section of HR6100 with a sampling rate of up to 20 Msps per channel, which is particularly suitable for high-speed dynamic data measurement. An internal memory of large capacity is installed in the HR6100 test system which has multiple trigger modes to ensure accurate capture of eligible data, especially the status data around the occurrence time of impulse current. The length of the negative delay can be set with a maximum length of 512 Kb. The 12–16 bit AD converters are used, to ensure sufficient magnitude accuracy. The following are the technical indicators of the acquisition section:
   - high-speed parallel acquisition with four data acquisition channels
   - maximum sampling rate per channel is 20 Msps
   - two trigger modes involved: manual mode and internal trigger mode
   - data storage length of each channel is 512 k data words.

7.4.3 Lightning Types

Channels of lightning directly striking the tower – The insulator flashover resulting from lightning current propagating along the tower can be accurately simulated by establishing the analytical model for electromagnetic transients during back flashover on the transmission line. The main factors that would influence back flashover include ground wire shunting, tower height, grounding resistance and conductor operating voltage (Figure 7.20).

Lightning channel in the case of shielding failures – The process of insulator flashover resulting from lightning current propagating along the tower can be accurately simulated by establishing the analytical model for electromagnetic transients in the case of shielding failures. The main factors influencing shielding failures include protection angle, terrain, conductor operating voltage and tower height (Figure 7.21).

7.4.4 Lightning Signal Acquisition of Dynamic Simulation Testbeds

Installation site of current transducer models – Lightning strike spots can be identified by installing lightning current transducers on ground wire supports and insulator string branches. When the shielding failure occurs on the line, the lightning current magnitude detected by the transducer installed on the insulator string branch would be far greater than those detected by the transducer installed on the ground wire support; when back flashover occurs, waveforms are recorded not only by the flashover phase of the insulator

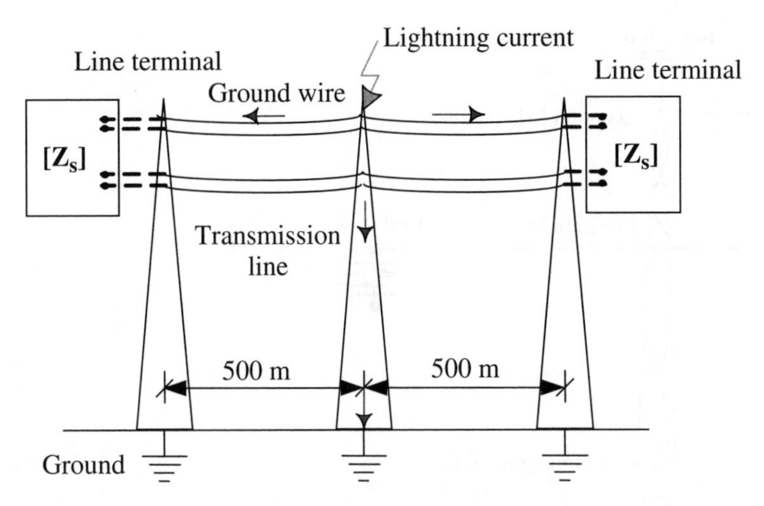

Figure 7.20 Schematic diagram of the channel of lightning directly striking the tower.

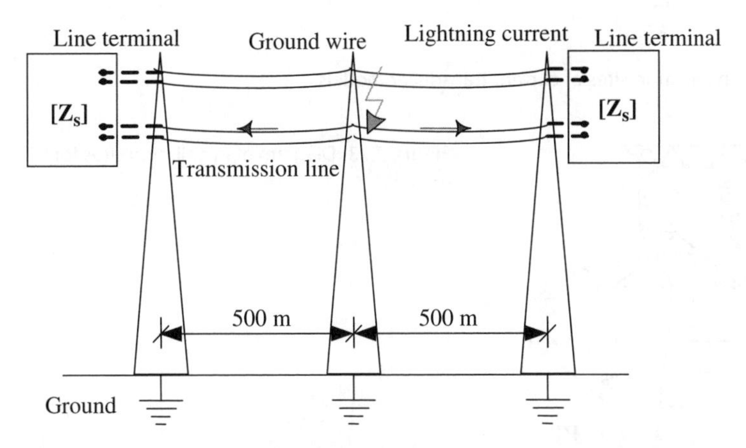

Figure 7.21 Schematic of the lightning channel in the case of shielding failures.

string but also by the transducers installed on the support of the ground wire of the tower (Figure 7.22).

Installation sites for PT models – When lightning accidents take place, the waveforms of the voltages monitored along the ground wire or at the transmission line tower can be used to identify the waveforms of the lightning overvoltage at fault sites by adopting the time difference location method and the inversion of the characteristics of the lightning channel attenuation (Figure 7.23).

7.4.5 Significance of Modeling

There will always be some degree of deviation between the simulation result and the practical situation, as the simulation cannot take all the factors into account. Lightning location and estimation of magnitudes and waveforms can be achieved by first building a miniature physical model for the selected line and then by differential processing of the signals measured at the substation terminals.

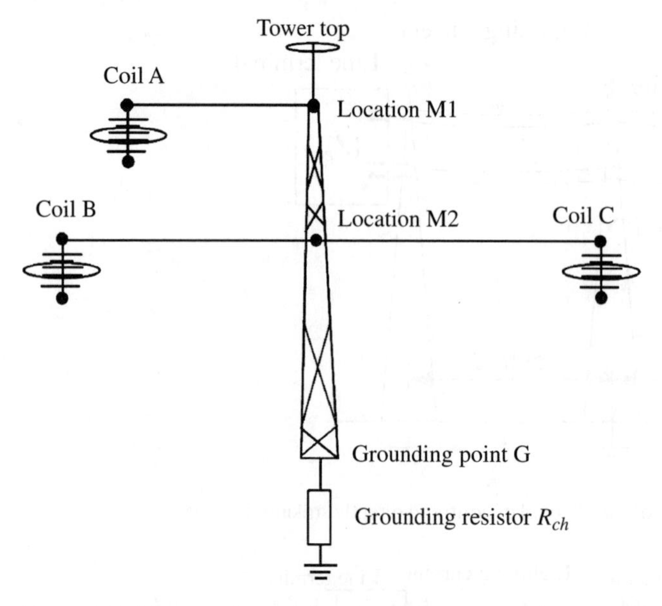

Figure 7.22 Diagram of installation sites of current transducer models.

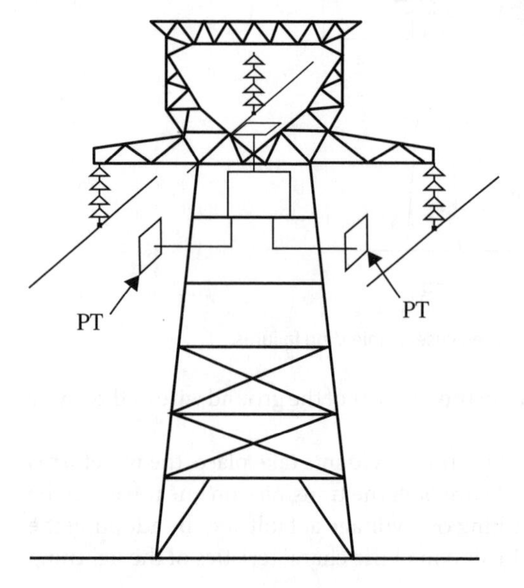

Figure 7.23 Diagram of installation sites for PT models.

The results presented by the study help to optimize transmission lines on the whole and upgrade the grounding connectors with higher lightning striking probability. Therefore, it is of great interest for secured operation in power grids and for low engineering cost.

8

Overvoltage Pattern Recognition in Power Systems

Online overvoltage monitoring devices have been widely applied in power grids of different voltage ratings. These devices are capable of recording overvoltage generation, waveforms and other parameters in a complete and accurate manner; however, analysis and identification functions are not provided, thus online overvoltage monitoring devices are impossible to identify overvoltage types and analyze accident causes. There is abundant information concerning power system operation status in overvoltage signals. Characteristic quantity extraction and recognition algorithms based on monitored overvoltage signals can achieve automatic identification and diagnosis, thereby not only providing scientific proof for quick overvoltage response suppression, but also helping to diagnose existing system weaknesses for improved insulation coordination. In a word, automatic overvoltage pattern recognition is important for safe operation of the power grid.

8.1 Selection of Characteristic Values

There are a wide variety of overvoltages in the power system that may be classified into internal and external overvoltages and further sub-classified into about 10 types of overvoltages according to the existing general classification rule. Overvoltage identification, based on overvoltage data acquisition, has become a world challenge. However, it is important to resolve this overvoltage identification issue so as to better understand system insulation weaknesses and to accumulate experience on system design and equipment improvement.

Currently, overvoltage waveform identification is mainly dependent on experts' experience, thus significantly increasing the time and labor cost. It is essential to establish a system that can recognize overvoltages intelligently. Tree-shaped hierarchical recognition adopted in this book not only improves recognition speed and accuracy but also better reflects overvoltage affiliations compared with the existing recognition approach based on unified characteristics. In order to make the recognition system better targeted and more efficient, prediction and estimation based on overvoltage mechanism (see Chapter 2) are performed on probability and damage of different overvoltages in this book. Linear resonance overvoltages and overvoltages caused by load rejection are not considered because they rarely occur because of suitable parameter coordination during the design phase or because they have little influence on normal system operation. Moreover, for lightning overvoltages frequently found on transmission lines, multiple

Measurement and Analysis of Overvoltages in Power Systems, First Edition. Jianming Li.

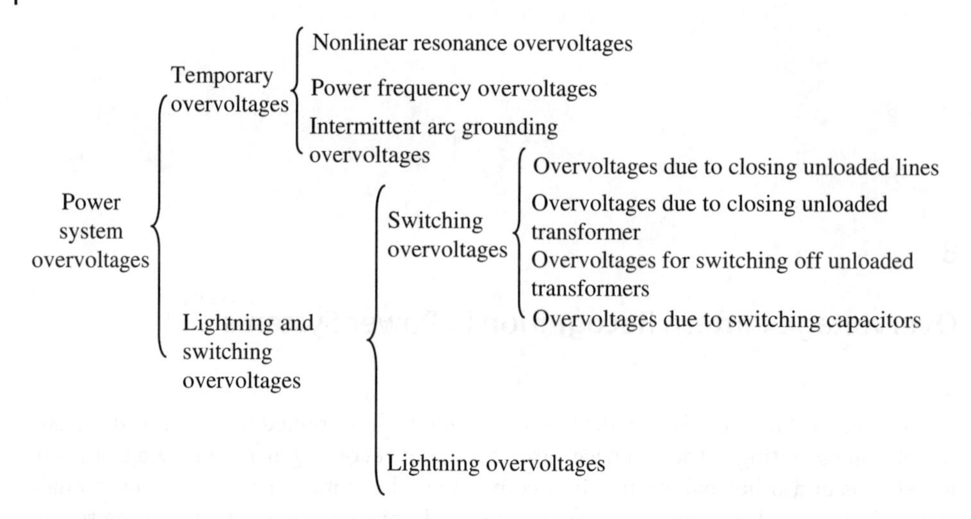

Figure 8.1 Overvoltage classification adopted in this book.

refraction and reflection during the long-distance propagation will dramatically distort their waveforms so that such waves reaching substations are hard to recognize. For this reason, lightning overvoltages are no longer subdivided into lightning bypassing overvoltages and back flashover overvoltages. In view of these principles, the tree-shaped classification of the recognition system established in this book is shown in Figure 8.1.

As can be seen from Figure 8.1, the overvoltage classification adopted in this book is different from the conventional idea of dividing overvoltages into internal and external overvoltages. This classification, which can recognize overvoltages more quickly and accurately, considers overvoltage magnitude, frequency, duration and other basic characteristics. For example, both switching overvoltages and temporary overvoltages belong to internal overvoltages; however, the switching overvoltage is separated from the temporary overvoltage due to its higher frequency and shorter duration. Also, the intermittent arc grounding overvoltage is essentially considered as a kind of switching overvoltages, but it is classified into temporary overvoltages as its duration is much longer than that of other kinds of switching overvoltages, and almost as long as that of temporary overvoltages. Selection of other characteristics will be detailed in the following content.

8.2 Time-Domain Characteristic Extraction

Time-domain characteristics are also called the statistical characteristics which mainly reflects the variation of signal amplitudes over time. Relevant mathematical methods can be adopted to extract characteristic quantities reflecting wave properties directly from acquired time-domain waveforms. The time-domain characteristic quantities may be described as dimensional or dimensionless according to different mathematical expressions.

The dimensional time-domain characteristic values include the average value, the root-mean-square value and the peak value.

Average value:

$$\overline{U} = \int_{-\infty}^{+\infty} up(u)du = \int_{0}^{T} udu = \frac{1}{N}\sum_{i=1}^{N} u_i \tag{8.2.1}$$

where $\breve{U}$ refers to the intermediate voltage amplitude over a certain period, which reflects the central tendency for overvoltage signal variation and is also deemed as the dc component in overvoltage oscillating waveforms.

Peak value:

$$U_p = \max(abs(u)) \tag{8.2.2}$$

The peak value U_P refers to the maximum voltage signal value.

Root-mean-square value:

$$U_{rms} = \sqrt{\int_{-\infty}^{+\infty} u^2 p(u)du} = \sqrt{\frac{1}{T}\int_{0}^{T} u^2 dt} = \sqrt{\frac{1}{N}\sum_{i=1}^{N} u_i^2} \tag{8.2.3}$$

The root-mean-square value U_{rms} is the signal effective value and may reflect the energy of the voltage over a certain time period.

Root-mean-square amplitude:

$$U_r = \left[\int_{-\infty}^{+\infty} \sqrt{|u|}p(u)du\right]^2 = \left[\frac{1}{T}\int_{0}^{T} \sqrt{|u|}dt\right]^2 = \left[\frac{1}{N}\sum_{i=1}^{N} \sqrt{|u_i|}\right]^2 \tag{8.2.4}$$

Average absolute value:

$$|\overline{U}| = \int_{-\infty}^{+\infty} |u|p(u)du = \frac{1}{T}\int_{0}^{T} |u|dt = \frac{1}{N}\sum_{i=1}^{N} |x_i| \tag{8.2.5}$$

Variance:

$$\sigma_u^2 = \int_{-\infty}^{+\infty} (u - \overline{u})^2 p(u)du = \frac{1}{T}\int_{0}^{T} (u - \overline{u})^2 dt$$

$$= \frac{1}{N}\sum_{i=1}^{N} (u_i - \overline{u})^2 = U_{rms}^2 - \overline{U}^2 \tag{8.2.6}$$

Variance $\sigma_u{}^2$ may reflect the dispersion degree of voltage waveforms. It refers to the degree of intensity or deviation of overvoltage signals fluctuating around the average value.

Skewness:

$$\alpha = \int_{-\infty}^{+\infty} u^3 p(u)du = \frac{1}{T}\int_{0}^{T} u^3 dt = \frac{1}{N}\sum_{i=1}^{N} u_i^3 \tag{8.2.7}$$

Skewness α refers to the asymmetry of amplitude probability density function with respect to the longitudinal axis; a larger α indicates greater asymmetry.

Kurtosis:

$$\beta = \int_{-\infty}^{+\infty} u^4 p(u)du = \frac{1}{T}\int_{0}^{T} u^4 dt = \frac{1}{N}\sum_{i=1}^{N} u_i^4 \tag{8.2.8}$$

Kurtosis may indicate overvoltage waveform steepness; it is highly sensitive, especially for signals with larger amplitudes. The peak value U_p, the effective value U_{rms} and the kurtosis β will all increase in the presence of overvoltages; however, β is the most effective in recognizing impulse signals, as it exhibits the fastest increase.

The dimensionless time-domain characteristic value includes the clearance factor, the impulse factor and the kurtosis value, etc.

Crest factor:

$$C = U_p/U_{rms} \tag{8.2.9}$$

Crest factor C may reflect the intensity degree of voltage oscillation, and a larger value suggests a larger oscillating voltage amplitude.

Impulse factor:

$$I = U_p/|\overline{U}| \tag{8.2.10}$$

Impulse factor I refers to the difference between the voltage wave signal and the average value, and a larger I means a larger difference and a greater deviation.

Clearance factor:

$$CL = U_p/U_r \tag{8.2.11}$$

Clearance factor CL refers to the difference between the maximum voltage amplitude and the root-mean-square amplitude.

Kurtosis value:

$$K = \beta/U_{rms}^4 \tag{8.2.12}$$

The kurtosis value β represents the curve steepness as mentioned earlier, and the fourth power within the kurtosis value K may refer to steepness or flatness of the voltage curve top.

8.3 Wavelet Transform Analysis

8.3.1 Basic Theory

A great many transients having non-fundamental frequency exist in the voltage and current of high-voltage power equipment that is subject to lightning or other faults. These transients vary with occurrence instants, lightning strike spots or fault locations, transition resistance and system composition; the resulting transient signals exhibit a non-stationery random process. The Fourier transform is limited in treating non-stationery signals as it does not possess frequency localization properties. It cannot tell when a frequency component occurs and hence loses the time information, which is essential to non-stationary transient overvoltage signals. The concept of the wavelet transform was first proposed in 1974 by J. Morlet, a French engineer working on signal processing in the oil industry. Time signals, when unfolded as the linear combination of wavelet functions, are localized in both time and frequency domains. Such a treatment is like providing an adjustable time-frequency window to enable simultaneous combined analysis of signals in time and frequency domains. Therefore, the wavelet transform can enable local transformation from time to frequency and extract information in different frequency ranges so as to extract the required signal features in a more effective manner.

Assuming that a base function (i.e. wavelet function) is given by

$$\omega_{s,\tau}(t) = \frac{1}{\sqrt{s}}\omega\left(\frac{t-\tau}{s}\right) \tag{8.3.1}$$

the inner products of the wavelet function at different scale s after displacement τ and the acquired overvoltage signal $f(t)$ are given by

$$WT_x(s,\tau) = \frac{1}{\sqrt{s}}\int_{-\infty}^{+\infty} f(t)\omega_{s,\tau}{}^*(t)dt \tag{8.3.2}$$

where $W_{s,\tau}{}^*(t)$ and $W_{s,\tau}(t)$ are conjugate to each other, with the inverse transformation formula expressed as

$$f(t) = \frac{1}{C_\omega}\int_{-\infty}^{+\infty}\int_{-\infty}^{+\infty}\frac{\alpha(s,\tau)\omega_{s,\tau}(t)}{s^2}dsd\tau \tag{8.3.3}$$

where the factor C_ω is determined by

$$C_\omega = \int_{-\infty}^{+\infty}\frac{|W_{s,\tau}(\omega)|^2}{|\omega|}d\omega \tag{8.3.4}$$

$W_{s,\tau}(\omega)$ may be obtained from $\omega_{s,\tau}(t)$ via the Fourier transform with the restriction on C_ω expressed by

$$\int_{-\infty}^{+\infty}\frac{|W_{s,\tau}(\omega)|^2}{|\omega|}d\omega < \infty \tag{8.3.5}$$

Seen from the above equation, $\omega_{s,\tau}(t)$ must be provided with a small waveform, i.e. the so-called wavelet.

If the time window width of the wavelet function $\omega(t)$ is Δt, and the frequency spectrum $W(\omega)$ has a frequency window width of $\Delta\omega$ after Fourier transformation, $\omega(t/s)$ will have a frequency window width of $s\Delta t$, and its frequency spectrum $W(s\omega)$ will have a frequency window width of $\Delta\omega/s$. Therefore, the wavelet transform can excellently resolve low-frequency signals in the frequency domain and high-frequency signals in the time domain. By changing s and τ in the first equation, a family of wavelet functions will be obtained. The signal $f(t)$ to be analyzed can be decomposed according to the function family, so components of $f(t)$ at a local frequency range and time can be solved based on the expansion coefficients. In this way, local time and frequency analysis of the adjustable window can be achieved.

8.3.2 Characteristic Extraction Based on Wavelet Decomposition

8.3.2.1 Selection of Decomposition Scales

As mentioned before, each level of wavelets represents signal components in different frequency ranges after wavelet decomposition, and all the frequency ranges are distributed over the whole frequency axis without any overlapping; therefore, frequency-domain local analysis may be achieved through wavelet decomposition. One key problem is to determine the coefficient of each wavelet component. Regarding the current research level, the Mallat algorithm is deemed to be the most successful algorithm, which in nature derives the orthogonal relations between each coefficient matrix on the basis of wavelet orthogonality and filters each level of wavelets from the high to low level. In short, suppose that the signal $f(t)$ has eight sampling points:

$$f(t) = \alpha_{\varphi}\varphi(t) + \alpha_0\omega(t) + \alpha_{1,0}\omega(2t) + \alpha_{1,1}\omega(2t-1) + \alpha_{2,0}\omega(4t) + \alpha_{2,1}\omega(4t-1) + \alpha_{2,2}\omega(4t-2) + \alpha_{2,3}\omega(4t-3) \tag{8.3.6}$$

its wavelet decomposition formula includes Level 0, 1 and 2 wavelets, recorded as f_0,f_1 and f_2, respectively, of which f_0 only includes $\alpha_0\omega(t)$, f_1 is superposed by shift wavelets $\alpha_{1,0}\omega(2t)$ and $\alpha_{1,1}\omega(2t-1)$, f_2 is superposed by shift wavelets $\alpha_{2,0}\omega(4t)$, $\alpha_{2,1}\omega(4t-1)$ and $\alpha_{2,3}\omega(4t-3)$; including the constant $f_{\varphi} = \alpha_{\varphi}\varphi(t)$, there are altogether eight items in the decomposition formula which is identical to the number of signal sampling points. The first step of the Mallat Algorithm is to filter the level 2 wavelet f_2, determine the coefficient of each shift wavelet $\alpha_{2,0}$, $\alpha_{2,1}$, $\alpha_{2,2}$ and $\alpha_{2,3}$ and decompose the signal into superposition of f_2 and $f_{\varphi} + f_0 + f_1$. The above process plays the role of a low-pass filter, and the high-frequency signal *Hf* of the level 2 signal is separated, while the low-frequency signal *Lf* is entirely retained. The second step is to filter the level 1 wavelet from *Lf* and determine its wavelet coefficients until all the coefficients are determined. The process of the algorithm is shown in Figure 8.2.

The monitoring sampling rate adopted in this book is 5M, i.e. 200ns/pt. Taking 2.5 kHz as a division value for signal frequency, most medium to high frequency overvoltage signals are above this value, while low-frequency oscillation is below 2.5 kHz. The zero-sequence overvoltage signal will be subject to 15-level decomposition combined with the Mallat algorithm theory. The db4 wavelet is selected as the mother wavelet as it is easier to obtain transient signals with wider frequency range, compared with other wavelets (such as Haar, Coiflet and Symlets). The corresponding frequency range of each level is shown in Table 8.1.

8.3.2.2 Case Study of Decomposition

Taking measured zero-sequence overvoltage data in the case of single-phase grounding faults as an example, the initial signal is subject to a four-level decomposition. When

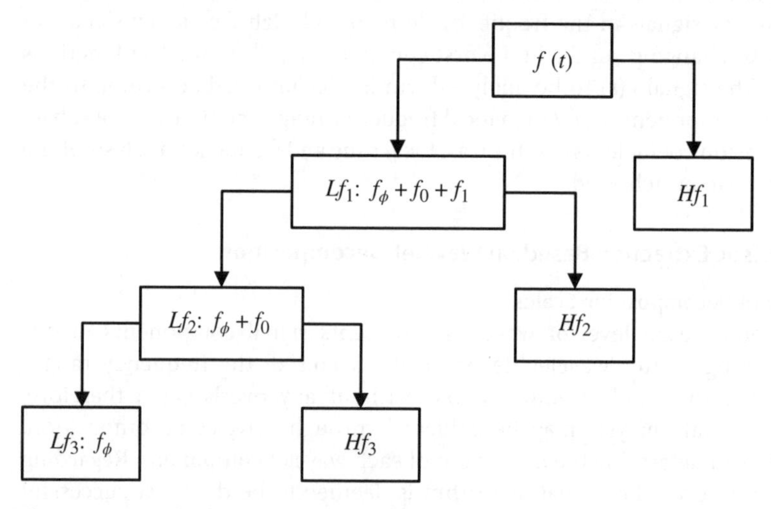

Figure 8.2 Wavelet decomposition process.

Table 8.1 Range of each frequency band.

Scale	d1	d2	d3	d4	d5	d6	d7	d8
Frequency band	1.25–2.5M	625k–1.25M	312–625k	156–312k	78–156k	39–78k	19–39k	9–19k
Scale	d9	d10	d11	d12	d13	d14	d15	a15
Frequency band	5–9k	2.5–5k	1.25–2.5k	610–1250	305–610	152–310	76–152	0–76

the system frequency reaches 160 kHz, the high-frequency signals contains a lot of background noise and interference. Hence, the frequency range over 160 kHz will not be considered here, so as to avoid the large impact of noise and interference and retain high-frequency signals to the greatest possible extent. As a result, only the a4 signal is retained after the four-level decomposition. The detailed results obtained from a ten-level decomposition are shown in Figure 8.3.

As seen in Figure 8.3(a), levels d_1–d_4 (160 k–2.5 MHz) are deemed as the background noise, and only level a_4 (0–160 kHz) contains the overvoltage signal after the first four-level decomposition. Such a decomposition is similar to denoising, and the level a_4 signal is subject to a ten-level decomposition for the second time. As seen in Figure 8.3(b), details around the power frequency are given in the level a_{10} (0–75 Hz) waveform which can be extracted as low-frequency signal characteristics. Similarly, levels d_1–d_5 (2.5 k–160 kHz) may be utilized to extract high-frequency signal characteristics, while levels d_6–d_{10} (75–2.5 kHz) may be used to extract low-frequency signal characteristics.

The wavelet energy within each frequency band sums up to the total energy of the original signal, and different frequency bands for different types of overvoltages account for different energy; for instance, switching and lightning overvoltages account for more energy than temporary voltages in the high-frequency range. Therefore, the energy within each frequency band may be taken as a characteristic value, with its

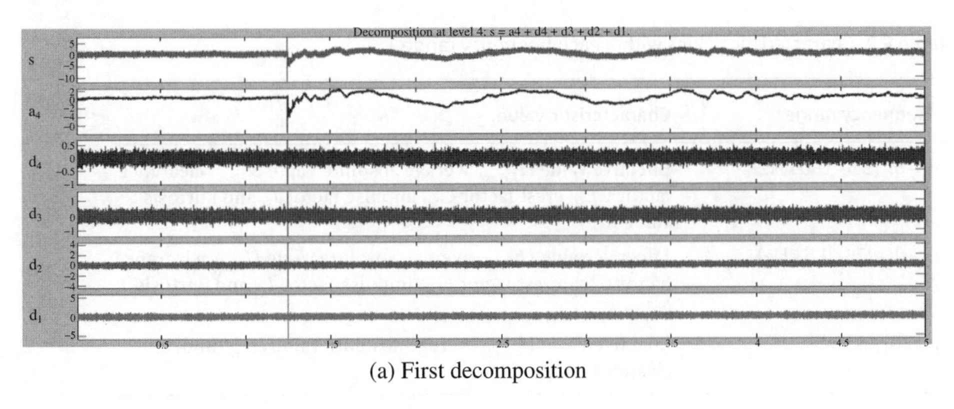

(a) First decomposition

Figure 8.3 Wavelet decomposition results.

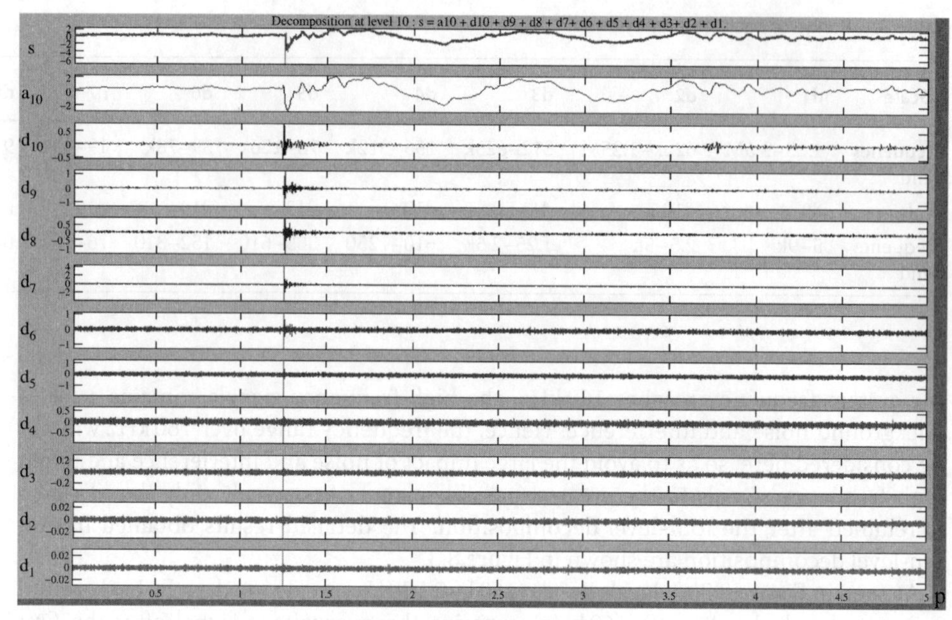

(b) Second decomposition

Figure 8.3 (*Continued*)

equation given by

$$E_L = \sum_{n=1}^{N} |d_L(n)|^2 \tag{8.3.7}$$

where L refers to the number of signal energy levels to be calculated, $d_L(n)$ refers to the coefficient of the level L high-frequency signal after wavelet decomposition, and E_L refers to the energy within each frequency range after wavelet decomposition. In addition, corresponding characteristic values may be extracted for signals within each frequency range using similar time-domain analysis, as shown in Table 8.2.

Table 8.2 Characteristic values within each frequency range.

Frequency range	Characteristic value
d_1–d_5 (2.5–160 kHz)	Effective value U_{Hrms}, average absolute value U_{Have}, clearance factor CL_H, crest factor C_H, impulse factor I_H and kurtosis value K_H
d_6–d_{10} (75–2.5 kHz)	Effective value U_{Mrms}, average absolute value U_{Mave}, clearance factor CL_M, crest factor C_M, impulse factor I_M and kurtosis value K_M
a_{10} (0–75 Hz)	Effective value U_{Lrms}, average absolute value U_{Lave} and clearance factor CL_L

8.4 Singular Value Decomposition (SVD) Theory

The concept of singular value decomposition was first proposed by Beltrami in 1873 for the real matrix. Through considerable research, Eckart and Yong have extended this concept into any rectangular matrix. Its definition is given below.

In the presence of m × n matrix A, a set of orthogonal matrixes $U = [u_1, u_2, \cdots u_m] \in R^{m\times m}$, $V = [v_1, v_2, \cdots v_m] \in R^{n\times n}$ are certain to exist to allow $U^TAV = diag[\sigma_1, \sigma_2, \cdots \sigma_p] = \sum, p = \min(m, n)$, that is

$$A = U \sum V^T \tag{8.4.1}$$

The equation is referred to as the singular value decomposition of matrix A, where $\sigma_1 \geq \sigma_2 \geq \ldots \geq \sigma_p \geq 0, \sigma_i$ $(i = 1, 2, \ldots p)$ are the singular values of A, and also the square roots of A^TA characteristic values.

The singular value is excellent in terms of stability. Its variation will not exceed the disturbance matrix norm when matrix A is subject to a slight disturbance. The signal will enjoy resistance to environmental interferences and background noise when singular values are used as overvoltage characteristic quantities. In addition, the singular value also enjoys constant scale. The singular value will remain constant in case of any change in rows or columns after the matrix is standardized. Moreover, the singular value is characterized by compression during dimensional reduction, meaning that the matrix will omit the singular values with little influence. The effective rank finally obtained is the number of reserved singular values. This process is deemed as a best rank approximation that can effectively improve the calculation speed. On account of these characteristics, singular value decomposition has already become one of the most effective methods in numerical analysis, and has been widely applied and developed in fields such as signal processing and statistical analysis.

For power system overvoltage recognition, the decomposition level is set to m, and the time sequence is divided into n scale(s) during wavelet decomposition of transient overvoltages to form an $m \times n$ matrix. Singular value decomposition is then used to significantly reduce the matrix dimension to display more abundant detailed signal characteristics within each scale.

The sampling frequency used in both simulation and practical measurement is 5 MHz in this book. The signal is subject to a 15-level decomposition with the detailed process described in Section 8.3. The wavelet coefficients for levels d_1–d_{15} and level a_{15} is thus obtained; the signal at each frequency band is then divided into ten scales to construct the matrix used for singular value decomposition. All the data used here consists of overvoltage zero-sequence signals. Singular value decomposition is performed on lightning overvoltages, intermittent arc grounding overvoltages and high-frequency resonance overvoltages, as examples. The results are presented in Figures 8.4–8.6.

As can be seen from Figures 8.4–8.6, the magnitude of each order of the lightning overvoltages is large after singular value decomposition, illustrating that there are many components within each frequency range. The second order is observed to be reduced by approximately 130% in magnitude compared with the first order; the other orders also show obvious numerical differences with rapid decline. Compared with lightning overvoltages, the singular value of each order for intermittent arc grounding overvoltages shows a significant decline, with the maximum high-frequency value of 56.53; similarly

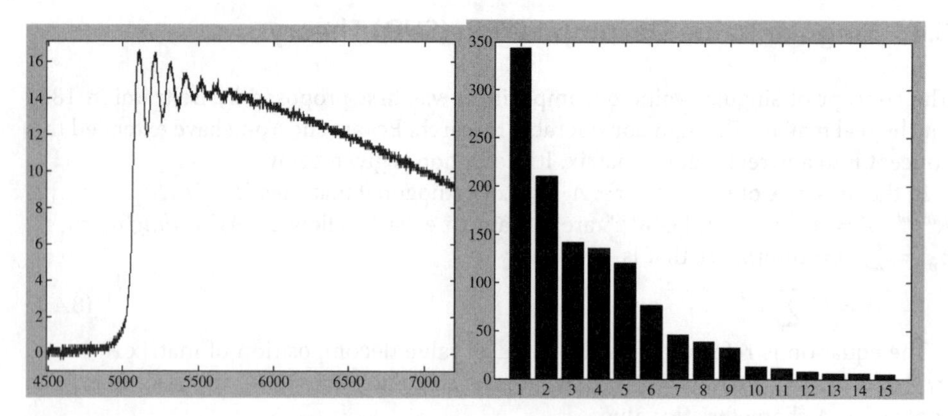

Figure 8.4 Singular value decomposition of lightning overvoltages.

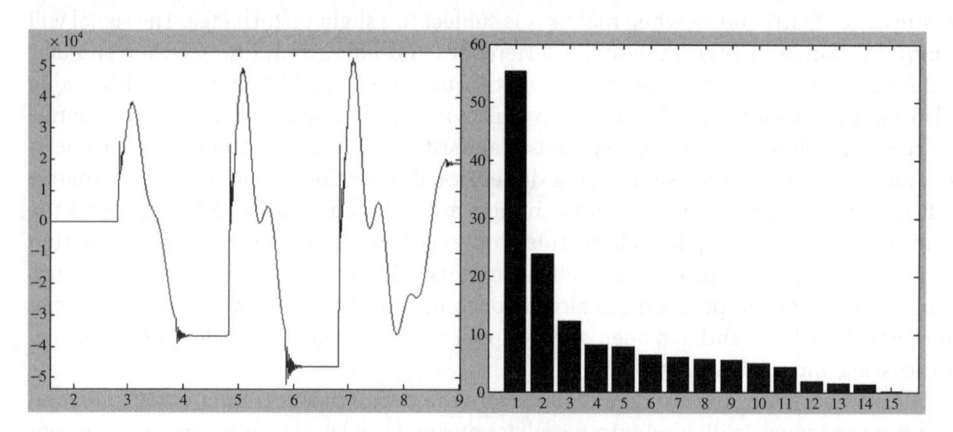

Figure 8.5 Singular value decomposition of intermittent arc grounding overvoltages.

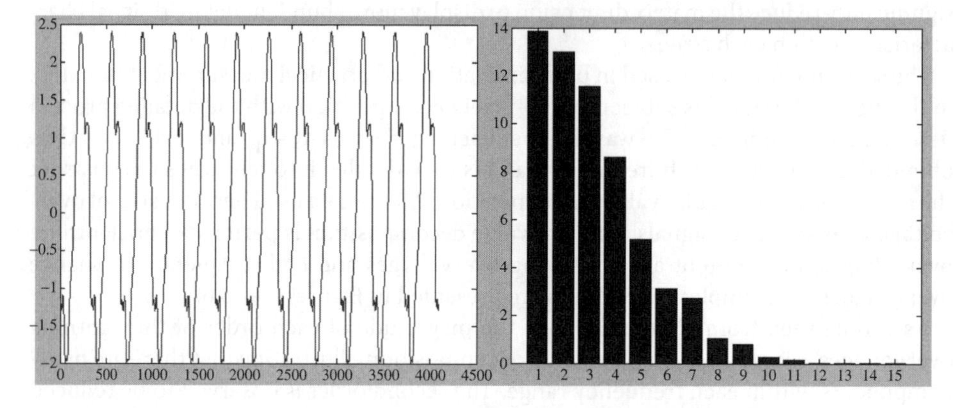

Figure 8.6 Singular value decomposition of high-frequency resonance overvoltages.

with other switching overvoltages, higher values are centralized in high-frequency orders that decline rapidly. For temporary overvoltages such as resonance overvoltages, the maximum singular value is 13.87 without obvious changes between each order, meaning a slow decline. According to the above characteristics, the following two characteristic values may be used in addition to the extracted singular value of each order.

Average singular value:

$$\gamma_{ave} = \frac{1}{15}\sum_{i=1}^{15}\gamma_i \tag{8.4.2}$$

where γ_{ave} is the average value of the singular spectrum, and $\gamma_i(i=1,2,\ldots 15)$ is the singular value of each order. The singular value γ shows the energy distribution over the time-frequency space, and the wider the energy distribution, the larger the value γ, therefore the average singular value may reflect the energy magnitude of each overvoltage.

Difference between the singular values of the first and second order:

$$\gamma_{1-2} = \gamma_1 - \gamma_2 \tag{8.4.3}$$

According to the previous analysis, the difference between γ_1 and γ_2 regarding different overvoltage types is clearly different, so it may be taken as the characteristic value.

8.5 Characteristic Value Selection for Sorters

8.5.1 Characteristic Value Selection for First-Level Sorters

As shown in Figure 8.7, the main objective of first-level sorter 1 is to distinguish temporary overvoltages from lightning and switching overvoltages.

Switching and lightning overvoltages are characterized by short duration, high amplitude and high frequency in terms of waveforms compared with temporary overvoltages. Generally, most switching and lightning overvoltages are millisecond/microsecond-level impulse overvoltages, while the duration of temporary overvoltages

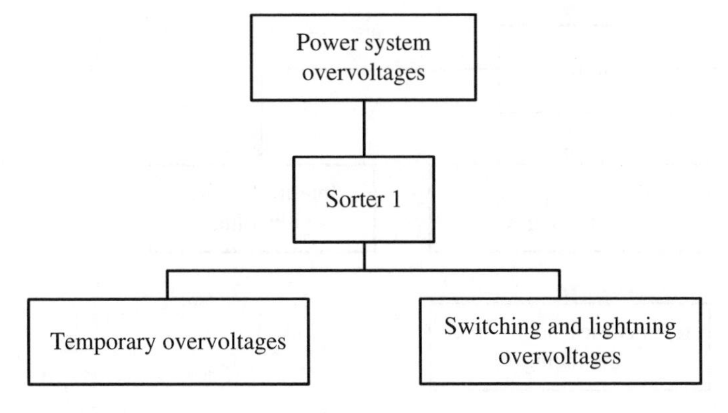

Figure 8.7 Structure of sorter 1.

far exceeds that order of magnitude, varying from dozens of microseconds to hundreds of seconds. In addition, most temporary overvoltages belong to low-frequency oscillation. Hence, time-domain characteristic values can help to recognize these two types of overvoltages.

In order to avoid split-phase recognition, characteristic values are extracted based on zero-sequence voltage signals in this book. According to the descriptions in Chapter 3, the root-mean-square value U_{0rms}, the average value U_{0ave}, the kurtosis β_0, the clearance factor CL_0, the kurtosis value K_0 and the impulse factor I_0 may be taken as characteristic parameters for sorter 1.

8.5.2 Characteristic Value Selection for Second-Level Sorters

The second-level sorter includes two sorters. The first sorter 2.1 is used to recognize nonlinear resonance overvoltages, power frequency overvoltages and intermittent arc grounding overvoltages, as shown in Figure 8.8. According to the analysis in Section 8.3, although all these three overvoltages have long durations, high-frequency resonance overvoltages contain more frequency components of integral multiple. Therefore, it is possible to extract the proportion of each frequency range to distinguish this overvoltage. Compared with the other two signals, intermittent arc grounding overvoltages exhibit more dramatic oscillation with more energy and with larger first-order singular value γ_1. Thus, the effective value U_{Hrms}, the average value U_{Have}, the clearance factor CL_H, the crest factor C_H, the impulse factor I_H and the kurtosis value K_H are taken as characteristic parameters to distinguish arc grounding overvoltages from power frequency and high-frequency resonance overvoltages. In the case of asymmetric grounding short-circuit, single-phase or two-phase voltages are considerably reduced to nearly zero. Therefore, the effective value $\min(U_{rms})$ of the minimal phase is taken as the characteristic parameter. To sum up, sorter 2.1 contains the following characteristic values: the effective value U_{Hrms}, the average value U_{Have}, the clearance CL_H, the crest factor C_H, the impulse factor I_H the kurtosis value K_H and the effective value $\min(U_{rms})$ of the fault phase and the first-order singular value γ_1.

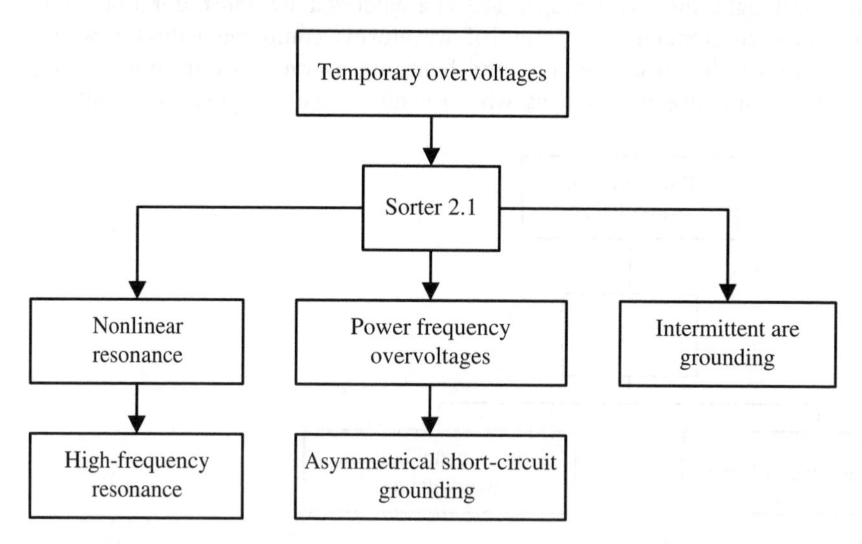

Figure 8.8 Structure of sorter 2.1.

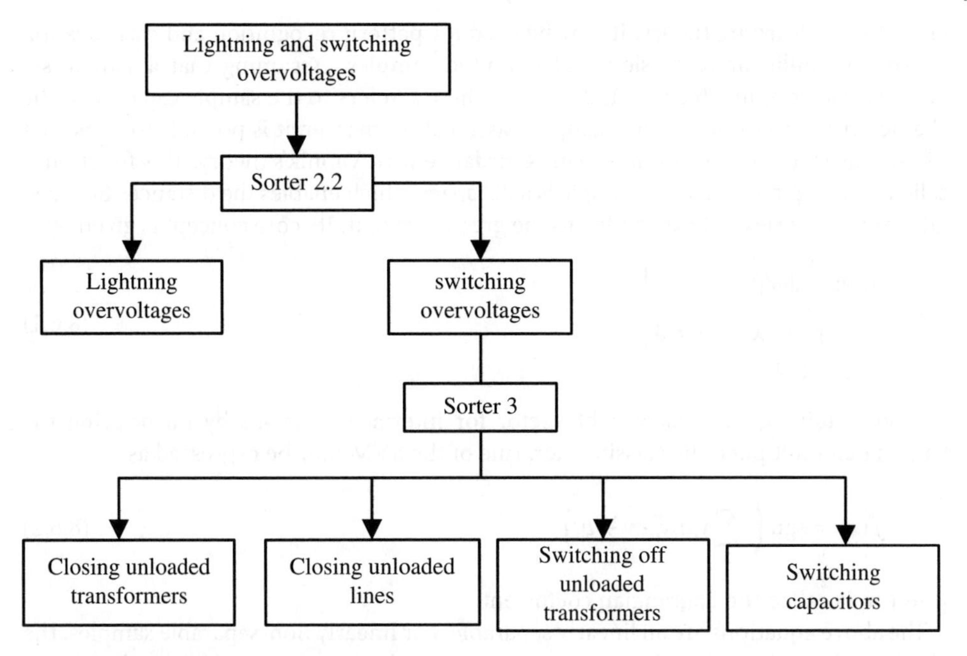

Figure 8.9 Structure of sorters 2.2 and 3.

Sorter 2.2 is used to distinguish between lightning overvoltages and switching overvoltages, while sorter 3 is used to distinguish between overvoltages caused by closing unloaded transformers, closing unloaded lines, opening (switching off) unloaded transformers and switching capacitors, as shown in Figure 8.9. According to the above analysis, characteristic quantities obtained through singular value decomposition can be used to effectively distinguish lightning and switching overvoltages; the maximum singular value γ_1, the average singular spectrum value γ_{ave} and the difference between first-order and second-order singular values $\gamma_{1\text{-}2}$ can be preferably selected by sorters 2.2 and 3 as the characteristic values. Also, each level of energy after wavelet decomposition E_L can be taken as a characteristic value.

Although it is difficult to distinguish lightning overvoltages and switching overvoltages by time-domain analysis, wavelet decomposition and singular value decomposition is able to magnify signal details locally and eliminate interference from background noise. Each frequency range and the singular value of each order regarding transient overvoltage signals can thus be extracted to identify switching overvoltages and lightning overvoltages. Finally, a suitable sorter at each level of the recognition system will be selected.

8.6 SVM-Based Transient Overvoltage Recognition System

8.6.1 Overview of Support Vector Machine (SVM)

The support vector machine (SVM) theory, similar to multi-level transducer networks and radial basis function networks, was first proposed in 1995 by Vapnik et al., based on statistical learning theory. It may be used for pattern recognition and classification as well as nonlinear regression under limited samples. Assuming that a sample set (x_i,y_i) is available, in which $i = 1, 2, 3, \ldots n$, where n refers to the sample capacity, x_i the characteristic vector and y_i the sample classification marker, it is possible to construct a best linear function $g(x) = \omega x + b$ in accordance with Vapnik's theory. This function is called a hyperplane when the sample is a 3D space which enables the distances between different categories to be extended to the greatest extent. Its core concept is given by

$$\left.\begin{aligned} &\min\ \frac{1}{2}\|\omega\|^2 \\ &s.t.\ \ y_i(\omega \cdot x + b) = 0 \\ &\quad i = 1, 2, \cdots n \end{aligned}\right\} \tag{8.6.1}$$

After obtaining the best weight vector for interface solutions by introducing the Lagrangian multiplier, the classification rule of the SVM may be expressed as

$$f(x) = \mathrm{sgn}\left(\sum_S y_i a_i x_i \cdot y + b\right) \tag{8.6.2}$$

where α_i is called the Lagrangian coefficient.

The above equations are all linearly separable. For linearly non-separable samples, the SVM may be made separable via higher-dimensional space mapping by introducing the kernel function $K(x_i,x)$, given by

$$g(x) = \sum_{i=1}^{n} a_i y_i K(x_i, x) + b \tag{8.6.3}$$

Currently, the most commonly used kernel functions include:

- linear kernel function: $K(x, x_i) = x^T x_i$;
- polynomial kernel function: $K(x, x_i) = (\gamma x^T x_i + r)^2,\ \gamma > 0$;
- radial basis function: $K(x, x_i) = \exp(-\gamma x - x_i^2),\ \gamma > 0$;
- two-level transducer kernel function: $K(x, x_i) = \tanh(\gamma x^T x_i + r)$.

If the SVM is still non-separable, after being mapped to a space with higher dimension, the slack variable ξ_i and the penalty factor c are introduced, expressed as

$$\left.\begin{aligned} &\min\ \frac{1}{2}\|\omega\|^2 + C\left(\sum_{i=1}^{n} \xi_i\right) \\ &s.t.\ \ y_i(\omega \cdot \varphi(x_i) + b) \geq 1 - \xi_i \\ &\quad \xi_i \geq 0 \quad i = 1,2, \cdots n \end{aligned}\right\} \tag{8.6.4}$$

where the slack variable ξ_i may be interpreted as introducing fault tolerance for threshold values of the classification plane, i.e. abandon all outliers in the sample characteristic vectors and no longer move towards them. The advantage of this is that the classification plan may gain larger geometric margins and become smoother with a non-negative value. When it is necessary to evaluate the extent of the loss of the target function caused by the outliers, the penalty factor c is required to be introduced. A larger c indicates less tolerance for the outlier losses.

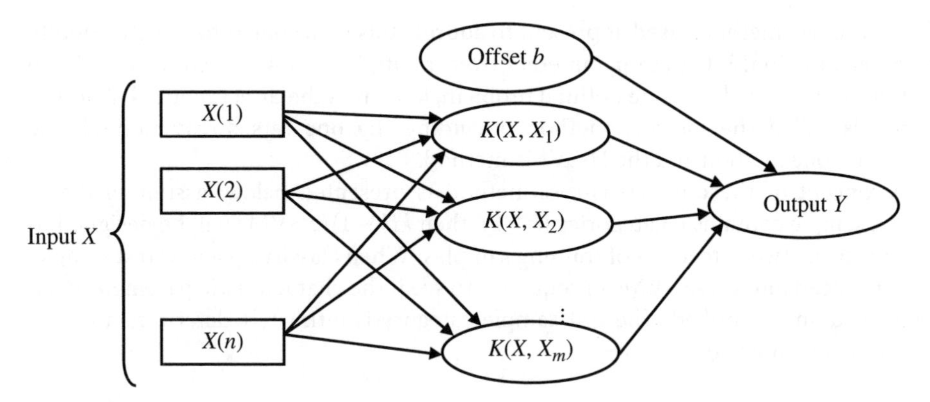

Figure 8.10 Schematic of the SVM Structure.

In view of pattern recognition, the support vector machine has been widely applied in fault diagnosis, energy quality estimation, security prediction and other fields, owing to its excellent performance in generalization, popularity and robustness as well as its advantages in simple operation and sound theory.

The SVM system structure is shown in Figure 8.10.

There are now many toolkits in Matlab that can be used to realize SVM, such as LIBSVM, LSSVM, SVMlight and the SVM toolkit integrated into the new Matlab version, each of which possesses its own strengths and weaknesses concerning functions and their extensibility.

LIBSVM, designed and developed by Professor Lin Zhiren et al. from Taiwan University is adopted in this book because of the following advantages:

1. Flexibility: LIBSVM provides the toolkit with open source code for the convenience of improvement and modification regarding recognition and classification of different issues.
2. Simplicity: compared with the conventional SVM, LIBSVM is optimized and upgraded in its algorithm to provide faster arithmetic speed, higher calculation accuracy and better memory usage.
3. Universality: LIBSVM includes multilingual versions, and can be used in about ten language environments, such as Java, Matlab, Labview and C#.

8.6.2 Multi-Class SVM

The support vector machine is practically designed for binary classification, namely it is only intended for two categories. However, there are a wide variety of transient overvoltages in the power system, and an overvoltage genre may be subdivided into four or five classes. Tree-shaped hierarchical classification adopted in this book divides all the categories into two classes, and then divides each class into two subclasses. However, a subclass may still include a number of overvoltage types; for instance, sorter 3 includes overvoltages caused by closing unloaded transformers, by closing unloaded lines, by opening (switching off) unloaded transformers and by switching capacitors, each of which is the minimal subclass that cannot be recognized by the conventional support vector machine.

There are two methods used at present to address this issue: one is to directly modify the target function, integrate parameters from multiple planes and convert multiple functions to be solved into one optimal problem, known as the direct method. Another method is called the indirect method, including the one-versus-rest method, the one-versus-one method and the H-SVMs method.

The concept of the one-versus-rest method is presented below: assuming that a training sample contains k categories ($k \geq 2$), then $k(k-1)/2$ SVMs will be designed to distinguish any two categories of training samples. When classifying a new test sample, it is introduced into these SVMs in sequence to see if the characteristic parameters can be matched and identified. The test sample category is ultimately determined by the one wining the most votes.

8.7 Data Preprocessing

8.7.1 Dimension Reduction

The characteristic parameters involved in transient overvoltage recognition system developed in this book have a large dimension, for instance, sorter 2.1 contains more than 10 characteristic parameters such as the effective average value U_{Hrms}, the average value U_{Have}, the clearance CL_H, the crest factor C_H, the impulse factor I_H, the kurtosis value K_H, the effective value $\min(U_{rms})$ under fault phase and the first-order singular value γ_1. Such complex sorter settings may lead to slow convergence speed. Because each characteristic parameter makes a different contribution during recognition, some characteristic parameters may have little impact on the accuracy. Principal component analysis (PCA) is adopted in sorter research which removes poor quality characteristic parameters and eliminates redundant information so as to effectively reduce the characteristic dimension and improve sorter learning efficiency and speed.

The basic concept of the PCA algorithm is to recombine correlated characteristic parameters through means such as linear conversion and mapping, to form a set of new comprehensive characteristic parameters with no correlation between them. For example, a certain question contains n characteristic parameters $T_1, T_2 \ldots T_n$, and it is typical to linearly recombine these characteristic parameters. Provided that X_1 refers to the characteristic parameter of the first linear combination, the larger its variance, i.e. the larger the $\mathrm{Var}(X_1)$, the more information it contains. When the variance reaches the maximum, X_1 is deemed as the first principal component of the new combination. If the first principal component is not enough to represent the information contained in the previous n indicators, the second linear combination X_2 has to be considered, and X_2 does not contain the information included in X_1. In this case, X_2 is referred to as the second principal component, and $1, 2, \cdots P$ principal components can be constructed in a similar fashion.

The general steps for PCA solution are:

1. Convert the acquired data into an $n \times T$ mixed data matrix, where n is the observed signal number and T is the sampling point number.
2. Calculate the average value of each observed signal (matrix row vector), and subtract the average value from the signal to get matrix X.

3. Construct a covariance matrix, and perform characteristic decomposition for XX^T to obtain its characteristic values and characteristic vectors λ_i.
4. Arrange characteristic values in a descending order, and select the first P maximum characteristic vectors according to their contribution rates. Among others, the contribution rate refers to the ratio between the selected characteristic value sum and the sum of all the characteristic values, and is generally greater than or equal to 95%. The equation for the contribution rate may be expressed as

$$\varphi = \frac{\sum_{i=1}^{p} \lambda_i}{\sum_{i=1}^{n} \lambda_i} \geq 95\% \tag{8.7.1}$$

Taken as an example, sorter 2.1 is selected and subjected to PCA dimension reduction. The results are shown in Figure 8.11:

It is observed from Figure 8.11 that the dimension reduces to 9 on the basis of little impact on the ultimate accuracy when the contribution rate is set to 95%, which effectively removes the useless characteristic parameters and redundancy.

8.7.2 Normalization

Normalization, as one of the most common methods during data preprocessing, is able to convert dimensional data into dimensionless data by simple calculation. Normalized mapping may significantly improve convergence and classification accuracy of the training data.

Normalized mapping in the range [0,1] may be expressed by

$$f : x \rightarrow y = \frac{x - x_{\min}}{x_{\max} - x_{\min}} \tag{8.7.2}$$

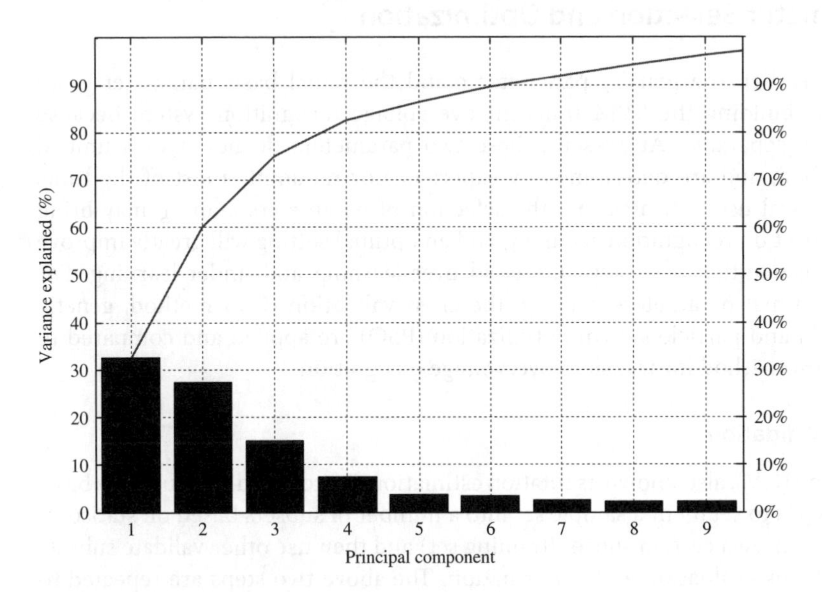

Figure 8.11 Results of PCA dimension reduction.

Table 8.3 Comparison of accuracy under different normalization approaches.

Normalization approach	Accuracy	c, g option
Not utilizing normalization	14.286%	'−c 2 −g 1'
[−1,1] normalization	71.429%	'−c 2 −g 1'
[0,1] normalization	85.714%	'−c 2 −g 1'

where $x, y \in R^n; x_{min} = \min(x); x_{max} = \max(x)$. Normalization helps to adjust original data to the range [0, 1], so as to effectively normalize the large value and the small value of data into a smaller set, i.e. $y_i \in [0, 1], i = 1, 2, \cdots n$.

In addition to the above normalization approach, another approach is known as the normalization within the range [−1,1] with its mapping given by

$$f : x \rightarrow y = 2 \times \frac{x - x_{min}}{x_{max} - x_{min}} \pm 1 \tag{8.7.3}$$

The classification accuracy of the prediction set in the presence/absence of normalization is shown in Table 8.3, in which the radial basis function is adopted as the kernel function, and the penalty factor and kernel function parameter are set to the default values.

Seen from the above table, normalization is able to improve the accuracy of transient overvoltage recognition by 5 times for [−1,1] normalization and 6 times for [0,1] normalization. In view of this, [0,1] normalization is adopted in this book for data preprocessing.

8.8 Parameter Selection and Optimization

As described earlier, the penalty parameter c and the radial basis parameter g are required while building the SVM transient overvoltage recognition system because it is linear non-separable. At present, these two parameters do not have a unified selection method, and are determined by experience or randomly most of the time. However, practical use indicates that the selection of parameters c and g may bring significant impact on recognition accuracy, and an optimal setting will greatly improve the ultimate recognition accuracy and avoid over-learning and under-learning. For selection of optimal parameters c and g, the cross validation (CV) method, genetic algorithm (GA) and particle swarm optimization (PSO) are applied and compared to find out the best method for transient overvoltage recognition.

8.8.1 Cross Validation

Cross validation (CV), also known as rotation estimation, is proposed by Skutin. Its basic optimizing step is to segment a sample set into a number of subsets based on statistics, calculate and analyze a certain subset (training set) and then use other validate subsets (validation set) for evaluation and confirmation. The above two steps are repeated to gain parameters c and g with the optimal accuracy.

The common CV methods include hold-out validation, k-fold cross validation (K-CV) and leave-one-out cross validation (LOO-CV); in this book, LOO-CV is adopted. When the original data contains N samples, the samples will be extracted one by one as the validation set, while the remaining $N-1$ samples are taken as the training set; therefore, LOO-CV finally contains n models. The average value of the classification accuracy of the n models is ultimately taken as the index to validate sorter performance. Compared with the other two methods, LOO-CV possesses the following two advantages:

1. Each set of data from the sample data set has been used as the training model; the final estimation is the closet to the original data distribution, and the conclusion is more reliable.
2. There will be no impact on the experimental data by random factors due to simple experimental process.

Results for parameter optimization of sorter 2.1 by the cross validation method are shown in Figure 8.12, which shows that the best value for transient overvoltage recognition system sorter 2.1 is $c=4$ and $g=0.70711$, and the final classification accuracy reaches 92.8571%.

8.8.2 Genetic Algorithm

The concept of the genetic algorithm (GA) is inspired by biological evolution processes, based on computer simulation technology. GA was invented by Holland, Michigan et al. in the USA in the 1960s, and is considered as a kind of adaptive optimizing probability algorithm that aims at parameter coding and is characterized by parallel operation from numerous points rather than just one point. Thus, local convergence of calculation results can be avoided. GA enjoys little dependence upon specific problems due to its adaptability. As a result, it has attracted much attention and has been developed in many subjects such as function optimization, pattern recognition and information processing.

The steps of genetic algorithm are shown in Figure 8.13.

Compared with cross validation, GA can find out the optimal global solution without circulating all the points in the grid, thus reducing memory usage and evaluation time. Hence, GA may be applied to search for the global optimal solution within a greater range in terms of transient overvoltage recognition research. The GA results are as shown in Figure 8.14.

In Figure 8.14, the red line (upper) represents the best fitness while the blue line (lower) represents the average fitness. If the genetic generation is set to 100 and the population is set to 20, the accuracy will be stable at 92.8571% with $c=5.8954$, $g=0.64735$.

8.8.3 Particle Swarm Optimization Algorithm

The particle swarm optimization algorithm (PSO algorithm) was first proposed by Kennedy and Eberhart in 1995 as a swarm intelligence optimization algorithm. Different from the ant colony algorithm and the fish swarm algorithm, the PSO algorithm is inspired by bird predatory behaviors. This algorithm first initializes a swarm of individuals known as particles in the solution space. Each particle can be seen as a potential optimal solution for the extremum optimization issue and has its own velocity, location and fitness in the space, of which the particle velocity shows the direction and distance of particle movement in the space. Particles will continuously move to find

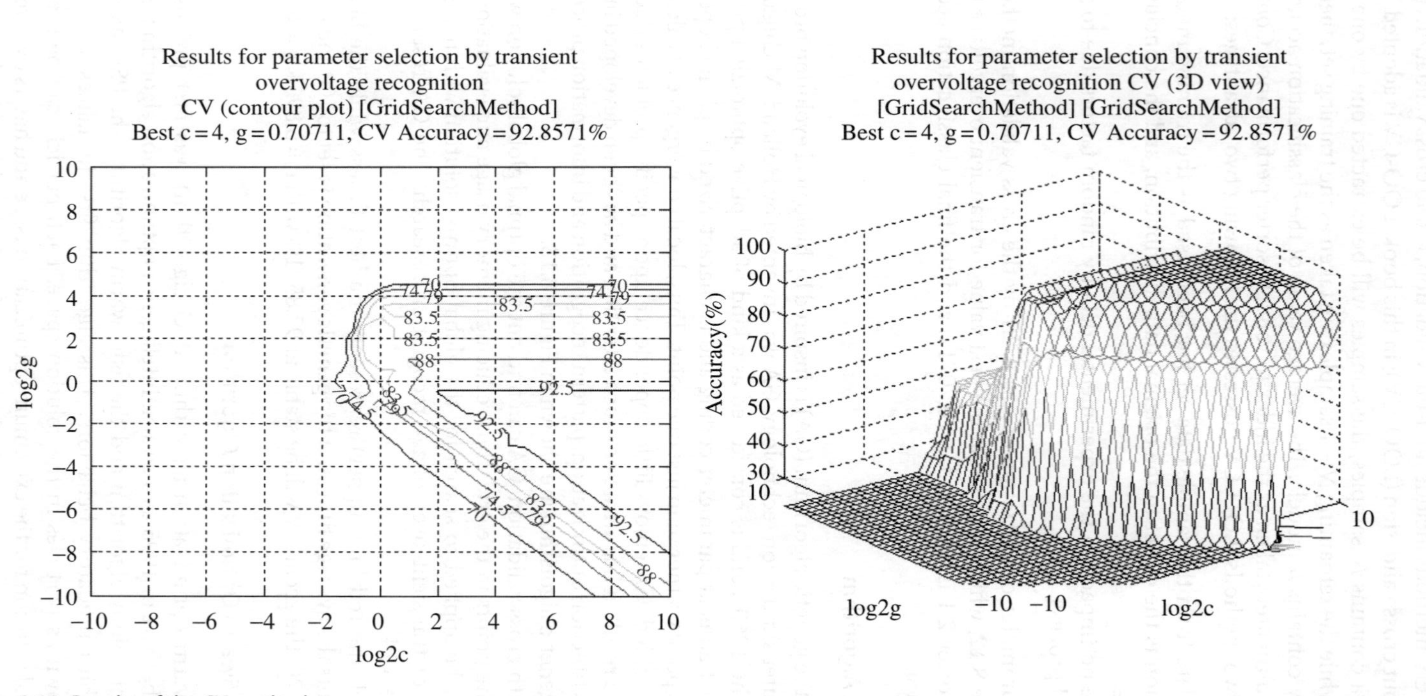

Figure 8.12 Results of the CV method.

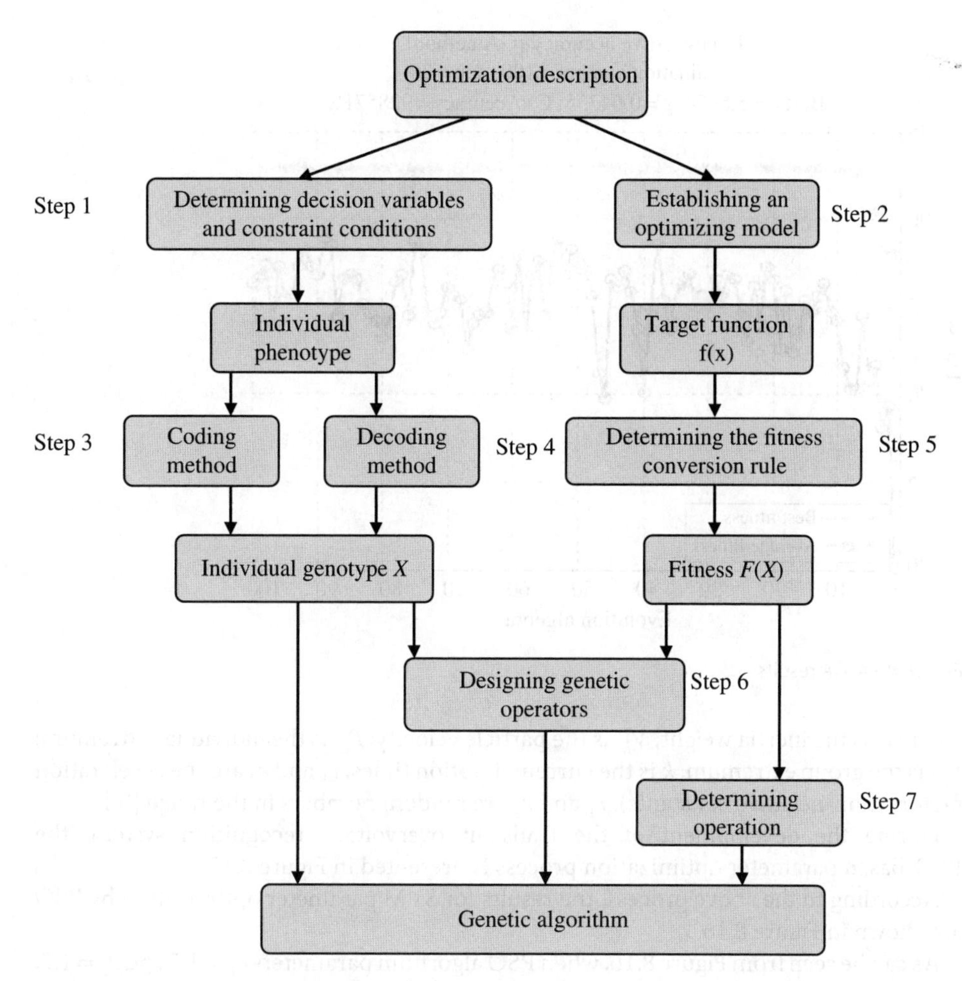

Figure 8.13 Basic steps of the genetic algorithm.

out the location with the best fitness by tracking extremums of individuals and groups. Specifically, the main concept of PSO is similar to the process where a bird judges the direction and distance of its next movement by observing the predatory locations of other birds and its own location.

It is supposed that a population $X = (X_1, X_2, \ldots, X_n)$ appears in a D-dimensional search space, where $X_1, X_2, \ldots X_n$ are the particles and n the particle number. The location of the ith particle in the D-dimensional space can be expressed by the vector $X_i = [X_{i1}, X_{i2}, \ldots, X_{iD}]^{\mathrm{T}}$ and the particle also represents a potential optimal solution to the question at the same time. During each iteration, the particle will update its velocity and location by judging individual and group extremums.

The updating formula is

$$V_{id}^{k+1} = \omega V_{id}^{k} + c_1 r_1 (P_{id}^{k} - X_{id}^{k}) + c_2 r_2 (P_{gd}^{k} - X_{id}^{k}) \tag{8.8.1}$$

with

$$X_{id}^{k+1} = X_{id}^{k} + V_{id}^{k+1} \tag{8.8.2}$$

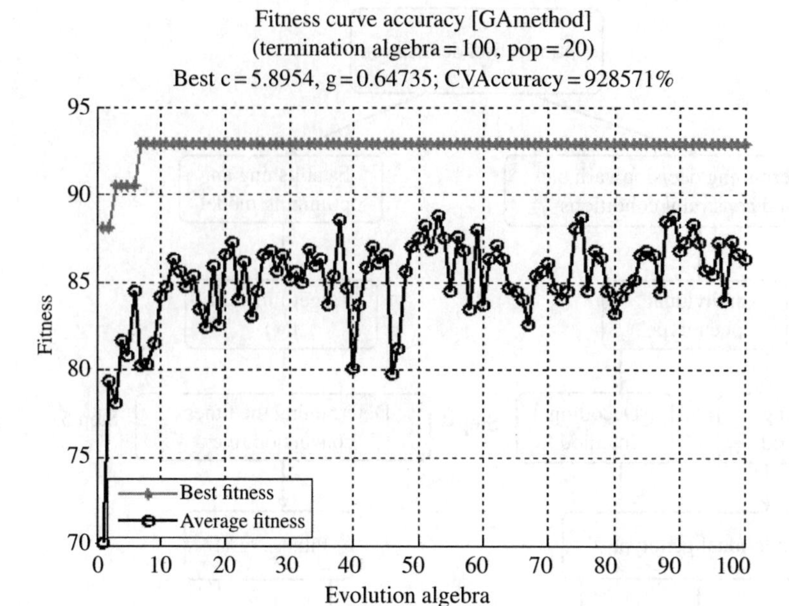

Figure 8.14 GA results.

where ω is the inertia weight; V_{id} is the particle velocity; P_{id} is the individual extremum; P_{gd} is the group extremum; k is the current iteration times; c_1 and c_2 are the acceleration factors (non-negative constants); r_1 and r_2 are random numbers in the range [0,1].

During the development of the transient overvoltage recognition system, the PSO-based parameter optimization process is presented in Figure 8.15.

According to the above process, the results for SVM parameter optimization by PSO are shown in Figure 8.16.

As can be seen from Figure 8.16, when PSO algorithm parameters $c_1 = 1.5$ and $c_2 = 1.7$, c and g reach the optimum, 28.1725 and 0.1, respectively, and the accuracy reaches 92.8571%.

According to the sorter 2.1 evaluation, based on the above three parameter optimization methods, the accuracies can all be improved to 92.8571%. Although the three methods all converge to the best fitness, we can find by comparison between Figures 8.14 and 8.16 that the average fitness exhibits more greater oscillation when adopting the PSO algorithm compared with that of the genetic algorithm, and the final penalty factor reaches 28.1725. As described before, the penalty factor impacts greatly upon sorter performances, and a too-large penalty factor will lead to machine over-learning. As described before, the penalty factor impacts greatly upon sorter performances, and a c value too large will lead to machine over-learning, while a value too small will result in machine under-learning. In order to avoid these two problems, median parameters are used to set sorters after analysis and comparison in this book. For example, the genetic algorithm is adopted for parameter optimization of sorter 2.1, parameters c and g are 5.8954 and 0.64735, respectively in this case.

The parameter optimization approach and its accuracy for each level of sorter are shown in Table 8.4, from which it can be seen that the accuracy of recognition of

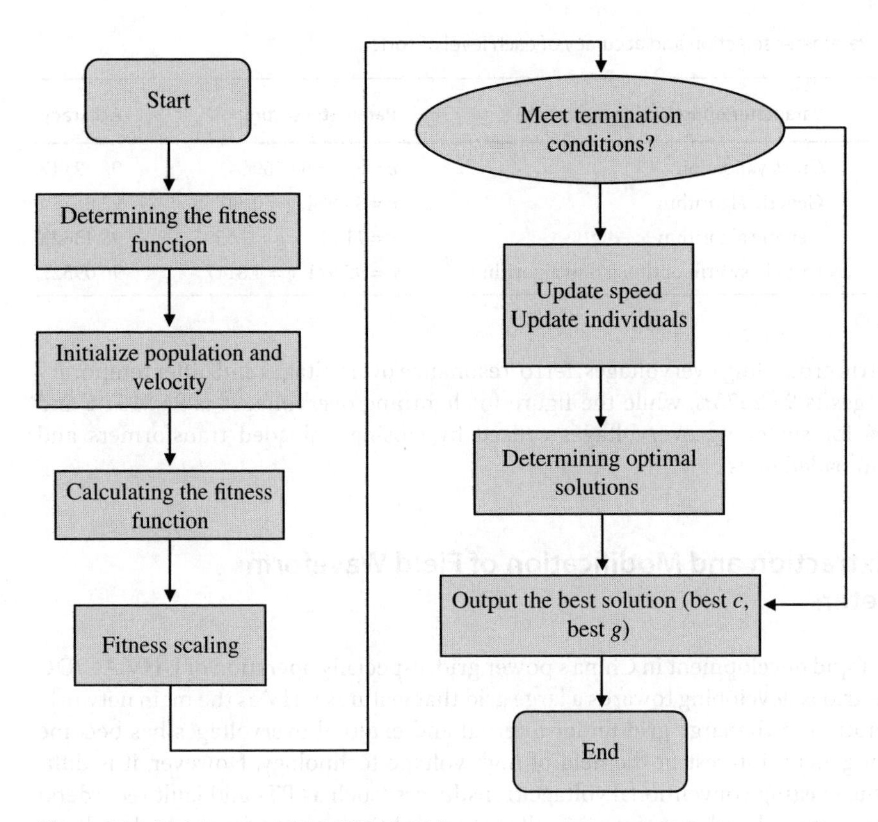

Figure 8.15 Basic process of the PSO algorithm.

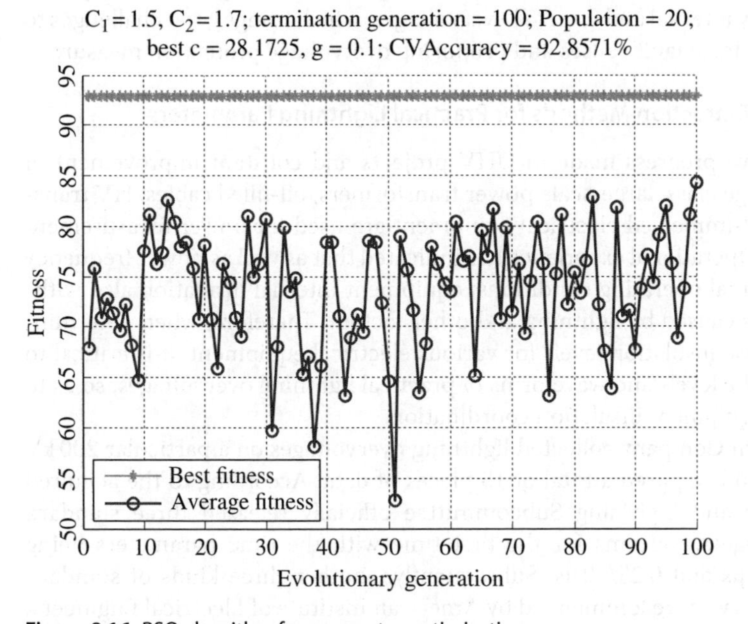

Figure 8.16 PSO algorithm for parameter optimization.

Table 8.4 Parameter selection and accuracy of each level of sorter.

Sorter	Parameter optimization approach	Parameter setting	Accuracy
1	Cross validation	$c = 8, g = 5.6569$	96.3214%
2.1	Genetic algorithm	$c = 5.8954, g = 0.64735$	92.8571%
2.2	Genetic algorithm	$c = 11.3173, g = 5.6595$	98.4362%
3	Particle swarm optimization algorithm	$c = 9.7971, g = 4.8117$	96.0952%

asymmetric grounding overvoltages, ferro-resonance overvoltages and other temporary overvoltages is 94.2573%, while the figure for lightning overvoltages is 94.8151% and 91.1128% for switching overvoltages caused by closing unloaded transformers and closing unloaded lines.

8.9 Extraction and Modification of Field Waveform Parameters

With the rapid development in China's power grid, especially operation of UHV AC/DC lines, the grid is developing towards a large grid that features UHV as the main network. Safe operation of the large grid under internal and external overvoltages has become a topic of general interest in the field of high voltage technology. However, it is difficult for the existing conventional voltage transducers (such as PTs and fault recorders) to monitor internal and external overvoltages in real time due to frequency band and intrinsic restrictions, and it is impossible to track original overvoltage data after power grid faults caused by internal and external overvoltages, which brings great challenges to determining overvoltage fault causes and proposing overvoltage protection measures.

8.9.1 Innovative Extraction Methods for Practical Lightning Parameters

With comprehensive progress made in UHV projects and constant improvement of power system voltage class, large-scale power transformers, oil-filled cables, HV transducer and other oil-immersed electrical equipment are used for power measurement and transmission. Operational experience has indicated that as well as power frequency voltages under normal operating conditions, equipment internal insulation also withstands overvoltages caused by lightning and other factors. Therefore, when determining lightning impulse insulation levels for various electrical equipment, it is critical to accurately master the levels and waveforms of practical lightning overvoltages, so as to reasonably select equipment insulation coordination.

In 1929, the Sporn Company collected lightning overvoltages on a particular 200 kV transmission line and acquired a total of 15 pieces of data. According to the acquired data, the Insulator and Lightning Subcommittee officially released three standard lightning overvoltage waveforms for the first time with the time parameters being 0.25/10 μs, 0.25/30 μs and 0.25/90 μs. Subsequently, another three kinds of standard lightning waveforms were recommended by American Institute of Electrical Engineers, namely, the 0.5/5 μs, 1/10 μs and 1.5/40 μs standard lightning waveforms, of which the

1.5/40 μs waveform is widely applied and written into the American standard. Later, the 10–90% method was adopted to determine the front time of the double exponential standard lightning wave and the standard lightning waveform was revised as 1/40 μs. In 1943, the Transformer Subcommittee initiated lightning impulse withstand tests on transformers and proposed a preliminary test method. As for the test waveform, the front time was 0.5–2.5 μs, and the half amplitude decay time was 40–50 μs. In 1962, the IEC recommended that a 1.2/50 μs standard lightning waveform be uniformly applied and the 30–90% method was adopted to determine the front time of lightning waves. It is thus clear that the standard waveform for lightning impulse tests was determined based on the direct lightning data of the transmission line acquired 80 years ago.

In this book, long-term online overvoltage monitoring data is studied in terms of statistical lightning waveform parameters and transient analysis on wave process. Specifically, characteristic parameters regarding lightning impulse waves such as the front and tail time, amplitude and steepness and their distribution characteristics are studied to obtain the transient changes during the wave process due to lightning waves reaching various electrical equipment and structures. Detailed research contents are as follows:

1. Online overvoltage monitoring devices are installed at the substation to monitor lightning waves reaching lines, transformers and other electrical equipment in a long term so as to count and analyze waveforms and transient processes of overvoltages including direct lightning overvoltages, induced lightning overvoltages and mixed overvoltages. Reasonable recognition methods for induced lightning overvoltages and direct lightning overvoltages are studied as well. Transient changes of the wave process regarding lightning waves in power systems are studied by comparison between the lightning waveform data from different substation sites in combination with relevant simulation and evaluation.
2. As for the lightning waveforms collected by the online overvoltage monitoring system in the long term, typical lightning waveforms are summarized; a feasible processing method in accordance with the definitions of front and tail time is proposed for non-standard oscillating waveforms; lightning wavefront/tail time parameters are counted, to study their probability distribution pattern and to obtain parameter modification data that can reflect the wave process of the lightning incoming waves.
3. Time-domain and frequency-domain characteristics of different lightning incoming waveforms are studied to get the equivalence and differences of different waveforms during the transient wave process, and the impacts of different waveform characteristic quantities and electrical environment parameters on the transient wave process of lightning incoming waves are analyzed.
4. Further research is made on the winding wave process after the lightning reaches the transformer. Waveforms and characteristic parameters regarding practical lightning waves reaching the transformer winding are analyzed; the potential and gradient distribution of the lightning wave reaching the transformer winding are studied in combination with simulation and evaluation, so as to further study the transient wave processes on major and minor insulation.

Characteristic quantities and their distribution law regarding practical lightning waves are obtained by counting measured lightning incoming wave parameters so as to modify the standard lightning waveform, and to offer proof for insulation coordination and lightning protection design of the electrical equipment. Detailed contents are as follows:

1. Filter the measured waveform characteristics of power system lightning waves to obtain typical waveforms and other parameters at each site of the substation. Compare the transient wave processes under the influence of various electrical equipment and stray reactance/capacitance after the lightning wave reaches the substation.
2. Evaluate time-related characteristic parameters of measured non-standard lightning waveforms on the basis of long-term measurement by the online monitoring system, and count these parameters to obtain the probability density model that can indicate statistical regularity of the incoming wave time parameters. Obtain quantitative waveform characteristics for lightning incoming waves in time and frequency domains so as to learn the impacts of different waveform characteristics on transient wave processes.
3. Determine typical waveforms and characteristic parameters of lightning incoming waves; modify the front time and tail time of the standard lightning wave according to its distribution regularity and probability density model, so as to obtain time parameters and typical waveforms regarding practical lightning incoming waves in the power system.
4. Establish the transient wave process model for lightning waves reaching transformer windings, and analyze the specific transient lightning overvoltage wave process inside the transformer, based on acquired waveform parameters for lightning reaching the transformer by the online monitoring system.

8.9.2 Modification of Practical Lightning Impulse Test Parameters

Safe operation of the electrical equipment is the first line of defense against major accidents in the power grid. Under practical operation, the internal insulation withstands impulse overvoltages caused by lightning and other factors in addition to power frequency voltages in normal operation. Lightning hazards at the substation are mainly of two kinds: equipment damage caused by lightning directly striking the substation (direct lightning stroke, for short) and insulator flashover (back flashover) caused by lightning striking transmission line towers or ground wires or equipment insulation failure resulting from lightning overvoltages generated by direct lightning striking transmission lines (shielding failure) and propagating along the line to reach substations (lightning incoming wave, for short). Operational experience in China indicates that direct lightning, back flashover and shielding failures seldom occur at substations that are properly equipped with lightning rods, ground wires and grounding devices, according to relevant standards. Owing to frequent lightning strikes on the lines, incoming lightning waves propagating along the line is considered as the main threat for safe equipment operation. Insulation coordination between existing electrical equipment is still limited for a lack of in-depth research on electric field distribution inside the electrical equipment under lightning impulse voltages and waveform parameters regarding practical lightning incoming waves. According to statistics, inadequate structure design, poor control over the manufacturing process and material quality and inherent defects of transformer insulation are considered as the second cause of transformer damage in the State Grid Corporation China in 2002–2004. In 2004, a total of 13 accidents regarding transformer rated 110 kV or above occurred in the State Grid, with a capacity of

1043.0 MVA, accounting for 24.4% of the total accidents and 24.7% of the total damage capacity.

In order to reduce the threat of lightning incoming waves to the substation transformers, the incoming terminal protection and arresters are provided, which are important lightning protection measures to reduce the amplitude and steepness of the incoming wave. In addition, in order to check whether the insulation impulse strength of the electrical equipment is up to national standards and to study improvement of insulation structure and design, lightning impulse withstand voltage tests have to be performed on electrical equipment. The test waveform specified by relevant standards is a (1.2±30%/50±20%) μs standard lightning waveform. However, in consideration of factors such as incoming section attenuation, in-station refraction and reflection, transformer winding resonance and LC oscillation, the real lightning waves reaching electrical equipment are no longer 1.2/50 μs standard lightning waves but non-standard lightning waves of all shapes, such as unidirectional and bidirectional oscillating overvoltage waveforms. Lightning waves of different parameters will lead to clear differences in dielectric breakdown characteristics, so replacing practical non-standard lightning waves with standard lightning waves in lightning impulse tests are not appropriate, which was pointed out by S. Berlijn et al.

Therefore, according to the research on wave process of lightning incoming waves in the power system, revising the lightning impulse test standard for electrical equipment in accordance with revised lightning wave parameters helps to give more insights into electrical aging mechanisms of electrical equipment internal insulation with lightning impulse voltages, and provides a good guide for insulation design, coordination and life evaluation.

In recent years, research institutes, colleges and universities have started to study real-time overvoltage monitoring successively and developed overvoltage transducers, signal conditioning units and transmission units valid for various voltage classes and applicable to substations and 10–500 kV ac and 800 kV UHV dc transmission lines. The achievements of these are of great theoretical and engineering interest in instructing overvoltage protection and constructing a strong smart grid.

Power grid online overvoltage monitoring has now become a research focus, and has attracted great attention from research institutions and power grid companies. The online overvoltage monitoring technology, however, will encounter the following opportunities and challenges.

R&D of high-precision non-contact passive overvoltage transducers – Most existing online overvoltage monitoring devices have direct contact with primary/secondary equipment, which can easily threaten the safe operation of the system. Meanwhile, online overvoltage monitoring devices require a power supply, which limits its application in outdoor transmission lines. For these reasons, high-precision non-contact passive overvoltage transducers and electromagnetic interferences for non-contact measurement need to be studied based on advanced optical and material technologies, which are of great concern for promotion of online overvoltage monitoring devices applied in the power grid.

Construction of fast intelligent recognition and large data analysis systems for power system overvoltages – Large amounts of power grid overvoltage monitoring data will be acquired with the wide application of online overvoltage monitoring devices. To address this issue, modern artificial intelligence methods and big data theory are adopted to investigate overvoltage data characteristics, establish overvoltage fingerprint databases and achieve fast intelligent overvoltage recognition, which are of essential engineering values in constructing sound overvoltage analysis systems, analyzing overvoltage causes and taking corresponding measure for overvoltage suppression.

Accurate measurement of special overvoltages such as VFTOs – The VFTO in GIS substations has always been the main threat to the safe operation of the substation, and accurate measurement of VFTO has always been a challenging issue in this field. It is of significant theoretical and engineering values to study nanosecond VFTO transducers and acquisition units, tackle electromagnetic compatibility issues during measurement and realize VFTO online monitoring so as to ensure safe operation of the GIS substation.

In recent years, worldwide research institutions, colleges and universities have made remarkable achievements in construction and research concerning power grids. We shall seize the opportunity brought by the grid's rapid development, make breakthroughs in challenges of online overvoltage monitoring, and get more involved with IEEE, CIGRE and IEC to lead the development of online overvoltage monitoring technology.

Bibliography

1 Qigong Shen, Yu Fang, Zecun Zhou et al. *High Voltage Technology*, China Electric Power Press, 2012.
2 Leonard L. Grigsby. *Electric Power Generation, Transmission and Distribution*. CRC Press, 2012.
3 Hector J. Altuve Ferrer, Edmund O. Schweitzer III, *Modern Solutions for Protection, Control and Monitoring of Electric Power Systems*, Quality Books, Inc., 2010.
4 Zhenya Liu. *Ultra-High Voltage AC/DC Grids*. China Electric Power Press, 2013.
5 Gomez D.F.G., Saens E.M., Prado T.A., Martinez M. "Methodology for Lightning Impulse Voltage Divisors Design". 2009 7(1). 71–77.
6 Wenxia Sima, Haitao Lan, Lin Du, et al. "Research on Response Characteristics of Voltage Transducers at Bushing Taps". Proceedings of the Chinese Society for Electrical Engineering. 2006 26(21). 5.
7 Birtwhistle D., Gray I.D. "A new technique for condition monitoring of MV metal-clad switchgear" 1998. 91–95.
8 Xiufang Jia, Shutao Zhao, Baoshu Li, Wei Zhao. "A new method of transient overvoltage monitoring for substation". North China Electr. Power Univ., Beijing. 2002(2). 994–997.
9 Yin Zhang, Yadong Gao, Bin Du, et al. "New Tower Models in Lightning Protection Evaluation for Transmission Lines". Journal of Xi'an Jiaotong University. 2004 (4).
10 Hongtao Li. "Investigations on Lightning Waves Reaching a 500 kV Substation". Chongqing University, 2006.
11 Yajun Li, Changming Wei, Shiyu Tang, et al. "Application of Lightning Location System in Chongqing". High Voltage Engineering. 2002 (2).60.
12 Jianming Li, Kang Zhu. "Test Methods for High Voltage Electrical Equipment". China Electric Power Press, 2001.
13 Gang Yang, Yanxia Zhang, Chaoying Chen. "Overvoltage Evaluation and Arrester Digital Simulation". High Voltage Engineering. 2001 (3) 64–66.
14 Ningping Tang, Shaoyu Rou, Fuwang Liao. "High Voltage Measuring Devices Based on Spatial Electric Field Effect". Advanced Technology of Electrical Engineering and Energy. 2009 28(1) 5.
15 Ziqiang Ming. "Research on Online Overvoltage Monitoring Systems in High Voltage Grids". Chongqing: Chongqing University. 2008.
16 Wenhui Wu, Xianglin Cao. "Electromagnetic Transient Evaluation and EMTP Applications in Power Systems". China Water&Power Press, 2012.

Measurement and Analysis of Overvoltages in Power Systems, First Edition. Jianming Li.

17 Chang C.C., Lin C.J. LIBSVM: "A library for support vector machine". http://www.csie.ntu.edu.tw/~cjlin/libsvm. 2009.
18 Xin Li. "Overvoltage Hierarchical Pattern Recognition and Its Applications in Power System". Chongqing: Chongqing University, 2012.
19 Jianming Li, Youquan Yin. "Transient Overvoltage Detection and Insulation Status Analysis of Electrical Equipment". Sichuan Electric Power Technology, 2005 (2) 15–38.
20 Juan Wu. "Research and Development of Comprehensive Dynamic Simulation Test System in Power System". Zhongnan University, 2008.
21 Shaoxun Ping. *Case Study: Internal Overvoltage Protection in Power System*. China Electric Power Press, 2006.
22 Zishu Zhu. *Power System Overvoltages*. Shanghai Jiao Tong University Press, 1995.
23 Weihan Wu, Fangliu Zhang, et al. *Overvoltage Numerical Computation in Power System*. Science Press, 1989.
24 Wei Shi. *Overvoltage Evaluation in Power System*. Xi'an Jiao Tong University Press, 1988.
25 Yin Xu, Shiyan Xu. *Overvoltage Protection and Insulation Coordination in AC Power Systems*. China Electric Power Press, 2006.
26 Weixian Chen. *Fundamentals of Internal Overvoltages*. Water Resources and Electric Power Press, 1981.
27 J.M. Li, T. Luo, Yu Zhang, et al. "The Response Characteristics of Lightning Waves on Power System Propagation". 2015 International Symposium on Lightning Protection (VIII SIPDA2015).
28 Yi Wen, Jianming Li, Chun Li. "Design and Analysis of Online Overvoltage Monitoring Devices in a 500kV Grid". Electrical Measurement & Instrumentation, 2013 (11) 92–95, 114.
29 Shaoqing Chen, Hanyu Wang, Lin Du, Jianming Li, "Research on Characteristics of Noncontact Capacitive Voltage Divider Monitoring System Under AC and Lightning Overvoltages". IEEE Transactions on Applied Superconductivity, 24 (5) 1–3, Oct. 2014.
30 Naihui Wang. "Investigations of Power System Overvoltage Monitoring Based on LabVIEW". Xihua University, 2008.
31 Cui'e Xu, Yi Wen, Jianming Li, et al. "Power Frequency Electromagnetic Environment Analysis in Ultra High Voltage Transmission Lines"[J]. Sichuan Electric Power Technology, 2014 (1) 63–67.
32 Shaoqing Chen, Hanyu Wang, Lin Du, Jianming Li. "Research on a New Type of Overvoltages Monitoring Sensor and Decoupling Technology". IEEE Transactions on Applied Superconductivity, 24 (5) 1–4, Oct. 2014.
33 Gao Luo. "Research on Online Transient Overvoltage Monitoring and Recording System in Power Systems". Xihua University, 2012.
34 Shao Qing Chen, Jian Ming Li, Tao Luo, Hui Ying Zhou, Qi Xiao Ma, Lin Du, "Study on the impulse characteristics of capacitive voltage transformer". 2013 IEEE International Conference on Applied Superconductivity and Electromagnetic Devices (ASEMD), pp. 65–68, 25–27 Oct. 2013.
35 Wei Zhang. "Switching Overvoltage Analysis and Protection of Vacuum Switchgear". Xihua University, 2010.

36 Juan Zhou. Power System Overvoltage Analysis Based on Virtual Instrument. Xihua University, 2007.

37 Shao Qing Chen, Jian Ming Li, Tao Luo, Hui Yang Zhou, Qi Xiao Ma, Lin Du. "Performance of overvoltage transducer for overhead transmission lines". 2013 IEEE International Conference on Applied Superconductivity and Electromagnetic Devices (ASEMD), pp. 48–51, 25–27 Oct. 2013.

38 S.Q. Chen, J.M. Li, T. Lao, H.Y. Zhou, Q.X. Ma, et al. "Performance of Overvoltage Transducer for Overhead Transmission Lines". Proceedings of 2013 IEEE International Conference on Applied Superconductivity and Electromagnetic Devices (ASEMD2013).

39 S.Q. Chen , J.M. Li, T. Lao, H.Y. Zhou, Q.X. Ma, et al. "Study on the Impulse Characteristics of Capacitive Voltage Transformer". Proceedings of 2013 IEEE International Conference on Applied Superconductivity and Electromagnetic Devices (ASEMD2013).

40 S.J. Xie, J.M. Li, et al. "Analysis of the Charge Distribution in Branched Downward Leaders using the Charge Simulation Method". 2015 Asia-Pacific International Conference on Lightning (APL2015).

Index

Measurement and Analysis of Overvoltages in Power Systems, First Edition. Jianming Li.

g

h

i

q

r